## Interactive Activities

As you read each chapter of *Life*, be sure to visit the text website to take advantage of a wide variety of study tools—eLearning, flash cards, case studies, and other engaging activities designed to reinforce learning.

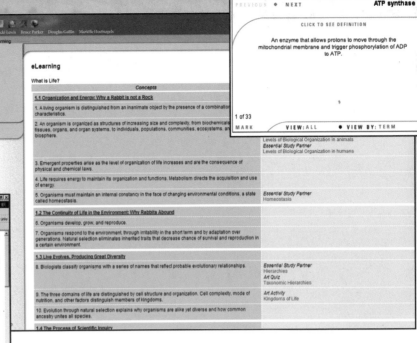

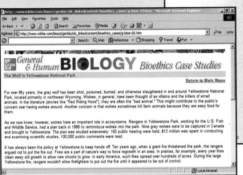

## Access to Premium Learning Materials

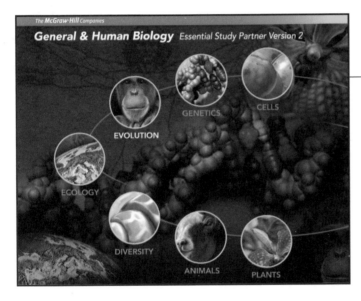

The *Life* text website is your portal to exclusive interactive study tools like McGraw-Hill's Essential Study Partner.

The Essential Study Partner offers art activities, animations, quizzes, and other activities to support your learning of the chapter material.

## Test Yourself

Take a quiz at the *Life* text website to gauge your mastery of chapter content. Along with multiple-choice quizzing, you'll find Thinking Scientifically and Testing Your Knowledge. Each chapter quiz is specifically constructed to test your comprehension of key concepts. Immediate feedback on your responses explains why an answer is correct or incorrect. You can even e-mail your quiz results to your professor!

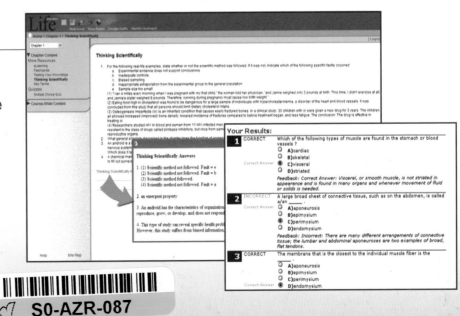

# Life

*sixth edition*

**Ricki Lewis**
*Contributing Editor,* The Scientist

**Bruce Parker**
*Utah Valley State College*

**Douglas Gaffin**
*The University of Oklahoma*

**Mariëlle Hoefnagels**
*The University of Oklahoma*

Boston   Burr Ridge, IL   Dubuque, IA   Madison, WI   New York   San Francisco   St. Louis
Bangkok   Bogotá   Caracas   Kuala Lumpur   Lisbon   London   Madrid   Mexico City
Milan   Montreal   New Delhi   Santiago   Seoul   Singapore   Sydney   Taipei   Toronto

*The McGraw·Hill Companies*

**McGraw Hill** **Higher Education**

LIFE, SIXTH EDITION

Published by McGraw-Hill, a business unit of The McGraw-Hill Companies, Inc., 1221 Avenue of the Americas, New York, NY 10020.

Some ancillaries, including electronic and print components, may not be available to customers outside the United States.

 This book is printed on recycled, acid-free paper containing 10% postconsumer waste.

1 2 3 4 5 6 7 8 9 0 DOW/DOW 0 9 8 7 6

ISBN-13  978–0–07–293112–9
ISBN-10  0–07–293112–4

Publisher: *Janice Roerig-Blong*
Sponsoring Editor: *Thomas C. Lyon*
Developmental Editor: *Fran Schreiber*
Director of Development: *Kristine Tibbetts*
Marketing Manager: *Tamara Maury*
Senior Project Manager: *Sheila M. Frank*
Senior Production Supervisor: *Kara Kudronowicz*
Senior Media Project Manager: *Jodi K. Banowetz*
Media Producer: *Eric A. Weber*
Senior Coordinator of Freelance Design: *Michelle D. Whitaker*
Cover/Interior Designer: *Maureen McCutcheon*
Senior Photo Research Coordinator: *John C. Leland*
Photo Research: *Mary Reeg*
Supplement Producer: *Melissa M. Leick*
Compositor: *Precision Graphics*
Typeface: *10/12 Minion*
Printer: *R. R. Donnelley Willard, OH*

United States Edition (USE) Cover Images:
Main cover image: © *David Fleetham/Visuals Unlimited*
Inset images (top to bottom): Bacteria Credit: © *Steven P. Lynch;* Diatoms in Fresh Water Credit: *E. Pollard/PhotoLink/Getty Images;* Amanita Parcivolvata Credit: *Photolink/Getty Images;* Leaves Credit: © *Comstock/PunchStock;* Elephant Credit: © *Photodisc, African Wildlife, vol. EP073*

The credits section for this book begins on page 976 and is considered an extension of the copyright page.

**Library of Congress Cataloging-in-Publication Data**

Life / contributing editor, Ricki Lewis . . . [et al.]. — 6th ed.
    p. ; cm.
  Includes index.
  ISBN 978–0–07–293112–9 — ISBN 0–07–293112–4 (hard copy : alk. paper)
  1. Biology. 2. Human biology. 3. Life sciences.  I. Lewis, Ricki.

  QH308.2.L485    2007
  570—dc22                                            2005034234
                                                      CIP

www.mhhe.com

# Life

## *About the Authors*

### Ricki Lewis

**Ricki Lewis** is a science writer who earned her Ph.D. in developmental genetics from Indiana University in 1980. She is author of *Human Genetics: Concepts and Applications,* original author of this textbook, co-author of Hole's *Human Anatomy & Physiology* and Hole's *Essentials of Human Anatomy & Physiology,* all from McGraw-Hill Higher Education; and author of *Discovery: Windows on the Life Sciences* published by Blackwell Science. She has authored thousands of articles in a wide range of scientific, medical, and consumer publications, and is a frequent public speaker. She currently writes for *The Scientist, Applied Neurology,* and *Nature.* Ricki has been a genetic counselor for a private medical practice in Schenectady, NY, since 1984. In 2005 she began serving as a hospice volunteer, visiting with nursing home patients daily. Ricki has a chemist husband, three daughters, and many felines. ralewis@nycap.rr.com

### Bruce Parker

**Bruce Parker** received his Ph.D. in molecular biology/biochemistry from Utah State University in 1988. His areas of expertise include virology, molecular cell biology, and biochemistry. He spent two years in London working on research into viruses that cause cancer, followed by another two years on the same project at St. Jude Children's Research Hospital in Memphis. He has taught general biology for nonmajors and majors and courses in cell biology and biochemistry at Utah Valley State College since 1992. He has been nominated for Faculty of the Year several times and has been included in *Who's Who Among America's Teachers* twice. Bruce currently serves as the associate Vice President for Academic Affairs at Utah Valley State College but will always consider himself to be a teacher. His hobbies include computer programming and amateur radio, when he is not fishing somewhere. parkerbr@uvsc.edu

### Douglas Gaffin

**Douglas Gaffin** holds a bachelor of science degree from the University of California at Berkeley, and he earned his Ph.D. in zoology from Oregon State University in Corvallis in 1994. His research interests are in sensory neurobiology, where his special focus is on the behavior and sensory physiology of sand scorpions. He has extensive biology teaching experience and has taught students in courses ranging from junior high school to graduate school levels. Doug is currently Dean of University College and associate professor in the Department of Zoology at the University of Oklahoma, and he has the privilege of teaching introductory zoology to thousands of undergraduates each year. His innovative teaching style and ability to inspire students have been recognized with awards both regionally and nationally. Among other organizations, he is a member of the Society for Neuroscience, the International Society for Neuroethology, and a life member of the American Arachnological Society. In his spare time he enjoys traveling, riding his bike, playing volleyball, and picking the banjo. One of his favorite activities is going to the desert each summer to observe and conduct field research on sand scorpions in their native habitat. ddgaffin@ou.edu

### Mariëlle Hoefnagels

**Mariëlle Hoefnagels** was raised near San Francisco, and received her B.S. in environmental science (1987) from the University of California at Riverside. After working in a soil analysis lab in Oregon for two years, she earned her master's degree in soil science from North Carolina State University (1991). Her research, on interactions between beneficial fungi and salt marsh plants, led her to return to Oregon to complete her Ph.D. in plant pathology (Oregon State University, 1997). Mariëlle's dissertation work focused on the use of bacterial biological control agents to reduce the spread of fungal pathogens on seeds. She is now assistant professor at the University of Oklahoma, where she teaches nonmajors courses in biology and microbiology, and a course on fungi for advanced botany and microbiology majors. She is a member of the National Association of Biology Teachers, the Association of Biology Laboratory Education, and the Mycological Society of America. Her hobbies include reading, traveling, gardening, and playing volleyball. hoefnagels@ou.edu

# Life
## *Brief Contents*

# Life

## *Detailed Contents*

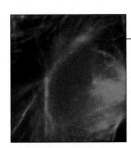

# UNIT 2
## *Genetics and Biotechnology*

# UNIT 3
## *Evolution*

# Chapter 15
## The Evolution of Evolutionary Thought 271

# Chapter 16
## The Forces of Evolutionary Change: Microevolution 287

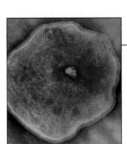

# UNIT 4
## *The Diversity of Life*

# UNIT 5
## *Plant Life*

## Chapter 27
## Plant Form and Function  515

## Chapter 28
## Plant Nutrition and Transport  537

## Chapter 29
## Reproduction of Flowering Plants  551

## Chapter 30
## Plant Responses to Stimuli  569

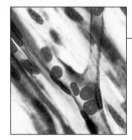

# UNIT 6
## *Animal Life*

# Chapter 31
## Animal Tissues and Organ Systems 589

# Chapter 32
## The Nervous System 609

# Chapter 33
## The Senses 637

# UNIT 7
## *Behavior and Ecology*

# Chapter 42
## Animal Behavior  825

# Chapter 43
## Populations  849

# Chapter 44
## Communities and Ecosystems  865

# Life
## *Preface*

The goal of *Life* has always been to engage the non-scientist or possible scientist in understanding how life works. We try to do that in a way that is fun, meaningful, and valuable in helping students make informed decisions on an ever-growing array of issues, from teaching evolution to human cloning and stem cells. This sixth edition continues our presentation of cutting edge life science and the technology that it spawns, with timely additions, such as discussions of animal behavior during the recent tsunami and removal of drugs such as Vioxx™ from the market.

The sixth edition of *Life* reunites four biologists whose areas of expertise complement beautifully. Geneticist and science journalist Ricki Lewis originated the text, using a unique style that weaves solid biology content with intriguing tales of scientific discovery and real-life cases and applications. Veterans from the fifth edition of *Life* are Douglas Gaffin and Mariëlle Hoefnagels of the University of Oklahoma and Bruce Parker of Utah Valley State College. Ricki has taught a variety of courses and has provided genetic counseling to a private medical practice since 1984. Doug, Mariëlle, and Bruce are active instructors who teach undergraduate biology to hundreds of majors and nonmajors each semester, and all have earned recognition on their campuses as outstanding teachers.

Devotion to, and passion about, teaching and communicating unite our team. We thoroughly enjoy telling interesting stories that are so easy to find at all levels of biology, from molecules to ecology—the stories that, when told correctly, mesmerize even the most reluctant students, causing them to perk up and think, "Wow, I never knew that! So that's why . . . !" We all love to watch students get excited about learning a subject they once viewed as too intimidating. Our areas of research expertise mesh beautifully, including animal physiology and behavior, plant-microorganism interactions, genetics, developmental biology, molecular biology, and biochemistry.

Together, the *Life* authors cover the breadth of general biology and make it easy for you to present

the latest information to your students, using the most accessible multimedia tools available. The unique mix of scientific expertise and journalistic experience results in a textbook that is not only substantial, but also accessible.

An interesting feature of the human mind is that it organizes and comprehends information better if first shown an overview of what it is about to encounter—much like knowing the overall route and destination before beginning a cross-country road trip. An expanded Chapter Preview at the beginning of each chapter gives a concise introduction to the key principles and ideas to be encountered in that chapter—giving students the chance to preview the information and understand where they are going in that chapter.

Responding to reviewer requests, this new edition splits the chapter on chemistry into two separate chapters, adds a chapter on Genetic Technology, and significantly enhances the narrative into one voice.

New chapter construction designed to assist the student includes a set of Reviewing Concepts summaries at the end of each major heading topic, followed by a Connections discussion at the end of the chapter. The Connections piece is intended to help the students connect topics previously discussed with topics yet to come. This section also stresses the major themes of *Life* and helps the student to understand why the information they have just covered is pertinent to their lives and to what else they have learned.

Some researchers have said that a student needs to repeat something at least three times to be able to remember it. The new layout of the textbook provides the students with a clearer opportunity to understand and retain the concepts of each chapter. The Chapter Preview is followed not only by the text, but also by Reviewing Concepts at the end of each section, and by a detailed summary at the end of each chapter. Students will have encountered the major ideas of each chapter at least four times when they finish each chapter!

The new elements that made the fifth edition friendlier and less cluttered were preserved. Figure and table references appear in bold type so that students can easily find related discussion. Scale bars have been added to micrographs so that students can appreciate the relative size of structures and organisms. New in-text cross-references aid recall and catalyze the conceptual connections that underlie biology.

For those looking for the end-of-chapter pieces of the fifth edition, the "Thinking Scientifically" and "References and Resources" sections have been retained as web-based resources. Likewise most of the Boxed Readings have been moved to the web and the Biotechnology Boxes found in the text of the fifth edition have been moved to the web or incorporated into the text of the new chapter on Genetic Technology. This has streamlined the text and allowed for more focus on the information in the text without losing the overall public interest level of the material.

The end-of-chapter material has been completely revised to provide the students with a built-in study guide for the textbook. The study guide questions are directly linked to the material covered in the text to help the student relate the questions to the sections of the text containing the answers. The study guide includes a much more extensive list of the key terms introduced and used in that chapter. Students taking biology for the first time are practically learning a new language in their effort to converse fluently with some 800 or more new terms. By expanding the key terms list, the "second language" element of the biology course is emphasized for the student.

The remainder of the end-of-chapter material consists of three types of brand-new questions written specifically for this edition. By including these questions at the end of each chapter, the student has an opportunity to immediately review what they have learned, or to refer to the questions while they read to guide their study of the concepts. The answers to all of the study questions are provided as an appendix. By using one of the textbook authors to write both the revisions of the sixth edition and the study guide questions and answers, students are assured of continuity in terminology and content covered by the questions and their answers.

## Overview of What's New for *Life*, Sixth Edition

- **Chapter Preview** introduces each chapter.
- **Reviewing Concepts** appear at the end of each major heading.
- Chemistry chapter divided into two chapters to help clarify for students.
- New **Genetic Technology** chapter.
- **Connections** help students tie everything together.
- Built-in **Student Study Guide** at the end of each chapter.
- New end-of-chapter questions.
- Extensive list of key terms.

## Today's Science for Today's Students

Each chapter begins with a brief essay designed to stimulate the student's interest. Many of the essays are new to this edition and add a personal touch, with stories from the authors' own experiences as well as topics that are found in news headlines. Throughout the text the authors have worked to provide the most current and up-to-date insights into the world of biology and biological research from a wide array of disciplines.

Unit 4, The Diversity of Life, is always a work in progress, as new data and new ways of looking at the living world reveal new classification possibilities. *Life's* "tree of life" illustrations, developed by Douglas Gaffin and found in this unit and others, encapsulate species relationships and evolutionary trends in an easy-to-follow form. While *Life's* classification schemes combine traditional and molecular approaches to taxonomy, we are careful to explain that new molecular data also throw the classification of life into upheaval—and acknowledge that the classification schemes in this book are provisional. It is important for students to realize that biological facts and concepts are not written in stone; that despite popular notions of scientific "proof," science is constantly changing to embrace new information. We would rather present the current state of taxonomic thought, uncertainties and all, than perpetuate dated classifications. For example, we revised and rearranged the animal diversity chapters to reflect the currently accepted split of protostomes into lophotrochozoans and ecdysozoans. The traditional scheme is included for comparison, but the book adopts the new classification scheme.

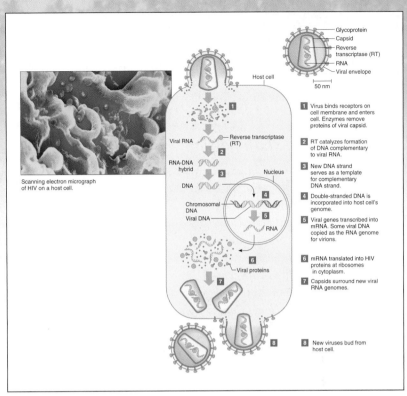

Scanning electron micrograph of HIV on a host cell.

Glycoprotein
Capsid
Reverse transcriptase (RT)
RNA
Viral envelope

50 nm

Host cell

Viral RNA — Reverse transcriptase (RT)

RNA-DNA hybrid

DNA

Nucleus

Chromosomal DNA
Viral DNA

RNA

Viral proteins

1. Virus binds receptors on cell membrane and enters cell. Enzymes remove proteins of viral capsid.
2. RT catalyzes formation of DNA complementary to viral RNA.
3. New DNA strand serves as a template for complementary DNA strand.
4. Double-stranded DNA is incorporated into host cell's genome.
5. Viral genes transcribed into mRNA. Some viral DNA copied as the RNA genome for virions.
6. mRNA translated into HIV proteins at ribosomes in cytoplasm.
7. Capsids surround new viral RNA genomes.
8. New viruses bud from host cell.

## The Art Program—A Visual Voyage

*Life's* art is not only visually spectacular, but also pedagogically sound, providing a consistent look from cover to cover. Repeated themes provide continuity, from biochemical reactions to life cycles to feedback loops in animal physiology to evolutionary tree diagrams. Use of color, arrows, and symbols is standardized throughout the text, easing learning and remembering. For example, DNA, membranes, and other cell structures have a consistent look and color throughout. A student can learn that every time a molecular structure appears that is purple, for instance, proteins are the heart of that structure. This approach vividly illustrates the organizational nature of life.

We have also selected unusual and interesting photos to show students glimpses of the natural world that they may never have seen before. The art and photos are combined in page layouts that are attractive and inviting—and above all, help students to learn.

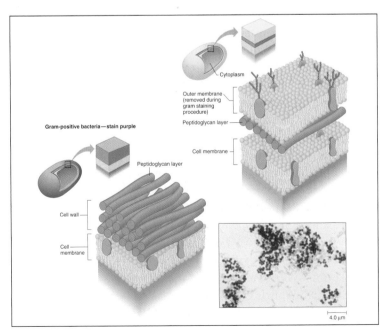

Cytoplasm

Outer membrane (removed during gram staining procedure)

Peptidoglycan layer

Cell membrane

Gram-positive bacteria—stain purple

Peptidoglycan layer

Cell wall

Cell membrane

4.0 µm

INSTRUCTOR

# A Word of Thanks

We offer special thanks to the reviewers who spent hours poring over chapter drafts in meticulous detail, spotting errors and inconsistencies, confirming what works and gently critiquing what doesn't, and pointing out sections that we could clarify.

## Reviewers of the Sixth Edition

Elaine Ashby, *Hagerstown Community College*

Mohammad Ashraf, *Olive-Harvey College*

Diane B. Beechinor, *Palo Alto College*

Penny L. Bernstein, *Kent State University – Stark*

Renee E. Bishop, *Penn State University*

Judy Bluemer, *Morton College*

Bradley S. Bowden, *Alfred University*

Debra Bretz, *Surry Community College*

Matthew Rex Burnham, *Jones County Junior College*

Nickolas A. Butkevich, *Schoolcraft Community College*

Claire Carpenter, *Yakima Valley Community College*

Thomas Chen, *Santa Monica College*

Barry Chess, *Pasadena City College*

Elisabeth Ciletti, *Pasadena City College*

Pamela Anderson Cole, *Shelton State Community College*

Jerry L. Cook, *Sam Houston State University*

Susan J. Cook, *Indiana University – South Bend*

Sarah Cooper, *Arcadia University*

Lee Couch, *University of New Mexico*

Jean DeSaix, *University of North Carolina – Chapel Hill*

Mary E. Dominiecki, *Slippery Rock University of Pennsylvania*

Cathy A. Donald-Whitney, *Collin County Community College*

Teresa H. Doscher, *Valdosta State University*

Donald S. Emmeluth, *Armstrong Atlantic State University*

Cheryld L. Emmons, *Alfred University*

James Engman, *Henderson State University*

Wm. Bruce Ezell, Jr., *University of North Carolina at Pembroke*

Lisa M. Flick, *Alfred University*

Wayne McCrady Forbes, *Slippery Rock University of Pennsylvania*

Donald P. French, *Oklahoma State University*

David Gerkensmeyer, *Ivy Tech State College*

Gordon L. Godshalk, *Alfred University*

Roberto B. Gonzales, *Northwest Vista College*

John Grew, *New Jersey City University*

John Griffis, *Joliet Junior College*

David J. Grisé, *Radford University*

Richard Hanke, *Rose State College*

Frankie L. Harriss, *Independence Community College*

Michael T. Harves, *Yakima Valley Community College*

Bernard Hauser, *University of Florida*

Karen R. Hickman, *Fort Hays State University*

Joyce D. Johnson, *Alcorn State University*

Walter S. Judd, *University of Florida*

Arnold J. Karpoff, *University of Louisville*

Jennifer B. Katcher, *Pima Community College*

Paul Kelly, *Salem State College*

Susan Keys, *Springfield College*

Stephen T. Kilpatrick, *University of Pittsburgh at Johnstown*

Dennis J. Kitz, *Southern Illinois University – Edwardsville*

Todd A. Kostman, *University of Wisconsin – Oskhosh*

Thomas G. Lammers, *University of Wisconsin – Oshkosh*

Denis A. Larochelle, *Clark University*

Lynn Larsen, *Portland Community College*

Ellen J. Lehning, *Jamestown Community College*

Peggy L. Lepley, *Cincinnati State Technical and Community College*

Tammy J. Liles, *Lexington Community College*

John F. Logue, *University of South Carolina – Sumter*

Fordyce G. Lux III, *Lander University*

Kevin Lyon, *Jones County Junior College*

Heather Miller Woodson, *Armstrong Atlantic State University*

V. Christine Minor, *Clemson University*

Archana Nair, *Tomball College (NHMCCD)*

Usana Nava, *Ivy Tech State College*

David H. Niebuhr, *College of William and Mary*

Shreekumar R. Pillai, *Alabama State University*

Eric Rabitoy, *Citrus College*

Robert J. Ratterman, *Jamestown Community College*

H. Bruce Reid, *Kean University*

Kathleen Richardson, *Portland Community College – Sylvania*

Laura H. Ritt, *Burlington County College*

Harry Roy, *Rennsselaer Polytechnic Institute*

Albert S. Rubenstein, *Ivy Tech State College*

Katherine N. Schick, *San Joaquin Delta Community College*

Jerred Seveyka, *Yakima Valley Community College*

Blair S. Shean, *Yakima Valley Community College*

Jennifer L. Siemantel, *Cedar Valley College*

John D. Sollinger, *Southern Oregon University*

Anthony J. Stancampiano, *Oklahoma City Community College*

Peter Svensson, *West Valley College*

Jeff H. Taylor, *Slippery Rock University*

John R. Taylor, *Southern Utah University*

Richard E. Trout, *Oklahoma City Community College*

Palaniswamy Vijay, *Indiana University School of Medicine*

Winfred E. Watkins, *McLennan Community College*

Lisa H. Weasel, *Portland State University*

Jason R. Wiles, *Portland Community College*

Kenneth Wunch, *Sam Houston State University*

Christopher J. Yahnke, *University of Wisconsin – Stevens Point*

M. Carol Yeager, *Grove City College*

ACKNOWLEDGMENTS

# Digital Content Manager

The Digital Content Manager provides illustrations, photos and tables from the textbook, in addition to supplemental media materials. With the Digital Content Manager instructors will have access to:

## Art Library

Color enhanced digital files of all the illustrations in the book, plus the same art saved in unlabeled versions, are included in the Art Library.

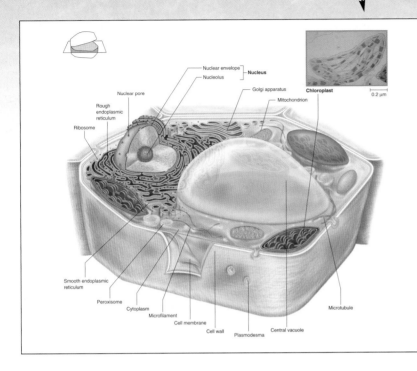

## Photo Library

Like the Art Library, digital files of all photographs from the book are available.

## Table Library

Every table that appears in the book is provided in electronic form.

## Additional Photo Library

Over 700 photos not found in the text are available for use in creating lecture presentations.

## PowerPoint Lecture Outlines

These ready-made presentations combine art and lecture notes for each of the chapters of the book. The presentations can be used as they are or customized to reflect your preferred lecture topics and organization.

All line art, photos, and tables are also pre-inserted into blank PowerPoint slides for ease of lecture presentations.

### TABLE 42.1

**Learning in Humans and Pets**

| Type of Learning | Example |
|---|---|
| Habituation | Being able to concentrate on written work with music playing in the background but jumping when the phone rings |
| Classical conditioning | Craving popcorn in a movie theater |
| Operant conditioning | A student learns what to do to earn high grades in school |
| Imprinting | Infant responds to mother's breast, voice, and face |
| Insight learning | A child places a box on roller skates to build a vehicle |
| Latent learning | A cat explores its new home and darts to safety when the doorbell rings |

# NEW! McGraw-Hill: Biology Digitized Video Clips

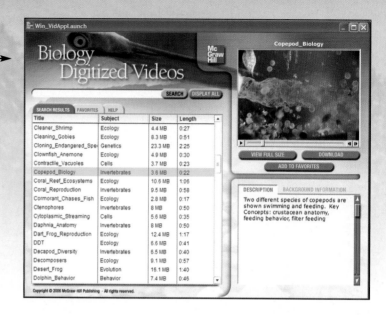

McGraw-Hill is pleased to offer adopting instructors a new presentation tool—digitized biology video clips on DVD! Licensed from some of the highest quality science video producers in the world, these brief segments range from about five seconds to just under three minutes in length and cover all areas of general biology from cells to ecosystems. Engaging and informative, McGraw-Hill's digitized biology videos will help capture students' interest while illustrating key biological concepts and processes such as mitosis, how cilia and flagella work, and how some plants have evolved into carnivores.

## Instructor's Testing and Resource CD-ROM

This cross-platform CD provides a wealth of resources for the instructor. Among the supplements featured on this CD is a computerized test bank that uses testing software to quickly create customized exams. The user-friendly program allows instructors to search for questions by topic, format, or difficulty level; edit existing questions or add new ones; and scramble questions for multiple versions of the same test. Word files of the test bank questions are provided for those instructors who prefer to work outside the test-generator software.

Other assets on the Instructor's Test and Resource CD include an Instructor's Manual with comprehensive teaching outlines and lecture suggestions.

## Transparencies

All of the illustrations from the text are included in this set of 750 transparencies. Approximately 50 illustrations from which the labels have been removed are also included. Be sure to visit the text website at www.mhhe.com/life6 for a correlation guide.

## Student Interactive CD-ROM

This CD includes chapter-based quizzes, animations of complex processes, and PowerPoints of all the images found in the textbook. It is organized chapter-by-chapter and provides a link directly to the text's Online Learning Center. This Interactive CD-ROM offers an indispensable resource for enhancing topics covered within the text.

**Interactive Laboratories and Biological Simulations**

## Ask your McGraw-Hill sales representative for more information about iLaBS.

# iLabs

**No setup. No mess. No limits!** Interactive Laboratories and Biological Simulations, or iLaBS, teach students real-life biomolecular applications and techniques using accurate and intriguing programs. Students can explore cutting-edge technologies like DNA fingerprinting and restriction mapping, repeating lab procedures until they are comfortable with the techniques. Because they can generate multiple sets of data, students gain practice in data interpretation and problem solving. iLaBS also provide the opportunity for students to virtually experience time consuming or complicated labs without the necessity for physical facilities.

New Features:

Now available on CD-ROM makes iLaBS more accessible. Accompanying student workbook gives detailed instructions on how to complete each lab, as well as providing background information and test questions for each.

Three new simulations have been added! Students can now view simulations on The Polymerase Chain Reaction, Sequencing, and Viruses. Instructor's CD-ROM contains all of the content from the student CD, PLUS a preview of iLaBS, tips on how to use iLaBS in the course, answers to the workbook questions, a low text version of the DNA Replication and Regulation of the Lac Operon simulations to make presentation in lecture easier, and bonus materials—simulations of cloning and Southern blotting and bioinformatics exercises on endosymbiosis and sickle cell anemia.

# Instructor Website
## www.mhhe.com/life6

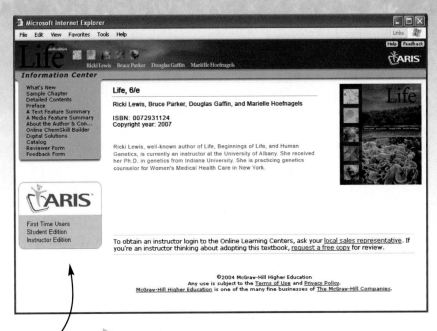

Through the *Life* text website, everything you need for effective, interactive teaching and learning is at your fingertips. Moreover, this vast McGraw-Hill resource is **easily loaded into course management systems such as WebCT or Blackboard.** Contact your local McGraw-Hill representative for details.

Instructor tools located on the enhanced website include:

**Instructor's Manual** This valuable resource includes chapter objectives and interesting tidbits not found in the textbook, lists areas of difficulty for students along with suggestions for resolving those difficulties effectively, and provides teaching suggestions for various content areas.

## ARIS

McGraw-Hill's ARIS for *Life* is a complete electronic homework and course management system, designed for greater ease of use than any other system available. Free on adoption of *Life*, instructors can create and share course materials and assignments with colleagues with a few clicks of the mouse. Instructors can edit questions, import their own content, and create announcements and due dates for assignments. ARIS has automatic grading and reporting of easy-to-assign quizzing, and testing. Once a student is registered in the course, all student activity within McGraw-Hill's ARIS is automatically recorded and available to the instructor through a fully integrated grade book that can be downloaded to Excel.

## Text Website—For the Student
## www.mhhe.com/life6

Resources on the *Life* website support each chapter in the text. Some of the learning tools available through the website include:

**E-learning sessions** with animations of key processes, art quizzes, outline summaries, and more featuring McGraw-Hill's Essential Study Partner 2.0.

**Thinking Scientifically** critical thinking questions challenge you to use concepts of the chapter to solve problems.

**Testing Your Knowledge** tests your recall of chapter material.

**Self-quizzing** with immediate feedback.

**Electronic flashcards** to review vocabulary.

Turn to the inside covers of this text to learn more about the exciting features provided for students through the *Life* website.

- **Fascinating photos** and stunning micrographs.

- **Overview Figures** simplify complex interactions and provide a sound study tool.

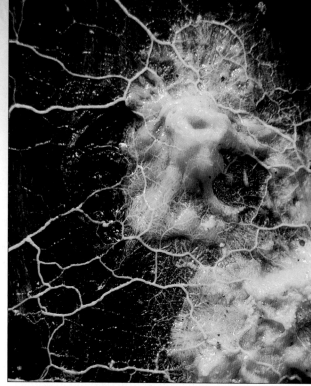

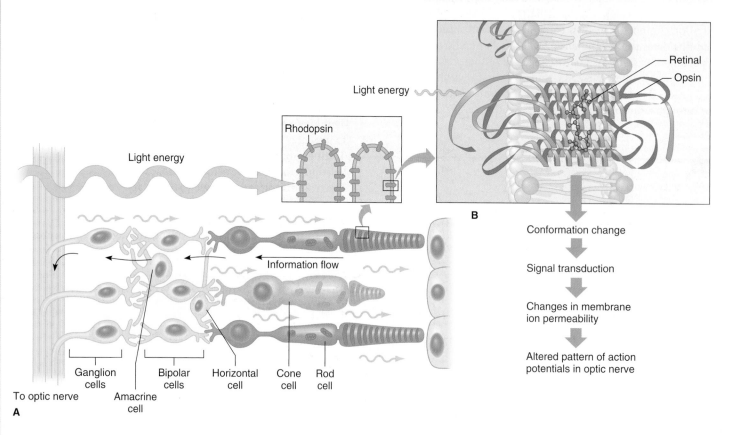

Light energy

Rhodopsin

Light energy

Retinal

Opsin

Information flow

Ganglion cells

Bipolar cells

Horizontal cell

Cone cell

Rod cell

To optic nerve

Amacrine cell

A

B

Conformation change

Signal transduction

Changes in membrane ion permeability

Altered pattern of action potentials in optic nerve

- **Process Figures** include step-by-step descriptions that walk the reader through a compact summary of important concepts.

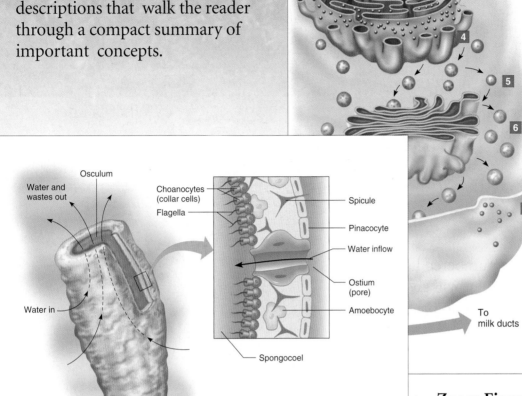

1 Milk protein genes transcribed into mRNA

2 mRNA exits through nuclear pores

3 mRNA forms complex with ribosomes and moves to surface of rough ER where protein is made

4 Enzymes in smooth ER manufacture lipids

5 Milk proteins and lipids are packaged into vesicles from both rough and smooth ER for transport to Golgi

6 Final processing of proteins in Golgi and packaging for export out of cell

7 Proteins and lipids released from cell by fusion of vesicles with cell membrane

To milk ducts

- **Zoom Figures** put structures into context by providing a macroscopic to microscopic view of life and its processes.

- **Combination figures** show the features that can be illustrated by an artist along with the appearance of structures and organisms in the real world.

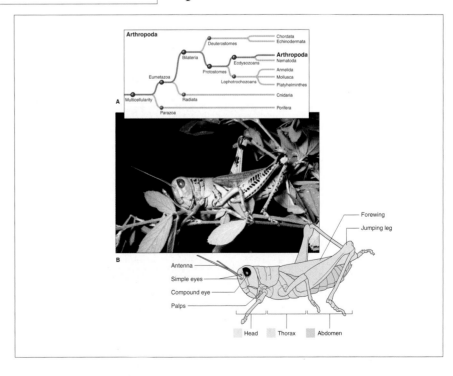

INSTRUCTIVE ART PROGRAM

- **Chapter Opening Vignettes**—Each chapter begins with a compelling vignette describing a real-life scientific issued related to the chapter topic.

- **Chapter Preview**—This numbered list at the beginning of each chapter will give you a quick preview of what is to come.

Plant or Animal? The leafy sea dragon is exquisitely adapted for its environment, its form mimicking the surrounding seaweed.

### The Surprising Sea Dragon—A Metaphor for Life

Welcome to biology, the study of life. Our journey will take us from atoms to biomes, to the familiar and past the strange. A recurrent lesson will surface as we travel: Life is not always what it seems. Take, for example, the leafy sea dragons (scientifically named *Phycodurus eques*). These surprising animals are good metaphors for life.

Sea dragons are not dragons, and though they are closely related to sea horses, they are not horses. They are fish. Yet, in common with horses, kelp, and all other organisms, they are made of cells, which are in turn composed of molecules made mostly of four common atoms. One such molecule is DNA, the molecule of inheritance, which dictates where a snout is to develop and the structure of a leaflike fin.

With other fish a leafy sea dragon shares many features including gills, gill coverings, and a swim bladder. Less obvious are the many characteristics in common with other vertebrates, such as sharks, snakes, birds, and horses. A skull encases a brain in the animal's head; a backbone encloses a nerve cord. All species share some features of life, but may manifest them in very different ways.

Incongruities in nature are provocative. They spark our curiosity and lead to bigger questions. What makes a fish a fish? How can a fish come to resemble kelp? Is behavior inherited? Why is there sex? How do parental roles evolve? Why do sea dragons live where they do? How do species influence each other? Some of the answers are in this book, and others remain to be discovered.

## What Is Life?

1

## Investigating Life    15.1

### Evolution in a Polluted River

The rapid adaptation of certain small aquatic invertebrates (animals without backbones) to environmental pollution illustrates microevolution in action. Animals whose genetic makeups render them vulnerable to pollution die out or do not produce fertile offspring, and their genes eventually disappear from the population. The allele(s) for pollution resistance, already present before exposure to the pollution, become(s) more common in the population. This happened in Foundry Cove, part of the Hudson River, about 50 miles (80 kilometers) north of New York City.

Foundry Cove has a toxic history. During the Revolutionary War, a forge at the site manufactured metal chains that were placed in the river to stop British ships. During the Civil War, bullets were produced near the cove. Then, in the 1950s, manufacturing facilities at Foundry Cove made batteries. With such a record of heavy metal manufacture, it's not surprising that the sediments at the bottom of the river contained up to 25% cadmium.

Curiously, the polluted river area swarmed with invertebrate life, as did neighboring coves. But when animals from nearby regions with cleaner sediments were moved to the polluted areas, they died. Had the Foundry Cove animals inherited the ability to survive heavy metal poisoning? That is, did the population once include animals with differing abilities to survive and reproduce in the presence of certain chemicals? Did years of exposure to pollution select the resistant animals, who came to constitute most of the population? To find out, biologists devised an experiment. They took animals from the areas with polluted sediments and bred them in the laboratory

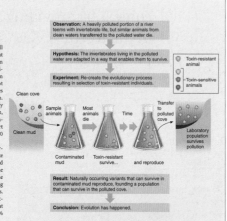

**Observation:** A heavily polluted portion of a river teems with invertebrate life, but similar animals from clean waters transferred to the polluted water die.

**Hypothesis:** The invertebrates living in the polluted water are adapted in a way that enables them to survive.

**Experiment:** Re-create the evolutionary process resulting in selection of toxin-resistant individuals.

○ — Toxin-resistant animal
○ — Toxin-sensitive animals

Clean cove / Clean mud / Sample animals / Most animals die / Time / Transfer to polluted cove / Laboratory population survives pollution

Contaminated mud / Toxin-resistant survive... / and reproduce

**Result:** Naturally occurring variants that can survive in contaminated mud reproduce, founding a population that can survive in the polluted cove.

**Conclusion:** Evolution has happened.

**FIGURE 15.A    Observing Evolution.** Evolution can occur rapidly enough for us to observe it, as the animal life in Foundry Cove illustrates.

for several generations. They then returned the descendants to the polluted cove, where they survived—indicating that the toxin resistance is genetic.

Next, researchers hypothesized that the population had evolved in a way that enabled the animals to withstand the pollution. This may have taken as little as 30 years. To test their hypothesis, researchers tried to re-create the cove's history in the laboratory (**figure 15.A**). They sampled

invertebrates from a clean cove and exposed them to cadmium-rich mud in the laboratory. Then, they bred the survivors with each other, building a population where most individuals could tolerate cadmium. When placed in polluted waters, the animals survived. Evolution had occurred, and was occurring, due to the survival and reproduction of animals that had inherited alleles enabling them to produce fertile offspring in the presence of heavy metal.

- **Boxed Readings**—These readings highlight the relevance of chapter contents to health and scientific inquiry.

  - *Health* readings discuss health issues of interest to the student.

  - *Investigating Life* features help removing some of the mystique of science, leading the reader through ways that scientists think when carrying out real experiments and investigations.

## Connections

The processes of cellular respiration are at the heart of everything we do. We can understand much of how the body functions by understanding its biochemical processes. We often think of energy as a somewhat abstract idea. When we recognize that we need energy to repair body damage, we rarely connect where that energy is actually used. If you consider that proteins need to be used in wound repair and that amino acids are needed for the protein, you can see how pulling a molecule out of the Krebs cycle to make that protein would result in less energy being extracted from that molecule. All of us have felt low on energy, but what causes those feelings? The body cannot send signals because it lacks the signal molecule. Or the movement of muscles is slower, caused by a delay in getting the ATP needed. Energy, in cellular terms, often means building molecules or moving them around. Chapters 9 and 10 explore some of these cellular processes.

## Student Study Guide

### Key Terms

| | | | |
|---|---|---|---|
| acetyl CoA  131 | aerobic respiration  127 | coenzyme A  131 | Krebs cycle  128 |
| anaerobes  127 | alcoholic fermentation  139 | cristae  128 | lactic acid fermentation  139 |
| aerobe  126 | anaerobic respiration  127 | electron transport chain  133 | matrix  128 |
| aerobic cellular | chemiosmotic | glycolysis  127 | pyruvic acid  128 |
| respiration  127 | phosphorylation  128 | intermembrane | substrate-level |
| | | compartment  128 | phosphorylation  128 |

### Chapter Summary

**8.1 Cellular Respiration: An Introduction**

1. All living cells use cellular respiration to extract energy from the bonds of nutrient molecules.
2. Aerobic organisms use oxygen as a final electron acceptor during cellular respiration. Anaerobic organisms use different molecules and pathways in the absence of oxygen.

**8.2 An Overview of Glucose Utilization**

3. The general equation for aerobic respiration is $C_6H_{12}O_6 + 6O_2 \longrightarrow 6CO_2 + 6H_2O + 30ATP$
4. Glycolysis is the most common energy-extracting process. It extracts some energy from splitting glucose into two molecules of pyruvate.
5. In eukaryotic cells, cellular respiration, except for glycolysis, occurs in the mitochondria.
6. Cells use phosphorylation to transfer energy from one molecule to another.

**8.3 Glycolysis: From Glucose to Pyruvic Acid**

7. Cells use enzyme systems to extract energy in a controlled manner by carefully rearranging energy-rich molecules.
8. Glycolysis splits glucose into two molecules of pyruvate while transferring energy to NADH.

**8.4 Aerobic Respiration: The Krebs Cycle**

9. Enzyme cycles use the product of one reaction as the substrate of the next, then re-form the starting molecule. The Krebs cycle transfers energy to ATP and NADH as the remaining carbons from glucose are rearranged.
10. The enzymes of the first energy-capturing step in the mitochondria release carbon dioxide and reduces $NAD^+$ to NADH. A carrier coenzyme transfers the remaining carbons to the Krebs cycle.

**8.5 The Electron Transport Chain**

11. Electron transport uses the energy from NADH and $FADH_2$ to form a proton gradient in the mitochondrial matrix, which is then used to phosphorylate ATP.
12. Oxygen is required at this step as a final acceptor of the energy-depleted electrons.
13. Each NADH yields 2.5 ATP, and each $FADH_2$ yields 1.5 ATP via this process. The net energy extracted from glucose is 30 to 32 ATP.
14. Feedback mechanisms regulate the activity of enzymes to ensure the cell has the energy it needs.

---

- **End-of-Chapter Material** will include ALL new features for the sixth edition:

  - The **Student Study Guide** will no longer need to be purchased as a separate study guide. It is now included at the end of each chapter!

  - **Connections** provides a short summary at the end of each chapter.

  - **Key terms** are now included. Page numbers are provided for easy reference.

- **Reviewing Concepts**—New to the 6th edition, this short summary appears at the end of each major heading.

---

organisms may have been unicellular algae. Australian fossils consisting of an organic residue 1.69 billion (1,690 million) years old are chemically similar to eukaryotic membrane components and may come from a very early unicellular eukaryote.

About 1.2 billion years ago, multicellular life appeared. Exactly how life proceeded from the single-celled to the many-celled is a mystery. Perhaps many unicellular organisms came together and joined, then took on different tasks to form a multicellular organism. Alternatively, a large single-celled organism may have divided into many subunits and diverged in gene expression and function. The earliest fossils of multicellular life are from a red alga that lived 1.25 billion to 950 million years ago in Canada. Abundant fossil evidence of multicellular algae comes from Siberia, dating from a billion years ago.

FIGURE 19.12 Complexity Increases. A "typical" Ediacaran organism, *Dickinsonia,* grew to a meter in diameter, but only 3 millimeters thick. It had segments and two different ends and internal detail that paleontologists have interpreted as remnants of a simple circulatory or digestive system. But just what it was remains unclear. One group of paleontologists couldn't agree on whether it most resembled a jellyfish, a flatworm, or a fungus! Like other Ediacarans, *Dickinsonia* might not have been anything like the species we are familiar with today.

**Reviewing Concepts**

- Fossil evidence suggests the first prokaryotic cell arose 3.8 billion years ago.
- The first photosynthetic cell followed at 3.5 billion years ago.
- This was followed by the diversification of cells to form the precursors of today's different cell types.

### 19.4 The Ediacarans

Paleontologists divide natural history into the Precambrian and everything that came during and after the Cambrian (590 until 505 million years ago), which was the first period of the Paleozoic era. The reason for this distinction is that fossils of all the major phyla of animals appear rather suddenly in sediments from the Cambrian. It seems highly unlikely that the late Precambrian was lifeless; instead, whatever lived then did not leave fossils. Most animal forms from the Precambrian were soft-bodied and did not readily fossilize. Charles Darwin wrote that the Cambrian explosion was misleading and that the Precambrian seas must have "swarmed with living creatures."

As the Proterozoic era became the Paleozoic, from about 600 to 544 million years ago, many parts of Earth were home to organisms called Ediacarans, which have no known modern descendants. The Ediacarans have long puzzled biologists because of their appearances, preserved in sandstones and shales as casts and molds, that are notably different from life as we know it today. Their bodies were apparently soft and very flat—figure 19.12 shows one representative organism, *Dickinsonia,* that was a meter in diameter but less than 3 millimeters in body thickness. This flattened form may have been an adaptation for extracting maximal oxygen from the ancient seas. Ediacaran bodies were built of sections of tubing, branches, ribs, discs, and fronds. Ediacarans had no complex internal organ systems, no

obvious body openings, and they probably could not move very far. Biologists have interpreted Ediacaran fossils to be everything from worms to ferns to fungi, as **figure 19.13** demonstrates.

The Ediacarans vanished from the fossil record about 544 million years ago, by which time the diversity of life in the Cambrian was exploding. Perhaps their disappearance resulted from change in the very specific conditions under which they were preserved. The height of the Ediacaran reign—from 570 to 580 million years ago—was also the time that phosphatization preserved large numbers of algae, sponges, and unidentified animal embryos in the seas (see figure 18.7E). Instead of Ediacaran life suddenly giving way to familiar animal phyla, these organisms may have more gradually vanished, opening up habitats for the organisms that left the phosphatized remains, and others. As the Ediacarans disappeared, the Cambrian seas filled with a spectacular diversity of life.

**Reviewing Concepts**

- The first multicellular life is represented by fossils of the Ediacarans, which were simple organisms without any obvious complex organs or systems, just an association of cells to form tissues.

Plant or Animal? The leafy sea dragon is exquisitely adapted for its environment, its form mimicking the surrounding seaweed.

## The Surprising Sea Dragon—A Metaphor for Life

Welcome to biology, the study of life. Our journey will take us from atoms to biomes, to the familiar and past the strange. A recurrent lesson will surface as we travel: Life is not always what it seems. Take, for example, the leafy sea dragons (scientifically named *Phycodurus eques*). These surprising animals are good metaphors for life.

Sea dragons are not dragons, and though they are closely related to sea horses, they are not horses. They are fish. Yet, in common with horses, kelp, and all other organisms, they are made of cells, which are in turn composed of molecules made mostly of four common atoms. One such molecule is DNA, the molecule of inheritance, which dictates where a snout is to develop and the structure of a leaflike fin.

With other fish a leafy sea dragon shares many features including gills, gill coverings, and a swim bladder. Less obvious are the many characteristics in common with other vertebrates, such as sharks, snakes, birds, and horses. A skull encases a brain in the animal's head; a backbone encloses a nerve cord. All species share some features of life, but may manifest them in very different ways.

Incongruities in nature are provocative. They spark our curiosity and lead to bigger questions. What makes a fish a fish? How can a fish come to resemble kelp? Is behavior inherited? Why is there sex? How do parental roles evolve? Why do sea dragons live where they do? How do species influence each other? Some of the answers are in this book, and others remain to be discovered.

# What Is Life?

**1.1   Organization and Energy: Why a Rabbit Is Not a Rock**
- Organisms Exhibit Dynamic Organization
- All Organisms Utilize Energy
- Organisms Maintain Internal Constancy

**1.2   The Continuity of Life in the Environment: Why Rabbits Abound**
- Organisms Reproduce, Grow, and Develop
- Organisms Respond to the Environment

**1.3   Life Evolves, Producing Great Diversity**
- Evolution Is Essential to Life
- Domains and Kingdoms Describe and Categorize Life's Diversity

**1.4   The Process of Scientific Inquiry**
- A Theory Summarizes Knowledge
- Controlled Experiments Verify Theories
- Experimental Design Seeks to Ensure Validity

**1.5   The Limitations of Scientific Inquiry**
- Results Can Be Interpreted Different Ways
- Unexpected Results May Be Discounted

## Chapter Preview

1. Life has unique characteristics that distinguish it from nonliving matter.

2. Organization, from atoms to ecosystems, is vital to the functions of organisms.

3. Organisms utilize energy to maintain life, grow, and reproduce. Diversity arises from the various ways organisms solve life's challenges.

4. All life shares common functions and structures, coming from a common ancestor. Evolution drives the formation of new species.

5. Biologists use increasingly detailed descriptions to classify life.

6. Science uses specific approaches to asking questions and learning about the natural world.

7. Science produces conclusions that are reliable due to controlled experiments and careful objectivity.

# 1.1    Organization and Energy: Why a Rabbit Is Not a Rock

We all have an intuitive sense of what life is: If we see a rabbit on a rock, we "know" that the rabbit is alive and the rock is not. But it is difficult to state just what makes the rabbit alive and not the rock. We recognize organisms are objects that have the characteristics of life, but what are those characteristics?

Throughout history, the question, "What is life?" has stymied thinkers in many fields. Eighteenth-century French physician Marie François Xavier Bichat poetically, but imprecisely, defined life as "the ensemble of functions that resist death." Others less eloquent and no more precise hypothesized that life is a kind of mysterious "black box" that endows a group of associated biochemicals with the qualities of life. But black-box thinking tends to assume that some things cannot be understood; by contrast, science seeks explanations.

One scientific approach to defining life reduces organisms to the smallest parts that still exhibit the characteristics of life and then identifies what makes those units different from their non-living components. Today's researchers seek the minimal requirements for life by probing the simplest organisms for those genes that must be present in all life. But many people believe life to be much more. Nineteenth-century philosophers called *vitalists* did, stating that some indefinable "essence or force" was what made life unique. To their way of thinking, the vital force entered just before birth to make something alive and left only at death. The medical profession usually defines death as cessation of brain

activity. But what constitutes death for the millions of other types of organisms, particularly those without brains?

After years of asking the question, science now recognizes five qualities that, in combination, constitute life:

1. organization;
2. energy use and metabolism;
3. maintenance of internal constancy;
4. reproduction, growth, and development;
5. irritability and adaptation (response to stimuli).

Some of these characteristics may also occur in inanimate objects: a rock in the sun absorbs sunlight and passes heat to the rabbit sitting on it, like the rabbit absorbs energy from food. A computer or self-flushing toilet responds to stimuli, as does a startled rabbit. A supermarket is highly organized, like the anatomy of a rabbit with all its organs and cells. Organisms must have all five characteristics, but each can be reflected in so many different ways that millions of species have emerged. The challenges of being alive have produced a tremendous range of diversity as each species evolves its own answers.

## Organisms Exhibit Dynamic Organization

Biological organization is apparent in all life. **Figure 1.1** summarizes the levels of this organization. Humans, eels, and evergreens, although outwardly very different, are all nonetheless comprised of structures organized in a particular three-dimensional relationship, often following a pattern of structures within structures within structures. Bichat was the first to notice this pattern in the human body. During the bloody French Revolution, as he performed autopsies, Bichat noticed that the body's largest structures, or **organs,** were sometimes linked to form collections of organs, or **organ systems,** but were also composed of simpler structures, which he named **tissues** (from the French word for "very thin"). Had he used a microscope, Bichat would have seen that tissues themselves are comprised of even smaller units called **cells.** Within complex cells, such as those that make up plants and animals, are structures called **organelles** that carry out specific functions. Bacteria, although smaller and less complex than animal or plant cells, also use highly organized structures. A cell is the simplest structure that can support life.

Life also functions beyond the level of the individual. A **population** is two or more members of the same type of organism, or species, living in the same place at the same time. A **community** includes the populations of different species in a particular region, and an **ecosystem** includes both the living and nonliving components of an area. Ultimately, the parts of the planet that can support life, and all of the organisms that live there, constitute the **biosphere**.

All life consists of cells, and cells are organized groups of **molecules.** In fact, cells may be considered as organized combinations of molecules that interact in controlled, highly specific ways (figure 1.1). Although some chemicals abundant in organisms, such as water, are also widespread in the nonliving world,

**Biosphere**
Parts of the planet and its atmosphere where life is possible.

**Ecosystem**
The living and nonliving environment. (The community of life, plus soil, rocks, water, air, etc.)

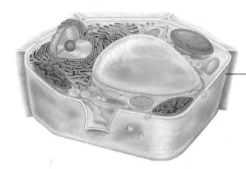

**Community**
All organisms in a given place and time.

**Population**
A group of the same type of organism living in the same place and time.

**Multicellular organism**
A living individual.

**Organ system**
Organs connected physically or chemically that function together.

**Organ**
A structure consisting of tissues organized to interact to carry out specific functions.

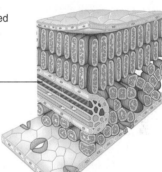

**Tissue**
A collection of specialized cells and the substances they secrete that function in a coordinated fashion.

**Cell**
The fundamental unit of life.

**Organelle**
A membrane-bounded structure within a complex cell that has a specific function.

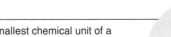

**Molecule**
A small group of joined atoms. (An amino acid is a building block of a protein.)

**Atom**
The smallest chemical unit of a type of pure substance (element). Includes protons, neutrons, and electrons.

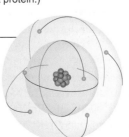

**FIGURE 1.1    Levels of Biological Organization Reveal Common Features of All Life.**

those that are uniquely found in cells are called **biochemicals.** One of the most important molecules, DNA uses four types of building blocks to store information, like a language of life that tells cells how to construct other vital molecules, such as proteins. Each cell's use of genetic instructions—as encoded in DNA— enables them to specialize and to function in tissues, organs, and organ systems. The organization of topics in this book reflects this ordered nature of life.

At all levels, and in all organisms, structural organization is closely tied to function: disrupt a structure, and function ceases. Shaking a fertilized hen's egg stops the embryo within from developing. Conversely, if function is disrupted, a structure will eventually break down. Unused muscles, for example, begin to atrophy (waste away). Biological function and form are interdependent.

An organism, however, is more than a collection of smaller parts, just as a car is much more than would be evident from a pile of its parts. Different levels of organization impart distinct characteristics that cooperate to form even more characteristics and maintain life. For example, nutrient molecules cannot harness energy directly from sunlight, but other molecules embedded in an organelle within a plant cell can use solar energy to synthesize nutrients. Functions that arise as complexity grows are called **emergent properties** (**figure 1.2**). They arise from physical and chemical interactions among components, much like flour, sugar, butter, and chocolate become brownies—something not evident from the parts themselves. The concept of emergent properties makes life, and all its diversity, possible from just a few types of starting materials.

## All Organisms Utilize Energy

Life's organization may seem contrary to the natural tendency of matter to become random or disordered. Maintaining order requires effort, so organisms must be able to acquire and use energy. This enables them to build new structures, repair or break down old ones, and reproduce. The term **metabolism** refers to the chemical reactions within cells that maintain life. Some reactions synthesize (build up) and others degrade (break down) the molecules of life. Metabolic processes occur at the whole-body level, as well as at the organ, tissue, and cellular levels.

All organisms extract energy from their environment (**figure 1.3**). **Producers** (also called autotrophs) extract energy from the nonliving environment, such as plants capturing light energy from the sun or bacteria deriving chemical energy from rocks. **Consumers** (also called heterotrophs), in contrast, obtain energy by eating nutrients made by other organisms, such as a cow eating grass. **Decomposers** are consumers that obtain nutrients from dead organisms. Fungi, such as mushrooms, are decomposers.

Energy requirements link life into chains and webs of "who eats whom." Producers are at the beginning of all such webs, continuing through several levels of consumers and decomposers. This interdependency of life is another type of organization in the living world.

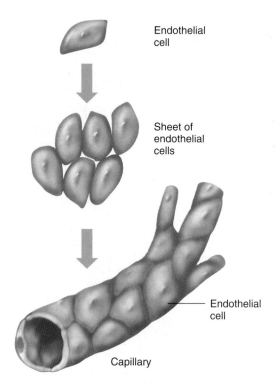

**FIGURE 1.2    An Emergent Property—From Tiles to Tubes.** Endothelial cells look like tiles. They adhere to one another to form a sheet. This sheet folds to form a tiny tubule called a capillary, which is the smallest type of blood vessel. The function of these cells does not "emerge" until they aggregate in a specific way. In other words, for an emergent property, the whole is greater than the sum of the parts.

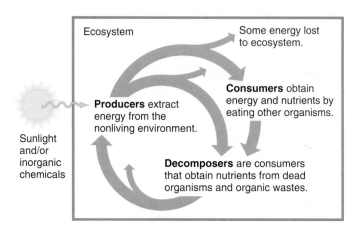

**FIGURE 1.3    Life Is Connected.** Organisms extract energy from each other (consumers) and, ultimately, the sun or inorganic chemicals (producers). Decomposers ultimately recycle nutrients to the nonliving environment.

## Organisms Maintain Internal Constancy

A rabbit is composed of many of the basic chemical elements found in a stream, yet it is obviously quite different. A stream can have a variety of different substances within its waters, depending on the season or pollution, and remain a stream. A rabbit, on the other hand, must maintain a precise composition, even in the face of drastic changes in the outside environment. The rabbit must take in nutrients, excrete wastes, and regulate its many chemical reactions to prevent deficiencies or excesses of specific substances. To keep metabolic processes running smoothly, organisms must maintain a certain temperature and the proper amount of water, salts, minerals, and so on. This ability to keep conditions constant is called **homeostasis.** An important characteristic of life, then, is the ability to sense and react to environmental change, thus keeping conditions within cells (and bodies) constantly compatible with being alive. Homeostasis in the presence of an ever-changing environment presents a significant challenge, requiring most of the energy consumed by an organism.

### Reviewing Concepts

- A living organism is distinguished from an inanimate object by the presence of a combination of characteristics.
- An organism is organized as structures of increasing size and complexity, from molecules, to cells, to tissues, organs, and organ systems, to individuals, populations, communities, ecosystems, and the biosphere.
- Emergent properties arise as the level of organization of life increases and are the consequence of physical and chemical laws.
- Metabolism utilizes energy to support homeostasis. Energy to fuel metabolism comes from the sun and other organisms.

## 1.2  The Continuity of Life in the Environment: Why Rabbits Abound

Most people, when asked to define life, will immediately think of abilities such as reproduction, movement and responses to stimuli. Ask anyone what they know about rabbits and most people will comment on their legendary reproductive capabilities. In fact, as we consider the qualities of life, reproduction is one of the most obvious. Most people will also think of the things that rabbits do to avoid predators. These are key characteristics found in one form or another in all species.

### Organisms Reproduce, Grow, and Develop

To maintain a population, or even life on a planet, organisms must reproduce—that is, make other individuals like themselves. Offspring then grow (increase in size) and develop, adding anatomical detail and taking on specialized functions until they resemble the parent and reproduce. Organisms have developed a dizzying array of approaches to reproduction, from the division of a bacterium, to the germination of a tree seed, or the concep-

Newly divided bacteria

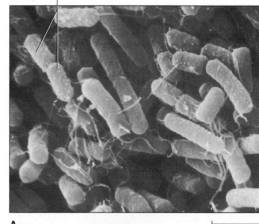

A
|—| 2 μm

B

C

**FIGURE 1.4    Reproduction Is One of the Most Obvious of Life's Characteristics.** **(A)** These *Escherichia coli* bacterial cells reproduce asexually every 20 minutes under ideal conditions. **(B)** A mighty oak tree begins with a small seedling, and each tree produces hundreds of acorns. **(C)** Mammals such as these deer usually produce only a few offspring per year.

tion, growth, and birth of a mammal (**figure 1.4**). Importantly, reproduction also transmits genetic instructions (DNA), which determine the defining characteristics from one generation to the next. Any group of organisms that can successfully reproduce over several generations is identified as a **species.**

Life employs two basic approaches to reproduction, depending on the organism and the environment. Offspring genetically identical to the parent are the product of **asexual reproduction.** Single-celled, or **unicellular,** organisms often reproduce asexually, particularly if the living conditions are nearly ideal. Some **multicellular** (composed of more than one cell) organisms can also reproduce asexually. For instance, a potato growing on an underground stem can sprout leaves and roots that form a new plant. Some simple animals, such as sponges and sea anemones, reproduce asexually when a fragment of the parent animal detaches and develops into a new individual. In a way, these new individuals are clones of the parent.

Regardless of the type of organism, asexual reproduction maintains the characteristics from generation to generation. By contrast, **sexual reproduction** mixes genetic material, usually from two individuals, to form genetically unique offspring. The number of parents does not define this type of reproduction, since some organisms can reproduce sexually from a sin-

gle parent by combining two genetically rearranged cells. By mixing genetic traits at each generation, sexual reproduction results in tremendous diversity in a population. Diversity ensures the survival of the species and also provides the raw material for new species.

Reproduction is a characteristic of life, but it is not restricted to life. A fire or a virus can "reproduce" very rapidly, but each lacks at least some of the other characteristics of life.

## Organisms Respond to the Environment

To stay alive, living organisms must sense and respond to certain environmental stimuli while ignoring others. **Irritability** is the tendency to respond immediately to stimuli. A person touching a thorn and jerking back in pain or a plant growing toward the sun are examples of irritability.

In contrast to the rapid, transient nature of irritability, **adaptation** is a response that develops over time. Individuals benefit from irritability, but adaptation is a population-level response that enables successive generations to successfully reproduce in a given environment. For example, "extremozymes" are enzymes produced by microorganisms that live in deep-sea hydrothermal vents, permitting metabolism at very high temperatures. Both a human and a dog will pull an extremity away from fire (irritability). But on a hot summer day, the human sweats, while the dog pants (adaptations).

Adaptations can be diverse and very striking. The color patterns of many organisms that enable them to literally fade into the background, such as the snake in **figure 1.5** and the similarity of a leafy sea dragon to sea plants, are adaptations. Such camouflaged organisms can hide from predators or await prey unnoticed.

Although irritability is not obvious in every species, adaptation is vital in all organisms, including plants. Most trees, for example, can survive strong winds because of their wide and sturdy trunks; flexible branches that sway without snapping; and strong, well-spread roots. **Figure 1.6** shows another adaptation to wind not readily apparent during the violence of a storm. A researcher who noticed that the fronds of certain algae close into cones and cylinder shapes in turbulent streams studied several types of trees during simulated "wind-tunnel storms" to see if similar reactions might occur. Indeed, he found the leaves of some types of trees do close into tighter shapes, which minimizes wind damage.

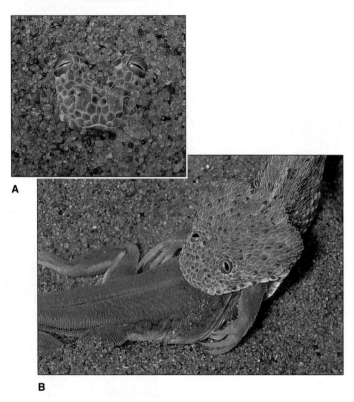

**FIGURE 1.5    An Adaptation to Acquire Food.    (A)** The superb camouflage of the adder snake, *Bitis peringueyi*, makes it virtually undetectable buried in the sand in the Namib Desert, Namibia. **(B)** It is little wonder that the sand lizard, *Aporosaura anchietae*, soon became the meal of the snake.

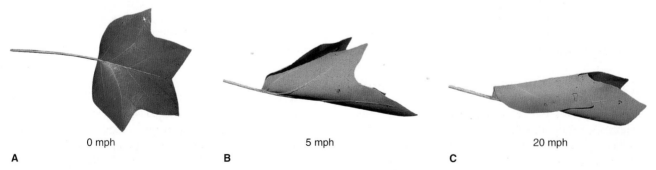

| 0 mph | 5 mph | 20 mph |
| --- | --- | --- |
| **A** | **B** | **C** |

**FIGURE 1.6    Leaves of Many Trees Are Adapted to Minimize Wind Damage.**    Researcher Steven Vogel subjected various leaves to a wind tunnel and snapped these photos, which reveal a previously unrecognized adaptation to survive a storm. Leaves **(A)** characteristically fold into a more compact form—under moderate **(B)** and high wind **(C)** conditions.

Adaptations accumulate in a population of organisms when individuals with certain inherited traits are more likely than others to survive in a particular environment. Trees with leaves that curl to resist wind are more likely to survive for the 20 or more years it takes for most trees to reproduce—even in the face of severe windstorms that strike, on average, every 5 years. Because trees with curling leaves survive more often than trees without this adaptation, more of them survive to reproduce and pass the advantageous trait to future generations. Trees with the adaptation eventually predominate in the population, unless environmental conditions change and protection against the wind becomes less important.

## Reviewing Concepts

- A species is a group of organisms that can successfully reproduce through several generations. Organisms use a variety of approaches to reproduction. Asexual reproduction maintains genetic information; sexual reproduction introduces new genetic changes.
- Organisms respond to the environment through irritability in the short term and by adaptation over generations. Natural selection eliminates inherited traits that decrease the chance of survival and reproduction in a certain environment.

# 1.3    Life Evolves, Producing Great Diversity

Evolution has been called the central principle of biology because of its ability to explain the origins of life's tremendous diversity and we will revisit this principle many times in this book. Evolution is often misunderstood because it occurs only over long periods of time. It is not the same as responding to stimuli, although prolonged exposure to some stimuli can lead to the formation of new species. As scientists have recognized these principles, they have formulated methods for classifying all species that show how they are related to each other.

## Evolution Is Essential to Life

Totally different species exhibit incredible similarities at the biochemical and cellular levels. The most logical explanation for these similarities is that the many species that have existed all descend from a common ancestral organism. Within each population, some individuals are better able to survive each new change in the environment due to inherited differences in their DNA. As individuals that have inherited particularly adaptive traits contribute more offspring, they come to make up more of the population. **Natural selection** is the enhanced survival and reproductive success of certain individuals from a population based on inherited characteristics. Over time, adaptation can mold the characteristics of a population as the genetic composition of the population changes, or **evolves.**

DNA not only encodes the information that cells require to function, but also may be changed, or **mutated.** Mutations in the DNA have provided, and continue to provide, the variation upon which natural selection acts. New species have emerged as genetic changes accumulate in response to an ever-changing environment. Evolution may be as subtle as the increasing prevalence of a particular trait or as profound as the extinction of a species. Largely because of natural selection, less than 1% of the species that have ever existed on Earth are alive today. Life is always changing.

## Domains and Kingdoms Describe and Categorize Life's Diversity

The mechanisms of evolution are the source of the many species that have populated the Earth. Life is united in its basic defining characteristics, yet **biodiversity** results from the many adaptive strategies different species use to maintain those characteristics. Biodiversity, all the varieties of life, makes the world a healthier place by providing support and connections among species that depend upon each other for survival, including humans.

We humans love to classify things. We organize large stores by departments; sort laundry; and assign students to grades and classes based on their ages, abilities, and interests. Classification helps us to make order out of huge amounts of information.

Similarly, the biological science of **taxonomy** classifies life according to what we know about the evolutionary relationships of organisms—that is, how recently two different species shared a common ancestor. Researchers compare anatomical, behavioral, cellular, and biochemical characteristics to identify similarities. The more similarities, the more closely related the species.

Biologists use several levels with increasingly restrictive criteria to describe and classify organisms. Just as a student is assigned to a particular school district, school, grade, and class, an organism is assigned to a specific **domain, kingdom, division** (or **phylum**), **class, order, family, genus,** and **species.** Each category contains organisms that share a set of defined characteristics.

Following the relatively recent discoveries of many types of unicellular organisms called Archaea, taxonomists added the domain as the broadest category. Basic differences in cellular constituents and organization distinguish three domains: Bacteria, Archaea, and Eukarya. Members of Bacteria and Archaea are all single-celled without nuclei (organelles that house the genetic material). All eukaryotes have nuclei, and many are multicellular like us. Bacteria and Archaea differ from each other in the types of molecules used to do similar tasks.

Cell complexity, mode of energy use and acquisition, and reproductive mechanisms distinguish the kingdoms. **Figure 1.7** illustrates the relationship between domains and kingdoms, which are more familiar and specific. Organisms within a kingdom demonstrate the same general strategies for staying alive, but they differ from each other in the details of how they do so.

The members of a domain or kingdom share many characteristics, yet they are quite diverse. A human, squid, and fly, for example, are all members of the animal kingdom, but they are clearly very different from each other and do not share features in the other more specific taxonomic levels (phylum to species). A human, rat, and pig are more closely related—all belong to the same kingdom, phylum (Chordata), and class

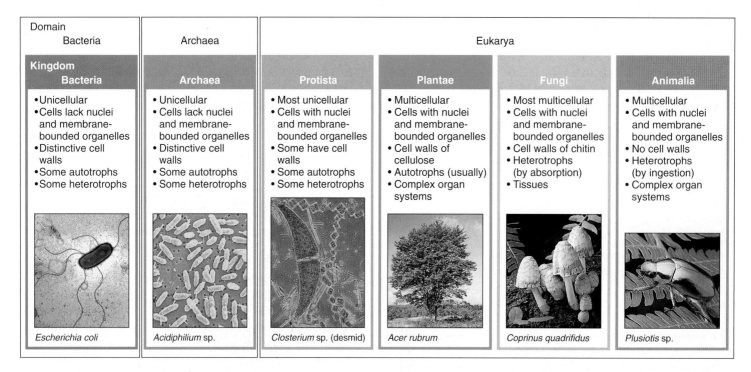

| Domain | | | | | |
|---|---|---|---|---|---|
| Bacteria | Archaea | Eukarya | | | |
| **Kingdom** **Bacteria** | **Archaea** | **Protista** | **Plantae** | **Fungi** | **Animalia** |
| • Unicellular<br>• Cells lack nuclei and membrane-bounded organelles<br>• Distinctive cell walls<br>• Some autotrophs<br>• Some heterotrophs | • Unicellular<br>• Cells lack nuclei and membrane-bounded organelles<br>• Distinctive cell walls<br>• Some autotrophs<br>• Some heterotrophs | • Most unicellular<br>• Cells with nuclei and membrane-bounded organelles<br>• Some have cell walls<br>• Some autotrophs<br>• Some heterotrophs | • Multicellular<br>• Cells with nuclei and membrane-bounded organelles<br>• Cell walls of cellulose<br>• Autotrophs (usually)<br>• Complex organ systems | • Most multicellular<br>• Cells with nuclei and membrane-bounded organelles<br>• Cell walls of chitin<br>• Heterotrophs (by absorption)<br>• Tissues | • Multicellular<br>• Cells with nuclei and membrane-bounded organelles<br>• No cell walls<br>• Heterotrophs (by ingestion)<br>• Complex organ systems |
| *Escherichia coli* | *Acidiphilium* sp. | *Closterium* sp. (desmid) | *Acer rubrum* | *Coprinus quadrifidus* | *Plusiotis* sp. |

**FIGURE 1.7    Organizing Life's Diversity.**    Before microscopes revealed the microbial world, it was easy to describe life as either plant or animal; the fungi were considered plants. With increasing discoveries and descriptions of microbes, it became clear that two, three, or even four kingdoms (animals, fungi, plants, and bacteria) would not suffice. A five-kingdom system, splitting microorganisms into bacteria and protista, led for a long time, until some years after the 1977 discovery of the Archaea. Identified first in extreme habitats like those of early Earth and therefore considered to be ancient (hence the name), the Archaea share some characteristics with the Bacteria, but also with organisms that have cells with nuclei. Many of their genes have never been described before. The descriptions of Bacteria and Archaea are identical here because most of their differences are at the molecular level. Today, biologists have reorganized the classification of life into three domains and are still determining the numbers of kingdoms in each domain and their relationships to each other. Overall, biological classification strives to depict evolutionary relationships among organisms, both past and present. Figure 4.4 lists more specific distinctions among the three domains of life. The criteria listed here will become more meaningful in later chapters.

(Mammalia). A human, orangutan, and chimpanzee are even more closely related, sharing the same kingdom, phylum, class, and order (Primates).

As part of a full classification system, taxonomists give each species a name that consists of two descriptive words: a human is *Homo sapiens.* Each category is based on highly descriptive common features of increasing complexity. For example, our full classification is Eukarya Animalia Chordata Mammalia Primates Hominidae Homo *Homo sapiens.* Another familiar organism is classified as Eukarya Plantae Angiospermophyta Monocotyledoneae Liliales Liliaceae Allium *Allium sativum*—otherwise known as garlic.

### Reviewing Concepts

• All life shares characteristics that emerged from a common ancestor. Evolution is the foundation of life, giving rise to all species.

• The tremendous diversity of organisms may be classified on the basis of increasingly complex characteristics.

• A species name is composed of two descriptive terms.

## 1.4    The Process of Scientific Inquiry

Anybody can be a scientist—it basically requires observing and asking questions about nature. Science seeks objective evidence that explains how nature works, and many scientific investigations are based on discovery. A microbiologist finds a never-before-seen microorganism in a hot spring; an ecologist catalogs the species that survived severe weather; a cell biologist discovers a molecule that signals a cell to divide. Based on some observation or discovery, a scientist formulates a prediction, called a **hypothesis,** which describes the processes behind the observation. The hypothesis is then tested through an **experiment,** which confirms or refines the hypothesis and adds additional observations. The hypothesis is strengthened as other scientists conduct their own experiments and confirm the results.

Often, new scientific information comes from connecting observations, which can lead to experimentation, as Investigating Life 1.1 describes. Through observations, many links have been established between environmental damage and the development of specific cancers. For example, researchers began comparing

## Elephants Calling—The Powers of Observation

Humans aren't the only animals to use language to communicate. Consider an elephant's extended family. An elephant clan, led by an elder female, or matriarch, and consisting of females and young males, has quite a vocabulary. Elephants communicate using infrasound, which is sound too low for human ears to detect.

In 1984, Cornell University researcher Katy Payne was standing near caged elephants at a zoo when she felt a "throbbing" in the air. The sensation brought back memories of singing in the church choir as a child where the church organ pipes made a similar throbbing. "In the zoo I felt the same thing . . . , without any sound. I guessed the elephants might be making powerful sounds like an organ's notes, but even lower in pitch," she writes in her children's book, *Elephants Calling*.

To test her hypothesis, Payne and two friends used equipment that could detect infrasound to record elephants in a circus and a zoo. Finding infrasound, she moved her study to the Amboseli Plain, a salty, dusty stretch of land at the foot of Mount Kilimanjaro in Tanzania. Her lab was a truck sitting among the elephants, who grew so used to its presence that they regarded it as part of the scenery.

Living among the elephants further sharpened Payne's already highly developed powers of observation. She and her fellow elephant-watchers soon became attuned to the subtle communications between mother and calf; the urges of a male ready to mate; and messages to move to find food or water. Writes Payne, "It is amazing how much you can learn about animals if you watch for a long time without disturbing them. They do odd things, which at first you don't understand. Then gradually your mind opens to what it would be like to have different eyes, different ears, and different taste; different needs, different fears, and different knowledge from ours."

One day, two bulls were fighting for dominance when the youngest family member, baby Raoul, slipped away (**figure 1.A**). Finding a hole, Raoul stuck his trunk inside,

**FIGURE 1.A**    Raoul rarely ventures far from his mother, Renata.

then leapt back, bellowing, as a very surprised warthog bounded out of his invaded burrow. Raoul's mother, Renata, responded with a roar. Payne described the scene:

*Elephants in all directions answer Renata, and they answer each other with roars, screams, bellows, trumpets, and rumbles. Male and female elephants of all sizes and ages charge past each other and us with eyes wide, foreheads high, trunks, tails, and ears swinging wildly. The air throbs with infrasound made not only by elephants' voices but also by their thundering feet. Running legs and swaying bodies loom toward and above us and veer away at the last second.*

The family reunited, all the animals clearly shaken. Unable to resist comparing the pachyderms to people, Payne writes, "Renata does not seem angry at Raoul. Perhaps elephants don't ask for explanations."

The kind of observation that leads to new knowledge or understanding, Payne says, happens only rarely. "You have to be alone and undistracted. You have to be concentrating on what's there, as if it were the only thing in the world and you were a tiny child again. The observation comes the way a dream—or a poem—comes. Being ready is what brings it to you."

Here is how the steps of scientific inquiry can be applied to Payne's research.

**Observation:** The air near the elephant's cage at the zoo seems to vibrate.

**Background knowledge:** Organ pipes make similar vibrations when playing very low notes.

**Hypothesis:** Elephants communicate by infrasound, making sounds that are too low for the human ear to hear.

**Experiment:** Record the elephants with equipment that can detect infrasound.

**Results:** Elephants make infrasound.

**Further observations:** elephants emit infrasound only in certain situations involving communication.

**Conclusion:** Elephants communicate with infrasound.

**Further question:** Do different patterns of sounds communicate different messages?

characteristics shared by people at higher-than-normal risk for cancer: smokers are more likely than nonsmokers to develop lung cancer; people exposed to atomic-bomb test blasts in the southwestern United States in the 1950s face increased risk of developing blood cancers; young women who painted a radioactive chemical onto watchfaces in the 1920s were more likely to develop bone cancer. Experiments on nonhuman animals and cells or tissues growing in culture then provided the details of how the environmentally induced cancer develops. The study of disease-related data from real life is called **epidemiology,** and it is often a starting point for experimental approaches.

## A Theory Summarizes Knowledge

When asked to comment on the idea of biological evolution, former U.S. president Ronald Reagan dismissed its validity, saying, "It's only a theory, anyway." In an informal sense, a theory may very well be little more than an opinion. But in many fields of study, "theory" has a distinct meaning, particularly in science.

Unlike a hypothesis, which is applied to initial observations, a **theory** is a systematically organized body of knowledge that applies to a variety of situations. Although theories attempt to logically explain natural phenomena, we can never know something scientific with absolute certainty. Although common in advertising, a scientist would never use the term "scientific proof." There is always more to learn. And theories change as knowledge accumulates. The history of science is full of long-established ideas changing as we learned more about nature, often thanks to new technology. People thought that Earth was flat and at the center of the universe before inventions and data analysis revealed otherwise. Similarly, biologists thought all life was plant or animal until microscopes unveiled a world of organisms invisible to our eyes. As we increase in our knowledge over many years, the strength of the evidence for a theory often grows to the point that it comes to be considered a law. This is the case for the principles that underlie heredity, called Mendel's laws. And, although many nonscientists scoff at evolution as "just a theory," most biologists, familiar with the abundant evidence, consider it law.

## Controlled Experiments Verify Theories

Theories build on the results of experiments and interpretations of observations. Those interpretations arise from application of the **scientific method,** which is a general way of thinking and of organizing an investigation. The scientific method is a framework in which to consider ideas and evidence in a way that can be repeated with the same results. Scientists can take several approaches to answering a particular question. An ecologist, a chemist, a geneticist, and a physicist conduct very different types of investigations.

Scientific inquiry consists of everyday activities—observing, questioning, reasoning, predicting, testing, interpreting, and concluding (**figure 1.8**). It includes thinking, detective work, and seeing connections between seemingly unrelated events. To explore the method of scientific inquiry and the development of a scien-

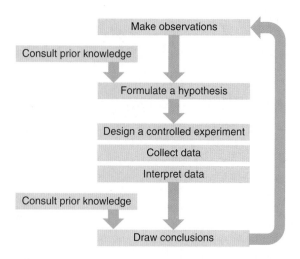

**FIGURE 1.8**    The Scientific Method Is a Means of Careful Discovery.

tific theory, we will consider several aspects of one environmental problem—the effects in animals of chemicals that resemble the hormone estrogen. An animal hormone is a biological messenger molecule that is produced in a gland and is transported in the bloodstream to where it exerts a specific effect on a particular organ. The idea that estrogen-like chemicals in the environment can harm health is called the estrogen mimic theory (**figure 1.9**).

## Observations

The scientific method begins with observations. These observations may be historical incidents or accidents:

- From 1949 to 1971, 2 million pregnant women in the United States took an estrogen-based drug called DES to prevent miscarriage. Years later, some of their daughters developed a rare vaginal cancer or less serious reproductive abnormalities.

- In 1980, a large amount of a now-banned pesticide, DDT, was dumped in Lake Apopka in Florida. Exposed male alligators had stunted penises. DDT is broken down to DDE, which functions as an estrogen. Alligators of both sexes had excess estrogen.

Observations can also be based on experimental results or may arise from making mental connections:

- Estrogen applied to human breast or uterine cells growing in culture stimulates them to divide.

- Women are more likely to develop breast cancer if they begin menstruating early and cease menstruating late—factors that expose them to more estrogen over their lifetimes than women who begin menstruating later and end earlier.

- Since 1938, human sperm counts have dropped, and the incidence of birth defects of the male reproductive system and abnormal sperm have increased. During the same time, use of pesticides containing estrogen-like chemicals has increased.

**Background Information**    Considering existing knowledge is important in scientific inquiry. To understand how estrogen

**FIGURE 1.9   Theories Are Built on Evidence.** The estrogen mimic theory proposes that estrogen-like chemicals in pesticides cause reproductive abnormalities. **(A)** DDT was sprayed on crops and on some people. **(B)** Bird populations exposed to DDE (a breakdown product of DDT) and other environmental estrogens experience many problems, such as this eagle with a malformed beak. **(C)** Nests containing broken eggs are evidence of the effect of DDT on the proper formation of eggshells, which are too fragile to bear the weight of the parent during incubation. **(D)** Abnormal sperm, such as this one with two extra tails, may also be the result of exposure to estrogen-based pesticides. More observations and experiments are needed to confirm the estrogen mimic theory.

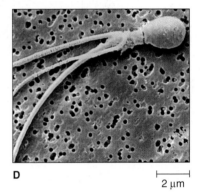

2 μm

might cause the observed effects, it is important to understand its normal role.

Estrogen molecules enter certain cells and bind to proteins called receptors, which fit them in much the same way that a catcher's mitt fits an incoming baseball. The binding occurs in the nucleus, the part of the cell that houses the genetic material. A cell that is sensitive to estrogen's stimulation may have many thousands of such receptor proteins. Estrogen molecules bound to receptors then activate certain genes, producing the hormone-associated effects, such as cell division. The proportion of estrogens to other sex hormones determines whether the reproductive system develops normally. Chemicals acting as estrogen mimics may disrupt this crucial hormonal balance in several ways.

**Formulating a Hypothesis** Observations in the scientific method lead first to a general question, such as

> Does excess estrogen exposure cause reproductive problems in animals, including humans?

The next step is to formulate a more specific statement to explain an observation, based on previous knowledge. This hypothesis is testable and should examine only one changeable factor, or variable. A hypothesis is more than a question or a hunch; it is based on known facts. Often a hypothesis is posed as an "if . . . then" proposition, which sets up a testable prediction:

> If exposure to large amounts of estrogen-like chemicals causes reproductive problems in animals, then eggs exposed to estrogen or similar substances should not hatch, or they will hatch to yield abnormal offspring.

**Devising an Experiment** Once a hypothesis is posed, a researcher collects information to test it. An experiment can disprove the hypothesis but can never prove it, because there is always the possibility of discovering additional information.

To determine whether DDE in the environment caused the reproductive abnormalities in the alligators, experiments recreated conditions that exposed the eggs to estrogen-like chemicals (**figure 1.10**). Investigators collected alligator eggs from a clean lake and incubated them at high temperatures, which in alligators ensures that the hatchlings are male. The researchers painted some eggs with estrogen, some with DDE, and some with an inert substance. This experimental design would test whether just painting an egg can alter development.

Among the estrogen-treated eggs, only 20% produced male hatchlings—a drastic change from the expected 100%. Only 40% of the eggs painted with DDE hatched males, which had excess hormone disturbances in the same proportions as did male alligators from Lake Apopka. Another 40% of the DDE-treated eggs hatched intersexes, which had both male and female reproductive structures. Intersexes are not seen among alligators in clean lakes. The remaining 20% of the DDE-treated eggs produced female

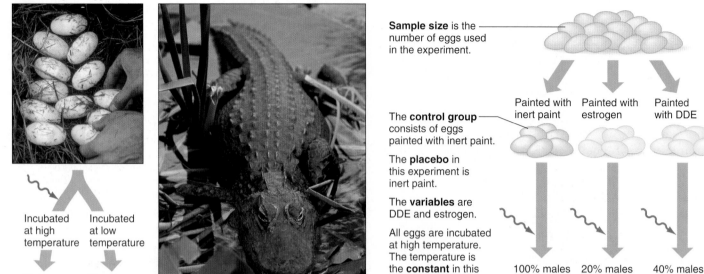

**FIGURE 1.10    Experiments Follow Rules.    (A)** In nature, the sex of alligators is determined by the incubation temperature of the eggs. Normally, alligator eggs incubated at high temperature hatch males. **(B)** Painting eggs with estrogen or DDE drastically lowers the percentage of male hatchlings. An experiment using alligator eggs to test the estrogen mimic theory must account for possible alternative explanations. The experiment was set up so that researchers did not know which eggs had received which treatment until after results had been tallied. This experimental design prevented unintentional bias, such as looking more carefully for abnormalities in animals that a researcher knows has received estrogen or DDE.

hatchlings. All of the eggs painted with the inert substance hatched normal male alligators.

These experimental results support the hypothesis and lead to the conclusion that DDE alters reproductive structures in alligators. But for a conclusion to become widely accepted, the results must be repeatable—other scientists must be able to perform the same experiments and observe the same results.

**Scientific Inquiry Continues**    Scientific inquiry does not end with a conclusion, because each discovery leads to further questions. Once experiments in painting alligator eggs provided evidence that DDE and estrogen cause reproductive abnormalities, a new question arose:

How do estrogen mimics exert their effects?

This new question suggests a new hypothesis:

If estrogen mimics produce the same or similar effects as estrogen, then the mimics should bind to estrogen receptors.

Chemical experiments support this hypothesis—estrogen receptors bind estrogens and DDE. Various insecticides, herbicides, pesticides, plastics, and fuels also bind to estrogen receptors, suggesting further roads of inquiry.

The cycle of scientific inquiry continues, as puzzle pieces are further connected to reveal a fuller picture of the estrogen mimic story. Because estrogen exposure correlates to increased risk of reproductive abnormalities and cancer, and pesticides introduce

estrogen-like compounds into the environment, we can link these facts to pose yet another hypothesis:

Are pesticide residues responsible for increases in breast cancer and reproductive problems in some populations of humans?

The experiments and observations discussed so far do not directly address this new hypothesis. The simultaneous increases in incidence of certain reproductive problems and pesticide use do not demonstrate cause-and-effect but rather a correlation—events occurring at the same time that may or may not be causally related. It will take experiments to investigate whether pesticide exposure causes breast cancer and other reproductive problems. If many tests continue to support a particular hypothesis, then that hypothesis attains the level of a theory. A scientific theory, then, is an idea, based on hypotheses that have survived strong testing, that has gained acceptance among scientists.

We now take a closer look at experimental design.

## Experimental Design Seeks To Ensure Validity

Scientists use several approaches when they design experiments in order to make data as valid as possible. The number of individuals or structures that are experimented on is the sample size. A useful experiment examines dozens of alligator eggs, or hundreds of cells, because a large sample helps ensure meaningful results. For example, sampling, by chance, only intersexes among the DDE-treated alligator eggs might support the conclusion that DDE exposure causes alligators to always develop as intersexes. Larger samples would reveal that DDE causes intersexes only 40% of the time.

Obtaining a large sample is not always practical or possible. When medical researchers study very rare disorders, only a few patients may be available. So they conduct small-scale (called pilot) experiments instead, which are valuable because they may indicate whether continuing research is likely to yield valid, meaningful results.

**Experimental Controls**  Distinguishing the unusual requires comparison to the usual. This is why well-designed experiments compare a group of "normal" individuals or components to a group undergoing treatment. Ideally, the only difference between the normal group and the experimental group is the one factor being tested. The normal group is called an experimental **control** and provides a basis of comparison. Designing experiments to include controls helps ensure that a single factor, or **variable,** is the actual cause of the observed effect.

Experimental controls may take several forms. A placebo presents the control group with the same overall experience as the experimental group. In the alligator egg experiment depicted in figure 1.10, the eggs painted with an inert substance received a **placebo.** In medical research, a placebo is often a stand-in for a drug being tested—a sugar pill or a treatment already known to be effective.

**Double-Blind Tests**  Another safeguard used in medical experiments is a **double-blind** design, in which neither the researchers nor the participants know who received the substance being evaluated and who received the placebo. The researchers break the "code" of who received which treatment only after the data are tabulated or if one group does so well that it would be unethical to withhold treatment from the placebo group. A famous advertising campaign used this approach to claim that Pepsi tastes better than Coke. In the alligator egg study, researchers didn't know which eggs had received which of the three treatments until they compiled the results. This avoided bias, such as looking more carefully for reproductive problems in alligators known to have been exposed to estrogen or DDE. However, it wasn't double-blind—unless the occupants of the eggs could know which treatment they received!

## Reviewing Concepts

- Scientific theories are ideas based on hypotheses that have survived rigorous testing. They attempt to explain natural phenomena.
- Scientific inquiry, which uses the scientific method, is a way of thinking that involves observing, questioning, reasoning, predicting, testing, interpreting, concluding, and posing further questions.
- Scientific inquiry begins when a scientist makes an observation; raises questions about it; and uses reason to construct an explanation, or hypothesis.
- Experiments test the validity of the hypothesis, and conclusions are based on data analysis.
- Experimental controls ensure the data reflect variables in the experiment, and not other factors.

## 1.5  The Limitations of Scientific Inquiry

The generalized scientific method is neither foolproof nor always easy to implement. **Table 1.1** lists some areas that are sometimes confused with science but are not science because they rely on beliefs and feelings, rather than evaluation of objective data such as measurements. But even data can be difficult to understand.

### Results Can Be Interpreted Different Ways

Experimental evidence may lead to multiple interpretations or unexpected conclusions, and even the most carefully designed experiment can fail to provide a definitive answer. Consider the observation that animals fed large doses of vitamin E live significantly longer than similar animals that do not ingest the vitamin. Does vitamin E slow aging? Possibly, but excess vitamin E causes weight loss, and other experiments associate weight loss with longevity. Does vitamin E extend life, or does the weight loss? The experiment of feeding animals large doses of vitamin E does not distinguish between these possibilities. Can you think of further experiments to clarify whether vitamin E or weight loss extends life?

Another limitation of implementing the scientific method is that researchers may misinterpret observations or experimental results. For example, scientists once concluded that life can arise by heating broth in a bottle and then corking it shut and observing bacteria in the brew a few days later. The correct explanation was that the cork did not keep bacteria out.

We cannot directly study natural phenomena that occurred only long ago. Many experiments have attempted to re-create the sequence of chemical reactions that might, on early Earth, have

### TABLE 1.1
### Distinguishing Science from Nonscience

| Science is | Science is NOT |
|---|---|
| Knowledgeable observations | Art Astrology Religion |
| Testable hypothesis | Creationism Extrasensory perception Philosophy |
| Controlled experiments | Fortune-telling Healing crystals Telepathy |
| Data-supported theories | Therapeutic touch Psychic phenomena Telekinesis |
| Understanding how nature works, based on objective evidence | Deciding how nature works through popular voting or subjective "feeling" |

formed the chemicals that led to life (see figures 18.3, 18.4, and
18.5). Although the experiments produce interesting results and
reveal ways that these early events may have occurred, we cannot
really know if they accurately re-created conditions at the beginning of life.  •  **origin of life, p. 344**

## Unexpected Results May Be Discounted

An investigator should try to keep an open mind toward observations, not allowing biases or expectations to cloud interpretation
of the results. To do so, scientists must expect the unexpected.
But it is human nature to be cautious in accepting an observation
that does not fit existing knowledge. The careful demonstration
that life does not arise from broth surprised many people who
believed that mice sprung from garbage, flies from rotted beef,
and beetles from cow dung.

Often, scientific discoveries depend very much upon our
willingness to accept unusual ideas. By 1796, people had
observed that young girls who milked cows and were exposed to
cowpox did not develop the devastating illness smallpox. An English country doctor, Edward Jenner, used this observation to
invent a vaccine for smallpox. He rubbed material from cowpox
lesions into the scratched upper arm skin of young human vol-
unteers, then exposed them to the related smallpox virus and
hoped this would alert their immune systems to prevent infection (**figure 1.11**). His actions would be considered unethical
today, and "volunteer" isn't the best term—it implies they were
informed of risks. A more recent example of accepting an
unusual explanation at the expense of a well-established one is
the common belief that stress causes ulcers. Today, we know that
most ulcers are caused by bacterial infection (see Investigating
Life 37.1).  •  **vaccines, p. 784**

Scientific research seeks to understand nature. Because
humans are part of nature, we sometimes tend to view scientific
research, and particularly biological research, as aimed at improv-

**FIGURE 1.11  Unusual Ideas Are Important in Science.**
Edward Jenner was at first ridiculed for his idea of using infection
with one virus (cowpox) to prevent infection by another (smallpox).
Many people at the time were afraid of receiving a smallpox vaccine
because they believed it would cause them to grow cow parts, as
this cartoon suggested. Some countries are reinstituting smallpox
vaccination to protect against use of the virus as a bioweapon.

ing the human condition. But knowledge without any immediate
application or payoff is valuable in and of itself—because we can
never know when information will be useful.

## Reviewing Concepts

- The scientific method does not always yield a complete
  answer or explanation or may produce ambiguous
  results.
- Discoveries may be unusual or unexpected.

## Connections

As we begin our study of life, it is important to remember that rules govern nature. Science also follows set rules as it seeks to
understand how nature works. We learn much about life by studying its organization, from atoms to worlds. The principles that govern interactions among species are ultimately based on principles that govern atoms and molecules, as we will explore in chapter 2.

# Student Study Guide
# Key Terms

| | | | |
|---|---|---|---|
| adaptation  6 | domain  7 | kingdom  7 | population  4 |
| asexual reproduction  6 | double-blind  13 | metabolism  4 | producer  4 |
| biochemical  4 | ecosystem  4 | molecule  4 | scientific method  10 |
| biodiversity  7 | emergent property  4 | multicellular  6 | sexual reproduction  6 |
| biosphere  4 | epidemiology  10 | mutate  7 | species  6 |
| cell  2 | evolve  7 | natural selection  7 | taxonomy  7 |
| class  7 | experiment  8 | order  7 | theory  10 |
| community  4 | family  7 | organ  2 | tissue  2 |
| consumer  4 | genus  7 | organelle  4 | unicellular  6 |
| control  13 | homeostasis  5 | organ system  2 | variable  13 |
| decomposer  4 | hypothesis  8 | phylum  7 | |
| division  7 | irritability  6 | placebo  13 | |

# Chapter Summary

## 1.1 Organization and Energy: Why a Rabbit Is Not a Rock

1. A living organism is distinguished from an inanimate object by the presence of a combination of characteristics.

2. An organism is organized as structures of increasing size and complexity, from biochemicals, to cells, to tissues, organs, and organ systems, to individuals, populations, communities, ecosystems, and the biosphere.

3. Emergent properties arise as the level of organization of life increases and are the consequence of physical and chemical laws.

4. Life requires energy to maintain its organization and functions. Metabolism directs the acquisition and use of energy.

5. Organisms must maintain an internal constancy in the face of changing environmental conditions, a state called homeostasis.

## 1.2 The Continuity of Life in the Environment: Why Rabbits Abound

6. A species is a group of organisms that can successfully reproduce through several generations.

7. Organisms develop, grow, and reproduce.

8. Organisms use a variety of approaches to reproduction. Asexual reproduction maintains genetic information; sexual reproduction introduces new genetic changes.

9. Organisms respond to the environment through irritability in the short term and by adaptation over generations. Natural selection eliminates inherited traits that decrease the chance of survival and reproduction in a certain environment.

## 1.3 Life Evolves, Producing Great Diversity

10. All life shares characteristics that emerged from a common ancestor. Evolution is the foundation of life, giving rise to all species.

11. The tremendous diversity of organisms may be classified on the basis of increasingly complex characteristics.

12. Biologists classify organisms with a series of names that reflect probable evolutionary relationships. A species name is composed of two descriptive terms.

13. The three domains of life are distinguished by cell structure and organization. Cell complexity, mode of nutrition, and other factors distinguish members of kingdoms.

## 1.4 The Process of Scientific Inquiry

14. Scientific theories are ideas based on hypotheses that have survived rigorous testing. Theories attempt to explain natural phenomena.

15. Scientific inquiry, by way of the scientific method, is a way of asking and answering questions in a rigorous and reproducible manner.

16. Scientific inquiry begins when a scientist makes an observation, raises questions about it, and uses reason to construct an explanation, or hypothesis.

17. Experiments test the validity of the hypothesis, and conclusions are based on data analysis. Experimental controls remove alternative explanations and minimize bias.

## 1.5 The Limitations of Scientific Inquiry

18. The scientific method does not always yield a complete answer or explanation or may produce ambiguous results. Discoveries may be unusual or unexpected.

19. Science is restricted to studying phenomena that yield reproducible data and can be empirically analyzed.

# What Do I Remember?

1. List the characteristics of life.
2. Cite two ways asexual and sexual reproduction differ.
3. How does natural selection act on adaptations in a way that causes evolution to occur?
4. Why do biologists assign taxonomic names to organisms, and what do those names reflect?
5. Describe the relationship between domains and kingdoms.
6. How is a theory more than an opinion and a hypothesis more than a guess?
7. Describe the necessary components of a scientific experiment.
8. What characteristics of life are also found in fire?
9. How is scientific inquiry a continuous process?
10. What are some limitations of scientific inquiry and experimentation?

### Fill-in-the-Blank

1. The presence of _____ provides the raw material for evolution.
2. Species with cell walls in multicellular, nonphotosynthetic bodies belong to the _____ kingdom.
3. The _____ of an experiment is a prediction that guides its design.
4. _____ is the variety of different approaches to the problems of being alive.
5. _____ is all of the chemical processes that maintain life.

### Multiple Choice

1. Which of the following is NOT a characteristic of life?
   a. the ability to convert sunlight energy into chemical energy
   b. a structure based upon cells and a hierarchy of organization
   c. the ability to establish a constant temperature
   d. a DNA molecule that stores information
2. Evolution is best described as
   a. the ability to respond to a stimulus.
   b. changes in an individual due to changes in environment.
   c. converting one form of energy to another.
   d. one species giving rise to a different species.
3. A scientific theory
   a. is just a guess, without supporting data.
   b. summarizes many years of accumulated experimental results.
   c. changes often as more experiments are conducted.
   d. is the foundation of an experimental design.
4. Which of the following best describes a species?
   a. a group of identical organisms
   b. a group of organisms that can reproduce
   c. two related groups of organisms
   d. all organisms that share a single characteristic
5. Which of the following sets is in the correct order, from smallest to largest?
   a. cell, molecule, organelle, organ, tissue
   b. cell, tissue, organ, organism, population
   c. kingdom, species, phylum, order, class
   d. phylum, class, order, family, genus

A lake can be deadly when it suddenly emits large volumes of carbon dioxide.

## Killer Lakes Spew $CO_2$

On an August morning in 1984, 37 people walking near Lake Monoun, near the village of Mjindoun in Cameroon, were suddenly enveloped by a white cloud coming from the lake and dropped dead. Perhaps because their numbers were small, from a remote place in the midst of political upheaval, their deaths did not capture media attention. Two years later, when 1,800 people died when nearby Lake Nyos similarly belched forth deadly gas, the world paid attention.

The geology of the region revealed what had happened. The two lakes are sandwiched among ancient, small volcanic cones ringed by tall grasses. Over the years, the cones emitted carbon dioxide ($CO_2$), which dissolved in the lakes and sank under the weight of the water. Then, a geologic upheaval—likely an underwater landslide triggered by seismic activity—released the gas trapped in the lakes' lower levels.

When a person inhales too much $CO_2$, the respiratory center in the brain increases breathing rate and the volume of air passed in and out of the lungs, in an attempt to obtain more oxygen. Normally, the person would exhale more $CO_2$ for awhile until blood gases normalized, but the output from the lakes was overwhelming. The respiratory system shut down.

Once geologist identified the source of the problem, plans began to sink large pipes into both lakes to release the $CO_2$ gradually. But the volcanic action continues to generate the gas, and the pipes are addressing the problem very slowly. Human activity is long gone from Lake Nyos, where so many perished, and the government now forbids settlement.

# Of Atoms and Molecules: Chemistry Basics

**2.1 Matter: The Basis for Rabbits and Rocks and Everything Else**
- Elements Have Unique Properties
- An Atom Is the Smallest Unit of an Element
- Molecules Are Atoms Bound Together

**2.2 When Atoms Interact: Chemical Bonds**
- Electrons Determine Whether Atoms Bond
- In Covalent Bonds, Electrons Are Shared
- In Ionic Bonds, Electrons Are Donated and Accepted
- Weak Chemical Bonds Provide Vital Links

**2.3 Water's Importance to Life**
- All Life Processes Take Place in Solutions
- Solutions May Be Acidic or Basic
- Water's Properties Regulate Temperature

# Chapter Preview

1. Any substance that occupies space is called matter. Matter can be broken down into pure substances called elements. An atom is the smallest unit of an element.

2. Subatomic particles include the positively charged protons and neutral neutrons, which form the nucleus, and the negatively charged, much smaller electrons, which circle the nucleus.

3. Compounds are formed when atoms join together as molecules. The characteristics of compounds are different from their constituent atoms.

4. Electrons move constantly, and their location is related to their energy level. The number of electrons in an atom determines how it will bond to other atoms, forming covalent, ionic, or hydrogen bonds.

5. The chemical reactions of life occur in aqueous environments. Water is cohesive and adhesive and has a high capacity for absorbing heat.

6. Life must control the amount of hydrogen ions released in solution to maintain chemical processes. The pH scale is an indicator of the concentration of hydrogen ions in a solution.

## 2.1 Matter: The Basis for Rabbits and Rocks and Everything Else

All solid objects are made of matter, often referred to as chemicals. Organisms consist of **matter** (material that takes up space) and **energy** (the ability to do work). Chapter 6 discusses the energy of life in detail; this chapter concentrates on the composition of living matter. Life is based on chemical principles. Without understanding these, we cannot understand life.

## Elements Have Unique Properties

All matter can be broken down into pure substances called **elements.** An element is a type of atom, and there are 92 known, naturally occurring elements and at least 17 synthetic ones. Each element has unique properties, and when we arrange elements according to their composition and properties, we have a chart called the **periodic table.**

It is interesting to look at the periodic table positions of the elements that make up organisms. These elements of life appear throughout the table, indicating they represent a diverse sampling of all chemicals (**figure 2.1**). Appendix D contains a complete periodic table.

Twenty-five elements are essential to life. Elements required in large amounts—such as carbon, hydrogen, oxygen, nitrogen, sulfur, and phosphorus—are termed **bulk elements;** those required in small amounts are called **trace elements.** Many trace elements are important in ensuring that vital chemical reactions occur fast enough to sustain life and are often the components found in vitamin supplements.

## An Atom Is the Smallest Unit of an Element

What gives each element a different character? It is the composition of the atom that comprises that element. An **atom** is the smallest possible "piece" of an element that retains the characteristics of the element. An atom is composed of three major types of subatomic particles: **protons** and **neutrons,** which form a centralized core called the **nucleus,** and **electrons,** which surround the nucleus. Atoms of each element have a characteristic number of

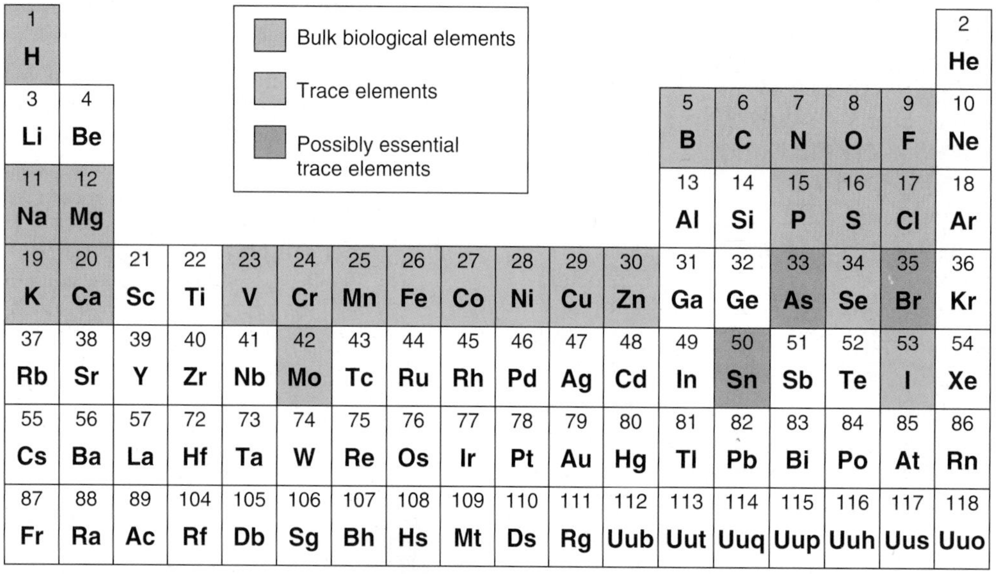

FIGURE 2.1  The Periodic Table of Elements.   Elements 58–71 and 90–103 are omitted for clarity.

protons. And, ultimately, it is the number of protons that determines the size and character of each atom. Hydrogen, the simplest of atoms, has only 1 proton and 1 electron; in contrast, an atom of uranium has 92 protons, 146 neutrons, and 92 electrons.

Protons carry a positive charge, electrons carry a negative charge, and neutrons are electrically neutral. Charge is the attraction between opposite types of particles. Within most atoms, the number of protons equals the number of electrons, and the atom is electrically neutral—that is, without a net charge. An electron is vanishingly small compared to a proton or a neutron, yet it exists far away from the nucleus of the atom. If the nucleus of a hydrogen atom were the size of a meatball, the electron belonging to that atom would be about 1 kilometer (0.62 mile) away from it! The term **orbital** refers to the most likely location for an electron relative to its nucleus. **Figure 2.2** summarizes the characteristics of the three types of subatomic particles.

The periodic table also depicts, in shorthand, the structures of the atoms of each element. Each element has a symbol, which can come from the English word for that element (**He** for helium, for example) or from the word in another language (**Na** for sodium, which is natrium in Latin). The **atomic number** above the element symbol and name shows the number of protons in the atom, which also establishes the identity of the atom. The elements are arranged sequentially in the periodic table by atomic number.

The **mass number** reflects the total number of protons and neutrons in the nucleus of the atom. (Weight and mass are related; weight is mass taking gravity into account.) For biological purposes, we can approximate the mass of a single proton (and that of a neutron) to be one. Since the contribution of an electron to an atom's mass is negligible, we can approximate the atomic mass by adding the number of protons and neutrons in each atom.

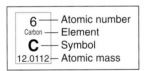

Often the atoms of an element can have a variable number of neutrons. These different forms, called **isotopes,** all have the same charge and chemical characteristics, but different masses. Often one isotope of an element is very abundant, and others are rare.

**FIGURE 2.2    Atoms Have Structure.**   An atom is composed of a nucleus, made of protons and neutrons, surrounded by a cloud of electrons.

For example, the most common isotope of carbon atoms has six neutrons, and 1% are isotopes with seven or eight neutrons. An element's **atomic mass**, presented beneath the element symbol in the periodic table, is the average mass of its isotopes. Since the vast majority of carbon atoms contain six neutrons, its mass is very close to 12 in the periodic table.

Many of the known isotopes are unstable, which means they tend to break down into more stable forms; often into different elements. When unstable radioactive isotopes break down, they emit energy that is relatively easy to detect, even from very small quantities of isotopes. Molecules containing these isotopes can be very easily tracked by their energy emissions. Each radioactive isotope has a characteristic half-life, which is the time it takes for half of the atoms in a sample to emit radiation or "decay" to a different, more stable form. Since radioactive isotopes behave chemically as their stable versions, they have a variety of uses in biomedical and life science research, and later chapters present them in context.  • **radiometric dating, p. 331**

## Molecules Are Atoms Bound Together

Atoms of two or more elements often exist joined together as a **compound,** which is a chemical substance with properties distinct from those of its constituent elements. A **molecule** is the smallest unit of a compound. A molecule usually consists of atoms of different elements, but a few exceptions are "diatomic," consisting of two atoms of the same element, such as hydrogen, oxygen, or nitrogen.

A compound's characteristics can differ strikingly from those of its constituent elements. Consider table salt, which is the compound sodium chloride. A molecule of salt contains one atom of sodium (Na) and one atom of chlorine (Cl). Sodium is a silvery, highly reactive solid metal, while chlorine is a yellow, corrosive gas. But combine them and an explosive reaction produces a white crystalline solid—table salt. Other compounds also combine with different characteristics than the starting elements. Carbon, a black, sooty solid, and hydrogen, a light, combustible gas, combine to make methane, a colorless gas. Likewise, two gases, hydrogen and oxygen, combine to form a liquid: water.

Scientists use symbols to indicate the kinds and numbers of different elements found in each molecule. For example, methane is written $CH_4$, which denotes 1 carbon atom bonded (joined) to 4 hydrogen atoms. A molecule of the sugar glucose, $C_6H_{12}O_6$ has 6 atoms of carbon (C), 12 of hydrogen (H), and 6 of oxygen (O). A coefficient indicates the number of molecules—6 molecules of glucose is written $6C_6H_{12}O_6$.

In **chemical reactions,** two or more molecules interact with each other to yield different molecules. Chemical reactions allow us to move, rebuild cells, manipulate energy, and generally make life possible. Chemical reactions are depicted as equations with the starting materials, or **reactants,** on the left and the end

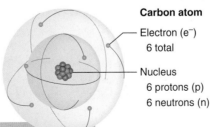

**Carbon atom**

Electron (e⁻)
6 total

Nucleus
6 protons (p)
6 neutrons (n)

| Subatomic particles | | | | | |
|---|---|---|---|---|---|
| **Particle** | **Charge** | **Mass** | **Function** | **Symbol** | **Location** |
| Electron | – | 0 | Bonding | e⁻ | Orbitals |
| Neutron | 0 | 1 | Nuclear stability | n | Nucleus |
| Proton | + | 1 | Identity | p | Nucleus |

**products** on the right. This allows us to write the equation for the formation of glucose in a plant cell:

$$6CO_2 + 6H_2O \longrightarrow C_6H_{12}O_6 + 6O_2$$

In words, this means "six molecules of carbon dioxide and six molecules of water react to produce one molecule of glucose plus six molecules of diatomic oxygen." The total number of atoms of each element must always be the same on either side of the equation. Note that each side of the equation indicates 6 total atoms of carbon, 18 of oxygen, and 12 of hydrogen.

Organisms are composed mostly of water and carbon-containing molecules. Most of the carbon-based molecules are **organic molecules,** which contain carbon and hydrogen. In addition to carbon and hydrogen, many organic molecules also include oxygen, nitrogen, phosphorus, and/or sulfur, which provide a variety of different functions. Some organic molecules of life are so large that they are termed **macromolecules,** which is the subject of chapter 3. The plant pigment chlorophyll, for example, contains 55 carbon atoms, 68 hydrogens, 5 oxygens, 4 nitrogens, and 1 magnesium. **Molecular mass** is an indication of a molecule's size. It is calculated by adding the atomic masses of the constituent atoms. • **chlorophyll, p. 109**

Many seemingly simple molecules play highly significant roles in cell chemistry. Carbon monoxide (CO), nitric oxide (NO), and carbon dioxide ($CO_2$) are not considered organic but can function as biological messenger molecules. How molecules form and how they are shaped determines how they can react in the key processes of life.

### Reviewing Concepts

- Atoms, the fundamental unit of matter, are composed of electrons surrounding a nucleus containing neutrons and protons.
- The different characteristics of elements come from differing numbers of protons.
- Chemical reactions bind atoms together as compounds that consist of unique molecules.
- Compounds have different characteristics than their constituent atoms.

## 2.2 When Atoms Interact: Chemical Bonds

Atoms can combine to form complex molecules of an infinite variety. The type of molecule that is formed is determined by the types of elements it contains. The electrons in atoms are the key to forming bonds between atoms. Molecules become the building blocks of life, and molecules interact to provide the functions of life.

### Electrons Determine Whether Atoms Bond

The more complex an atom, the more electrons it contains. More electrons occupy more space as they move about the nucleus like a swarm of flies around food. With more electrons in an atom, some are forced to occupy space farther from the nucleus. The farther an electron is from the nucleus, the more energy it has. So we use the particular distance from the nucleus, called an **energy shell,** to describe electrons. Energy level also helps approximate an electron's location since electrons occupy the lowest energy shell available to them and almost always fill one shell before beginning another. If an electron absorbs energy, it shifts to a higher energy level. Upon releasing that energy, the electron returns to the original level. We see this principle used by substances that "glow in the dark." Their electrons absorb energy and are boosted to higher energy levels. To return, the electrons release the energy as a burst of light.

Often electrons are illustrated as dots moving in concentric circles around a nucleus, much like planets moving around a sun, with the orbits symbolizing the energy shells (**figure 2.3**). These depictions, called Bohr models, are useful for visualizing the interactions between atoms to form bonds. However, they do not accurately portray the three-dimensional structure of atoms. In a Bohr model, the innermost shell, the one closest to the nucleus, can hold only two electrons. The outermost shell in an atom is called the **valence shell.** This shell is most actively involved in chemical reactions because it is most likely to be only partially filled with electrons. The electrons occupy the space in each shell in very characteristic shapes. The unique features of molecules are intimately linked to their shapes, which are determined by the configurations of the electrons involved in bonding.

Atoms are the most stable—that is, least likely to combine with other atoms—when their valence shells are filled. Since most of the valence shells contain spaces for eight electrons, atoms will become most stable with those eight vacancies filled. This tendency to require eight electrons in the valence shell is referred to as the **octet rule.** To fill that valence shell with exactly eight electrons, atoms can share or take electrons from other atoms. Strong chemical bonds are formed when atoms share

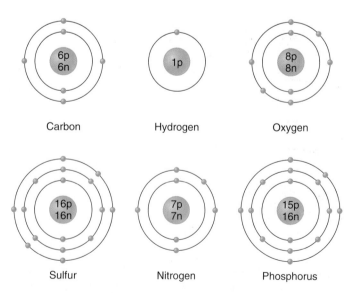

**FIGURE 2.3     Structures of the Atoms Prevalent in Life.** Shown here are Bohr models of the six most common atoms that make up organisms.

electrons. Weaker bonds form from the attraction between two atoms that have, respectively, given up or taken electrons from each other.

## In Covalent Bonds, Electrons Are Shared

**Covalent bonds** hold together most of the molecules of life. These strongest of bonds form when two atoms share pairs of valence electrons, and they tend to form among atoms that have three, four, or five valence electrons. Sharing electrons enables both atoms to fill their outermost shells with eight electrons. One electron from each atom spends time around each nucleus, strongly connecting the atoms.

As an example, carbon has four electrons in its outermost shell, and, therefore, it requires four more electrons to satisfy the octet rule. A carbon atom can attain the stable eight-electron configuration in its outer shell by sharing electrons with four hydrogen atoms, each of which has one electron in its only shell (**figure 2.4**). The resulting molecule is the swamp gas methane ($CH_4$). **Figure 2.5** shows several ways to represent its chemical structure.

Each single covalent bond contains two electrons, one from each atom, depicted with lines between the interacting atoms, with each line representing one bond. By sharing four electrons with four hydrogen atoms, the octet rule is satisfied for carbon, while also filling the shell of each hydrogen atom. Carbon atoms can also bond with each other and share two or three electron pairs, forming double and triple covalent bonds, respectively (**figure 2.6**). The fact that a carbon atom may form four covalent bonds, which satisfies the octet rule, allows this element to assemble into long chains, intricate branches, and rings. Carbon can also bond to many other elements, yielding a great variety of biological molecules. Methane is an example of sharing single electron pairs between carbon and hydrogen.

Since the electrons are shared equally in methane, these bonds are **nonpolar covalent bonds.** In contrast, in a **polar covalent bond,** electrons draw more toward one atom's nucleus than the other. The term **polar** means there is a difference between opposite ends of a molecule, usually formed by opposite charges. Water ($H_2O$), which consists of two hydrogen atoms bonded to an oxygen atom, is a molecule with polar covalent bonds (**figure 2.7**). The nucleus of each oxygen atom attracts the electrons on the hydrogen

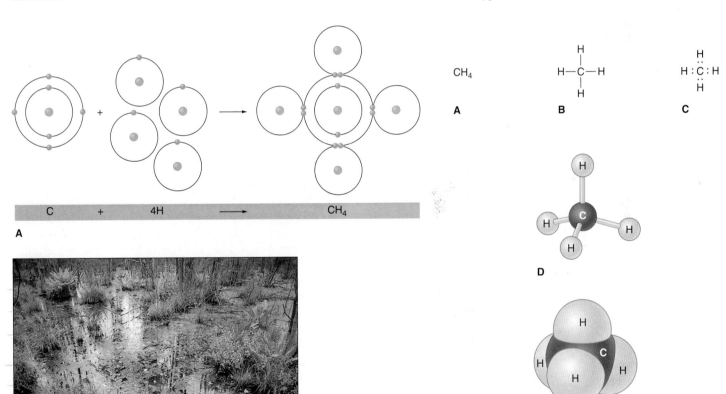

**FIGURE 2.4   Covalent Bonds Form Molecules.   (A)** Methane ($CH_4$) is a covalently bonded molecule. One carbon and four hydrogen atoms complete their outermost shells by sharing electrons. Note that the first electron shell is complete with two electrons. **(B)** Methane, also known as swamp gas, was once a major constituent of Earth's atmosphere.

**FIGURE 2.5   Different Types of Diagrams Are Used to Represent Molecules.   (A)** The molecular formula $CH_4$ indicates methane consists of one carbon atom bonded to four hydrogen atoms. **(B)** The structural formula shows single bonds as single lines. **(C)** The electron dot diagram shows the number and arrangement of shared electrons. **(D)** A ball-and-stick model reveals the angles of the bonds between the hydrogen and the carbon atoms. **(E)** A space-filling model shows bond relationships as well as the overall shape of a molecule.

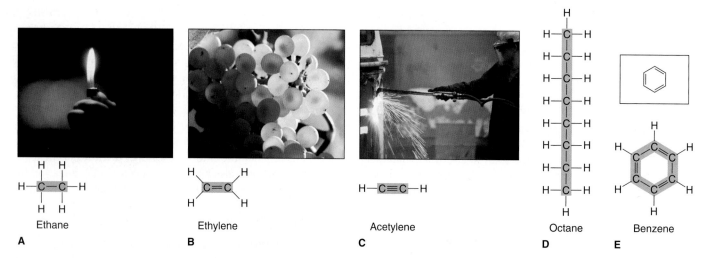

**FIGURE 2.6    Carbon Atoms Form Four Covalent Bonds.**    Two carbon atoms can bond, forming single, double, or triple bonds. Note that as the number of bonds between carbon atoms increases, the number of bonded hydrogens decreases. (**A**) Ethane, a component of natural gas, is a hydrocarbon built around two singly bonded carbon atoms. Breaking the atoms apart releases enough energy to heat a home or light a fire. (**B**) Ethylene consists of two carbon atoms linked by a double bond. Ethylene is a plant hormone, triggering flowers to drop and fruit to ripen. (**C**) Acetylene, consisting of two carbons held by a triple bond, is a flammable gas used in torches because tremendous heat energy is released when a triple bond is broken. (**D**) Octane is a key ingredient in gasoline, and (**E**) benzene is a common solvent. Benzene shows how a ring structure is possible. The inset shows an abbreviation of the ring that assumes carbon atoms are at the apexes, each with an implied hydrogen atom attached.

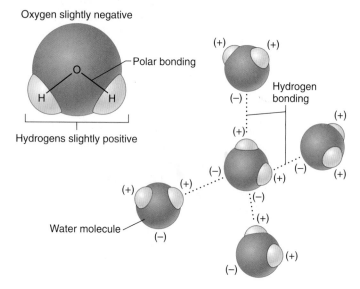

**FIGURE 2.7    Polar Covalent Bonds.**    Polar covalent bonds hold together the two hydrogen atoms and one oxygen atom of water ($H_2O$). Because the oxygen attracts the negatively charged hydrogen electrons more strongly than the hydrogen nuclei do, the oxygen atom bears a partial negative charge, and the hydrogens carry a partial positive charge. The resulting partial charges attract one molecule to another in hydrogen bonds.

atoms more than the hydrogen nuclei do. As a result, the area near the oxygen carries a partial negative charge (from attracting the negatively charged electrons), and the area near the hydrogens is slightly positively charged (as the electrons draw away).

The tendency of an atom to attract electrons is termed **electronegativity.** Oxygen is highly electronegative. Its ability to accept electrons is crucial in helping organisms extract energy from nutrients. Polar covalent bonds form when less electroneg-

ative atoms, such as carbon and hydrogen, bond with electronegative atoms, such as oxygen or nitrogen.

## In Ionic Bonds, Electrons Are Donated and Accepted

Atoms share electrons in covalent bonds. But sometimes, the electronegativity of one atom is so great that the atom takes an electron from the other atom. The two atoms then have opposite charges and attract. This attraction is an **ionic bond** and can be a very strong attraction, though not as strong as a covalent bond. An atom with one electron in the outermost shell might lose it to an atom with seven valence electrons in an attempt to fill all of the valence shells, or eliminate them. A sodium (Na) atom, for example, has one valence electron. When it donates this electron to an atom of chlorine (Cl), which has seven electrons in its outer shell, the two atoms bond ionically to form NaCl (**figure 2.8**). Other combinations are also possible between atoms with fewer than four electrons in the valence shell and those with six or more.

Once an atom loses or gains electrons, it has an electric charge and is called an **ion.** Atoms that lose electrons lose negative charges and thus carry a positive charge. Atoms that gain electrons become negatively charged. In NaCl, the oppositely charged ions $Na^+$ and $Cl^-$ attract in such an ordered manner that a crystal results (figure 2.8B). A compound composed of oppositely charged ions, like NaCl, is often called a **salt.** A salt can consist of any number of ions.

Atoms behave very differently as ions than they do normally. Native sodium is lethal if ingested but is vital to life as an ion. When salts are placed in water, the ionic bonds are often disrupted by the charges in water. The ions are then independent of each other and are important in many biological functions. Nerve

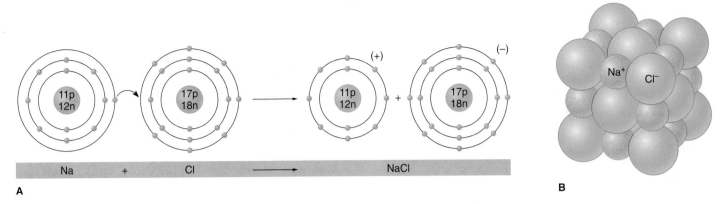

**FIGURE 2.8    Table Salt, an Ionically Bonded Molecule.    (A)** A sodium atom (Na) can donate the one electron in its valence shell to a chlorine atom (Cl), which has seven electrons in its outermost shell. This satisfies the octet rule for both atoms. The resulting ions (Na$^+$ and Cl$^-$) bond to form the compound sodium chloride (NaCl), better known as table salt. **(B)** The ions that constitute NaCl occur in a repeating pattern that produces crystals.

transmission depends upon passage of Na$^+$ and K$^+$ in and out of nerve cells (see figure 32.4). Muscle contraction depends upon movement of calcium (Ca$^{2+}$) ions. In many types of organisms, egg cells release calcium ions, which clear a path for the sperm's entry.

## Weak Chemical Bonds Provide Vital Links

In addition to covalent and ionic bonds, there are weak chemical attractions that may also be thought of as a type of bond. These can be very important in the molecules of life since they can readily reverse their connections, much like Velcro or a zipper. When two molecules contain polar covalent bonds, parts of each atom have small partial charges. The opposite charges on two molecules attract each other and form **hydrogen bonds** (see figure 2.7). In these bonds, the partially charged hydrogen on one molecule attracts the partial negative charges on other molecules. These bonds always form between molecules, not individual atoms, and the molecules can separate and rejoin in the processes of life. Hydrogen bonds are relatively weak, but in large molecules, such as DNA, the numbers of bonds are so high that they contribute significantly to holding the molecule together. They allow DNA to be more dynamic than covalent bonds

would. Water also contains many hydrogen bonds, which provide some of its interesting characteristics.

Different parts of the large and intricately shaped chemicals of life may be temporarily charged because their electrons are always in motion. Dynamic attractions between molecules or within molecules that occur when oppositely charged regions approach one another are called **van der Waals attractions.** They help shape the molecules of life. van der Waals attractions are also important on a larger scale because they account for attractions between surfaces that come into close contact. Geckos (a type of lizard) rapidly scale walls and ceilings and mussels cling to ship hulls using these forces.

Interactions with water produce another type of chemical attractive force that helps determine the three-dimensional shape of molecules. Like a drop of oil in a glass of water, if part of a molecule lacks any kind of charge, it cannot interact with water and is **hydrophobic** (water-fearing). **Hydrophilic** (water-loving) parts of molecules have charge and are attracted to water. Large molecules, such as proteins, contort to move hydrophobic regions away from water. Interactions with water also help shape the surfaces of cells by forming different compartments. **Table 2.1** summarizes chemical bonds. Without these, life would be impossible.

## TABLE 2.1

### Chemical Bonds and Attractive Forces

| Type | Chemical Basis | Strength | Example |
|------|----------------|----------|---------|
| Covalent bonds | Atoms share electron pairs | Strong | Glucose |
| Ionic bonds | Electrons donated; ions attract | Moderate | Sodium chloride |
| Hydrogen bonds | Partial charges on molecules due to uneven electron sharing with hydrogen; oppositely charged regions attract | Weak | Water, DNA |
| van der Waals | Partial charges on molecules due to random electron motion; oppositely charged regions attract | Weak | Proteins, animal adhesion |
| Hydrophobic | Hydrophobic regions of molecules combine to minimize contact with water | Strong | Cell membrane, protein structure |

# 2.3    Water's Importance to Life

Water has been called the "mater and matrix" of life. "Mater" means mother, and indeed life as we know it could not have begun were it not for this unusual substance. Water is also the matrix, or medium, of life, because it is vital in most biochemical reactions.

## All Life Processes Take Place in Solutions

A molecule of water consists of two atoms of hydrogen and one atom of oxygen. But water mostly exists in organisms as trillions of such molecules interacting with each other in tiny "pools." Every second, the hydrogen bonds between a single water molecule and its nearest neighbors form and re-form some 500 billion times! This constant changing of the hydrogen bonding makes water fluid. Fluidity is the ability of a substance to flow. The attraction between identical molecules, which accounts for water's constant rebonding, is called **cohesion** (**figure 2.9**). Water also readily forms hydrogen bonds to many other compounds, a property called **adhesion,** which is important in many biological processes.

Cohesion and adhesion are important in plant physiology. Movement of water from a plant's roots to its highest leaves depends upon cohesion of water within the plant's water-conducting tubes. Water entering roots is drawn up as water evaporates from leaf cells. Water movement up trees also depends upon the adhesion of water to the walls of the conducting tubes. Water's adhesiveness accounts for **imbibition,** the tendency of substances to absorb water and swell. Rapidly imbibed water swells a seed so that it bursts through the seed coat, stimulating further growth of the embryo within.

Biologists use several terms to describe water's ability to carry other chemicals. A **solvent** is a chemical in which other chemicals, called **solutes,** dissolve. A **solution** consists of one or more chemicals dissolved in a solvent. An **aqueous** solution is one that uses water as the solvent. Water molecules are polar; they have a partial positive charge at one end and a partial negative charge at the

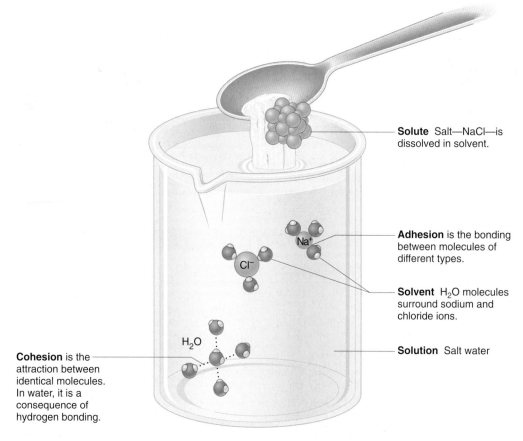

**Solute** Salt—NaCl—is dissolved in solvent.

**Adhesion** is the bonding between molecules of different types.

**Solvent** $H_2O$ molecules surround sodium and chloride ions.

**Solution** Salt water

**Cohesion** is the attraction between identical molecules. In water, it is a consequence of hydrogen bonding.

**FIGURE 2.9    Solutions Are Mixtures of Molecules.**    Interactions between the solutes and solvents allow solutes to dissolve by forming bonds with the solvent.

In the complex, three-dimensional world of biological molecules, shape is vitally important.

## Discriminating Drugs: They May Not Always Do What We Intend

Doctors working to develop new antibiotics had discovered one that seemed very promising. But something odd happened as they tested the drug: They found a tremendous variation in potency of the drug from one batch to the next.

Testing the compound in the lab did nothing to explain this finding—each batch behaved in a chemically similar way. What could cause the difference in potency?

The puzzle was solved only when the three-dimensional structure of the molecule was examined. As it turned out, the molecule came in two forms that were identical—except that one was a mirror image of the other. The "right-handed" form showed modest antibacterial action, but the "left-handed" form was extremely potent against bacteria.

The investigators found a way to separate the two mirror-image forms of the drug, and the drug called Levaquin (levofloxacin) was born. Levaquin is a potent antibiotic against anthrax as well as other bacterial infections.

Three-dimensional molecular shape can be crucial to biological function. What seem to be inconsequential differences in molecules can often cause radically different responses in living organisms. Drug testing seeks to establish the range of responses and side effects of therapeutic molecules, but unanticipated effects may sometimes occur. One recent example has been the withdrawal of certain pain medications known as cox-2 inhibitors from the market while their potential for increasing the risk of heart attack or stroke is reviewed.

# Life's Chemistry

3.1 **Carbon: The Basis of Life's Chemistry**
- Molecules Can Be Functional Groups
- Life Depends on Four Major Types of Molecules
- Polymers Expand Properties of Monomers
- Building Polymers Relies on a Common Chemical Reaction

3.2 **Carbohydrates: Energy and Structure**
- Monosaccharides and Disaccharides Are Simple Carbohydrates
- Polysaccharides Provide Energy Storage and Structure

3.3 **Lipids: Energy-Rich and Oily**
- Fatty Acids Combine to Form Triglycerides and Phospholipids
- Sterols Make Up Hormones and Cholesterol
- Waxes Protect Cells

3.4 **Proteins: Highly Diverse Molecules**
- Amino Acids Are the Monomers of Proteins
- Proteins Must Fold to Function
- Enzymes Are Life's Catalysts

3.5 **Nucleic Acids: Carriers of the Genetic Blueprint**

# Chapter Preview

1. Carbon atoms form an infinite variety of molecules whose functions differ with changes in structure and the presence of other elements.

2. Cells can provide new functions by making complex molecules from small subunit molecules called monomers, which possess characteristics distinct from the resulting polymers.

3. The major molecules of life are carbohydrates, proteins, lipids and nucleic acids.

4. Carbohydrates are mostly soluble molecules composed of carbon, oxygen and hydrogen that provide structure, energy storage, cell recognition and connection. The most common carbohydrates are sugars.

5. Lipids are insoluble compounds composed mostly of carbon and hydrogen. These diverse organic compounds store energy and help separate the interior of the cell from the environment.

6. The most diverse type of biomolecules are the proteins, which are comprised of monomers called amino acids.

7. Life makes use of 20 types of amino acids, each of which consists of a central carbon atom bonded to a hydrogen, an amino group, a carboxyl group, and a different functional group.

8. Each protein has a unique three-dimensional shape which is vital to its function. Any change in the cellular environment has the potential to change the shape of proteins.

9. Enzymes are proteins that accelerate specific chemical reactions under specific conditions and are vital to life.

10. Nucleic acids are long polymers of nucleotides that form either DNA or RNA, molecules that store and carry instructions for building and repairing the cell.

11. ATP, a nucleotide found in RNA, is the key energy carrier in cells.

## 3.1 Carbon: The Basis of Life's Chemistry

Life on Earth uses **carbon** as the main component of the molecules that make up living systems. A carbon is able to form bonds with four other atoms, including itself, resulting in an infinite variety of shapes and functions. Therefore, living cells can use carbon to make some very complicated molecules that act as the framework of even more complicated structures. Carbon can be formed in many shapes: long chains, ring molecules, branching molecules, and so on. (see figure 2.7).

## Molecules Can Be Functional Groups

As we learned from chapter 2, organic molecules contain at least carbon and hydrogen. The most basic organic molecules are **hydrocarbons,** which contain only carbon and hydrogen. The gasoline in your car is a hydrocarbon, but the peanut butter on the seat is something more. The differences come from what else is found in the molecules of peanut butter. By adding **functional groups** to a carbon skeleton, life can change the functions of the carbon molecule. Functional groups, then, are atoms or groups of atoms that add functions by combining oxygen, phosphorus, sulfur, and nitrogen to larger carbon skeletons. A tremendous variety of functions and reactions are possible by mixing different combinations of atoms and molecules from a basic set, as we will see.

## Life Depends on Four Major Types of Molecules

All living organisms are composed of four major classes of carbon molecules: carbohydrates, lipids, proteins, and nucleic acids and, for the most part, we acquire these molecules as we eat. Cells use these molecules directly or rearrange them to provide the properties that are needed to maintain a living organism. Vitamins are another group of vital molecules, but we require them in smaller amounts.

## Polymers Expand Properties of Monomers

Many of the biologically important molecules are large compounds made from small, single-unit molecules called **monomers.** Linked monomers form **polymers,** which usually possess very different characteristics than those of the monomers alone (**figure 3.1**). In discussing these versatile molecules, we often classify them by the numbers of unit monomers they contain by adding to a descriptive name a prefix that indicates the number: mono-, for one; di-, two; tri-, three; oligo-, 5 to 100; and poly-, for more than 100 units.

## Building Polymers Relies on a Common Chemical Reaction

Cells use a type of chemical reaction called **dehydration synthesis** ("made by losing water") to join monomers into larger molecules (figure 3.1). In the opposite reaction, **hydrolysis** ("breaking with water"), cells separate monomers by the addition of water. Much of what we call digestion is the release of monomers through hydrolysis. Monomers can then be recycled or further broken down. For instance, the proteins we eat are broken down and the monomers are used to build our own proteins. **Table 3.1** on p. 32 summarizes the major organic compounds of life, and we will address each in turn.

**Monosaccharides**—simple sugars composed of carbon, hydrogen, and oxygen in the proportions 1:2:1.

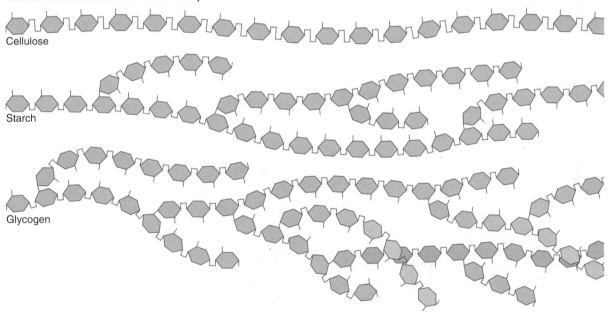

Glyceraldehyde $C_3H_6O_3$ — Ribose $C_5H_{10}O_5$ — Glucose $C_6H_{12}O_6$ — Fructose $C_6H_{12}O_6$

**Disaccharides**—molecules composed of two monosaccharides joined by dehydration synthesis. Hydrolysis converts disaccharides into their component monosaccharides. (The structures of the molecules are simplified to emphasize the joining process.)

Glucose $C_6H_{12}O_6$ + Fructose $C_6H_{12}O_6$ ⇌ Sucrose $C_{12}H_{22}O_{11}$ + Water $H_2O$

**Polysaccharides**—also known as complex carbohydrates, composed of long chains of simple sugars, usually glucose. Their chemical characteristics are determined by the orientation and location of the bonds between the monomers.

Cellulose

Starch

Glycogen

**FIGURE 3.1    Carbohydrates Can Be Simple or Complex.** Monosaccharides are composed of single molecules, such as glucose or fructose. Monosaccharides generally have the equivalent of one $H_2O$ for every carbon atom. Disaccharides are formed by dehydration synthesis, which binds two monosaccharides and removes water. For instance, glucose and fructose bond to form sucrose. Polysaccharides are long chains formed in a similar way from monosaccharides, such as glucose. Different orientations of these bonds produce different characteristics in the molecules. Recall from figure 2.6 that rings represent carbon atoms at the apexes.

## TABLE 3.1

### The Macromolecules of Life

| Type of Molecule | Structure | Functions | Examples |
| --- | --- | --- | --- |
| Carbohydrates | Sugars and polymers of sugars | Energy storage, structure, identity | Glucose, sucrose, starch, cellulose |
| Lipids | Fatty acids, triglycerides, sterols | Membranes, energy storage, waterproofing, hormones | Fat, oil, wax, cholesterol, testosterone |
| Proteins | Polymers of amino acids | Enzymes, structure, communication, recognition, transport | Hemoglobin, enzymes, keratin |
| Nucleic acids | Polymers of nucleotides | Genetic information, energy | DNA, RNA, ATP |

### Reviewing Concepts

- Carbon forms an infinite variety of molecules.
- Functional groups are molecules that add function to other molecules.
- The major molecules of life are carbohydrates, lipids, proteins and nucleic acids.
- Most large biological molecules are polymers, which have different properties than the unit monomers.
- Polymers are made through dehydration synthesis and recycled through hydrolysis.

## 3.2    Carbohydrates: Energy and Structure

**Carbohydrates** consist of carbon, hydrogen, and oxygen, often in the proportion 1:2:1. That is, a carbohydrate has twice as many hydrogen atoms as either carbon or oxygen atoms $(CH_2O)_n$. Sugars and starches are familiar examples. Carbohydrates store energy, which is released when their bonds are broken. Some carbohydrates physically support cells and tissues, and others distinguish different cell types. The generic term "saccharide" or sugar is used when we describe carbohydrates.

### Monosaccharides and Disaccharides Are Simple Carbohydrates

**Monosaccharides** are the smallest carbohydrates (figure 3.1). Monosaccharides differ from each other by how many carbons they contain, usually from three to seven, and how their atoms are bonded. For example, three monosaccharides with the same molecular formula $(C_6H_{12}O_6)$ but different chemical structures are glucose (blood sugar), galactose, and fructose (fruit sugar).

Monosaccharides containing only three carbons $(C_3H_6O_3)$ are key molecules in photosynthesis (discussed in chapter 7) and are also formed as cells extract energy from glucose in a process discussed in chapter 8.

Monosaccharides and disaccharides are sometimes called simple carbohydrates or sugars. The smallest complex carbohydrate is a **disaccharide,** which forms when two monosaccharides link through dehydration synthesis. Figure 3.1 shows how sucrose (table sugar) is made from a molecule of glucose and a molecule of fructose. Many plants, including sugarcane and sugar beets, contain abundant sucrose. Maltose, formed from two glucose molecules, provides energy in sprouting seeds and is used to make beer. Lactose, or milk sugar, is a disaccharide containing glucose and galactose.

### Polysaccharides Provide Energy Storage and Structure

**Oligosaccharides** are moderately sized carbohydrates often used by cells for identification and forming complex structures. Oligosaccharides are attached by the cell to proteins or lipids on the cell's surface and give the cell unique functions and identity (**figure 3.2C**). These **glycoproteins** and **glycolipid**s on cell surfaces are important in immunity, which is based on the distinctiveness of cell membranes among individuals. Oligosaccharides are also important in enabling proteins called antibodies to assume their characteristic three-dimensional shapes, which is essential to their function in protecting an animal's body from infection.

Complex carbohydrates are familiar to dieters and runners as food but are found in other organisms as structural elements. Chemically, they are usually made of **polysaccharides** containing hundreds of glucose monomers (see figure 3.1) or modified sugars. The most common polysaccharides are cellulose, chitin, starch, and glycogen. **Cellulose** and **starch** are long chains of glucose (figure 3.2A), but they differ from each other by the orientation of the bonds that link the monomers. Starch is a familiar food component, but cellulose forms wood and parts of plant

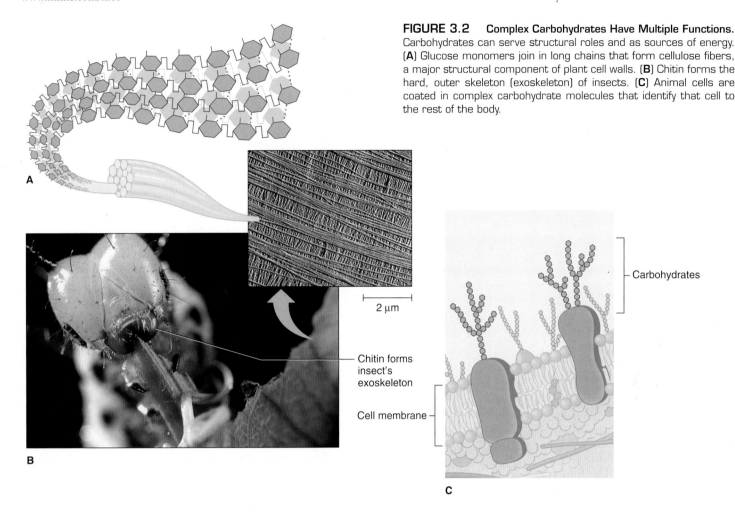

**FIGURE 3.2    Complex Carbohydrates Have Multiple Functions.** Carbohydrates can serve structural roles and as sources of energy. (**A**) Glucose monomers join in long chains that form cellulose fibers, a major structural component of plant cell walls. (**B**) Chitin forms the hard, outer skeleton (exoskeleton) of insects. (**C**) Animal cells are coated in complex carbohydrate molecules that identify that cell to the rest of the body.

2 μm

Chitin forms insect's exoskeleton

Cell membrane

Carbohydrates

cell walls and is the most common organic compound in nature. The seemingly minor chemical difference between them results in very different characteristics, including our ability to digest one and not the other. **Chitin,** the second most common polysaccharide in nature, resembles cellulose, but one OH group in each glucose molecule is replaced with a functional group that contains nitrogen. Chitin forms the flexible exoskeletons of insects, spiders, and crustaceans (figure 3.2B) and forms the cell wall of fungi.

## Reviewing Concepts

- Carbohydrates are based on the formula $(CH_2O)_n$. The most common monosaccharide (simple sugar) is glucose.
- Oligosaccharides are important components of identity molecules. Polymers of glucose form complex carbohydrates (polysaccharides) such as starch and cellulose.
- Changing the arrangement of the monomers and modifying glucose produces different characteristics for starch, cellulose, and chitin.

## 3.3    Lipids: Energy-Rich and Oily

All of us are familiar with oil and fat, which are composed of **lipids.** Like carbohydrates, lipids are diverse molecules containing carbon, hydrogen, and oxygen. But with fewer oxygen atoms than carbohydrates, lipids have very different properties. Most notably, lipid molecules dissolve in organic solvents but not in water.

Lipids are vital to life in many ways. They are necessary for growth and for the utilization of some vitamins. Fat is also an excellent energy source, providing more than twice as much energy as equal weights of carbohydrate or protein. Because they are hydrophobic, lipids are a major component of the membranes that enclose all cells. Within cells, they form compartments, separating one aqueous environment from another.

In our own bodies, nerve transmission is faster due to the lipid-rich cells that ensheath nerve cells. Lipids called waxes coat leaves, fur, and feathers, making them water-repellent. Human milk is rich in lipids, partly to suit the rapid growth of the brain in the first 2 years of life.

Fat cells aggregate as adipose tissue in animals. White adipose tissue forms most of the fat in human adults, cushioning organs and insulating against loss of body heat. A rare type, brown adipose tissue, is found in hibernating mammals and in newborn humans. Metabolic processes unique to these cells convert fat directly to heat, making hibernation possible and keeping infants warm.

## Fatty Acids Combine to Form Triglycerides and Phospholipids

By varying the kinds of molecules found in lipids, a variety of functions are possible. The simplest type of lipid in nature, **fatty acids** are long hydrocarbons of up to 36 carbon atoms with an acidic functional group at one end (**figure 3.3**). The character of fatty acids depends on their degree of saturation, which is a measure of their hydrogen content. A **saturated** fatty acid contains all the hydrogens it possibly can, which occurs when single bonds connect all the carbons. A fatty acid is **unsaturated** if it has at least one double bond and polyunsaturated if it has more than one double bond.

Unsaturation (double bonds) in the fatty acids causes them to form kinks (figure 3.3) and spread their "tails." This allows lipids to be more fluid and produces an oily consistency at room temperature. Lipids in plants are less saturated than those in animals. Olive oil is an example of a monounsaturated fat. The more saturated animal fats tend to be more solid, like butter or lard. A food-processing technique called hydrogenation, used to produce margarine, adds hydrogen to an oil to solidify it—in essence, saturating a formerly unsaturated fat.

Fatty acids combine to form more complex lipids. A **triglyceride** consists of three fatty acids joined to a three-carbon molecule called glycerol (figure 3.3). Dehydration synthesis combines the fatty acids and glycerol, releasing water. A triglyceride is what is commonly known as "fat" and is a compact way for cells to store energy.

The fundamental molecule of membranes, the **phospholipid** (**figure 3.4A**), is formed when enzymes replace one of the fatty acids in a triglyceride with **phosphate** (a functional group, $PO_4$). The oxygen-rich phosphate is highly negatively charged and, therefore, hydrophilic. This end of a phospholipid associates readily with water, while the other avoids it. We will discuss more of the function of membranes in chapter 4.

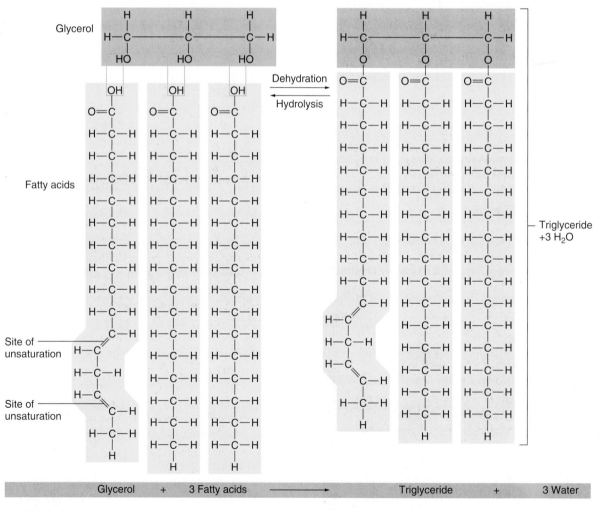

**FIGURE 3.3    Lipids Form Complex Molecules.**    A triglyceride is formed by bonding fatty acids to glycerol. This triglyceride is tripalmitin. The double bonds bend the fatty-acid tails, making the lipid more fluid. The bend is shown schematically here.

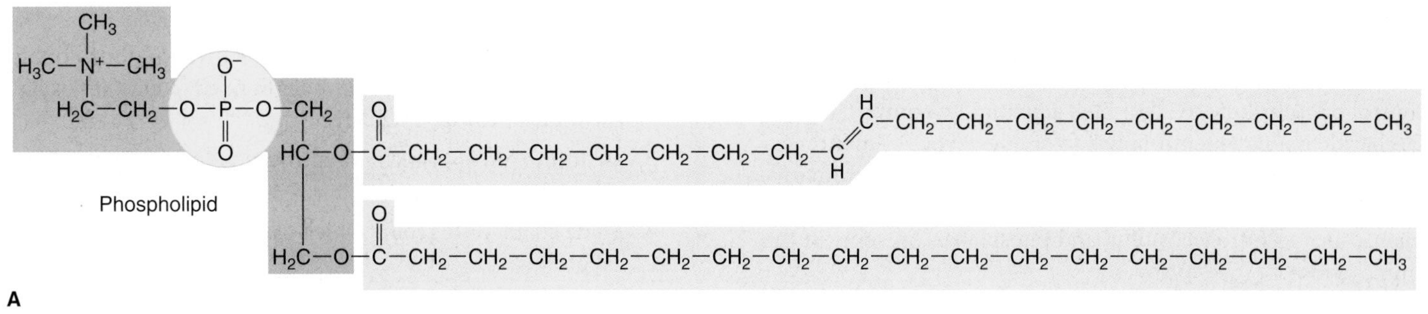

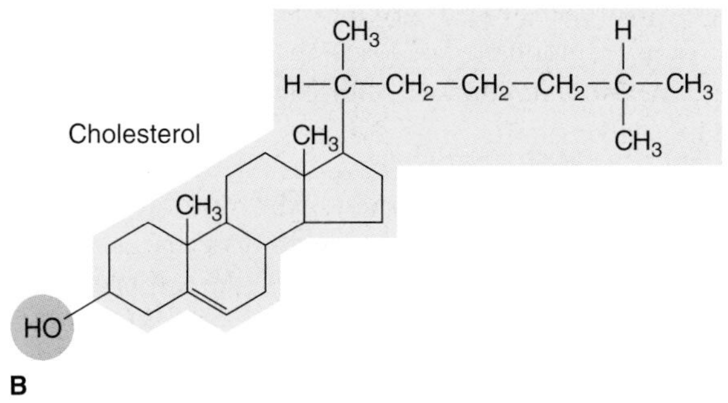

**A**

**Cholesterol**

**B**

**FIGURE 3.4**   **Lipids in Cell Membranes.**   Phospholipids (**A**) are the fundamental unit of all biological membranes. Double bonds create "kinks" that make membranes more fluid. In animal cells, cholesterol (**B**) also adds to membrane fluidity, in addition to providing raw material for hormone production.

**FIGURE 3.5**   **Lipids Come in Many Forms.**   Waxes waterproof the coat of the otter and the cuticles of the grasses growing in the background.

## Sterols Make Up Hormones and Cholesterol

**Sterols** are lipid molecules based on four interconnected carbon rings. Additions and modifications of this basic structure yield hormones, vitamins, and **cholesterol** (figure 3.4B). Cholesterol is vital for cells to maintain the fluidity of cell membranes and can be modified to make other lipids, including the sex hormones **testosterone** and **estrogen.** As virtually all cells need it, cholesterol is constantly being produced by specialized cells. In humans, this occurs in the liver, and it is this process that is inhibited by new drugs that lower cholesterol. Since it is so hydrophobic, proteins must surround cholesterol to carry it through the bloodstream. If this process is disrupted, cholesterol condenses on arteries, causing disease.

## Waxes Protect Cells

Waxes are fatty acids combined with either alcohols (molecules containing an OH functional group) or other hydrocarbons that usually form a hard, water-repellent covering. These lipids help waterproof fur, feathers, leaves, fruits, and some stems (**figure 3.5**). Jojoba oil, used in cosmetics and shampoos, is unusual in that it is a liquid wax.

- Attaching fatty acids to a glycerol in a variety of ways produces fats (triglycerides) or membrane components (phospholipids). Other modifications produce wax.
- Sterols are lipids based on an interconnecting ring structure. Examples are cholesterol and sex hormones.
- Cholesterol is vital to cells but needs special transport methods.

### Reviewing Concepts

- Lipids contain less oxygen than carbohydrates and are insoluble in water.
- The fundamental unit of most lipids is the fatty acid, a long hydrocarbon with an acidic functional group at one end.
- Fatty acids behave differently if they are unsaturated (containing double bonds).

## 3.4   Proteins: Highly Diverse Molecules

**Proteins** consist of monomers of **amino acids** linked to form one or more polypeptide chains. Proteins enable blood to clot, muscles to contract, oxygen to reach tissues, and nutrients to be broken down to release energy. **Enzymes** are proteins that allow biochemical reactions to proceed fast enough to sustain life. Structural proteins are the foundation of hair and bone. Proteins in membranes mark our cell surfaces as distinctly ours. Proteins control or contribute to all of life's activities. In contrast, a polysaccharide is a polymer of only one or a few types of monomers and has little diversity of function. Proteins also differ from carbohydrates and lipids due to the presence of a variety of elements, such as nitrogen, sulfur, and phosphorus, in some of the amino acids.

## Amino Acids Are the Monomers of Proteins

To build proteins, organisms make use of 20 types of amino acids, even though chemically many others exist. These 20 amino acids make possible a nearly infinite variety of proteins.

**Amino Acid Structure**    All amino acids (**figure 3.6A**) contain a central carbon atom bonded to four different functional components:

1. a hydrogen atom;

2. a **carboxyl** group (acid), which is a carbon atom double-bonded to one oxygen and single-bonded to another oxygen carrying a hydrogen (COOH);

3. an **amino group,** which is a nitrogen atom single-bonded to two hydrogen atoms ($NH_2$)

4. a side chain, or **R group,** which can be any of several chemical groups.

The nature of the R group distinguishes the 20 types of biological amino acids (see appendix E). An R group may be as simple as the lone hydrogen atom in glycine or as complex as the two organic rings of tryptophan. The R groups of two amino acids, cysteine and methionine, contain sulfur. Figure 3.6B shows three of life's amino acids.

**The Peptide Bond**    Two amino acids join by way of dehydration synthesis, just as two monosaccharides yield a disaccharide or fatty acids become triglyceride. To join amino acids, the carboxyl group

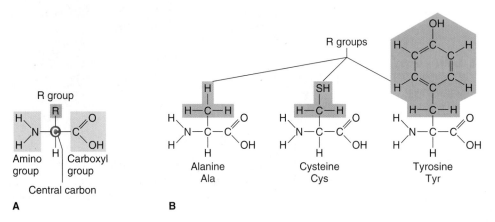

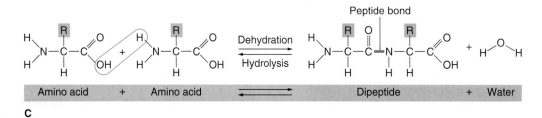

**FIGURE 3.6    Amino Acids Join to Form Peptides.**    Amino acids are the monomer subunits of proteins. (**A**) An amino acid is composed of an amino group, an acid (carboxyl) group, and one of 20 R groups attached to a central carbon atom. (**B**) The composition of the R groups contributes different functions to the final protein. (**C**) A peptide bond forms when an OH from a carboxyl group of one amino acid combines with a hydrogen from the amino group of another amino acid, creating a water molecule and linking the carboxyl carbon of the first amino acid to the nitrogen of the other. (**D**) Long chains of amino acids are polypeptides, which form proteins.

A

B

C

**FIGURE 3.7   A Gallery of Structures Based on Keratin.**
Alpha-keratin, a water-repellent protein, forms (**A**) the beak of a bird, (**B**) the scales of a snake, and (**C**) the horns of a ram.

of one amino acid combines with the nitrogen group of the other and forms a **peptide bond** (figure 3.6C). Two linked amino acids form a dipeptide; three, a tripeptide; larger chains (figure 3.6D) with fewer than 100 amino acids are oligopeptides (often just called peptides); and finally, those with 100 or more amino acids are polypeptides. The kind and number of amino acids joined together determines what the protein will be and how it is used (**figure 3.7**). This complex process of protein synthesis is the topic of chapter 13.

## Proteins Must Fold to Function

As a protein is synthesized in a cell, it folds into a three-dimensional structure, or **conformation** based on the order and kinds of amino acids it contains. The final shape of a protein arises from interactions with other proteins and water molecules and bonds that form within the protein itself.

**Levels of Protein Structure** The conformation of a protein may be described at four levels (**figure 3.8**). The simplest, or **primary** (1°), structure of a protein is just the amino acid

sequence of its polypeptide chain. Hydrogen bonds between parts of the peptide "backbone" form the **secondary** (2°) structure, folding the polypeptide into coils, sheets, loops, and combinations of these shapes. R groups can affect where these hydrogen bonds form but are not actually part of them. As a result, common patterns of structure, called **motifs,** will emerge. Figure 3.8B shows two of the more common motifs of secondary structure, alpha helices and beta-pleated sheets.

Proteins fold into their final **tertiary** (3°) structures through interactions between R groups and each other or water (figure 3.8C). Oppositely charged R groups bend the polypeptide as they attract and form ionic bonds. Other R groups attract and bind through hydrogen bonds. Since thousands of water molecules usually surround a protein, hydrophilic R groups move toward water, and hydrophobic R groups move away from water toward the protein's interior. The resulting hydrophobic interactions are a major contributor to the final shape of the protein. The entire structure is further stabilized by the formation of covalent bonds between sulfur atoms in some R groups. Called **disulfide bonds,** these are abundant in structural proteins such as keratin, which forms hair, scales, beaks, wool, and hooves (figure 3.7). A "permanent wave" curls hair by breaking disulfide bonds in hair keratin and re-forming those bonds in hair that has been wrapped around curlers.

Many proteins are functional at the tertiary level. However, some proteins are composed of more than one polypeptide, held together through hydrogen or ionic bonds (figure 3.8D). This level of organization is referred to as **quaternary** (4°) structure. Many of the critical enzymes of life are composed of dozens of different polypeptides. Even hemoglobin, the major functional protein of blood, exists as two sets of two different polypeptides. The quaternary structure is also important in controlling protein functions, as accessory polypeptides reversibly combine to either activate or inhibit a protein.

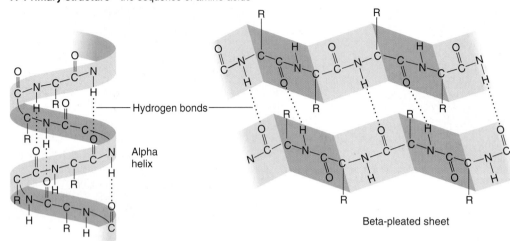

A  **Primary structure**—the sequence of amino acids

**FIGURE 3.8    Four Levels of Protein Structure.** (**A**) The amino acid sequence of a polypeptide forms the primary structure, while (**B**) hydrogen bonds between non-R groups create secondary structures such as helices and sheets. The tertiary structure (**C**) is formed when R groups interact, folding the polypeptide in three dimensions and forming a unique shape. (**D**) If different polypeptide units must interact to be functional, this forms the quaternary structure of a protein.

B  **Secondary structure**—hydrogen bonds between nonadjacent carboxyl and amino groups

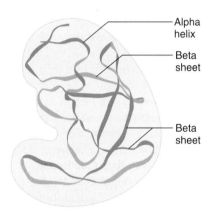

C  **Tertiary structure**—disulfide and ionic bonds between R groups, interactions between R groups and water

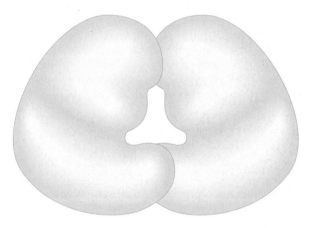

D  **Quaternary structure**—hydrogen and ionic bonds between separate polypeptides

**Denaturation**  A protein's conformation makes possible, and in large part determines, its function. A digestive enzyme holds a large nutrient molecule in just the right way to break the nutrient apart. Muscle proteins form long, aligned fibers that slide past one another, shortening their length to create muscle contractions.

All functions of a protein are ultimately determined by its primary structure, which also dictates the other levels of conformation. If we disrupt the tertiary structure, we destroy the function even if the primary structure of the protein remains intact. Any such dramatic disruption of structure, termed **denaturation,** destroys a protein's function. Soap cleans by interfering with the ionic bonds and hydrophobic interactions of all proteins. Methods of preserving food, such as salting or pickling or boiling, work in the same way: they disrupt protein structure. Living systems have evolved to constantly replace proteins as they are affected by the environment and to maintain the best possible environment for protein function.

## Enzymes Are Life's Catalysts

Among the most important of all biological molecules are enzymes. **Enzymes** are proteins that speed the rates of specific chemical reactions without being consumed in the process, a phenomenon called **catalysis.** Catalysts may be inorganic (lacking carbon) or organic (containing carbon and hydrogen). Platinum is a common inorganic component of an automobile's catalytic converter system. An enzyme is a biological catalyst that binds **reactants** (starting materials) in such a way that a reaction can readily and rapidly occur. Without enzymes, many biochemical reactions would proceed far too slowly to support life; some enzymes increase reaction rates a billion times.

Enzymes catalyze specific chemical reactions. The key to an enzyme's specificity lies in its **active site,** which is a region to which the reactants, also called **substrates,** bind. At its most basic level, an enzyme works by increasing the likelihood that two sub-

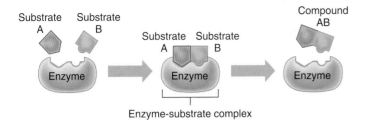

**FIGURE 3.9**   **Enzyme Action.**   In this highly schematic depiction, substrate molecules A and B fit into the active site of an enzyme. An enzyme-substrate complex forms as the active site moves slightly to accommodate its occupants. A new compound, AB, is released, and the enzyme is reused. Enzyme-catalyzed reactions can break down as well as build up substrate molecules.

strates will encounter each other in just the right way to react. A substrate combines with the active site of an enzyme as the enzyme contorts slightly around it, forming a short-lived enzyme-substrate complex (**figure 3.9**). An enzyme might hold two substrate molecules that react to form one product molecule or a single substrate that splits to yield two products. Once the reaction is complete, the complex breaks down to release the products (or product) of the reaction. The enzyme is unchanged, and its active site is empty and able to pick up more substrate.

Most enzymes function under very specific pH and temperature conditions, and cells continually work to provide and maintain those conditions. If the conditions are not optimal, most enzymes will simply stop working, often because the protein itself is denatured. The enzymes of bacteria found in hot springs, for example, can function under very high or low temperature or pH, but they stop working at room temperature. These "extremozymes" are providing scientists with many clues to life's origins and how proteins function.

## Reviewing Concepts

- Proteins are polymers of amino acids and provide the majority of life's functions.
- Twenty amino acids form the basis of all proteins. Amino acids have a common structure that facilitates their joining through dehydration synthesis and a unique R group that adds to protein function.
- The final conformation (shape) of a protein is based on four levels of interactions between components of the polypeptide. Proteins are only functional in their folded state. Changes in temperature, pH, or ions in the environment of a protein will disrupt (denature) the protein's shape and function.
- Vital to life, enzymes are proteins that facilitate (catalyze) chemical reactions. Enzymes bind very specific reactants (substrates) in their active sites and produce specific products.

## 3.5 Nucleic Acids: Carriers of the Genetic Blueprint

Synthesizing a protein is a more complex task than synthesizing a carbohydrate or fat because of the great variability of amino acid sequences. How does an organism "know" which amino acids to string together to form a particular protein? Like a blueprint or recipe, a protein's amino acid sequence is encoded in the sequence of chemical units in a molecule called a **nucleic acid.**

The two types of nucleic acids are **deoxyribonucleic acid (DNA)** and **ribonucleic acid (RNA).** Both are polymers of molecules called **nucleotides (figure 3.10A).** A nucleotide consists of a five-carbon sugar (**deoxyribose** in DNA and **ribose** in RNA), another example of sugar being used for something other than energy. To this sugar are added one or more phosphate groups ($PO_4$) and one of five types of nitrogen-containing compounds called **nitrogenous bases.** The nucleotides (often just "bases") adenine (A), guanine (G), thymine (T), cytosine (C), and uracil (U) form all nucleic acids. Both DNA and RNA use A, C, and G, but T is found only in DNA, and U is unique to RNA. DNA is a double helix resembling a spiral staircase, with alternating sugars and phosphates forming the rails and nitrogenous bases forming the rungs (figure 3.10B).

**DNA**   Long sequences of DNA contain information that is first copied to RNA molecules, then used by cells to guide assembly of amino acids into polypeptide chains. Each group of three bases in a row specifies a particular amino acid, in a correspondence called the **genetic code.** An entire polypeptide is encoded by sequence of DNA known as a **gene.** Every organism uses DNA to encode its proteins, using the same genetic code to decipher each unique gene. Different genes mean different proteins; different proteins mean different organisms.

The two strands of nucleotides that make up the DNA double helix are **complementary,** or "opposites," of each other (figure 3.10B). Hydrogen bonds form between the bases on opposite strands—A with T and C with G—holding the strands together. This pairing of bases is highly specific: if one strand contains an A at a particular location, the base on the complementary strand must be a T. Since every base has a complement on the opposite strand, each strand of DNA contains the information for the other, providing a mechanism for the molecule to replicate. Future chapters (12 and 13) will explore more of the detail of DNA function and protein synthesis and the relationships between proteins and DNA.

**RNA**   Although DNA is the genetic material, RNA is, in some ways, even more important. The various functions of RNA are possible because its single-stranded structure enables it to assume different shapes. In its various guises, RNA enables the cell to make use of the information in DNA without damaging the DNA itself. The complex shapes possible for RNA also enable some types of RNA molecules to function as enzymes. Even its nucleotides are important: one particular RNA nucleotide, **ATP,**

**Nucleotides**—consist of a sugar (ribose or deoxyribose), a phosphate, and one of five nitrogenous bases.

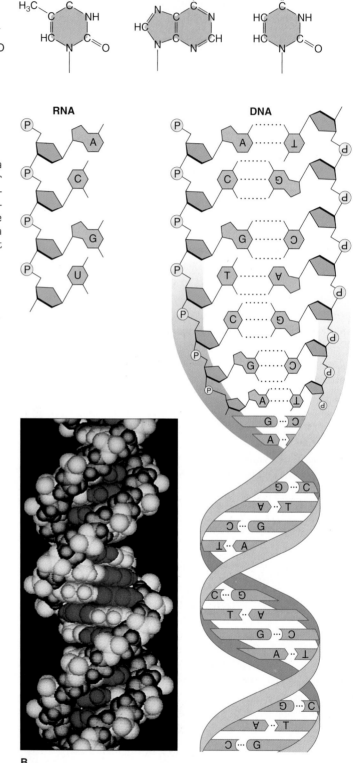

**FIGURE 3.10    DNA Structure.    (A)** The monomer unit of DNA is a nucleotide, which consists of a nitrogenous base (A, T, C, or G), a sugar (deoxyribose), and a phosphate. **(B)** Nucleotide pairs form as complementary bases attract—A with T and G with C. A DNA molecule is double-stranded, with the strands running in opposite orientations—this is why the labels on one strand are upside down. The two strands entwine to form a double-helix shape. RNA is usually single-stranded and contains a different sugar—ribose—and a different base—uracil—which replaces thymine.

serves a vital role in carrying energy that is used in nearly all biological functions. Because of its eclectic roles, RNA, or a molecule similar to it, may have been a bridge between complex groups of chemicals and the first organisms ● **RNA world, p. 348**

Life can be thought of as a complex mix of molecules that interact in myriad ways. Table 3.1 (p. 32) reviews the characteristics of the major types of organic molecules.

## Reviewing Concepts

- Nucleic acids are polymers of nucleotides and form DNA or RNA.
- Nucleotides (bases) share a common structure, with five varieties: A, C, T, G, and U.
- DNA and RNA use different bases (T and U, respectively) and a different sugar (deoxyribose and ribose, respectively) and have different cellular uses.
- DNA stores information in the sequence of its nucleotides. RNA helps retrieve and utilize that information, sometimes acting as an enzyme.
- Three bases correspond to a single amino acid, a relationship known as the genetic code which is the same for virtually all life on Earth.
- A gene is a sequence of nucleotides containing the information for an entire protein.
- DNA is a double helix comprised of two complementary strands held together via hydrogen bonds between base pairs. The pairing is highly specific and allows replication of DNA.
- ATP, a nucleotide found in RNA, is the key energy carrier in cells.

**Nucleic acids**—nucleotides joined together in long chains to form DNA or RNA. DNA is composed of the nucleotides A, C, T and G. RNA contains the sugar ribose and the nucleotide U instead of T.

# Connections

Although representing only four basic types, the carbon-based molecules of life can be used to produce a tremendous variety of structures and functions. Understanding the characteristics of each—carbohydrate, protein, lipid, and nucleic acid—helps us to understand the fundamentals of how life works. For instance, lipids reject water and can align to form a barrier. This gives us the beginnings of a cell membrane. Add the functions of enzymes and a method of storing information and we have a cell, the topic of our chapter 4.

# Student Study Guide

## Key Terms

| | | | |
|---|---|---|---|
| active site *38* | deoxyribonucleic acid | lipid *33* | protein *35* |
| amino acid *35* | (DNA) *39* | monomer *30* | quaternary *37* |
| ATP *39* | deoxyribose *39* | monosaccharide *32* | reactant *38* |
| carbohydrate *32* | disulfide bond *37* | motif *37* | R group *36* |
| carbon *30* | enzyme *35, 38* | nitrogenous base *39* | ribonucleic acid (RNA) *39* |
| carboxyl *36* | estrogen *35* | nucleic acid *39* | ribose *39* |
| catalysis *38* | fatty acid *34* | nucleotide *39* | saturated *34* |
| cellulose *32* | functional group *30* | oligosaccharide *32* | secondary *37* |
| chitin *33* | gene *39* | peptide bond *37* | starch *32* |
| cholesterol *35* | genetic code *39* | phosphate *34* | sterol *35* |
| complementary *39* | glycolipid *32* | phospholipids *34* | substrate *38* |
| conformation *37* | glycoprotein *32* | polymer *30* | tertiary *37* |
| dehydration synthesis *30* | hydrocarbon *30* | polysaccharide *32* | testosterone *35* |
| denaturation *38* | hydrolysis *30* | primary *37* | triglyceride *34* |
| | | | unsaturated *34* |

## Chapter Summary

### 3.1  Carbon: The Basis of Life's Chemistry

1. Carbon forms an infinite variety of molecules. Functional groups are molecules that add function to other molecules.
2. Most of the large biological molecules are composed of small subunit molecules called monomers, which possess characteristics distinct from the resulting polymers.
3. Polymers are made through dehydration synthesis and recycled through hydrolysis.

### 3.2  Carbohydrates: Energy and Structure

4. Carbohydrates consist of carbon, hydrogen, and oxygen, often in the proportion 1:2:1.
5. Carbohydrates provide structure, energy storage, cell recognition, and connections between cells.
6. Monosaccharides are single-molecule sugars such as glucose. A disaccharide is a molecule of two joined monosaccharides. Oligosaccharides are composed of 2 to 100 monomers, whereas polysaccharides are enormous molecules of hundreds of monomers.

7. The orientation of the covalent bonds that join the monomers in a polysaccharide provides the different properties of cellulose and starch.

### 3.3  Lipids: Energy-Rich and Oily

8. Lipids are diverse organic compounds that provide energy, slow digestion, waterproof the outsides of organisms, cushion organs, and preserve body heat.
9. Lipids include fats and oils; do not dissolve in water; and contain carbon, hydrogen, and oxygen but have less oxygen than carbohydrates.
10. Triglycerides consist of glycerol and three fatty acids, which may be saturated (no double bonds), unsaturated (at least one double bond), or polyunsaturated (more than one double bond). Double bonds make a lipid oily at room temperature, whereas saturated fats are more solid.
11. Sterols are lipids containing four carbon rings. Common examples are cholesterol and the sex hormones testosterone and progesterone.

## 3.4 Proteins: Highly Diverse Molecules

12. Proteins are polymers of amino acids that function in tremendously diverse ways.

13. Life makes use of 20 types of amino acids, each of which consists of a central carbon atom bonded to a hydrogen, an amino group, a carboxyl group, and an R group.

14. A peptide bond is a special type of covalent bond that connects amino acids into long chains through dehydration synthesis.

15. A protein's conformation, or three-dimensional shape, is vital to its function and is determined by the amino acid sequence (primary structure) and interactions between the non-R group atoms (secondary structure); ionic, covalent, and hydrophobic interactions between R groups (tertiary structure) in the sequence also contribute. A protein with more than one polypeptide has a quaternary structure as subunits join through hydrogen and ionic bonds.

16. Enzymes are proteins that accelerate specific chemical reactions under specific conditions and are vital to life.

## 3.5 Nucleic Acids: Carriers of the Genetic Blueprint

17. Nucleic acids are long polymers of nucleotides and form DNA or RNA.

18. Nucleotides consist of a phosphate, a sugar, and one of five nitrogenous bases: A, C, T, G and U. They are often referred to only by the base they contain.

19. DNA and RNA use different bases (T and U, respectively) and a different sugar (deoxyribose and ribose, respectively) and have different cellular uses. DNA stores information in the sequence of its nucleotides. RNA helps retrieve and utilize that information, sometimes acting as an enzyme.

20. Three bases correspond to a single amino acid, a relationship known as the genetic code, which is the same for virtually all life on Earth. A gene is a sequence of nucleotides containing the information for an entire protein.

21. DNA is a double helix comprised of two complementary strands held together via hydrogen bonds between base pairs. The pairing is highly specific and allows replication of DNA.

22. ATP, a nucleotide found in RNA, is the key energy carrier in cells.

# What Do I Remember?

1. What are the major types of molecules found in living organisms?

2. What are the chemical compositions and different types of carbohydrates, lipids, proteins, and nucleic acids?

3. What are the functions of carbohydrates, lipids, proteins, and nucleic acids?

4. Why are proteins extremely varied in organisms, but carbohydrates and lipids are not?

5. What is the significance of a protein's conformation?

6. How are DNA and RNA different in structure and function?

7. Compare and contrast the structures and functions of carbohydrates, lipids, proteins, and nucleic acids.

**Fill-in-the-Blank**

1. A(n)_____ is the unit molecule of all polymers.

2. The chemical formula _____ represents a carbohydrate.

3. The _____ structure of a protein is formed by interactions between R groups.

4. The RNA nucleotide _____ is used to supply energy for metabolic reactions.

5. An amino acid contains an amino group, a hydrogen, _____ and an R group.

**Multiple Choice**

1. A molecule with the chemical formula $C_2H_2$ must contain
   a. two double bonds
   b. a triple covalent bond
   c. two negative charges
   d. oxygen atoms

2. Which of the following is NOT a function of proteins?
   a. storing information
   b. catalyzing chemical reactions
   c. separating the inside of a cell from the environment.
   d. linking one cell to another

3. RNA contains _____, which is (are) never found in DNA.
   a. nucleotides
   b. protein
   c. ribose
   d. thymine

4. Dehydration synthesis forms covalent bonds in which of the following?
   a. proteins
   b. carbohydrates
   c. lipids
   d. all of these

5. Triglycerides, commonly known as "fat," contain
   a. glycerol and three fatty acids.
   b. one fatty acid and three amino acids.
   c. proteins and carbohydrates.
   d. nucleic acid and glycerol.

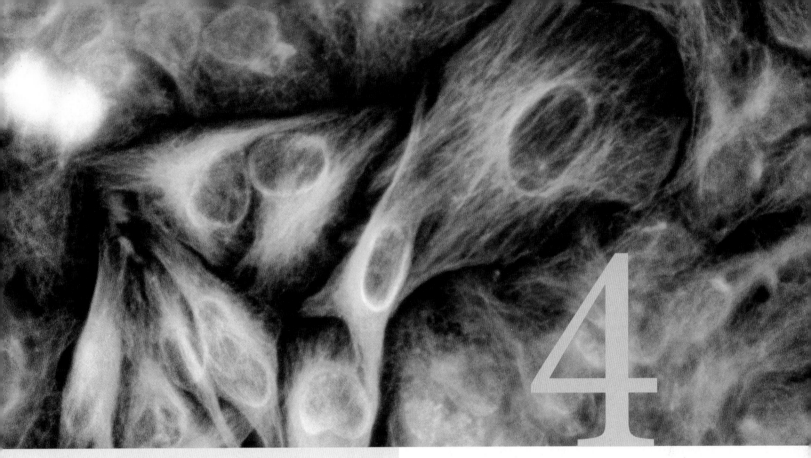

Cancer cells, when stained for the presence of proteins characteristic of cancer cells, look very different from surrounding healthy tissue. The orange cells are a melanoma (skin cancer) that is invading normal skin.

# Cells: Units of Life

## Cancer Cells: A Tale of Two Drugs

On August 4, 1993, a physician-friend, looking at me from across a room, said, "What's that lump in your neck?" So began my medical journey. A few days later, a specialist stuck seven thin needles into my neck to sample thyroid cells for testing, assuring me that thyroid tumors are benign (non-cancerous) 99% of the time.

The doctor's phone call came early on a Monday morning. Knowing that doctors don't like to deliver bad news on a Friday, I was panic stricken. I knew I had become a statistic.

I was in the 1 in 100 whose thyroid lump defied the odds, but fortunately, I had the "good" type of thyroid cancer. Surgery and radiation soon followed, and I was, and am, fine. But I will never forget the terror of discovering that I had cancer.

A healthy cell has a characteristic shape, with a boundary that allows entry to some substances, yet blocks others. Not so the misshapen cancer cell, with its fluid surface and less discriminating boundaries. The cancer cell squeezes into spaces where other cells do not. Cancer cells disregard the "rules" of normal cell division and multiply unchecked, forming tumors or uncontrolled populations of blood cells.

Unraveling how deranged signals prompt the out-of-control cell division of cancer has produced new drugs that target only the abnormal cells. These drugs interfere with an aspect of cell biology unique to the cancer cells—and as a result, have fewer adverse effects. The science of cell biology can save lives.

43

# Chapter Preview

1. Cells, the units of life, are the microscopic components of all organisms and are the ultimate source of the functions of multicelled organisms. The cell theory states that all life is composed of cells, that cells are the functional units of life, and that all cells come from preexisting cells.

2. Advances in lens technology permitted the discovery of cells and increased our understanding of their structures and functions.

3. There are three major cell types, representing three domains of life. Bacteria are unicellular, lack organelles, and usually have a rigid cell wall.

4. Archaea are unicellular and lack nuclei. They share some characteristics with bacteria and eukaryotes but also have unique structures and biochemistry.

5. Eukaryotic cells sequester certain biochemical activities in organelles. Organelles increase membrane surfaces and enclose unique chemical environments that allow cells to contain a greater variety of processes.

6. Organelles facilitate protein synthesis and processing, energy manipulation, recycling components, or protecting the cell.

7. Mitochondria and chloroplasts share characterstics not found in other organelles and may have originated from early primitive cells.

## 4.1 The Discovery of Cells

A rabbit, a rose, a mushroom, and a bacterium appear to have little in common other than being alive. However, on a microscopic level, these organisms are similar. All organisms consist of microscopic structures called **cells,** and some organisms are just a single cell. Within cells, highly coordinated biochemical activities carry on the basic functions of life, and, in some cases, provide specialized functions (**figure 4.1**). To understand life, we must understand the cell, for the functions of all organisms have their beginnings in actions of cells. Here we introduce the cell, and the chapters that follow delve into specific cellular events.

Cells, as the basic units of life, exhibit all of the characteristics of life. A cell requires energy, genetic information, and structures to carry out the activities of life. Movement occurs within living cells, and some, such as the sperm cell, can move about in the environment. All cell types have some structures in common that allow them to reproduce, grow, respond to stimuli, and obtain and manipulate energy. In all cells, a **cell membrane** separates the living matter from the environment and limits size. Complex cells house specialized structures, called **organelles,** in which particular activities take place. The remaining interior of cells, both complex and simple, consists of **cytoplasm.**

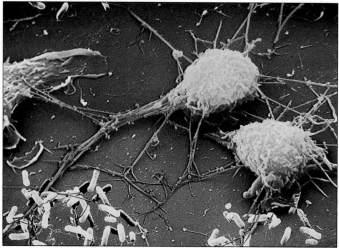

A

10 μm

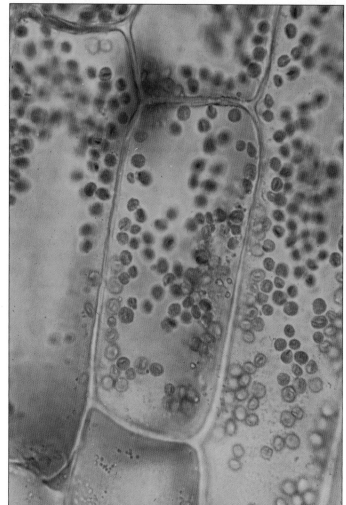

B

20 μm

**FIGURE 4.1    Specialized Cells.    (A)** Macrophages (stained blue) are specialized cells that engulf and destroy bacteria (yellow). Long extensions of the macrophages capture the bacteria and draw them inside the cell, where enzymes destroy the bacteria. Macrophages will consume bacteria until they wear out and become part of "pus" in an infected wound. **(B)** Specialized leaf cells are loaded with green chloroplasts, which contain the machinery for photosynthesis.

Cells can specialize to contribute specific functions to the whole organism. The unique chemical processes in these cells often depend on particular types of organelles. The unique composition of each cell makes it specialized. For example, an active muscle cell contains many more of the type of organelles that enable it to use energy than does an adipose cell, which is little more than a blob of fat. Specialized leaf cells are packed with chloroplasts, which are organelles that capture the sun's energy in a process called photosynthesis (figure 4.1B). A root cell has few, if any, chloroplasts, yet specializes in absorption. Multicelled organisms begin life with just one or two **stem cells** that have the capacity to become any of a variety of specialty cells. Humans start as a single fertilized cell that gives rise to more than 200 varieties to form a complete adult.

Our knowledge of the structures inside cells depends upon technology because most cells are too small for the unaided human eye to see. Cell biologists use a variety of microscopes to greatly magnify different types of images of cell contents. Since those contents are often transparent, they also use many types of stains to visualize the structures within the cell.

## Lenses Reveal the World of the Cell

The ability to make objects appear larger probably dates back to ancient times, when people noticed that pieces of glass or very smooth, clear pebbles could magnify small or distant objects. By the thirteenth century, the ability of such "lenses" to aid people with poor vision was widely recognized in the Western world.

Three centuries later, people began using paired lenses to increase magnification. Many sources trace the origin of a double-lens compound microscope to Dutch spectacle makers Johann and Zacharius Janssen. Reports claim their children were unwittingly responsible for this important discovery. One day in 1590, a Janssen youngster was playing with two lenses, stacking them and looking through them at distant objects. Suddenly he screamed—the church spire looked as if it was coming toward him! When the elder Janssens looked through both pieces of glass, the faraway spire indeed looked as if it was approaching. One lens had magnified the spire, and the other lens had further enlarged the magnified image. This observation led the Janssens to invent the first compound optical device, a telescope. Soon, similar double-lens systems were constructed to focus on objects too small for the unaided human eye to see. The compound microscope was born.

The study of cells—cell biology—began in 1660, when English physicist Robert Hooke melted strands of spun glass to create lenses that he focused on bee stingers, fish scales, fly legs, feathers, and any type of insect he could hold still. When he looked at cork, which is bark from a type of oak tree, it appeared to be divided into little boxes, which were remnants of cells that were once alive. Hooke called these units "cells," because they looked like the cubicles (cellae) where monks studied and prayed. Although Hooke did not realize the significance of his observation, he was the first person to see the outlines of cells.

In 1673, Antonie van Leeuwenhoek of Holland improved lenses further. He used only a single lens, but due to its quality, it was more effective at magnifying and produced a clearer image than most two-lens microscopes then available. One of his first objects of study was tartar scraped from his own teeth, and his words best describe what he saw there:

> To my great surprise, I found that it contained many very small animalcules, the motions of which were very pleasing to behold. The motion of these little creatures, one among another, may be likened to that of a great number of gnats or flies disporting in the air.

Over the next few years, Leeuwenhoek built more than 500 microscopes that opened up a vast new world to the human eye and mind (**figure 4.2**). He viewed bacteria and protista; life that people hadn't known existed. However, he failed to see the single-celled "animalcules" reproduce, and therefore he perpetuated the popular idea at the time that life arises from the nonliving or from nothing. Nevertheless, he described, with remarkable accuracy, microorganisms and microscopic parts of larger organisms,

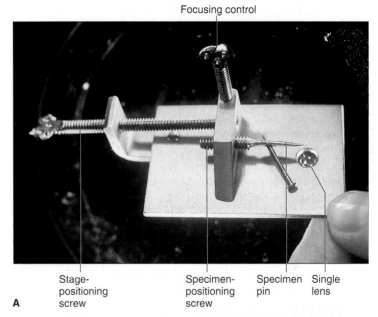

Focusing control

Stage-positioning screw    Specimen-positioning screw    Specimen pin    Single lens

**A**

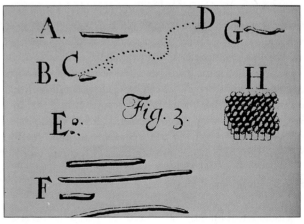

**B**

**FIGURE 4.2   A First Microscope.   (A)** Antonie van Leeuwenhoek made many simple microscopes such as this example, which opened the microscopic world to view. **(B)** He drew what he saw and, in doing so, made the first record of microorganisms.

including human red blood cells and sperm. His work set the foundation for modern fields of microbiology and cell biology. Investigating Life 4.1 describes some of the types of microscopes used in cell biology, and **figure 4.3** provides a sense of the size of objects that these microscopes can image.

## The Cell Theory Emerges

Despite the accumulation of microscopists' drawings of cells made during the seventeenth and eighteenth centuries, the **cell theory**—the idea that the cell is the fundamental unit of all life—did not emerge until the nineteenth century. Historians attribute this delay to poor technology, including crude microscopes and lack of procedures to preserve and study living cells without damaging them. Neither the evidence itself nor early interpretations of it suggested that all organisms were composed of cells. Hooke had not observed actual cells but rather what they had left behind. Leeuwenhoek made important observations, but he did not systematically describe or categorize the structures that cells had in common.

In the nineteenth century, more powerful microscopes, with better magnification and illumination, revealed details of life at the subcellular level. In the early 1830s, Scottish surgeon Robert Brown noted a roughly circular structure in cells from orchid plants. Finding the structure in every orchid cell, he then identified it in all cells from a variety of other organisms. He named it the **nucleus,** a term that stuck. Brown never realized the importance of the organelle he discovered, but today we know the nucleus houses DNA for complex cells.

The cell theory finally emerged in 1839, when German biologists Matthias J. Schleiden and Theodor Schwann made careful comparisons of plants and animals. Schleiden first noted that cells were the basic units of plants, and then Schwann compared animal cells to plant cells. After observing many different plant and animal cells, they concluded that cells were "elementary particles of organisms, the unit of structure and function." Schleiden and Schwann described the components of the cell as a cell body and nucleus contained within a surrounding membrane. Schleiden called a cell a "peculiar little organism" and realized that a cell can be a living entity on its own, but the new theory also recognized that in larger plants and animals, cells are part of a larger living organism.

Many cell biologists extended Schleiden and Schwann's observations and ideas. German physiologist Rudolph Virchow added the important corollary in 1855 that all cells come from preexisting cells, contradicting the still-popular idea that life can arise from the nonliving or from nothingness. Virchow's statement also challenged the popular concept that cells develop on their own from the inside out, the nucleus forming a cell body around itself, and then the cell body growing a cell membrane. Virchow's observation set the stage for descriptions of cell division in the 1870s and 1880s (see chapter 9).

Virchow's thinking was ahead of his time because he hypothesized that abnormal cells cause diseases that affect the whole body. Many new treatments for diverse disorders are based on understanding cellular processes.

Although we have known about the relationship between cells and life for many years, the cell theory is still evolving. For

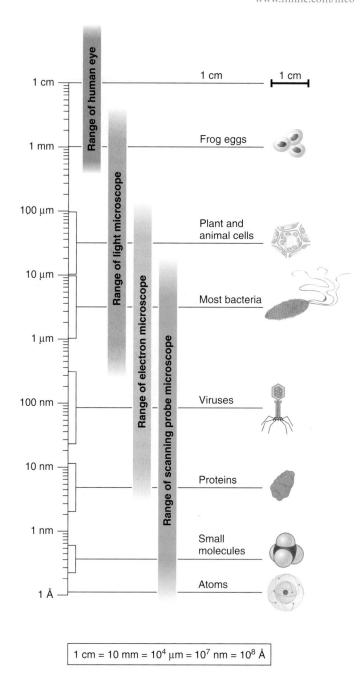

1 cm = 10 mm = $10^4$ μm = $10^7$ nm = $10^8$ Å

**FIGURE 4.3** **Ranges of the Light, Electron, and Scanning Probe Microscopes.** Biologists use the metric system to measure size (see appendix C). The basic unit of length is the meter (m), which equals 39.37 inches (slightly more than a yard). Smaller metric units measure many chemical and biological structures. A centimeter (cm) is 0.01 meter (about 2/5 of an inch); a millimeter (mm) is 0.001 of a meter; a micrometer (μm) is 0.000001 meter; a nanometer (nm) is 0.000000001 meter; an angstrom unit (Å) is 1/10 of a nanometer. In the scale shown, each segment represents only 1/10 of the length of the segment beneath it. The sizes of some chemical and biological structures are indicated next to the scale.

example, until recently, scientists viewed a complex cell as a structure containing a nucleus and jellylike cytoplasm, with organelles suspended in no particular organization—simply because we could not discern a pattern in their spatial distributions. Researchers are now learning that organelles have precise

| Domain Bacteria | Domain Archaea | Domain Eukarya |
|---|---|---|
|  | 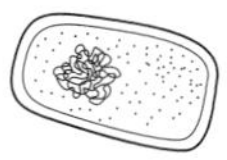  |  |
| • 1–10 micrometers | • 1–10 micrometers | • 10–100 micrometers |
| • Cell wall of peptidoglycan | • Cell wall of various molecules; pseudopeptidoglycan, protein | • Cell wall of cellulose or chitin in many |
| • No introns present | • Some introns present | • Introns present |
| • Membrane based on fatty acids | • Membrane based on non-fatty acid lipids (isoprenes) | • Membrane based on fatty acids |
| • No membrane-bounded organelles | • No membrane-bounded organelles | • Membrane-bounded organelles |
| • 4-subunit RNA polymerase | • Many-subunit RNA polymerase | • Many-subunit RNA polymerase |

**FIGURE 4.4** Cells of the Three Domains of Life. For many years, biologists considered cells to be two types—prokaryotic or eukaryotic—distinguished by absence or presence of a nucleus, respectively. Investigation at the molecular level, however, has revealed that not all prokaryotes are alike. Biologists now recognize three types of cells: a bacterium, the superficially similar archaean, and the eukaryotic cell.

locations in cells, near other structures with which they interact. These locations can change as the cell participates in different functions. We continue to learn there is much more complexity to cells than we ever suspected. Cell biology is emerging as one of the hottest areas of scientific discovery.

## Reviewing Concepts

- Cells are the source of all of the characteristics of life but can also specialize.
- With increased improvement in lenses, more details concerning cells emerged, resulting in the cell theory: all life is composed of cells, and cells are alive.
- Structures within cells support functions.

# 4.2 Variations on the Cellular Theme

Until recently, biologists recognized just two types of organisms, prokaryotes (whose cells lack organelles—small membrane-bounded compartments) and eukaryotes (whose cells have organelles). Recent discoveries have confirmed the existence of three basic cell types, representing three taxonomic domains. Eukaryotic cells, in domain Eukarya, are larger and more complex than prokaryotic cells. Prokaryotic cells, in the domains Archaea and Bacteria, lack nuclei and other membrane-bounded organelles. In 1977, University of Illinois physicist-turned-microbiologist Carl Woese detected differences in key molecules in some of the prokaryotes that were great enough to suggest they were a completely different form of life. He first named them Archaebacteria, which was changed to Archaea when it became apparent their resemblance to bacteria was only superficial.

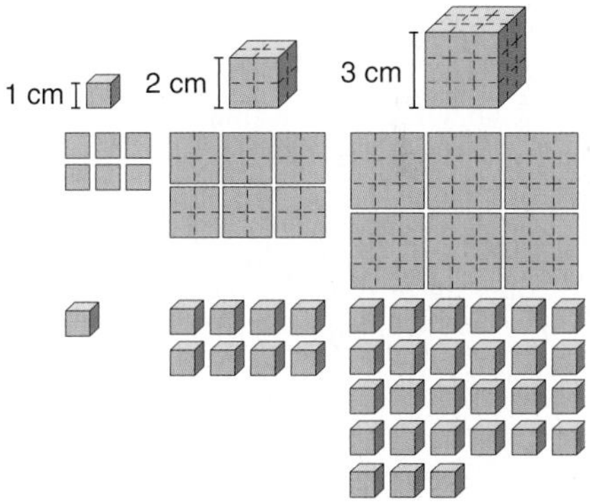

| | 1 cm | 2 cm | 3 cm |
|---|---|---|---|
| Surface area (cm$^2$) | 6 | 24 | 54 |
| Volume (cm$^3$) | 1 | 8 | 27 |
| Ratio of surface area to volume | 6.0 | 3.0 | 2.0 |

**FIGURE 4.5** The Important Relationship Between Surface Area and Volume. When an object enlarges, its volume grows faster than its surface area. Cells have limited sizes because if they grow too large, the surface areas would be too small to support the large volumes.

The archaea lack organelles and have several characteristics unique to that domain but also share features with bacteria and the Eukarya. Since much of their uniqueness lies at the molecular level, we do not have a complete enough picture of the archaea to even depict a "typical" cell, as we can for the other two domains (**figure 4.4**).

One way to distinguish Eukarya from the other domains is by cell size. Most eukaryotic cells are 100 to 1,000 times larger than those of Bacteria and Archaea. All cells require relatively large surface areas through which they interact with the environment. Nutrients, water, oxygen, carbon dioxide, and waste products must enter or leave a cell through its surfaces. As a cell grows, its volume increases at a faster rate than its surface area, a phenomenon demonstrable with simple calculations (**figure 4.5**).

Investigating Life

## Microscopes Reveal Cell Structure

Studying life at the cellular and molecular levels requires microscopes to magnify structures. A researcher must usually prepare specimens of tissue or single cells before they can be observed under a microscope. First, specimens are fixed. This means that certain organic chemicals are applied that stop enzyme action, solidify structures, and basically hold the cell and its constituents in place. Next, if the specimen is too thick to be imaged, it must be sectioned into thin slices. Dyes are typically added, alone or in combination, to provide contrast to cell parts that are naturally translucent or transparent.

All microscopes provide two types of power—*magnification* and *resolution* (also called resolving power). A microscope produces an enlarged, or magnified, image of an object. Magnification is the ratio between the size of the image and the object. Resolution refers to the smallest degree of separation at which two objects appear distinct. A compound light microscope can resolve objects that are 0.1 to 0.2 micrometer (4 to 8 millionths of an inch) apart. The resolving power of an electron microscope is 10,000 times greater. Figure 4.A compares these two types of microscopes.

Following is a survey of several types of microscopes.

### The Light Microscope

The *compound light microscope* focuses visible light through a specimen. Different regions of the object scatter the light differently, producing an image. Three sets of lenses help generate the image. The condenser lens focuses light through the specimen. The objective lens receives light that has passed through the specimen, generating an enlarged image. The ocular lens, or eyepiece, magnifies the image further. Multiplying the magnification of the objective lens by that of the ocular lens gives the total magnification. A limitation of light microscopy is that it focuses on only one two-dimensional plane at a time, diminishing the sense of depth.

A *confocal microscope* is a type of light microscope that enhances resolution by passing white or laser light through a pinhole and a lens to the object. This eliminates the problem of light reflecting from regions of the specimen near the object of interest, which can blur the image. The result is a scan of highly focused light on one tiny part of the specimen. "Confocal" refers to the fact that both the objective and condenser lenses focus on the same small area. Computers can integrate many confocal images of specimens exposed to fluorescent dyes to produce spectacular peeks at living structures. (Fluorescence is an optical phenomenon discussed in chapter 7.)

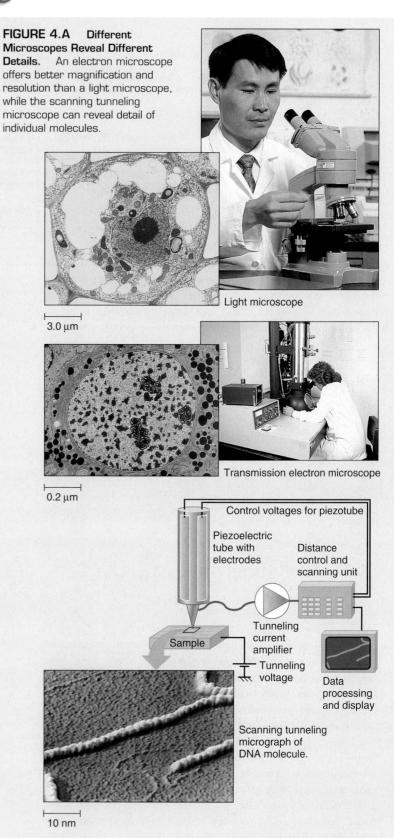

**FIGURE 4.A  Different Microscopes Reveal Different Details.** An electron microscope offers better magnification and resolution than a light microscope, while the scanning tunneling microscope can reveal detail of individual molecules.

Light microscope

$3.0 \ \mu m$

Transmission electron microscope

$0.2 \ \mu m$

Control voltages for piezotube

Piezoelectric tube with electrodes

Distance control and scanning unit

Sample

Tunneling current amplifier

Tunneling voltage

Data processing and display

Scanning tunneling micrograph of DNA molecule.

10 nm

## The Electron Microscope

Electron microscopes provide greater magnification, better resolution, and better depth than light microscopes. Instead of focusing light, the *transmission electron microscope* (TEM) sends a beam of electrons through a specimen, using a magnetic field rather than a glass lens to focus the beam. Different parts of the specimen absorb electrons differently. When electrons from the specimen hit a fluorescent screen coated with a chemical, light rays are given off, translating the contrasts in electron density into a visible image.

In TEM, the specimen must be killed, chemically fixed, cut into very thin sections, and placed in a vacuum, a treatment that can distort natural structures. The *scanning electron microscope* (SEM) eliminates some of these drawbacks. It bounces electrons off a metal-coated, three-dimensional specimen, generating a 3-D image on a screen that highlights crevices and textures.

## Scanning Probe Microscopes

These microscopes work on a different principle than light or electron microscopes. They move a probe over a surface and translate the distances into an image—a little like moving your hands over someone's face to get an idea of his or her appearance. There are several types of scanning probe microscopes.

A *scanning tunneling microscope* reveals detail at the atomic level. A very sharp metal needle, its tip as small as an atom, scans a molecule's surface. Electrons "tunnel" across the space between the sample and the needle, creating an electrical current. The closer the needle, the greater the current. An image forms as the scanner continually adjusts the space between the needle and specimen, keeping the current constant over the topography of the molecular surface. The needle's movements over the microscopic hills and valleys are expressed as contour lines, which a computer converts and enhances to produce a colored image of the surface.

Electrons do not pass readily through many biological samples, so *scanning ion-conductance microscopy* offers an alternative approach. It uses ions instead of electrons, which is useful in imaging muscle and nerve cells, which are specialized to use ions in communication. The probe is made of hollow glass filled with a conductive salt solution, which is also applied to the sample. When voltage passes through the sample and the probe, ions flow to the probe. The rate of ion flow is kept constant, and a portrait is painted as the probe moves. In yet another variation, the *atomic force microscope* uses a diamond-tipped probe that presses a molecule's surface with a very gentle force (figure 4.B). As the force is kept constant, the probe moves, generating an image. This variation on the theme is useful for recording molecular movements, such as blood clotting and cells dividing.

Figure 4.C contrasts images of human red blood cells taken with the light, electron, and scanning probe microscopes.

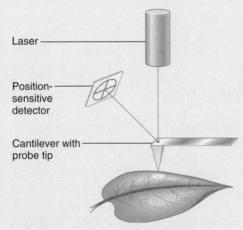

**FIGURE 4.B    An Atomic Force Microscope "Feels" a Biological Surface.** This device uses a microscopic force sensor called a cantilever, which is attached to a tip that assesses the space between itself and the sample. As the cantilever moves over the surface, the detected distances are converted into an image.

Laser

Position-sensitive detector

Cantilever with probe tip

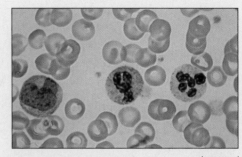

A

50 µm

B

25 µm

C

3.5 µm

**FIGURE 4.C    Three Views of Red Blood Cells.** **(A)** A "smear" of blood visualized under a light microscope appears flat. **(B)** The disc-shaped red blood cells appear with much greater depth with the scanning electron microscope, but material must be killed, fixed in place, and subjected to vacuum before viewing, thus limiting applications. **(C)** The scanning probe microscope captures images of cells and molecules in their natural state, as parts of living organisms.

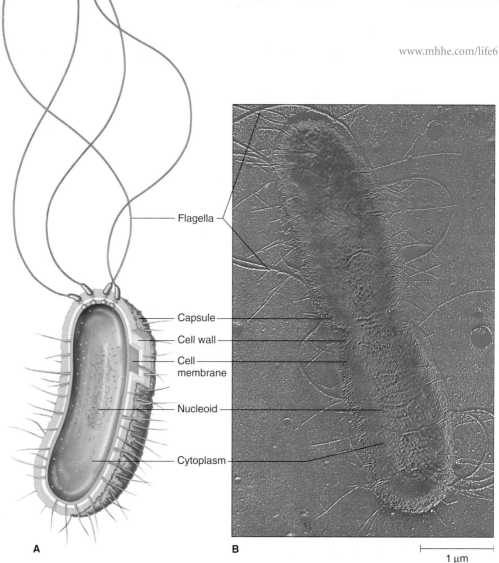

FIGURE 4.6    Anatomy of a Bacterium.
(A) Bacteria lack organelles such as a nucleus. Their DNA is suspended in the cytoplasm among the ribosomes, enzymes, nutrients, and fluids that make up the inside of a cell. Most bacteria use a rigid cell wall to maintain their shape, and some use flagella, long whiplike tails, to provide movement. (B) This common bacterium, *Escherichia coli,* is dividing.

The increased volume of a large cell means the interior components' cells are farther away from the cell's surface. The ultimate inability of a cell's surface area to keep pace with its volume can limit a cell's size, and this may be why bacterial and archaean cells are usually small. The evolution of organelles enabled the larger cells of eukaryotic organisms to have more overall membrane surface area to carry out the needed functions. Another adaptation to maximize surface area is for the cell membrane to fold, just as inlets and capes extend the perimeter of a shoreline. Eukaryotic cells have also evolved an internal network of "tracks" that allows them to rapidly move substances from one part of the cell to another, discussed in chapter 5.

## Bacterial Cells Lack Organelles

All bacterial cells lack distinct membrane-enclosed nuclei (**figure 4.6**). Once known as blue-green algae because of their characteristic pigments and their similarity to the true (eukaryotic) algae, cyanobacteria also lack nuclei. Like the cyanobacteria, many of these prokaryotic organisms are photo-synthetic, while others play a critical role in ecosystems as decomposers. Bacteria exhibit a wide variety of shapes and approaches to life (**figure 4.7**). Although a few bacteria cause illnesses, others are very valuable in food and beverage processing and pharmaceutical production. Bacteria are vital to life on Earth.

Most bacterial cells are surrounded by rigid **cell walls** built of peptidoglycan, a molecule consisting of amino acid chains and carbohydrates. Many antibiotic drugs, such as amoxicillin, halt bacterial infection by interfering with the microorganism's ability to construct its cell wall. In addition to the cell wall, some bacteria form a polysaccharide capsule that protects the cell or attaches the bacteria to specific types of surfaces. You see an example of this in the sticky film of bacteria found on your teeth each morning.

Members of this domain are extremely abundant and diverse—the species we are most familiar with are only a small subset of the entire group. To classify bacteria, microbiologists look at differences in cell wall structure; biochemical characteristics, such as metabolic pathways; and shape. Bacterial cells may be round (cocci), rod-shaped (bacilli), spiral (spirilla), comma-shaped (vibrios), or spindle-shaped (fusiform). Micro-

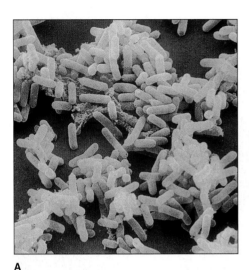

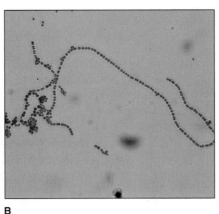

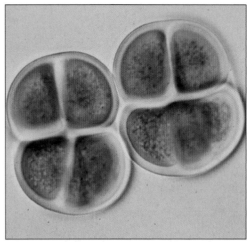

A

B

C

**FIGURE 4.7**    **A Trio of Bacteria.**    **(A)** *E. coli* (×35,000) inhabits the intestines of some animal species, including humans. **(B)** *Streptococcus pyogenes* infects humans (×900). **(C)** Cyanobacteria, such as this *Chroococcus furgidus* (×600), have pigments that enable them to photosynthesize.

biologists use a specific staining procedure (described in figure 21.4) known as Gram-staining to distinguish two types of cell wall structures. Gram-positive bacteria have cell walls that contain very thick layers of peptidoglycan. Gram-negative bacteria have a thinner peptidoglycan layer plus an outer membrane rich in protein and lipopolysaccharide.

Prokaryotic cells also differ from eukaryotes in the nature of their DNA. Unlike the linear chromosomes of most eukaryotes, the main genetic material of a bacterium is a single circle of DNA, associated with unique proteins. This circle of DNA is located in a fibrous region called the **nucleoid.** Nearby are regions of protein synthesis (covered in detail in chapter 13) composed of RNA molecules and **ribosomes;** protein synthesis complexes consisting of RNA and protein. The prokaryotic ribosomes are distinct from those in eukaryotic cells and are another target for antibiotics. The close arrangement of DNA, RNA, and ribosomes in prokaryotic cells makes protein synthesis more rapid than in eukaryotes. This adaptation allows bacteria to grow and reproduce rapidly in the challenging environments in which they live.

## Archaean Cells Represent a Distant Ancestor

The first members of Archaea to be described were microorganisms that use carbon dioxide and hydrogen from the environment to produce methane—hence, they are called methanogens. Originally, archaea were considered bacteria, but Woese delineated their differences from the better-known bacteria and eukarya:

- Their cell walls lack peptidoglycan, found in bacteria.
- They may have unique coenzymes to produce methane.

- The base sequence of two characteristic types of RNA (rRNA and tRNA) are distinctly different from similar molecules in members of the other domains.

The methanogens have a curious mix of characteristics. These archaea can transport ions within their cells like bacteria do, and some surface molecules are identical to those in bacteria. Yet, like eukaryotes, they have proteins, called histones, associated with their genetic material. Their protein synthesis machinery more closely resembles that of eukaryotes than prokaryotes. When researchers deciphered all of the genes of a methanogenic archaean in 1996, they found, as Woese expected they would, that more than half of the genes had no counterpart among bacteria or eukarya. But the fact that nearly half of the genes do correspond indicates that the three forms of life branched from a shared ancestor long ago (see figure 1.7).

Most of the first Archaea identified came from environments that have extremes of temperature, pressure, pH, or salinity. Researchers have since discovered them in a variety of habitats, including swamps, rice paddies, and throughout the oceans (**figure 4.8**). Yet, we still know hardly anything about Archaea compared to what we know of the other two domains. Chapter 21 discusses what we do know.

## Eukaryotic Cells Use Organelles

Plants, animals, fungi, and protista are composed of eukaryotic cells (**figures 4.9** and **4.10**). In these cells, organelles provide membrane surfaces and create specialized compartments to organize functions. To improve efficiency, some membranes contain different enzymes on their surfaces physically laid out in the order in which they participate in biochemical reactions (such as

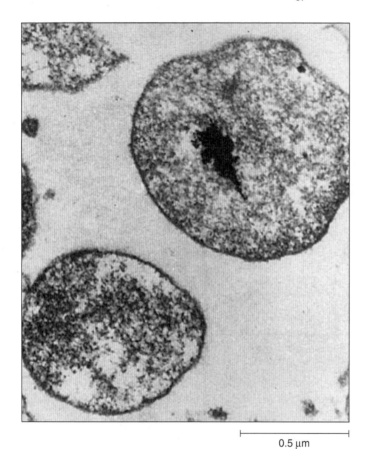

0.5 μm

**FIGURE 4.8    An Archaean.** *Thermoplasma* has characteristics of bacterial and eukaryotic cells. The DNA wraps around proteins much as it does in eukaryotic cells, but there is no nucleus. *Thermoplasma* thrives in the high heat and acidic conditions of smoldering coal deposits. When researcher Gary Darland first discovered these cells, he was so startled by the habitat and mix of features that he named it the "wonder organism."

those involved in photosynthesis, discussed in chapter 7). However, some cellular processes will not occur until the concentration of reactants is sufficiently high. But often the reactants, such as hydrogen ions, become increasingly destructive to other parts of the cell with increasing concentration. The compartmentalization provided by organelles makes these reactions possible without disrupting the cell. This also makes it unnecessary for the entire cell to maintain a high concentration of any particular biochemical. Reactions can be isolated from one another in different organelles, making some processes available to a cell that could not occur otherwise.

The most prominent organelle in most eukaryotic cells is the nucleus, which protects and organizes the cell's DNA. The eukaryotic DNA is combined with protein, forming threadlike chromosomes that condense and appear as the familiar rodlike structures when the cell divides. The remainder of the cell consists of other organelles and cytoplasm. In fact, nearly half of the volume of an animal cell is organelles. In contrast, some

plant cells contain up to 90% water, much of it within a large organelle called a **vacuole**. All cells, however, are very watery. On the science fiction program *Star Trek,* beings from another planet quite correctly called humans "ugly bags of mostly water."

In addition to the organelles, some eukaryotic cells also contain stored nutrients, minerals, and pigment molecules. Arrays of protein rods and tubules within plant and animal cells form the **cytoskeleton**, which helps to give the cell its shape. Protein rods and tubules also form appendages that enable certain cells to move, and they form structures that are important in cell division. (Chapter 5 examines the cytoskeleton in depth.)

## Reviewing Concepts

- Three basic types of cells are represented in all organisms.
- Bacteria do not contain organelles.
- Archaea are unique cells that share features of both bacteria and eukaryotes but have unique biochemical features as well.
- Eukaryotic cells use membranes to increase surface area to allow bigger cells and to provide compartmentalization for unique chemical processes.

# 4.3    An Introduction to Organelles

Organelles effectively compartmentalize a cell's activities, improving efficiency and protecting cell contents from harsh chemicals. Organelles enable cells to secrete substances, derive energy from nutrients, degrade debris, and reproduce.

Organelles divide the activities of life at the cellular level, like rooms in factories divide different aspects of the manufacturing process. Organelles also interact, providing basic life functions as well as specialized characteristics.

## Organelles Synthesize and Process Proteins

Proteins, including enzymes, are key to determining the functions of a cell and how it interacts with other cells in the body. Genes, instructions for building each protein, are stored in the nucleus, which protects them from degradation in the cytoplasm. Connected to the nucleus, the **endomembrane system** is a series of compartments formed from highly folded membranes. Within the first compartment, the **rough endoplasmic reticulum** (ER), many of the cell's proteins are manufactured. Subsequent compartments house unique enzymes that process these proteins. Each compartment is connected to the next through **vesicles**, which are small packages of proteins and other molecules surrounded by membrane.

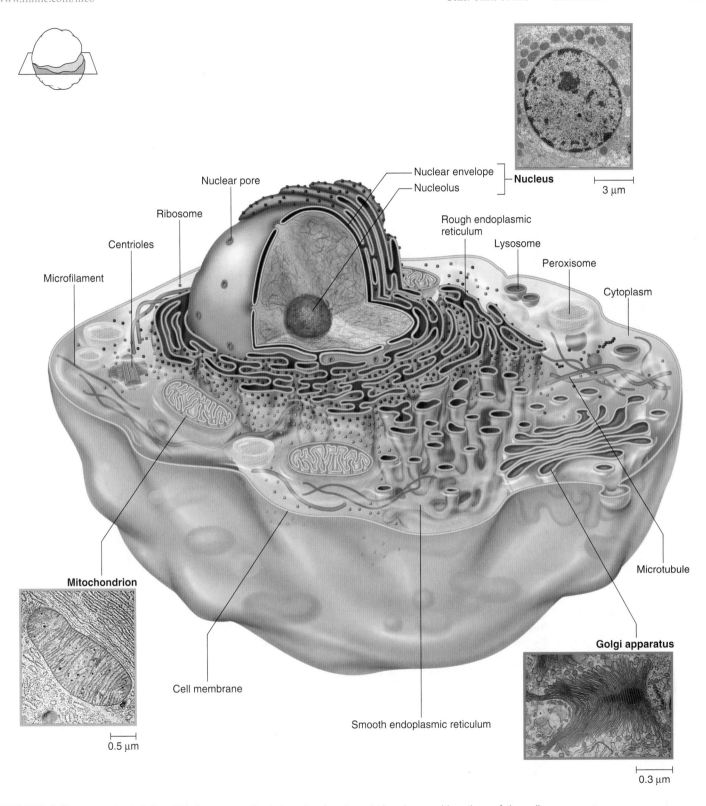

**FIGURE 4.9    An Animal Cell.**    This is a generalized view showing the relative sizes and locations of the cell components.

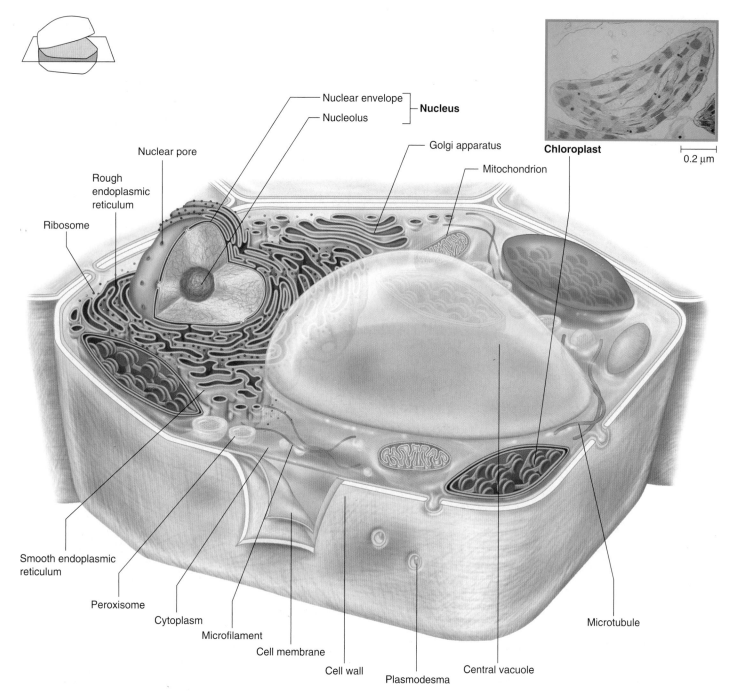

**FIGURE 4.10    A Plant Cell.**    This generalized view illustrates key features of the plant cell. Compare to the animal cell in figure 4.9.

In a related area, the **smooth endoplasmic reticulum,** lipids are synthesized and modified, and toxins are neutralized. The final compartment, the **Golgi apparatus,** finishes the processing and sorts proteins for export out of the cell, or into sacs called **lysosomes,** which contain digestive enzymes. Other compartments within the cell provide energy, contain special enzymes for oxidation reactions, or simply store useful molecules. Examining a coordinated function, milk secretion, illustrates how organelles interact to produce, package, and release from the cell a complex mixture of biochemicals.

## Organelles Interact to Secrete Substances

Special cells in the mammary glands of female mammals produce milk, which is a complex mixture of proteins, fats, carbohydrates, and water in a proportion ideal for development of the young of a particular species. Human milk is also rich in lipids, which the rapidly growing newborn's brain requires, and immune-system products, to protect from infections. Dormant most of the time, the special cells of the mammary glands increase their metabolic activ-

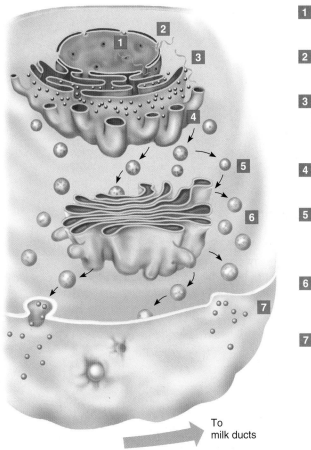

1 Milk protein genes transcribed into mRNA

2 mRNA exits through nuclear pores

3 mRNA forms complex with ribosomes and moves to surface of rough ER where protein is made

4 Enzymes in smooth ER manufacture lipids

5 Milk proteins and lipids are packaged into vesicles from both rough and smooth ER for transport to Golgi

6 Final processing of proteins in Golgi and packaging for export out of cell

7 Proteins and lipids released from cell by fusion of vesicles with cell membrane

To milk ducts

**FIGURE 4.11    Secretion.**    Milk production and secretion illustrate organelle functions and interactions in a cell from a mammary gland; (*1*) through (*7*) indicate the order in which organelles participate in this process.

ity during pregnancy and then undergo a burst of productivity shortly after the female gives birth. Organelles form a secretory network that enables individual cells to manufacture milk. We follow here the production and secretion of human milk (**figure 4.11**).

---

### Reviewing Concepts

- The endomembrane system is a site for protein synthesis and provides mechanisms for protein processing and packaging.
- The rough ER is involved in synthesis, the smooth ER and Golgi appartatus process proteins, and vesicles are used to package and transport proteins.

---

## 4.4    The Nucleus and the Cytoplasm

### The Nucleus Exports RNA Instructions

Secretion begins in the nucleus (figure 4.11, step 1), where specific genes are copied into another nucleic acid, messenger RNA,

or mRNA. These genes encode milk protein and enzymes required to synthesize the carbohydrates and lipids in milk. The mRNA molecules exit from the interior of the nucleus (figure 4.11, step 2) toward the cytoplasm through the **nuclear pores** (**figure 4.12**), holes in the two-layered **nuclear envelope** that separates the nucleus from the cytoplasm. Nuclear pores are not merely perforations but highly specialized channels composed of more than 100 types of proteins that span both membranes. Traffic through the nuclear pores is busy, with millions of proteins and mRNA molecules passing in or out each minute, but highly selective. Proteins tend to enter, and mRNA molecules leave. Proteins colorfully called "importins" and "exportins" help facilitate the passage of molecules between the nucleus and the cytoplasm.

### The Cytoplasm Is the Site of Protein Synthesis

Once in the cytoplasm, the mRNA binds (see figure 4.11, step 3) to one of the millions of protein-manufacturing **ribosomes** that are found in the cytoplasm. Ribosomes are complexes of dozens of proteins surrounding a core of three ribosomal RNA (rRNA) molecules, one of which functions as the key enzyme that forms the peptide bond (see figure 3.6c). The components of ribosomes assemble in a region of the nucleus known as the **nucleolus** and are then transported to the cytoplasm. Once attached to mRNA,

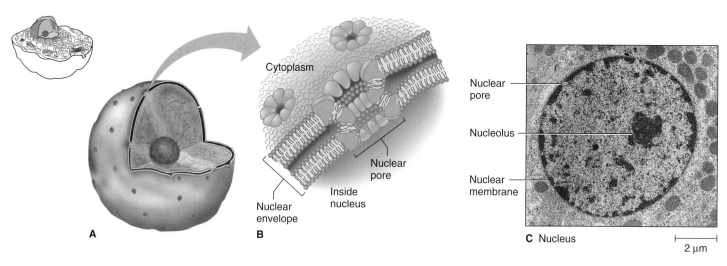

**FIGURE 4.12    The Nucleus.    (A)** The largest structure within a typical eukaryotic cell, the nucleus, is surrounded by two membrane layers, which make up the nuclear envelope **(B)**. Pores through the envelope allow specific molecules to move in and out of the nucleus. The darkly staining nucleolus **(C)** is the site of ribosome manufacture and assembly.

ribosomes immediately begin synthesizing the encoded protein. For secretory proteins, the entire complex of ribosomes, mRNA, and partially made protein becomes anchored to the surface of the rough endoplasmic reticulum (rough ER) shortly after protein synthesis begins (**figure 4.13**). (*Endoplasmic* means "within the cytoplasm," and *reticulum* means "network".) The proteins are inserted through the membrane as they are made and wind up in the interior compartment of the rough ER where they are folded, modified, and packaged for secretion from the cell. Membrane proteins are made in a similar fashion. Any proteins that fail to form properly are removed and destroyed.

The presence of so many ribosomes anchored to its surface gives the membrane of the rough ER its "rough" appearance and name. Adjacent to the rough ER, a section of the network called smooth ER (see figure 4.13) synthesizes lipids and other membrane components. The smooth ER also houses enzymes that detoxify certain chemicals. The lipids and proteins made by the ER (see figure 4.11, step 4) exit the organelle in vesicles that pinch off from the tubular endings of the ER membrane (see figure 4.11, step 5).

A loaded vesicle carries its contents to the next stop in the secretory production line, the Golgi apparatus (**figure 4.14**). This organelle is a stack of flat, membrane-enclosed sacs that functions as a processing center. As proteins from the ER pass through the series of Golgi sacs (see figure 4.11, step 6) in specific sequences, they complete their intricate processing and become functional. The Golgi distinguishes between proteins destined for secretion and proteins that will continue into a lysosome. The

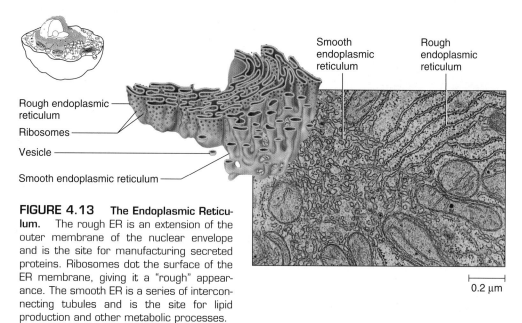

**FIGURE 4.13    The Endoplasmic Reticulum.**    The rough ER is an extension of the outer membrane of the nuclear envelope and is the site for manufacturing secreted proteins. Ribosomes dot the surface of the ER membrane, giving it a "rough" appearance. The smooth ER is a series of interconnecting tubules and is the site for lipid production and other metabolic processes.

Golgi apparatus, therefore, compartmentalizes the sequence of steps necessary to produce functional proteins. Finally, enzymes in the Golgi apparatus manufacture and attach complexes of carbohydrates to proteins to form glycoproteins or to lipids to form glycolipids.

Vesicles are cellular transports keyed to carry their cargo of proteins to locations specific to the proteins and their functions. Vesicles budding off of the Golgi apparatus in this example contain milk proteins and fats. The vesicles move toward the cell membrane, where they fuse to become part of the cell membrane and open out, facing the exterior of the cell, and release the proteins (see figure 4.11, step 7). It isn't surprising that cells that secrete copiously have large ER and numerous Golgi apparatuses.

When a baby suckles, hormones (chemical messengers) released in the mother's system stimulate muscle cells surrounding

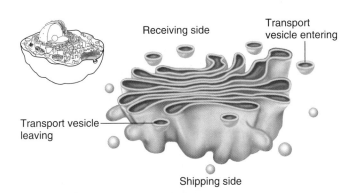

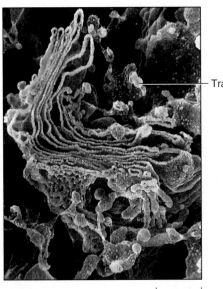

**FIGURE 4.14** **The Golgi Apparatus.** Seen here in a scanning electron micrograph, the Golgi apparatus is composed of a series of membrane vesicles and flattened sacs. Proteins are sorted and processed as they move through the Golgi apparatus on their way to the cell surface or lysosomes.

balls of glandular cells in the breast to contract and release milk into ducts that lead to the nipple.

All eukaryotic cells use secretory pathways in one way or another. But eukaryotic cells contain additional organelles that are not directly involved in the secretion of milk.

### Reviewing Concepts

- Proteins destined to be secreted from the cell are manufactured initially on the rough ER and processed through the Golgi.
- The ribosomes are manufactured in the nucleolus and transported to the cytoplasm.
- Nuclear pores facilitate controlled movement of RNA from the nucleus to the cytoplasm, where ribosomes attach and begin protein synthesis.

## 4.5 Lysosomes and Peroxisomes: Cellular Digestion Centers— and More

To recycle vital nutrients as well as make use of ingested ones, the cells of eukaryotes break down molecules and structures as well as produce them. Specialized compartments prevent these processes from interfering with each other.

### Lysosomes Are Cellular Recycling Centers

Lysosomes, named because their enzymes lyse (cut apart) their targets, dismantle captured bacteria, worn-out organelles, and debris. Lysosomal enzymes also break down large nutrients (fats, proteins, and carbohydrates) into the constituent monomer molecules the cell can use, releasing them through the organelle's membrane into the cytoplasm. Lysosomes fuse with vesicles carrying debris from outside or worn-out organelles within the cell, and the lysosomal enzymes then degrade the contents (**figure 4.15**).

Lysosomal enzymes are manufactured within the rough ER. The Golgi apparatus then detects and separates enzymes destined for lysosomes by recognizing a particular type of sugar attached to them and packaging them into vesicles that eventually become lysosomes. Lysosomal enzymes can function only in a very acidic environment, and the organelle maintains this environment without harming other cellular constituents.

Different types of cells have differing numbers of lysosomes. White blood cells, for example, have many lysosomes because the function of these cells is to engulf debris and bacteria to protect and maintain the larger organism. Liver cells require many lysosomes to process cholesterol and blood components.

In human cells, a lysosome contains more than 40 types of digestive enzymes. The correct balance of these enzymes is important to health. Absence or malfunction of just one type of enzyme can cause a lysosomal storage disease, in which the molecule that is normally degraded accumulates. The lysosome swells, crowding organelles and interfering with the cell's functions. In Tay-Sachs disease, for example, lack of an enzyme that normally breaks down lipid in cells surrounding nerve cells buries the nervous system in lipid. An affected infant gradually loses sight, hearing, and the ability to move, typically dying within 3 years. Even before birth, the lysosomes of affected cells become hugely swollen.

In plant cells, the large central vacuole serves a function similar to, but not as extensive as, the lysosomes. These enzyme-containing vacuoles soften cells in some fruits, making them more palatable to animals that eat them and spread the seeds.

### Peroxisomes Facilitate Oxidative Reactions

**Peroxisomes** are single-membrane-bounded sacs, present in all eukaryotic cells, that contain several types of enzymes, usually those involved in oxidizing (removing electrons) other molecules. Unlike lysosomes, peroxisomes contain enzymes made in the cytoplasm and then transported into vesicles, which become peroxisomes. In some species, exposure to environmental toxins triggers an explosive production of peroxisomes, which helps cells to survive the insult. Due to their ability to catalyze oxidizing reactions, peroxisomal enzymes catalyze a variety of biochemical reactions, including

- synthesizing bile acids, which are used in fat digestion;
- breaking down lipids called very-long-chain fatty acids;
- degrading rare biochemicals;
- metabolizing potentially toxic compounds that form as a result of oxygen exposure.

Some peroxisomal enzymes produce hydrogen peroxide ($H_2O_2$) as a by-product of their normal activity. Hydrogen peroxide releases highly reactive oxygen-free radicals (oxygen atoms with unpaired electrons), which can damage the cell. To counteract the free-radical buildup, peroxisomes contain abundant catalase, an enzyme that removes an oxygen atom from hydrogen peroxide and combines it with hydrogen to produce harmless water molecules. Liver and kidney cells contain many peroxisomes, which help dismantle toxins from the blood (**figure 4.16A**).

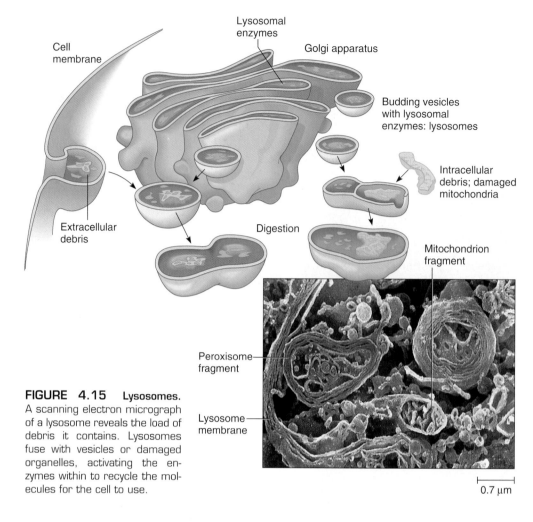

**FIGURE 4.15   Lysosomes.** A scanning electron micrograph of a lysosome reveals the load of debris it contains. Lysosomes fuse with vesicles or damaged organelles, activating the enzymes within to recycle the molecules for the cell to use.

0.7 µm

**FIGURE 4.16   Peroxisomes.   (A)** The high concentration of enzymes inside peroxisomes results in crystallization of the proteins, giving these organelles a characteristic appearance. **(B)** In plants, peroxisomes have enzymes that catalyze many oxidation-reduction reactions that assist in photosynthesis and defense.

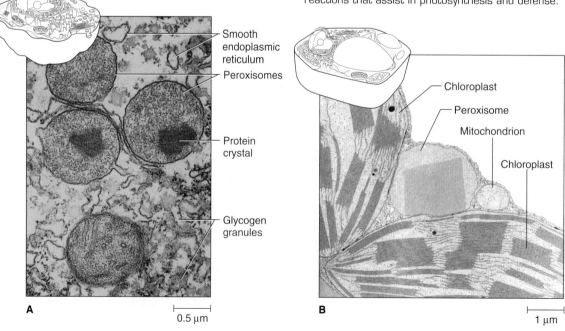

A    0.5 µm

B    1 µm

The leaf cells of plants contain many peroxisomes. The concentration of enzymes such as catalase reaches such high levels in these cells that the protein condenses into easily recognized crystalline arrays that are a hallmark of peroxisomes (figure 4.16B). Peroxisomes in plants help to break down and process organic molecules.

Abnormal peroxisomal enzymes can harm health. A defect in a receptor on the peroxisome's membrane may affect several of the enzymes and cause a variety of symptoms, or a single enzyme type may be abnormal. In adrenoleukodystrophy, for example, one of two major proteins in the organelle's outer membrane is absent. Normally, this protein transports an enzyme into the peroxisome, where it catalyzes a reaction that helps break down a type of very-long-chain fatty acid. Without the enzyme transporter protein, the fatty acid builds up in the cells of the brain and spinal cord, eventually stripping these cells of fatty coverings necessary for nerve conduction. Symptoms include weakness, dizziness, low blood sugar, darkening skin, behavioral problems, and loss of muscular control.

## Reviewing Concepts

- Lysosomes are packages of digestive enzymes that cells use to recycle cellular components and release nutrient molecules from ingested products.
- Peroxisomes store oxidative enzymes that help break down molecules and protect the cell from harmful chemicals.

# 4.6 Mitochondria and Chloroplasts: Organelles of Energy

## Mitochondria Extract Energy from Nutrients

The activities of secretion, as well as the many chemical reactions taking place in the cytoplasm, require a steady supply of energy. In cells of eukaryotes, organelles called **mitochondria** extract energy from nutrient molecules (**figure 4.17**). The number of mitochondria in a cell can vary from a few to tens of thousands. A typical liver cell has about 1,700 mitochondria; cells with high energy requirements, such as muscle cells, may have many thousands.

A mitochondrion has an outer membrane and an intricately folded inner membrane. The folds of the inner membrane, called **cristae,** contain enzymes that catalyze the biochemical reactions that acquire energy (see chapter 8). This organelle is especially interesting because it contains its own genetic material, a point we will return to at the chapter's end.

Another unique characteristic of mitochondria is that they are inherited from the female parent only. This is because mitochondria are found in the middle regions of sperm cells but not

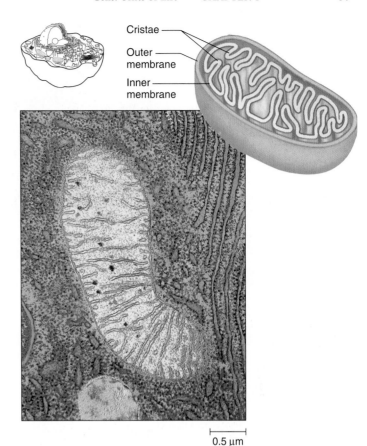

FIGURE 4.17   **Mitochondria Extract Energy from Nutrients.** A transmission electron micrograph of a mitochondrion. Cristae, infoldings of the inner membrane, increase the available surface area containing enzymes for energy reactions.

in the head region, which is the portion that enters the egg to fertilize it. In humans, a class of inherited diseases whose symptoms result from abnormal mitochondria are always passed from mother to offspring. These mitochondrial illnesses usually produce extreme muscle weakness, because muscle is a highly active tissue dependent upon the functioning of many mitochondria.

## Chloroplasts Provide Plant Cells with Nutrients

In plants, and many protista, an additional organelle, the **chloroplast** (**figure 4.18**), carries out photosynthesis, providing the cell with nutrients and useable energy in the form of glucose. Like a mitochondrion, this organelle contains additional membrane layers. Two outer membrane layers enclose a space known as the **stroma.** Within the stroma is a third membrane system folded into flattened sacs called **thylakoids.** The thylakoids are stacked and interconnected in structures called **grana.** The process of photosynthesis, the subject of chapter 7, occurs in these thylakoids.

Like the mitochondria, chloroplasts contain their own unique DNA. Chloroplast genes encode proteins and other molecules unique to photosynthesis and the structure of the chloroplasts.

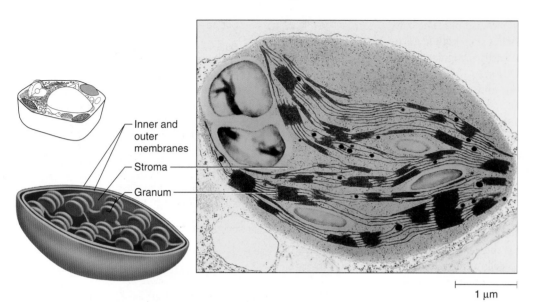

Inner and outer membranes

Stroma

Granum

**FIGURE 4.18    Chloroplasts Are the Sites of Photosynthesis.** A transmission electron micrograph of a chloroplast reveals the stacks of thylakoids that form the grana within the inner compartment, the stroma. Enzymes and light-harvesting proteins are embedded in the membranes of the thylakoids to convert sunlight to chemical energy.

1 μm

## TABLE 4.1

### Structures and Functions of Organelles

| Organelle | Structure | Function |
|---|---|---|
| Nucleus | Enveloped sac enclosing DNA | DNA storage |
| Ribosome | Protein enclosing rRNA | Protein synthesis |
| Endoplasmic reticulum | Membrane network; rough ER includes ribosomes | Protein folding and processing; lipid synthesis; detoxifying |
| Vesicle | Membrane sac | Temporary molecule storage, transport |
| Golgi apparatus | Stacks of membrane sacs | Protein processing; glycolipid and glycoprotein synthesis |
| Lysosome | Membrane sac containing digestive enzymes | Recycling of cellular molecules; digestion of ingested molecules |
| Peroxisome | Membrane sac containing oxidative enzymes | Oxidation of molecules; cellular protection |
| Mitochondrion | Two membranes; inner highly folded; unique DNA | Energy extraction from nutrients |
| Chloroplast | Three membranes; inner stacks of flattened sacs; unique DNA | Photosynthesis |

**Table 4.1** summarizes the organelles discussed in this chapter, but cells have other specialized organelles and structures. Some play a role in cell division; others are important in photosynthesis. These organelles are explored in later chapters. Chapter 5 considers the cell membrane, the cytoskeleton, and how cells interact.

### Reviewing Concepts

- Mitochondria and chloroplasts manipulate energy by manufacturing nutrients or extracting energy from them.
- They contain unique membrane structures that facilitate their biochemical functions.
- They each contain unique DNA.

## 4.7    Origins of Complex Cells

How did eukaryotic cells arise? The **endosymbiont theory** proposes that these complex cells formed as large, nonnucleated cells engulfed smaller and simpler cells. (An endosymbiont is an organism that can live only inside another organism, a relationship that benefits both partners.)

### Structure and DNA Sequences Provide Evidence

The compelling evidence supporting the endosymbiont theory is the striking resemblance between mitochondria and chloroplasts, which are present only in eukaryotic cells, and certain types of bacteria and archaea. These similarities include size,

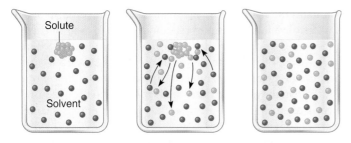

**FIGURE 5.4 Diffusion.** Molecules and atoms collide, then spread, so the same volume of space surrounds each one. Diffusion always results in molecules or atoms moving from regions of high concentration toward low until equilibrium is reached.

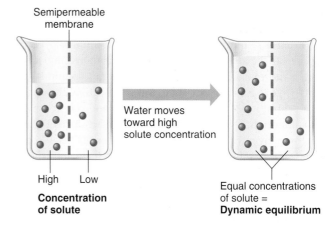

**FIGURE 5.5 Osmosis.** An artificial membrane dividing a beaker demonstrates osmosis by permitting water to pass from one chamber to another, but preventing large solute molecules from doing the same. Water will flow from an area of low salt (solute) concentration toward an area of high salt concentration. Eventually, the volume on each side of the membrane will be different, but the final concentrations (amount of solute per unit of volume) will be the same. Dynamic equilibrium is reached when there is no net tendency for water to flow in either direction.

protein-lined channels), while others cannot. For example, oxygen ($O_2$), carbon dioxide ($CO_2$), and water ($H_2O$) freely cross cell and other biological membranes. They do so by diffusion, without using energy (**figure 5.4**). Diffusion occurs because molecules are in constant motion, and they move so that two regions of differing concentration become equal. Heat increases diffusion by increasing the rate of collisions between molecules. An easy way to observe diffusion is to place a tea bag in a cup of hot water. Compounds in the tea leaves dissolve gradually and diffuse throughout the cup. The tea is at first concentrated near the bag, but the brownish color eventually spreads to create a uniform brew.

The natural tendency of a substance to move from where it is highly concentrated to where it is less so is called "moving down" or "following" its **concentration gradient.** A gradient is a general term that refers to a difference in some quality between two neighboring regions. Concentration, electrical, pH, and pressure differences all create gradients important to life. Ions such as sodium ($Na^+$) and potassium ($K^+$) establish electrical gradients. Hydrogen ions ($H^+$) produce gradients of pH, charge, and concentration that are vital in energy transfers within cells.

Simple diffusion (requiring no energy input) eventually reaches a point where the concentration of the substance is the same on both sides of the membrane. After this, molecules of the substance continue to flow randomly back and forth across the membrane at the same rate, so the concentration remains equal on both sides. This point of equal movement back and forth is called **dynamic equilibrium.**

To envision dynamic equilibrium, picture a party taking place in two rooms. Everyone has arrived, and no one has yet left. People walk between the rooms in a way that maintains the same number of partiers in each room, but the specific occupants change. If people leave the party from one room, the remaining partiers spread out over the available space until people are once again evenly distributed. The party is in dynamic equilibrium.

## Osmosis Is the Movement of Water

Cells must control solvent amounts as well as that of solutes, not only to control volume, but also to provide the best environment for enzymes. The fluids that continually bathe cells of multicellular organisms consist of molecules and ions dissolved in water. Because cells are constantly exposed to water, they must regulate

water entry. If water enters, a cell swells; if too much leaves, it shrinks. Either response may affect a cell's ability to function. Movement of water across biological membranes by simple diffusion is called **osmosis.** The concentration of dissolved substances inside and outside the cell determines the direction and intensity of movement.

In osmosis, water is driven to move because the membrane is impermeable to the solute, and the solute concentrations differ on each side of the membrane (**figure 5.5**). Water moves across the membrane in the direction that dilutes the solute on the side where it is more concentrated.

Variants of the word "tonicity" are used to describe osmosis in relative terms. Water concentration is critical, and the cytoplasm is so rich in nutrients that it is really quite salty. Tonicity refers to the differences in solute concentration in two compartments separated by a semipermeable membrane. A cell interior is **isotonic** to the surrounding fluid when solute concentrations are the same within and outside the cell. In this situation, there is no net flow of water, a cell's shape does not change, and salt concentration is ideal for enzyme activity.

Disrupting a cell's isotonic state changes its internal environment and shape as water rushes in or leaks out. If a cell is placed in a solution in which the concentration of solute is lower than it is inside the cell, water enters the cell to dilute the higher solute concentration there. In this situation, the solution outside the cell is **hypotonic** to the inside of the cell. The cell swells. In the opposite situation, if a cell is placed in a solution in which the solute concentration is higher than it is inside the cell, water leaves the cell to dilute the higher solute concentration outside. In this case, the outside is **hypertonic** to the inside. This cell shrinks.

Hypotonic and hypertonic are relative terms and can refer to the surrounding solution or to the solution inside the cell. It may help to remember that *hyper* means "over," *hypo* means "under," and *iso* means "the same." A solution in one region may be hypotonic or hypertonic to a solution in another region.

The effects of immersing a cell in a hypertonic or hypotonic solution can be demonstrated with a human red blood cell, which is normally suspended in an isotonic solution called plasma (**figure 5.6**). In this state, the cell is doughnut-shaped, with a central indentation. Placing a red blood cell in a hypertonic solution draws water out of the cell, and it shrinks. Placing the cell in a hypotonic solution has the opposite effect. Because there are more solutes inside the cell, water flows into the cell, causing it to swell. Size changes caused by osmosis in plant cells are less dramatic because of the rigid cell wall.

Because shrinking and swelling cells may not function normally, unicellular organisms must regulate osmosis to maintain their shapes. Many cells alter membrane transport activities, changing the concentrations of different solutes on either side of the cell membrane in a way that drives osmosis in a direction that

maintains the cell's shape. This enables some single-celled organisms that live in the ocean to remain isotonic to their salty environment, keeping their shapes. In contrast, the paramecium, a single-celled organism that lives in ponds, must work to maintain its oblong form. A paramecium contains more concentrated solutes than the pond, so water tends to flow into the organism. A special organelle called a contractile vacuole pumps the extra water out (**figure 5.7**).

Plant cells also face the challenge of maintaining their shapes even with a concentrated interior. Instead of expelling the extra water that rushes in, as the paramecium does, plant cells expand until their cell walls restrain their cell membranes. The resulting rigidity, caused by the force of water against the cell wall, is called **turgor pressure** (**figure 5.8**). A piece of wilted lettuce demonstrates the effect of losing turgor pressure. When placed in water, the leaf becomes crisp, as the individual cells expand like inflated balloons.

In the human body, osmosis influences the concentration of urine. Brain cells called osmoreceptors shrink when body fluids are too concentrated, which signals the pituitary gland to release antidiuretic hormone (ADH). The bloodstream transports ADH to cells lining the kidney tubules, where it alters their permeabilities so that water exits the tubules and enters capillaries (microscopic blood vessels) that entwine about the tubules. This conserves water. Without ADH's action, this water would remain in the kidney tubules and leave the body in dilute urine. Instead, it returns to the bloodstream, precisely where it's needed (see figure 39.18). • **kidney function, p. 773**

Blood cells in isotonic solution
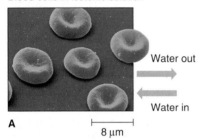

Blood cells in hypertonic solution
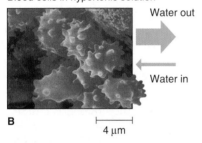

Blood cells in hypotonic solution
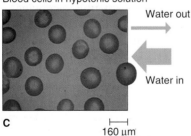

**FIGURE 5.6** **Diffusion Affects Cell Shape.** A red blood cell changes shape in response to changing plasma solute concentrations. (**A**) A human red blood cell is normally isotonic to the surrounding plasma. When water enters and leaves the cell at the same rate, the cell maintains its shape. (**B**) When the salt concentration of the plasma increases, water leaves the cells to dilute the outside solute faster than water enters the cell. The cell shrinks. (**C**) When the salt concentration of the plasma decreases relative to the salt concentration inside the cell, water flows into the cell faster than it leaves. The cell swells and may even burst.

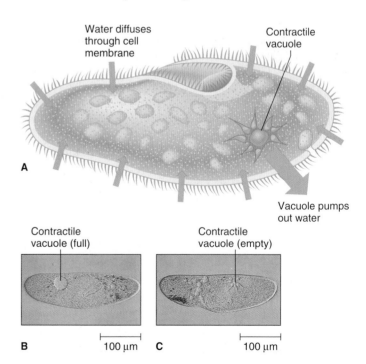

**FIGURE 5.7** **Aquatic Organisms Must Pump Water.** Paramecium keeps its shape with a contractile vacuole that fills and then pumps excess water out of the cell across the cell membrane (**A**). The contractile vacuole moves near the cell membrane as it fills (**B**), and then releases the water to the outside. The organelle then resumes its empty shape (**C**) and moves back to the interior of the cell.

Swallowing and digesting a meal almost as big as oneself requires a huge energy investment. This African rock python is consuming a Thomson's gazelle.

## Eating for Life

The African rock python lay in wait for the lone gazelle. When the gazelle came close, the snake moved suddenly, positioning the victim's head and holding it in place while it swiftly entwined its 30-foot-long body snugly around the mammal. Each time the gazelle exhaled, the snake squeezed, shutting down the victim's heart and lungs in less than a minute. Then the swallowing began.

How can a 200-pound snake eat a 130-pound meal? Thanks to adaptations of the reptile's digestive system, the snake can indeed swallow and digest its meal.

The snake begins to feed by opening its jaws at an angle of 130 degrees (compared to 30 degrees for most humans) and places its mouth over the gazelle's head, using strong muscles to gradually envelop and push along the carcass. Saliva coats the prey, easing its journey to the snake's stomach. After several hours, the huge meal arrives at the stomach, and the digestive tract readies itself for several weeks of dismantling the gazelle. When digestion is completed, only a few chunks of hair will remain to be eliminated.

Eating a tremendous meal once every few months places great energy demands on the snake. While most organisms that eat frequently invest 10 to 23% of a meal's energy in digesting it and assimilating its nutrients, snakes invest 32% in energy acquisition. The African rock python expends an equivalent of half the energy in the chemical bonds of its meal just to digest it.

# The Energy of Life

# Chapter Preview

## 6.1 Energy Definitions

A tortoiseshell butterfly expends a great deal of energy to stay alive. Not only does it fly from flower to flower, but it migrates seasonally over long distances. It also uses energy in a less obvious way to power the many biochemical reactions of its metabolism. This butterfly obtains energy from nectar that a flower produces (**figure 6.1**); the flowering plant gets its energy from the sun. In a related fashion, all animals ultimately extract energy from plants (or certain microorganisms) that capture the energy in sunlight (or inorganic chemicals) and convert it into the chemical energy of organic compounds. Energy is vital to life.

The term *energy* was coined about two centuries ago, when the Industrial Revolution redefined familiar ideas of energy—from the power behind horse-drawn carriages and falling water to the power of the internal combustion engine. As people began to think more about harnessing energy, biologists realized that understanding energy could reveal how life itself is possible. **Bioenergetics** is the study of how living organisms use energy to perform the activities of life.

**Energy** is the ability to do work—that is, to change or move matter against an opposing force, such as gravity or friction. Because energy is an ability, it is not as tangible as matter, which has mass and takes up space. Energy comes in different forms that can be converted from one to another, such as using the energy in gasoline to run a generator that recharges batteries. Since electrons can carry energy, energy can be transferred from molecule to molecule as energized electrons. As molecules are rearranged by enzymes, electrons can be pushed to contain more energy, and then transferred with that energy.

Calories are units used to measure energy. A **calorie** (cal) is the amount of energy required to raise the temperature of 1 gram of water from 14.5°C to 15.5°C. The most common unit for measuring the energy content of food and the heat output of organisms is the **kilocalorie** (kcal), which is the energy required to raise the temperature of a kilogram of water 1°C. The kilocalorie equals 1,000 calories. (Dietary calories are kilocalories.)

The energy transformations that sustain life are similar in all organisms. The two most important pathways are cellular respiration and photosynthesis, and they are intimately related. The *energy-requiring stage* of biological energy acquisition and utilization is usually photosynthesis, the subject of chapter 7. During this process, chlorophyll molecules (or similar molecules

**FIGURE 6.1  It Takes Energy to Get Energy.**  A butterfly expends great amounts of energy flying from flower to flower to obtain food and migrating sometimes thousands of miles. This tortoiseshell butterfly is feeding on nectar from a flower.

**FIGURE 6.2  Energy Can Take Many Forms**  Energy flows from the sun and is captured by plants to make energy-rich chemicals. Those chemicals, in turn, provide energy for the plants themselves and the organisms that eat plants. In this way, energy is continually flowing through one organism to another.

in certain microorganisms) absorb light energy. They use this energy to reduce carbon dioxide (a low-energy compound) to carbohydrate (a high-energy compound). Oxygen is released as a by-product. Carbohydrate, in turn, fuels the activities of the plant and ultimately other organisms. During **cellular respiration,** the energy-releasing stage of the biological energy process, energy-rich carbohydrate molecules are oxidized to carbon dioxide and water. Cellular respiration, explored in chapter 8, liberates the energy necessary to power life.

## Organisms Obtain Energy from the Sun, the Earth, and Other Living Things

Most organisms obtain energy from the sun (**figure 6.2**), either directly through photosynthesis or indirectly by consuming other organisms. Even the energy in fossil fuels and in organisms originated as solar energy.

Deep within the sun, temperatures of 10,000,000°C fuse hydrogen atoms, forming helium and releasing electrons and **photons,** which are packets of light energy that travel in waves and can be absorbed by electrons in certain biomolecules. Life depends upon the ability to transform this solar energy into chemical energy before it is converted to heat—the eventual fate of all energy.  • **What is light? p. 109**

The sun's total yearly energy output is about 3.8 sextillion megawatts of electricity, of which Earth intercepts only about two-billionths. Although this is the equivalent of burning about 200 trillion tons of coal a year, most of this energy doesn't reach life. Nearly a third of the sun's incoming energy is reflected back to space, and another half is absorbed by the planet, converted to heat, and returned to space. Another 19% of incoming solar radiation powers wind and other weather phenomena and drives photosynthesis. Of this 19%, only 0.05 to 1.5% is incorporated into plant material—and only about a tenth of that makes its way into the bodies of animals that eat plants. Far less of the original solar energy reaches animals that consume plant-eating animals. Life on Earth uses just a tiny portion of solar energy.

To acquire usable energy, most organisms depend, directly or indirectly, on organisms such as plants that photosynthesize. However, a few species can use geothermal energy and energy from certain inorganic chemicals. Bacteria thrive around dark cracks in the ocean floor, called deep-sea hydrothermal vents (**figure 6.3**), where hot molten rock seeps through, creating an intensely hot environment. These bacteria extract energy from

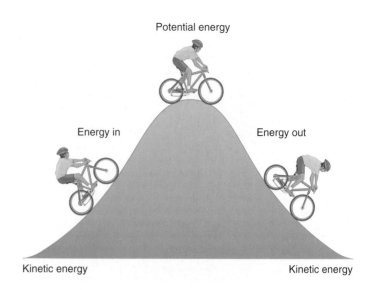

**FIGURE 6.5    Potential Energy and Kinetic Energy.**    Potential energy in food is converted to kinetic energy as muscles push the cyclist to the top of the hill. The potential energy of gravity provides a free ride by conversion to kinetic energy on the other side.

**FIGURE 6.3    Chemical Energy Supports Life.**    Chemoautotrophic bacteria support life in deep-sea hydrothermal vents. The bacteria derive energy from inorganic chemicals, such as hydrogen sulfide, that come from Earth's interior. Bacteria, in turn, support other organisms, such as crabs, clams, and the tube worms shown here. Each worm is about 1 meter long.

A

B

**FIGURE 6.4    Kinetic and Potential Energy**    **(A)** The living world abounds with illustrations of kinetic energy, the energy of action. Here a chameleon's tongue whips out to ensnare a butterfly. **(B)** The ball this pitcher is about to throw has potential energy.

the chemical bonds of hydrogen sulfide ($H_2S$), a gas present at the vents, and from other inorganic chemicals. The bacteria then synthesize organic compounds that are nutrients for them and the organisms that consume them. Organisms like this that use inorganic chemicals as the energy source to manufacture nutrient molecules are called **chemoautotrophs.** This term comes from the more general term **autotroph,** which refers to organisms that synthesize their own nutrients.

## Energy At Rest Is Potential—Energy on the Move Is Kinetic

To stay alive, an organism must continually take in energy, through eating or from the sun or the earth, and convert it into a usable form. Every aspect of life centers on converting energy from one form to another. In life, two basic types of energy are constantly being utilized and exchanged: potential energy and kinetic energy.

**Potential energy** is stored energy available to do work. When matter is placed in certain positions or arrangements, it contains potential energy. A tank of gas, a teaspoon of sugar, a snake about to strike at its prey, and a baseball player about to throw a ball illustrate potential energy. In organisms, potential energy is stored in the chemical bonds of nutrient molecules, such as carbohydrates, lipids, and proteins.

**Kinetic energy** is energy being used to do work. Burning gasoline, converting sugar, the snake striking, and the soaring baseball all demonstrate kinetic energy. A chameleon shooting out its sticky tongue to capture a butterfly uses kinetic energy, as does an elephant trumpeting and a Venus's flytrap closing its leaves around an insect. The adder snake in figure 1.5 demonstrates potential energy as it awaits the approach of a sand lizard. Somewhere between figures 1.5A and 1.5B, the snake uses kinetic energy to grab the lizard. **Figures 6.4** and **6.5** illustrate potential and kinetic energy.

Kinetic energy transfers motion to matter. The movement in the pitcher's arm transfers energy to the ball, which then takes flight. Similarly, flowing water turns a turbine, and a growing root can push aside concrete to break through a sidewalk. Put another way, kinetic energy moves objects. Heat and sound are types of kinetic energy because they result from the movement of molecules.

## Reviewing Concepts

- Energy, the capacity to do work, is vital to life and with few exceptions, ultimately comes from the sun.
- Potential energy is stored in matter and in chemical bonds.
- Kinetic energy is energy of motion, or under use.
- Organisms can extract energy from the sun or from molecules.

# 6.2 The Laws of Thermodynamics: Obeyed by Both Rabbits and Rocks

Like all aspects of nature, energy is governed by laws. The **laws of thermodynamics** regulate the energy conversions vital for life, as well as those that occur in the nonliving world. These laws apply to a system and its surroundings; the system is the collection of matter under consideration, and the surroundings are the rest of the universe. In **figure 6.6**, the elephant is the system, and its watery background is the surroundings. An open system exchanges energy with its surroundings, whereas a closed system is isolated from its surroundings—it does not exchange energy with anything outside the system. Thus, the open system that is the elephant gains energy from the sun and loses heat to its surroundings as it splashes about. The closed system is the universe: it cannot gain more energy.

The laws of thermodynamics apply to all energy transformations—gasoline combustion in a car's engine, a burning chunk of wood, or a cell breaking down glucose.

## Energy Is Neither Created nor Destroyed, but Changes Form

The laws of thermodynamics follow common sense. The first law of thermodynamics is the law of energy conservation. It states that energy cannot be created or destroyed but only converted to other forms. This means that the total amount of energy in a system and its surroundings remains constant; thus, on a grander scale, the amount of energy in the universe is constant. However, the energy in a living system is constantly changing.

In a practical sense, the first law of thermodynamics explains why we can't get something for nothing. The energy released when

**FIGURE 6.6    A System and Its Surroundings.**    The elephant obtains energy from the chemical bonds of the nutrient molecules in its food. When the animal moves in the water, it dissipates some energy as heat, sound, and motion into the surroundings.

a baseball hurtles towards the outfield doesn't appear out of nowhere—it comes from a batter's muscles. Likewise, green plants do not manufacture glucose from nothingness; they trap the energy in sunlight and store it in chemical bonds. The energy in sunlight, in turn, comes from nuclear reactions in the sun's matter.

According to the first law of thermodynamics, the amount of energy an organism uses cannot exceed the amount of energy it takes in through the chemical bonds contained in the nutrient molecules of food. Even when starving, an organism cannot use more energy than its tissues already contain. Similarly, the amount of chemical energy that a plant's leaves produce during photosynthesis cannot exceed the amount of energy in the light it has absorbed. No system can use or release more energy than it takes in.

## All Energy Transformations Increase Entropy

The second law of thermodynamics concerns the concept of **entropy,** which is a tendency toward randomness or disorder (**figure 6.7**). This law states that all energy transformations are inefficient because every reaction results in increased entropy and loses some usable energy to the surroundings as heat. Unlike other forms of energy, heat energy results from random molecule movements. Any other form of energy can be converted completely to heat, but heat cannot be completely converted to any other form of energy. Because all energy eventually becomes heat, and heat is disordered, all energy transformations head towards increasing disorder (entropy). In general, the more disordered a system is, the higher its entropy.

**FIGURE 6.7** **Entropy Represents Disorder.** The destruction caused by a violent storm symbolizes entropy—extreme disorder. It takes great energy to rebuild because specific objects must be placed in specific places.

Because of the second law of thermodynamics, events tend to be irreversible (proceed in one direction) unless energy is added to reverse them. Processes that occur without an energy input are termed **spontaneous.** Although spontaneous, a reaction may not be instantaneous, since speed is not relevant to energy transfers. In natural processes, irreversibility results from the loss of usable energy as heat during energy transformation. It is impossible to regain order in molecules that have dispersed as a result of heat.

The second law of thermodynamics governs cell energetics. Cells derive energy from nutrient molecules and use it to perform such activities as growth, repair, and reproduction. The chemical reactions that transfer this energy are (as the law predicts) inefficient and release much heat. The cells of most organisms are able to extract and use only about half of the energy in nutrients. Although organisms can transform energy—storing it in tissues or using it to repair a wound, for example—ultimately, much of the energy is dissipated as heat because a small amount of energy is lost with each transfer.

Because organisms are highly organized, they may seem to defy the second law of thermodynamics—but only when they are considered alone, as closed systems. Organisms remain organized because they are *not* closed systems. They use incoming energy and matter, from sources such as sunlight and food, in a constant effort to maintain their organization and stay alive. Although the entropy of one system, such as an organism or a cell, may decrease as the system becomes more organized, the organization is temporary; eventually the system (organism) will die. In addition, life connects, or couples, energy reactions, so that one reaction occurs at the expense of another. In a more general sense, **coupled reactions** mean that organisms can increase in complexity as long as something else decreases in complexity by a greater amount. Life remains ordered and

complex because the sun is constantly decreasing in complexity and releasing energy. The entropy of the universe as a whole is always increasing, even though the total energy in the universe remains the same.

We're familiar with energy transformations that release large amounts of energy at once—explosions, lightning, or a plane taking off. Cellular energy transformations, by contrast, release energy in tiny increments. Cells extract energy from glucose in small amounts at a time via the biochemical pathways of cellular respiration, the subject of chapter 8. If all the energy in the glucose chemical bonds was released at once, it would be converted mostly to heat and produce deadly high temperatures. Instead, cells extract energy from glucose by slowly reducing the overall organization of the molecule in several controlled steps. Each reaction transfers energy from molecule to molecule. Some of this energy is lost as heat, but much of it is stored in the chemical bonds of molecules within the cell.

## Reviewing Concepts

- Energy does not arise from nothing; it cannot be created or destroyed but can change form.
- The physical laws that explain energy transformation underlie energy use in life, too. With each transfer, some energy becomes unusable as heat.
- Cells extract energy from chemical bonds in carefully controlled steps to capture as much as possible in a usable form.

## 6.3 Metabolism and Energy Transformations: How Rabbits Run

The life of a cell is a complex and continual web of interacting biochemical reactions that build new molecules and dismantle existing ones. Synthesizing the new requires energy; breaking down the old releases energy.

**Metabolism** consists of the chemical reactions that change or transform energy in cells. The reactions of metabolism occur in step-by-step sequences called **metabolic pathways,** in which the product of one reaction becomes the starting point, or substrate, of another (**figure 6.8**). Pathways may branch or form cycles. Many of these reactions release energy and produce raw materials (intermediates) needed to make biomolecules. Enzymes enable metabolic reactions to proceed fast enough to sustain life, a point we return to later in the chapter. Metabolism in its entirety is an enormously complex network of interrelated biochemical reactions organized into chains and cycles.

# Investigating Life

## Firefly Bioluminescence

Photosynthesis transforms light energy into chemical energy. In **bioluminescence,** the reverse can happen—chemical energy is converted into light energy. Biolumines- cence is common in the oceans, where hordes of glowing microscopic organisms called dinoflagellates lend an eerie bluish cast to fishes, dolphins, or even ships that interrupt their movement (see figure 21.10).

More familiar, perhaps, is the glow of a firefly's abdomen in late summer. More than 1,900 species of fireflies are known, and members of each use a distinctive repertoire of light signals to attract a mate. Typically, flying males emit a pattern of flashes. Wing- less females, called glowworms, usually are on leaves, where they emit light in response to the male. In one species, *Photuris versi- color,* the female emits the mating signal of another species and then eats the tricked male who approaches her. Some frogs con- sume so many fireflies that they glow!

In the 1960s, Johns Hopkins University researchers William McElroy and Marlene DeLuca asked Baltimore schoolchildren to bring them jars of fireflies. They then used the insects to decipher the firefly bioluminescence reaction. McElroy and DeLuca found that light is emitted when a molecule called *luciferin* reacts with ATP, yielding the intermediate compound *luciferyl adenylate* (**figure 6.A**). The enzyme *luciferase* then catalyzes reaction

of this intermediate with molecular oxygen ($O_2$) to yield *oxyluciferin*—and a flash of light. Oxyluciferin is then reduced to luciferin, and the cycle starts over.

Chemical companies sell luciferin and luciferase, which researchers use to detect ATP. When ATP appears in a sample of any substance, it indicates the presence of an organism. For example, the manufacturers of Coca-Cola use firefly luciferin and luciferase to detect bacteria in syrups used to produce the bev- erages. Contaminated syrups glow in the presence of luciferin and luciferase because the ATP in the bacteria sets the biolumines- cence reaction into motion. Fire- fly luciferin and luciferase were also aboard the *Viking* spacecraft sent to Mars. Scientists sent the compounds to detect possible life—a method that would only succeed if Martian life-forms use ATP.

Although we understand the biochemistry of the firefly's glow, the ways animals use their biolu- minescence are still very much a mystery. This is particularly true for the bioluminescent synchrony seen in fireflies in the same trees. When night falls, first one firefly, then another, then more, begin flashing from the tree. Soon the

tree twinkles like a Christmas tree. But then, order slowly descends. In small parts of the tree, the lights begin to blink on and off together. The synchrony spreads. A half-hour later, the entire tree seems to blink on and off every second. Biologists studying animal behavior have joined mathematicians study- ing order to try to figure out just what the fireflies are doing—or saying—when they synchronize their glow.

**FIGURE 6.A   Fireflies Exhibit a Unique Use of Energy.** ATP is used to create flashes of light as energy is trans- ferred to an electron in a specialized molecule called luciferin. The energy is released as light as the electron drops back to its original position

from molecules broken down in other reactions. This reaction is represented as

$$ADP + P_i + energy \longrightarrow ATP + H_2O$$

Investigating Life 6.1 addresses the role of ATP in biolumi- nescence, the reaction that gives fireflies a "glow."

ATP is an effective biological energy currency for several rea- sons. First, converting ATP to ADP + $P_i$ releases about twice the amount of energy required to drive most reactions in cells. The extra energy is dissipated as heat, adding to the organism's body heat. Second, ATP is readily available. The large amounts of energy in the bonds of fats and starches are not as easy to access—they must first be converted to ATP before the cell can

use them. Finally, ATP's terminal phosphate bond, unlike the covalent bonds between carbon and hydrogen in organic mole- cules, is unstable—so it may be broken to release energy.

Just as you can use currency to purchase a great variety of different products, all cells use ATP in many chemical reactions to do different kinds of work. If you ran out of ATP, you would die instantly. Organisms require huge amounts of ATP. A typical adult uses the equivalent of 2 billion ATP molecules a minute just to stay alive. However, ATP is recycled so rapidly that only a few grams are available at any given instant. Organisms recycle ATP at a furious pace, adding phosphate groups to ADP to reconsti- tute ATP, using the ATP to drive reactions, and turning over the entire supply every minute or so.

## Cells Couple ATP Formation and Breakdown to Other Reactions

Cells couple the breakdown of nutrients to ATP production, and they couple the breakdown of ATP to other reactions that occur at the same time and place in the cell. Coupled reactions, as their name implies, are reactions that occur in pairs. One reaction drives the other, which does work or synthesizes new molecules. Consider once again the formation of sucrose and water from glucose and fructose. This reaction is not spontaneous; it requires a net input of energy. When ATP provides additional energy, the reactants now have more energy available than the products, and the reaction proceeds. ATP breakdown, an exergonic and spontaneous reaction, is therefore coupled to sucrose synthesis.

A cell uses ATP as an energy source in two major ways: to energize a molecule or to change the shape of a molecule. Both activities transfer the terminal phosphate to another molecule, a process called **phosphorylation.** The phosphate is released after the movement occurs or the new bond is formed. The net result is the change of ATP to ADP, but this always occurs in stages. About 7 kilocalories of energy are released in splitting $10^{23}$ molecules of ATP to ADP and phosphate. The presence of a phosphate on a target molecule can make that molecule more likely to bond with other molecules. In this way, ATP is often used to fuel anabolic reactions, such as assembling amino acids to make a protein. The other main mechanism of action using ATP changes the shape of a molecule. Adding phosphate to a protein can force that protein into a different shape, and removing phosphate allows that molecule to return to its original shape. The cell uses the change to a new shape to move molecules and structures throughout the cell. Muscle contraction is the large-scale effect of millions of small molecules changing shape in a coordinated way. ATP provides the energy.

## Cofactors and Coenzymes Move Electrons Through Chains

Several other compounds besides ATP participate in the cell's energy transformations (**table 6.1**). Nonprotein helpers called **cofactors** assist other chemicals in enabling certain reactions to proceed. Cofactors are often trace minerals in the form of ions. $Mg^{2+}$, for example, must be present to stabilize many important enzymes.

Organic cofactors, which are called **coenzymes,** usually carry protons or electrons. Many coenzymes are nucleotides, as is ATP. But a coenzyme's energy content, unlike ATP's, depends on its ability to donate electrons or protons, not on the presence or absence of a particular phosphate bond. Vitamins are a major source of coenzymes in cells and help to drive metabolic reactions. Vitamins, therefore, do not directly supply energy but make possible the reactions that extract energy from nutrient molecules.

$NAD^+$, $NADP^+$, and FAD are other important molecules that function as coenzymes. We introduce them here and will return to them in chapters 7 and 8.

Nicotinamide adenine dinucleotide ($NAD^+$), like ATP, consists of adenine, ribose, and phosphate groups (**figure 6.13**). However, $NAD^+$ also has a nitrogen-containing ring, called nicotinamide, which is derived from niacin (vitamin $B_3$). The nicotinamide is the active part of the molecule. $NAD^+$ is reduced when it accepts two electrons and two protons (hydrogens) from a substrate. Both electrons and one proton actually join the $NAD^+$, leaving a proton ($H^+$). This reaction is written as

$$NAD^+ + 2H^+ + 2e^- \longrightarrow NADH + H^+$$

$NADH + H^+$ is reduced with an electron at a higher energy shell and is therefore packed with potential energy. The cell uses that energy to synthesize ATP and to reduce other compounds.

### TABLE 6.1

#### Some Molecules Involved in Cellular Energy Transformations

| Molecule | Mechanism |
| --- | --- |
| ATP | High potential energy in the phosphate bonds is released when phosphate is transferred (phosphorylation) or cleaved; the negative charges drive changes in shape |
| Cofactors | Substances, usually ions, that stabilize proteins |
| Coenzymes | Vitamin-derived molecules that transfer electrons and protons; aid in enzyme shape and function |
| NADH | Coenzyme that carries high-energy electrons; powers ATP synthesis via chemiosmosis; source of electrons in enzymatic redox reactions |
| NADPH | Coenzyme that carries high-energy electrons; source of electrons in enzymatic redox reactions; used to reduce $CO_2$ in photosynthesis |
| $FADH_2$ | Coenzyme that carries high-energy electrons; powers ATP synthesis via chemiosmosis |
| Cytochromes | Iron-containing electron carriers |

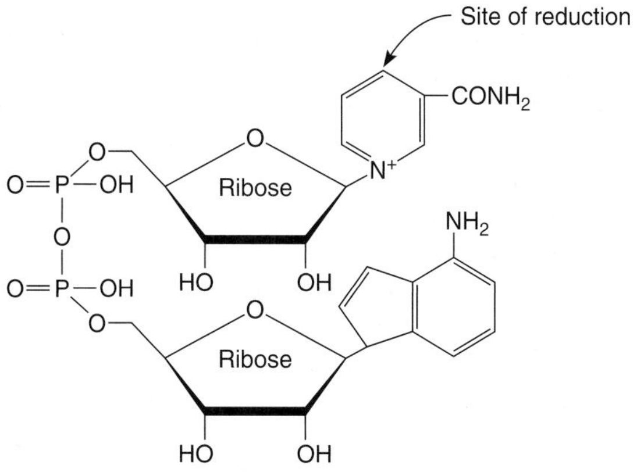

FIGURE 6.13 **NADH Carries Electrons.** NAD⁺ is a form of nucleotide that is used in many cellular reactions to store and transfer energy-rich electrons. As the electron is added to NAD⁺, it becomes reduced as NADH.

The structure of nicotinamide adenine dinucleotide phosphate ($NADP^+$) is similar to that of $NAD^+$ but with an added phosphate group. NADPH supplies the hydrogen and energy that reduce carbon dioxide to carbohydrate during photosynthesis. This process "fixes" atmospheric carbon into organic molecules that then serve as nutrients.

Flavin adenine dinucleotide (FAD) is derived from riboflavin (vitamin $B_2$). FAD, like $NAD^+$, carries two energy-rich electrons. However, it also accepts two protons to become $FADH_2$.

The cytochromes are iron-containing molecules that transfer electrons in metabolic pathways. When oxidized, the iron in cytochromes is in the $Fe^{3+}$ form. When the iron accepts an electron, it is reduced to $Fe^{2+}$. The several types of cytochromes all carry electrons in cells. In metabolic pathways, cytochromes align to form electron transport chains, with each molecule accepting an electron from the molecule before it and passing an electron to the next. **Figure 6.14** shows such a chain; chapters 7 and 8 show more. Small amounts of energy are released at each step of an electron transport chain, and the cell uses this energy in other reactions. Cytochromes take part in many energy transformations in life, suggesting that this cellular strategy for energy transformation is quite ancient.

### Reviewing Concepts

- ATP, through high-energy phosphate bonds, temporarily stores energy that a cell uses for a wide variety of activities.
- Breaking the bonds through transient phosphate transfers releases enough energy for most of a cell's reactions, with some left over as heat. ATP is constantly recycled.
- Other molecules assist the cell as carriers of energetic electrons.

## 6.5 Enzymes: Life's Catalysts

Many chemical reactions require an initial boost of energy, the energy of activation, to bring reactants together. Sometimes this energy comes from heat in the environment. For example, heat from a lighted match provides the energy of activation to ignite a piece of wood. The amount of heat required to activate most metabolic reactions in cells would swiftly be lethal were it not for enzymes (**figure 6.15**).

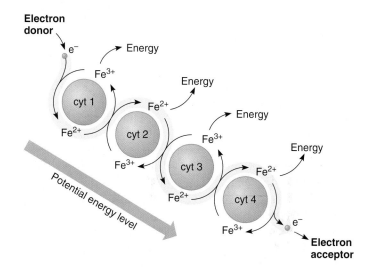

FIGURE 6.14 **Electron Transport Chains.** Different cytochrome (cyt) proteins align within membranes, forming electron transport chains. Electrons pass down the chain in a series of oxidation-reduction reactions, releasing energy in small, usable increments. Electrons are passed between iron molecules attached to the cytochromes. Iron in the ferric form ($Fe^{3+}$) gains an electron, becoming reduced to the ferrous ($Fe^{2+}$) form. Enzymes use the energy transfers to pump hydrogens across the membrane and transfer energy ultimately to ATP or other molecules.

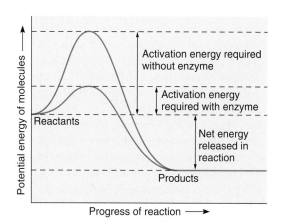

FIGURE 6.15 **Enzymes Lower the Activation Energy.** Enzymes speed chemical reactions by increasing the chance that reactants will come together. This decreases the energy required to start the reaction: the activation energy.

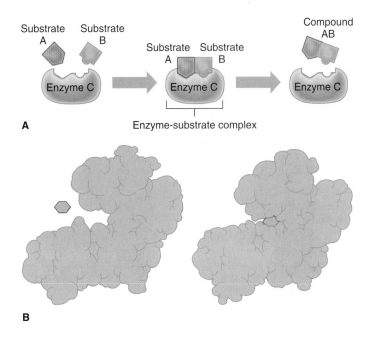

**A**

Enzyme-substrate complex

**B**

**FIGURE 6.16**  **Enzymes Are Specific.**    (A) The shape of the enzyme creates an active site that binds to one or more specific substrates. (B) When binding to its substrate, an enzyme changes shape very slightly, as is illustrated here with the binding of glucose to the enzyme hexokinase.

## Enzymes Speed Biochemical Reactions

Without enzymes, many biochemical reactions would not occur fast enough to support life. Recall from chapter 3 that an **enzyme** is a protein that catalyzes (speeds) specific chemical reactions without being consumed by them. An enzyme catalyzes a chemical reaction by decreasing the energy of activation. The enzyme binds the substrate in a way that helps the reaction to occur. The conformation of the enzyme is such that the substrate fits into its active site, but not as precisely as a key fits a lock. Rather, the active site contorts slightly, as if it is hugging the substrate (**figure 6.16**). This "induced fit" provides some additional controls to the enzyme.

● **enzymes, p. 38**

Because enzymes are proteins, they are very sensitive to conditions that would disrupt protein structure, known as **denaturation.** Increases in temperature, changes in pH, high or low salt concentrations, and toxins can change the shape of an enzyme. Most enzymes have a very narrow range of temperatures within which they function correctly. If the temperature is too high or too low, these enzymes stop working. Very high temperatures destroy most enzymes. This is one reason organisms work to maintain a uniform cellular temperature. Organisms adapted to high-temperature environments have special heat-tolerant enzymes, but they still function within a narrow range. Homeostasis is, in many respects, necessary for proper functioning enzymes.

## Cells Control Metabolic Pathways

Several biological mechanisms regulate cellular metabolism and preserve a delicate balance of thousands of reactions. Many chemical reactions are single-step, using a single enzyme to make a single product. More complex processes in cells make use of a sequence of enzyme-catalyzed reactions to form a final product needed by the cell. But cells need to control these processes to be efficient and have developed several ways of doing so.

Certain enzymes control metabolism by functioning as pacesetters at important junctures in biochemical pathways. The enzyme with the slowest reaction sets the pace for the pathway's productivity, just as the slowest runner on a relay team limits the overall pace for the whole team. This is because each subsequent reaction in the metabolic pathway (like each subsequent relay runner) requires the product of the preceding reaction to continue. The reaction this enzyme catalyzes is called the rate-limiting step because it regulates the pathway's pace and productivity.

**Negative Feedback**  Enzymes are highly sensitive to specific chemical cues. When enough of a product has been made, cells may need to "turn off" an enzyme for a time. One simple way to accomplish this is to have the final product bind to an enzyme and inhibit it from working. This type of regulation is called **negative feedback,** or feedback inhibition. Negative feedback prevents too much of one substance from accumulating—an excess of a particular biochemical effectively shuts down its own synthesis until its levels fall. At that point, the pathway resumes its activity. Negative feedback is somewhat like a thermostat—when the temperature in a building reaches a certain level, the thermostat shuts the heat off for a while. Falling temperature cues the thermostat to ignite the furnace again (see figure 30.6).

A product molecule can inhibit its own synthesis in two general ways. It may bind to the enzyme's active site, preventing it from binding substrate and temporarily shutting down the pathway. This is called competitive inhibition because a substance other than the substrate competes to occupy the active site. Alternatively, product molecules may bind to the enzyme at a site other than the active site, but in a way that alters the shape of the enzyme so that it can no longer bind substrate. This indirect approach is sometimes called noncompetitive inhibition because the inhibitor does not directly compete to occupy the active site. Both competitive and noncompetitive inhibition are forms of negative feedback. Other molecules may also function as either type of inhibitor. To control other enzymes, cells often use special **regulatory proteins** that control the activity of specific enzymes by binding them.

**Figure 6.17** illustrates negative feedback. It shows the sequential pathway that certain bacteria use to synthesize the amino acid histidine. When excess histidine accumulates, it binds (noncompetitively) to the junction between two subunits of the enzyme that regulates the pathway. This temporarily destabilizes the enzyme and impairs catalysis. For a time, histidine synthesis ceases. When levels of histidine in the bacterium fall, the block on the enzyme is lifted, and the cell can once again synthesize the amino acid.

Substances from outside the body, such as drugs and poisons, can also inhibit enzyme function. A foreign chemical similar in conformation to the substrate may block the enzyme's active site and prevent the substrate from binding. The drug sulfanilamide, for example, competitively inhibits certain enzymes in bacteria and is therefore useful to fight certain infections.

Organisms called **anaerobes** obtain energy in the absence of oxygen by using different pathways that are referred to as fermentation. To extract as much energy as possible, molecules such as nitric oxide ($NO_3$) are used instead of $O_2$ as the terminal electron acceptor. Since oxygen is not used, this is called **anaerobic respiration.**

Organisms use cellular respiration to extract energy from nutrients. In addition to glucose, the cell can use a variety of carbon compounds, such as amino acids or fats, as a source of energy. When nutrients are broken down, carbon-to-carbon bonds are cleaved, and each carbon combines with oxygen, forming $CO_2$, which is a metabolic waste. Cellular respiration, then, explains why our respiratory systems obtain oxygen and get rid of $CO_2$.

In animals, cellular respiration begins where digestion leaves off. Consider a chipmunk eating a nut (**figure 8.2**). The nut passes through the rodent's digestive system and is broken into clumps of cells. Digestive enzymes break the cells apart, releasing proteins, carbohydrates, and lipids. More enzymes in the chipmunk's stomach digest these macromolecules into their component amino acids, monosaccharides, and fatty acids. These monomers are small enough to enter the blood and be transported to the body's tissues. When these smaller nutrient molecules enter the animal's cells, they are used as precursors to manufacture other cellular components, such as protein or fat, or are broken down further to provide energy. The energy is transferred to the high-energy phosphate bonds of ATP. The chipmunk stays alive.

This chapter describes how cells extract energy from nutrients. The journey from food to energy entails several biochemical pathways with many chemical names, and it may appear overwhelming. But if we consider energy release in major stages, and also one step at a time, the logic emerges.

**FIGURE 8.2**　　**Eating Provides the Raw Materials for Cellular Respiration.**　　The chipmunk eats the peanuts, which provide fats and amino acids and some glucose. As the digestive system breaks the larger molecules down, cellular respiration gains the molecules it uses to make ATP.

## Reviewing Concepts

- Cellular respiration is a common biochemical pathway that extracts energy from the bonds of nutrient molecules.
- Aerobic organisms use their respiratory systems to provide oxygen as a final electron acceptor to extract energy. Anaerobic organisms use different molecules and pathways in the absence of oxygen.

# 8.2　An Overview of Glucose Utilization

As we learned in chapter 6, ATP is an ideal energy carrier for a cell. ATP functions like a rechargeable battery, alternating between a charged form (ATP) and an uncharged form (ADP) ready to receive energy. Charging ATP is done by phosphorylation—the addition of a high-energy phosphate group. Converting ATP to

ADP releases just enough energy—7.5 kcal per mole—to fuel most of the reactions of a cell. Other molecules such as NADH and $FADH_2$ act as electron carriers that transfer larger amounts of energy within the cell.

All organisms obtain energy from glucose using **glycolysis** (literally "breaking glucose") (**figure 8.3**). The glucose can come from complex nutrients such as starch or cellulose, other nutrients, or simply from ingested molecules of glucose. Even plants must first make glucose via photosynthesis, transport the sugar to another part of the cell, and then extract the energy through glycolysis. But glycolysis does not completely oxidize glucose to carbon dioxide and therefore doesn't yield as much energy as other processes that use oxygen.

**Aerobic respiration** refers to the entire breakdown of glucose to carbon dioxide in the presence of oxygen. The general equation for aerobic respiration is

$$\text{glucose} + \text{oxygen} \longrightarrow \text{carbon dioxide} + \text{water} + \text{energy}$$

$$C_6H_{12}O_6 + 6O_2 \longrightarrow 6CO_2 + 6H_2O + 30ATP$$

Interpreted, the equation says that energy in glucose is transferred to ATP in the presence of oxygen while being broken down to carbon dioxide, with water as a by-product.

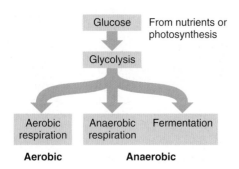

**FIGURE 8.3    All Cells Undergo Glycolysis.**    The first step in energy release is glycolysis. Subsequent pathways continue the release of energy from nutrient molecules. Three common pathways are shown.

The reactions that finish what glycolysis starts are called the **Krebs cycle** and the **electron transport chain.** The enzymes for these reactions are found in the membranes and cytoplasm of bacteria and in the membranes and matrix of mitochondria.

In animals, the reactions of cellular respiration begin with glucose. Other carbohydrates in food, such as sucrose (table sugar) or fructose (fruit sugar), must be digested or converted to glucose. Plants begin cellular respiration with compounds that may come from other sources, too. Later in the chapter, we will see how the other major nutrients—proteins and lipids—enter the energy pathways. But for now, we will focus on glucose.

## Respiration Reactions Take Place Both Outside and Inside Mitochondria

Burning a log releases the energy in the wood essentially all at once as heat. If cells were to release all of the energy in glucose (686 kcal per mole) in one uncontrolled step, enough heat would be released that the cell would die. To gain the energy in a usable form, cells carefully extract the energy in glucose in small amounts and transfer that energy to ATP. Due to the second law of thermodynamics, each step releases some energy as heat. But because the energy is never released all at once, the heat generated in this way can be used to maintain a cell's internal temperatures. The heat dissipates quickly into the surroundings, and no one part of a cell becomes hot enough to burst into flames.

Glycolysis, the first part of cellular respiration, is a series of enzyme-catalyzed steps that occur in the cytoplasm. To control the release of energy, enzymes first rearrange glucose in a way that concentrates the energy in the electrons involved in some of the bonds. When removed, the energized electrons of these bonds carry energy with them to molecules of NADH and ATP. In the process, glucose is split from one six-carbon molecule to two three-carbon molecules of **pyruvic acid.** (The ionized form is called pyruvate.) Next, pyruvic acid is transported into the mitochondria, where additional rearrangements and energy transfers take place. Glycolysis yields only a net of two ATP molecules.

The remainder of the energy must be extracted by the reactions of the mitochondria. Aerobic bacteria, which lack mito-

chondria, employ their cell membrane as the location of the respiratory enzymes. Recall from chapter 4 that a mitochondrion consists of an outer membrane and a highly folded inner membrane, an organization that creates two compartments (**figure 8.4**). The innermost compartment is called the **matrix,** and the area between the two membranes is called the **intermembrane compartment.** The structure of these membranes allows the cell to produce and manipulate a proton gradient.

The second and third parts of cellular respiration occur in the matrix of the mitochondria. The first step releases a molecule of $CO_2$ and transfers the other two carbons from pyruvic acid into the third part of cellular respiration: the Krebs cycle. The two carbons from the original glucose molecule are manipulated through the cycle while enzymes transfer the remaining energy to more molecules of NADH, $FADH_2$, and ATP. The carbons are finally released as carbon dioxide.

The final stage of cellular respiration, electron transport, transfers energy from NADH and $FADH_2$ to ATP. The inner mitochondrial membrane is folded into numerous projections called **cristae,** which are studded with the enzyme ATP synthase and electron carrier molecules. This extensive folding greatly increases the surface area on which the reactions of the electron transport chain can occur. The energy-rich electrons and hydrogen ions from NADH and $FADH_2$ are transferred through a series of membrane proteins, and their energy is used to create a gradient of hydrogen ions across the mitochondrial membrane. This gradient holds tremendous potential energy and is used to phosphorylate ATP. The electrons, now drained of energy, are transferred to oxygen.

The overall result of the energy-releasing pathways uses energy stored in organic molecules to phosphorylate ADP to ATP. Cellular respiration is quite efficient, capturing approximately 35% of the energy in glucose as 30 to 32 ATP, depending upon the tissue and organism. Just as we use ordinary currency, the cell can use its energy currency—ATP—in several ways, from synthesizing protein to contracting muscle fibers. **Figure 8.5** shows where all of these reactions occur relative to one another in a eukaryotic cell. We will explore these steps in greater detail later in the chapter.

## Two Types of Phosphorylation Make ATP

In acquiring useful energy, cells employ two methods for ATP synthesis. In keeping with the laws of thermodynamics, both mechanisms are possible because the resulting molecules contain less energy than the reactants. In the first mechanism, called **substrate-level phosphorylation,** an enzyme oxidizes an energy-rich substrate molecule, liberating enough energy to directly attach inorganic phosphate to a second substrate. A second reaction transfers this phosphate to ADP. In the example of this reaction that occurs in glycolysis, enough energy is liberated to reduce $NAD^+$ to NADH as well.

The second mechanism of making ATP, called **chemiosmotic phosphorylation** (or oxidative phosphorylation), uses the energy in a proton gradient to add phosphate to ADP. (A nearly identical mechanism operates in photosynthesis.) The proton gradient is

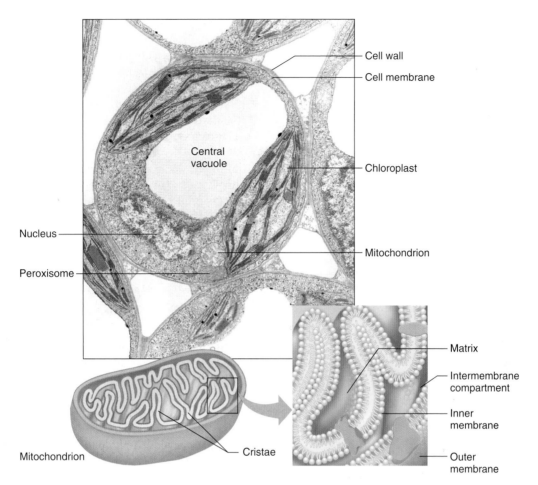

**FIGURE 8.4** **Cellular Respiration Occurs in Mitochondria.**      Mitochondria in eukaryotic cells (such as the leaf cell pictured) provide most of the ATP for cellular functions. Enzymes in the matrix and membrane oxidize pyruvic acid to carbon dioxide and transfer the released energy to ATP. The inner membranes of mitochondria are used to form the proton gradient of chemiosmotic phosphorylation, just as the thylakoid membrane does in photosynthesis (see figure 7.11).

formed as high-energy electrons are passed from carriers such as NADH through a series of membrane-bound proton pumps in the inner mitochondrial membrane. As the electrons drop from higher energy levels, they release energy that is used to pump hydrogen ions out of the mitochondrial matrix. The resulting gradient of protons contains a tremendous amount of potential energy due to the pressure of protons attempting to move across the membrane. The only available release of that pressure is an enzyme-linked channel that couples the movement of protons to the phosphorylation of ATP. The process of energy transfer used here is similar to creating electricity using water pressure. Water from rain flows downhill and is trapped by a dam. Tremendous pressure builds on the face of the dam. That pressure is released by forcing the water to turn blades that spin an electric generator. Electricity is made from rain! A mitochondrion uses a similar mechanism to transfer energy from NADH to ATP.

Substrate-level phosphorylation is the simpler and more direct mechanism for producing ATP, but it accounts for only a small percentage of the ATP produced—specifically, that from glycolysis and the Krebs cycle. In the electron transport chain

that follows these reactions, much more ATP is formed through chemiosmotic phosphorylation.

We will now explore each of the processes of respiration in more detail.

## Reviewing Concepts

- The general equation for aerobic respiration is
$C_6H_{12}O_6 + 6O_2 \longrightarrow 6CO_2 + 6H_2O + 30ATP$

- Glycolysis is found in all organisms and extracts some energy from splitting glucose into two molecules of pyruvate. The rest of cellular respiration in eukaryotic cells occurs in the mitochondria.

- Energy in cells comes from phosphorylation. There are two types—substrate level and chemiosmotic. Enzymes transfer phosphate directly from molecule to molecule in the first type. The second type uses a proton gradient as a source of potential energy to drive the phosphorylation of ATP.

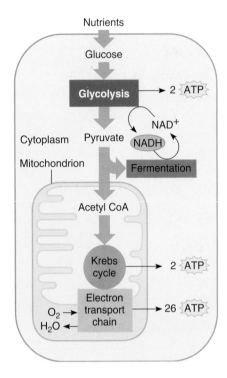

**FIGURE 8.5** **An Overview of Cellular Respiration.** Glucose is broken down to carbon dioxide through a series of enzyme-catalyzed reactions. The energy is captured as ATP. As the chapter progresses, each of the biochemical pathways will become more detailed. Look to the insets that repeat this diagram with different sections highlighted to follow the part of the overall pathway under discussion.

## 8.3    Glycolysis: From Glucose to Pyruvic Acid

Glucose contains considerable bond energy, but cells recover only a small portion of it during glycolysis. The entire process requires 10 steps, all of which occur in the cytoplasm (**figure 8.6**), to extract energy and split glucose into two three-carbon compounds. Each step is catalyzed by a unique enzyme, providing control and direction. The first half of the pathway activates glucose so that energy can be redistributed in the molecule. The second half of the pathway then extracts some of this energy.

### The First Half of Glycolysis Activates Glucose and Produces PGAL

The first step of glycolysis uses one molecule of ATP to phosphorylate one molecule of glucose, representing a net loss of energy to the cell. But phosphorylation activates glucose, enabling enzymes to rearrange the energy it contains, and traps it in the cell. Because phosphate is negatively charged and since charged compounds cannot easily cross the cell membrane, phosphorylation prevents glucose from leaving.

Step 2 rearranges the atoms of phosphorylated glucose. In step 3, the new molecule (fructose-6-phosphate) is phosphorylated again using another ATP, forming fructose-1,6-bisphosphate. In steps 4 and 5, an enzyme splits this compound into two three-carbon compounds, each containing one phosphate. One of the products, phosphoglyceraldehyde (PGAL), will continue through the steps of glycolysis. The other product, dihydroxyacetone phosphate, is converted first to PGAL and then is broken down along with the other PGAL. Formation of the two molecules of PGAL derived from each glucose molecule marks the halfway point of glycolysis (steps 1 through 5 in figure 8.6). So far, energy in the form of ATP has been invested, but no ATP has been produced. To maintain life, the cell must continue through the rest of the pathway. Poisons such as arsenic cause death by blocking enzymes at the next step in the pathway.

### The Second Half of Glycolysis Extracts Some Energy and Produces Pyruvic Acid

The first energy-obtaining step of glycolysis (step 6) occurs in the second half, when enzymes reduce $NAD^+$ to NADH through the coupled oxidation of PGAL to phosphoglyceric acid. Enough energy is also released to add a second phosphate group, forming 1,3-bisphosphoglyceric acid. Substrate-level phosphorylation is completed when one of the phosphates of 1,3-bisphosphoglyceric acid is then transferred to ATP (step 7). The three-carbon molecule that remains, 3-phosphoglycerate, is rearranged to form 2-phosphoglycerate (step 8). An enzyme removes water, producing phosphoenolpyruvate (PEP) (step 9). PEP then becomes pyruvic acid when phosphate is transferred to a second ADP (step 10). Each PGAL from the first half of glycolysis yields two ATP and one pyruvic acid.

Because one molecule of glucose yields two molecules of PGAL, glycolysis produces four ATP and two pyruvic acid molecules per glucose. However, because the first half of glycolysis requires two ATP, the net gain is two ATP per molecule of glucose.

Glycolysis transfers a small amount of the total chemical energy in glucose to molecules of ATP and NADH. However, most of the energy of glucose remains in the bonds of pyruvic acid. This energy is tapped in the mitochondrion to synthesize more ATP. Note that, to this point, we have not used oxygen in any part of cellular respiration.

### Reviewing Concepts

- Enzymes allow a cell to carefully rearrange molecules as they extract energy in a controlled manner.
- The first half of glycolysis splits glucose into two molecules of the three-carbon compound PGAL. Energy is invested, but none yet returned.
- In the second half of glycolysis, PGAL is oxidized as $NAD^+$ is reduced to NADH, and two ATP are formed. PGAL is rearranged to form pyruvic acid.

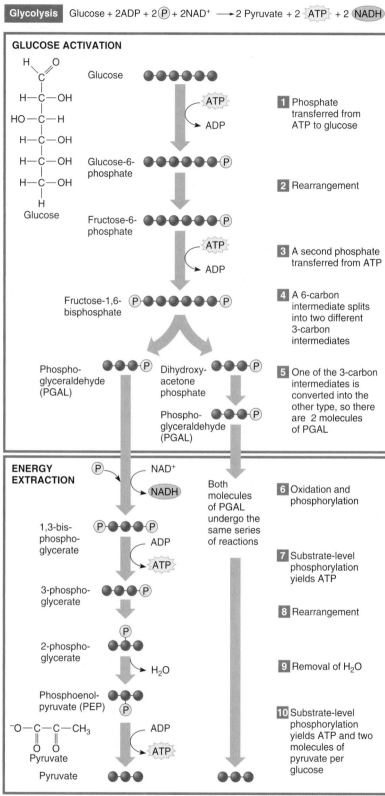

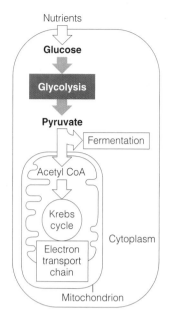

**Glycolysis** Glucose + 2ADP + 2(P) + 2NAD$^+$ ⟶ 2 Pyruvate + 2 ATP + 2 NADH

**GLUCOSE ACTIVATION**

Glucose

1 Phosphate transferred from ATP to glucose

Glucose-6-phosphate

2 Rearrangement

Fructose-6-phosphate

3 A second phosphate transferred from ATP

Fructose-1,6-bisphosphate

4 A 6-carbon intermediate splits into two different 3-carbon intermediates

Phospho-glyceraldehyde (PGAL)

Dihydroxy-acetone phosphate

Phospho-glyceraldehyde (PGAL)

5 One of the 3-carbon intermediates is converted into the other type, so there are 2 molecules of PGAL

**ENERGY EXTRACTION**

Both molecules of PGAL undergo the same series of reactions

6 Oxidation and phosphorylation

1,3-bis-phospho-glycerate

7 Substrate-level phosphorylation yields ATP

3-phospho-glycerate

8 Rearrangement

2-phospho-glycerate

9 Removal of $H_2O$

Phosphoenol-pyruvate (PEP)

10 Substrate-level phosphorylation yields ATP and two molecules of pyruvate per glucose

Pyruvate

Pyruvate

**FIGURE 8.6   Glycolysis.**   Glucose is rearranged and split into two three-carbon intermediates, each of which is rearranged further to eventually yield two molecules of pyruvic acid (pyruvate). Along the way, four ATP and two NADH are produced. Two ATP are consumed in activating glucose, so the net yield is two ATP molecules per molecule of glucose. (Each black sphere represents a carbon atom.)

## 8.4   Aerobic Respiration: The Krebs Cycle

### Pyruvic Acid Is Converted to Acetyl CoA

The pyruvic acid that is transported into the mitochondrial matrix goes through a processing step before entering the Krebs cycle (**figure 8.7**). This single step oxidizes pyruvic acid, removes a single carbon as carbon dioxide, and harvests enough energy to reduce NAD$^+$ to NADH. The remaining two-carbon molecule, called an acetyl group, is transferred to a carrier called **coenzyme A** (CoA). Together, they form **acetyl CoA.**

The conversion of pyruvic acid to acetyl CoA links glycolysis and the Krebs cycle. Pyruvic acid is the final product of glycolysis, and acetyl CoA is the compound that enters the Krebs cycle.

### The Krebs Cycle Produces ATP and NADH

Because the last step continually regenerates the reactants of the first step, the Krebs reactions are referred to as the Krebs cycle (**figure 8.8**). Since the product of one reaction becomes the reactant of the next, the enzymes for the cycle are positioned in the matrix of the mitochondrion to each other to improve the overall efficiency of the process. In addition to continuing the breakdown of glucose, the Krebs cycle forms molecules, known as intermediates, used to manufacture other nutrients such as amino acids or fats. Due to its involvement with energy and many key metabolic processes, the Krebs cycle forms the heart of metabolism for all cells.

**FIGURE 8.7    Transition to the Mitochondria.**    Acetyl CoA formation bridges glycolysis and the Krebs cycle. After pyruvic acid enters the mitochondrion, crossing both membranes, $CO_2$ is removed as $NAD^+$ is reduced to NADH. The remaining two carbons combine with coenzyme A to yield acetyl CoA. For every glucose molecule that entered glycolysis, two acetyl CoA molecules now enter the Krebs cycle.

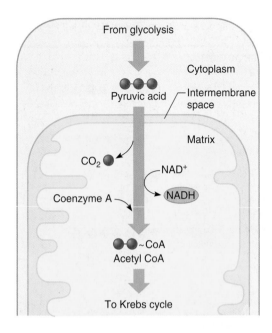

**FIGURE 8.8    Products of the Krebs Cycle.**    One glucose molecule yields two molecules of acetyl CoA. Therefore, one glucose molecule is associated with two turns of the Krebs cycle. Each turn of the Krebs cycle generates one molecule of ATP, three molecules of NADH, one molecule of $FADH_2$, and two molecules of $CO_2$, as the carbons of acetyl CoA are rearranged and oxidized.

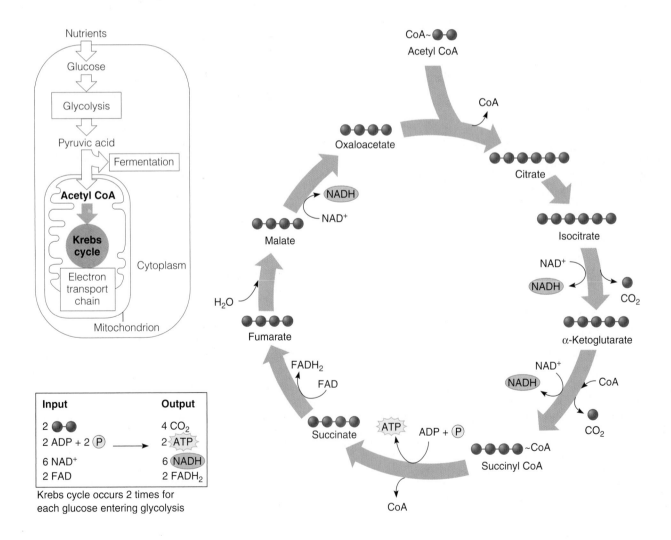

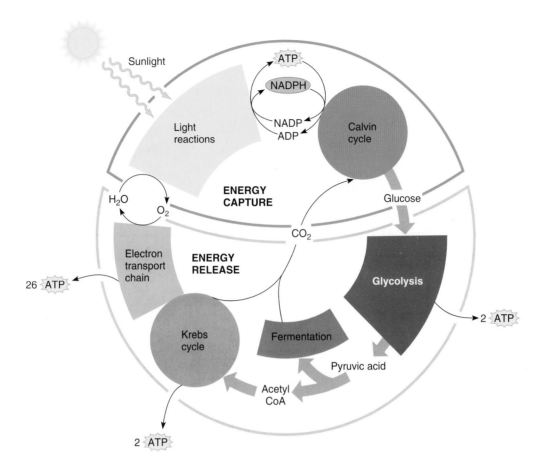

**FIGURE 8.16     The Energy Pathways and Cycles Connect Life.**     An overview of energy metabolism illustrates how biological energy reactions are interrelated.

constantly, from sunlight. Photosynthesis, over time, released oxygen into the primitive atmosphere, paving the way for an explosion of new species capable of using this new atmospheric component. In addition, electrical storms split water in the atmosphere, releasing single oxygen atoms that joined with diatomic oxygen to produce ozone ($O_3$). As ozone accumulated high in the atmosphere, harmful ultraviolet radiation was blocked, reducing genetic damage and allowing new varieties of life to arise.

Photosynthesis could not have debuted in a plant cell because such complex organisms were not present on the early Earth. The first photosynthetic organisms may have been anaerobic bacteria or archaea or an ancestor of both that used hydrogen sulfide ($H_2S$) in photosynthesis instead of the water that plants use. These first photosynthetic microorganisms would have released sulfur, rather than oxygen, into the environment. Eventually, evolutionary changes in pigment molecules enabled some of these organisms to use water instead of hydrogen sulfide. If a large cell engulfed an ancient photosynthesizing microorganism, it may have become a eukaryotic-like cell that may have been the ancestor of modern plants. (Recall the endosymbiont theory discussed in chapter 4.) Mitochondria might have evolved

in a similar way, when larger cells engulfed bacteria capable of using oxygen.   ●   **origins of complex cells, p. 60**

After endosymbiosis, different types of complex cells probably diverged, leading to evolution of a great variety of eukaryotic organisms. Today, the interrelationships of the biological reactions of photosynthesis, glycolysis, and aerobic respiration and the great similarities of these reactions in diverse species demonstrate a unifying theme of biology: all types of organisms are intimately related at the biochemical level.

## Reviewing Concepts

- The energy pathways are the core of all metabolic processes and evolved slowly, beginning with glycolysis when there was little oxygen available.
- The Krebs cycle added more levels of efficiency, but the best efficiency had to wait for photosynthesis to provide oxygen for use by electron transport.
- Endosymbiosis formed a variety of organisms with differing capabilities.

## Connections

The processes of cellular respiration are at the heart of everything we do. We can understand much of how the body functions by understanding its biochemical processes. We often think of energy as a somewhat abstract idea. When we recognize we need energy to repair body damage, we rarely connect where that energy is actually used. If you consider that proteins need to be used in wound repair and that amino acids are needed for the protein, you can see how pulling a molecule out of the Krebs cycle to make that protein would result in less energy being extracted from that molecule. All of us have felt low on energy, but what causes those feelings? The body cannot send signals because it lacks the signal molecule. Or the movement of muscles is slower, caused by a delay in getting the ATP needed. Energy, in cellular terms, often means building molecules or moving them around. Chapters 9 and 10 explore some of these cellular processes.

# Student Study Guide

## Key Terms

| | | | |
|---|---|---|---|
| acetyl CoA   *131* | alcoholic fermentation   *139* | coenzyme A   *131* | Krebs cycle   *128* |
| aerobe   *126* | anaerobes   *127* | cristae   *128* | lactic acid fermentation   *139* |
| aerobic cellular | anaerobic respiration   *127* | electron transport chain   *133* | matrix   *128* |
| respiration   *126* | chemiosmotic | glycolysis   *127* | pyruvic acid   *128* |
| aerobic respiration   *127* | phosphorylation   *128* | intermembrane | substrate-level |
| | | compartment   *128* | phosphorylation   *128* |

## Chapter Summary

### 8.1  Cellular Respiration: An Introduction

1. All living cells use cellular respiration to extract energy from the bonds of nutrient molecules.
2. Aerobic organisms use oxygen as a final electron acceptor during cellular respiration. Anaerobic organisms use different molecules and pathways in the absence of oxygen.

### 8.2  An Overview of Glucose Utilization

3. The general equation for aerobic respiration is $C_6H_{12}O_6 + 6O_2 \longrightarrow 6CO_2 + 6H_2O + 30ATP$
4. Glycolysis is the most common energy-extracting process. It extracts some energy from splitting glucose into two molecules of pyruvate.
5. In eukaryotic cells, cellular respiration, except for glycolysis, occurs in the mitochondria.
6. Cells use phosphorylation to transfer energy from one molecule to another.

### 8.3  Glycolysis: From Glucose to Pyruvic Acid

7. Cells use enzyme systems to extract energy in a controlled manner by carefully rearranging energy-rich molecules.
8. Glycolysis splits glucose into two molecules of pyruvate while transferring energy to NADH.

### 8.4  Aerobic Respiration: The Krebs Cycle

9. Enzyme cycles use the product of one reaction as the substrate of the next, then re-form the starting molecule. The Krebs cycle transfers energy to ATP and NADH as the remaining carbons from glucose are rearranged.
10. The enzymes of the first energy-capturing step in the mitochondria releases carbon dioxide and reduces $NAD^+$ to NADH. A carrier coenzyme transfers the remaining carbons to the Krebs cycle.

### 8.5  The Electron Transport Chain

11. Electron transport uses the energy from NADH and $FADH_2$ to form a proton gradient in the mitochondrial matrix, which is then used to phosphorylate ATP.
12. Oxygen is required at this step as a final acceptor of the energy-depleted electrons.
13. Each NADH yields 2.5 ATP, and each $FADH_2$ yields 1.5 ATP via this process. The net energy extracted from glucose is 30 to 32 ATP.
14. Feedback mechanisms regulate the activity of enzymes to ensure the cell has the energy it needs.

wear down, and the cells would cease to divide—and the plant would cease to grow.   •   **meristems, p. 516**

## A Variety of Signals Inhibit or Initiate Division

Signals from outside the cell and from within affect the cell cycle. Normal cells growing in culture stop dividing when they form a one-cell-thick layer (a monolayer) lining their container. If the layer tears, remaining cells bordering the tear grow and divide to fill in the gap but stop dividing once it is filled (**figure 9.12**). The term **contact inhibition** refers to the inhibiting effect of cell crowding on cell division. Lack of contact inhibition is one characteristic of cancer cells.

Certain hormones and growth factors are biochemicals that signal a cell to divide. In an animal, a **hormone** is a substance that is manufactured in a gland and travels in the bloodstream to another part of the body, where it exerts a specific effect (see chapter 34). Cell division is one such consequence of a hormonal signal, as we saw in chapter 1 for estrogen. At a certain time in the monthly hormonal cycle in the human female, estrogen levels

Cells form single layer in culture.

**A**

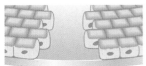

Cells removed.

**B**

Cells replace removed cells; division stops when single layer is repaired.

**C**

Cancer cells in culture will continue to divide and pile up haphazardly.

**D**

**FIGURE 9.12 Contact Inhibition.** **(A)** Normal animal cells in culture divide until they line their container in a one-cell-thick sheet (a monolayer). If the monolayer is damaged **(B)**, the cells bordering the removal site grow and divide, **(C)** filling the gap. However, the cells do not pile up on each other because of contact inhibition. In contrast, **(D)** cancer cells pile up on each other.

peak, stimulating the cells lining the uterus to divide and build tissue in which a fertilized egg can implant. If an egg is not fertilized, another hormonal shift triggers cell death that breaks the lining down, resulting in menstruation. In plants, hormones coordinate growth and development, often by stimulating cell division in roots, shoots, seeds, fruits, and young leaves (see chapter 30).

**Growth factors** are proteins that stimulate local cell division. Damaged cells at a wound site release growth factors that increase the rate of cell division, which replaces damaged tissue. For example, epidermal growth factor (EGF) stimulates epithelium (lining tissue) to divide, which fills in new skin underneath a scab. Salivary glands produce EGF, which aids healing when an animal licks its wounds.

Biochemicals inside cells also stimulate cell division. Proteins called **kinases** and **cyclins** activate genes whose protein products interact to carry out cell division. Cyclins are highly active at the $G_1$ checkpoint, when they determine whether or not a cell divides. A kinase is a type of enzyme that activates other proteins by adding phosphates to them. The kinase controlling the cell cycle is present in all cells of eukaryotes at all times. In contrast, levels of cyclin fluctuate, as its name implies. At one point in the cell cycle, cyclin levels plummet as if something is rapidly degrading it.

It is the relationship between kinase and cyclin that controls the cell cycle (**figure 9.13**). First, kinase molecules bind to cyclin molecules, which have accumulated during the previous interphase. These kinase-cyclin complexes activate other enzymes, making mitosis inevitable and stimulating production of enzymes that break down cyclin. As cyclin levels fall, levels of cyclin-degrading enzymes also drop, and cyclin accumulates again. When enough cyclin accumulates that it again combines with the always-present kinase, cell division begins anew. The name "cyclin" is derived from this cylic appearance and removal of these proteins. The control of the cell cycle by cyclin and kinase is an example of negative feedback. That is, when a substance accumulates to a certain level, an event occurs that stops further synthesis.   •   **negative feedback, fig. 6.17**

## Stem Cells Provide a Reservoir for New Tissues

Tissues must maintain their specialized natures, as well as retain the capacity to generate new cells as the organism grows or repairs tissues damaged by injury or illness. To achieve this, many tissues contain a few cells, called **stem cells,** that can divide and thereby replenish the tissue. When a stem cell divides to yield two daughter cells, one remains a stem cell, able to divide again, while the other specializes to perform certain functions. **Figure 9.14** shows stem cells in the basal (deepest) layer of the epidermal skin layer in a human. These basal cells readily divide, but the cells above them, which divided from the basal stem cells, do not. The cells in the upper skin layers die via apoptosis—they flake off after a brisk toweling. It is believed

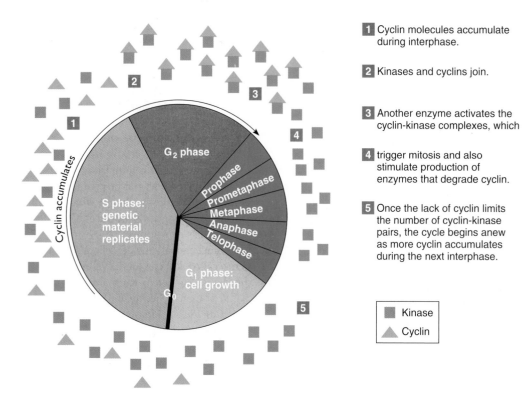

1 Cyclin molecules accumulate during interphase.

2 Kinases and cyclins join.

3 Another enzyme activates the cyclin-kinase complexes, which

4 trigger mitosis and also stimulate production of enzymes that degrade cyclin.

5 Once the lack of cyclin limits the number of cyclin-kinase pairs, the cycle begins anew as more cyclin accumulates during the next interphase.

Kinase
Cyclin

**FIGURE 9.13** **Cyclins and Kinases Regulate the Cell Cycle.** Kinase molecules are always present in the cell, and they join cyclin molecules that accumulate during interphase.

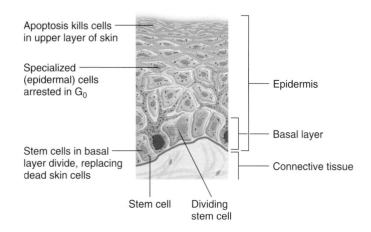

**FIGURE 9.14** **Stem Cells.** In some tissues, only cells in certain positions divide. The outer layer of human skin, the epidermis, has actively dividing stem cells in its basal (deepest) layer. As they divide, they push most of their daughter cells upward, yet they maintain a certain number of stem cells.

this is why tatoos remain for a lifetime—the pigments are stored in stem cells. This organization of stem cells pushing more specialized cells upward also occurs in the folds of the small intestinal lining. Researchers have known about the stem cells in the skin and small intestine for many years. More recently, stem cells have been discovered in places they were never thought to reside, such as the brain and heart.

Finding stem cells in the human brain was a great surprise because most neurons there are in $G_0$ and therefore do not divide. However, small patches of neural stem cells hug spaces in the brain called ventricles. These cells can divide to replace neurons and neuroglia (cells that support, nourish, and interact with neurons) as they die or to restore function following injury. Perhaps too few neural stem cells cause neurodegenerative disorders such as Alzheimer disease; too many may lead to cancer. Implants made of neural stem cells hold promise for treating spinal cord injuries and neurodegenerative disorders.

Stem cells in the heart were discovered in a very dramatic set of experiments. Using a technique that adds a fluorescent tag to a Y chromosome, which is found only in cells of a male, researchers discovered such cells in hearts that had been transplanted from eight women to men. In addition, the male cells sported surface proteins that identified them as stem cells. Those men who survived the longest after the transplant had the fewest stem cells in their hearts but had more Y-bearing

Researchers know of hundreds of oncogenes and tumor suppressor genes. Often a cancer results from a series of genetic changes. Some forms of colon cancer, for example, occur because of a sequence of genetic abnormalities that include oncogene activation and tumor suppressor inactivation. Environmental influences such as diet, exercise habits, sun exposure, and cigarette smoking also affect the activities of oncogenes and tumor suppressor genes. Exposure to certain chemicals, radiation, and viruses as well as nutrient deficiencies can raise cancer risks by damaging DNA.

Many cancers are treatable. They may be surgically removed or their growth slowed sufficiently to allow many years of relative health. Traditional cancer treatments are surgery, drugs (chemotherapy), and radiation. Since the 1990s, many new treatments that boost the immune response against cancer cells have become available. Often, treatment approaches work together. For example, a patient whose bone marrow has been dangerously depleted from chemotherapy may receive a hormone that bolsters red blood cell production (erythropoietin, or EPO) as well as colony stimulating factors (CSFs), which enable the bone marrow to produce more white blood cells. Taking EPO and CSFs can enable a patient to withstand higher doses of chemotherapy, which are more effective. New, more targeted drugs act on the receptors for growth factors that are overabundant on some cancer cells. Such new drugs are very successful in treating certain forms of breast cancer and leukemia. The future has never been brighter for new treatments—many of which are based on understanding the cell cycle.

### Reviewing Concepts

- Cancer is a derangement in cell cycle control that permits cells to continue to divide and often lose specificity of function.
- As the cells grow, they form tumors and may invade other tissues, causing death.
- Mutations in normal genes change vital proteins that regulate the cell cycle and apoptosis.

## Connections

The cell cycle, once believed to represent only living, functional cells, represents a series of decisions made by each cell to divide or remain in a resting, functional state. As scientists have studied the controls on the cell cycle, they have discovered fundamental concepts regarding cancer formation that could lead to breakthroughs in medical science. Understanding the cell cycle has enabled scientists to clone animals, which offer additional avenues for understanding the development of a body from a single cell. The exquisite mechanisms used by a cell to divide ensure that each cell has all of the genetic information needed to sustain life and respond to the environment. But life on planet Earth needs to be adaptable. Cells have provided a way to force changes in the genetic makeup of each generation of complex organisms, providing raw materials for evolution. This mechanism, called meiosis, is a variation on the steps of mitosis but provides the foundation for sexual reproduction.

## Student Study Guide
## Key Terms

| | | | |
|---|---|---|---|
| anaphase *151* | cleavage furrow *151* | karyokinesis *147* | prophase *149* |
| apoptosis *146* | contact inhibition *155* | kinase *155* | S phase *148* |
| aster *150* | cyclin *155* | kinetochore *150* | stem cell *155* |
| caspase *158* | cytokinesis *147* | M phase *149* | telomerase *154* |
| cell cycle *147* | $G_0$ phase *147* | metaphase *150* | telomere *154* |
| centromere *148* | $G_1$ phase *147* | metastasis *159* | telophase *151* |
| centrosome *149* | $G_2$ phase *148* | mitosis *146* | tumor suppressor gene *160* |
| checkpoint *147* | growth factor *155* | mitotic spindle *148* | |
| chromatid *148* | hormone *155* | oncogene *160* | |
| chromatin *149* | interphase *147* | prometaphase *150* | |

# Chapter Summary

## 9.1 The Balance Between Cell Division and Cell Death

1. Mitosis produces the cells for multicellular species, and binary fission reproduces single-celled organisms.

2. Apoptosis is the controlled elimination of cells, thereby forming structures and regulating growth. Cancer results when these processes run out of control.

## 9.2 The Cell Cycle

3. The cell cycle is a series of highly coordinated and controlled steps that ensure each daughter cell will be genetically identical.

4. Cells do not need to be constantly dividing to stay alive, but can enter a quiescent phase that prevents unwanted growth and allows for cell specialization.

## 9.3 Cell Cycle Control

5. Stem cells give rise to all other cell types and provide a reservoir for new tissues.

6. The cell cycle is controlled by proteins, the length of telomeres, and extracellular signals. Biochemicals coordinate the needs of the body by serving as signals that trigger growth and cell division. Stem cells provide continuity and maintain the growth and specialization of a tissue.

## 9.4 Cell Death

7. Apoptosis, programmed cell death, is used to form body structures by removing unwanted cells and is a way of destroying unregulated cells that could lead to cancer.

## 9.5 Cancer—When the Cell Cycle Goes Awry

8. Proteins help control checkpoints in the cell cycle. If mutations disrupt these proteins, cells can divide indefinitely, leading to cancer.

9. In addition to loss of cell cycle control, cancerous cells have specific characteristics, such as loss of contact inhibition and differentiation states. Malignant cells can invade other tissues.

# What Do I Remember

1. Describe each stage of the cell cycle and where the major checkpoints exist.

2. Describe the events that occur during each of the stages of the cell cycle.

3. How does telomere length serve as a cell division clock in some organisms?

4. What is the function of stem cells?

5. Which biochemicals outside and inside cells affect cell division rate?

6. Explain why some people can experience severe radiation exposure and never develop cancer, while others get cancer.

## Fill-in-the-Blank

1. The _____ _____ is the apparatus that actually pulls apart the sister chromatids.

2. During _____ phase, the DNA replication machinery is most active.

3. _____ is the process of physically dividing chromosomes.

4. _____ are proteins that activate regulators of the cell cycle.

5. A bacterium divides by means of the process known as _____.

6. A(n) _____ is part of a duplicated chromosome that joins the two halves.

## Multiple Choice

The most recognizable phase of mitosis is

    a. telophase.

    b. prophase.

    c. metaphase.

    d. interphase.

2. A human cell possesses _____ in a cell that has just finished cytokinesis.

    a. 46 pairs of chromosomes

    b. 23 DNA molecules

    c. 46 duplicated chromosomes

    d. 46 DNA molecules.

3. The mitotic spindle is primarily composed of

    a. cellulose.

    b. actin.

    c. intermediate filaments.

    d. microtubules.

4. The last step in plant cell division involves the formation of the

    a. cell plate.

    b. cleavage furrow.

    c. spindle.

    d. chromosomes.

5. To prevent shrinkage of the telomeres, cells employ the enzyme

    a. reverse transcriptase.

    b. phosphatase.

    c. telomerase.

    d. nucleotide transferase.

Certain bees can develop from unfertilized eggs.

# 10

# Meiosis

## Of Aphids and a Mosaic Child: Parthenogenesis

Parthenogenesis is Greek for "virgin birth" and refers to the unusual situation of an oocyte (immature egg cell) becoming activated, doubling its genetic material, and then repeatedly dividing mitotically to yield a viable offspring. This reproduction without a mate occurs in salamanders, lizards, snakes, turkeys, roundworms, flatworms, and various pond dwellers.

The most familiar parthenogenotes are the males (drones) in certain bee, wasp, and ant societies, which develop from unfertilized eggs. Various stimuli can provoke parthenogenesis, allowing calcium to enter an oocyte. For example, wasp eggs start developing when they are mechanically stimulated while exiting the female's abdomen. When certain female salamanders mate with males from a related species, the sperm activate their oocytes, but the male cells degenerate before they can contribute paternal nuclei. Researchers can induce parthenogenesis in some species by pricking an oocyte, which admits calcium.

Parthenogenesis is advantageous when an organism is well adapted to its environment, because it can lead to a sudden population explosion. Parthenogenesis explains how a pond forming overnight from a puddle becomes rapidly overrun with little swimming animals. In a changing environment, though, parthenogenesis becomes a liability. The organisms are genetically alike, and all perish if conditions become harsh. Because environments change, parthenogenesis is rare.

163

# Chapter Preview

1. Asexual reproduction, such as binary fission, can be successful in an unchanging environment.

2. Sexual reproduction mixes traits and therefore protects species in a changing environment. Sexual reproduction also increases the number of organisms. It occurs when haploid gametes fuse, restoring the diploid state.

3. Meiosis halves the genetic material, and fertilization is the joining of haploid nuclei. Meiosis and fertilization are two key events in sexual life cycles.

4. Sexual reproduction in plants involves an alternation of generations and generations with multicellular haploid and diploid phases.

5. Conjugation, a form of gene transfer in some microorganisms, is sexual because one individual transfers genetic material to another, but it is not reproduction because no additional individual forms.

6. Meiosis halves the number of chromosomes in somatic cells, producing haploid gametes. A species' chromosome number stays constant because in gamete-producing cells, the DNA replicates once, but the cells divide twice.

7. Meiosis provides genetic variability by partitioning different combinations of genes into gametes through independent assortment. Crossing over, which occurs in prophase I as homologous pairs synapse, further increases the variability.

8. Meiosis is different in males and females, producing the specialized cells needed for reproduction. The specialized structures and timing of meiosis in each are needed to ensure genetic diversity.

## 10.1 Reproduction: The Perpetuation of Species

Organisms must reproduce—generate other individuals like themselves—for a species to survive. The amoeba in **figure 10.1A** demonstrates a straightforward way to reproduce. A single cell first replicates its genetic material and then splits in two. The cellular contents, including the DNA, are apportioned into two identical daughter cells. Similar to mitosis, this ancient form of reproduction, called **binary fission,** is still common among single-celled organisms.

In an unchanging environment, the mass production of identical individuals, as in binary fission, makes sense since they will all have the characteristics needed to maintain the species.

But environmental conditions are rarely constant in the real world. If all organisms in a species were well suited to a hot, dry climate, the entire species might perish during a cooling period. Genetic diversity within a population of the same species enables that species to survive in a changeable environment. Binary fission cannot create genetic diversity, but **sexual reproduction** can as it mixes up and recombines inherited traits from one generation to the next. The persistence of sexual reproduction over time and in most species attests to its success in a changing world.

The kittens in figure 10.1B differ from each other because they were conceived sexually. In contrast are the identical amoebae in figure 10.1A, which are the products of **asexual reproduction,** or reproduction without sex. In another rare form of reproduction, **parthenogenesis,** an offspring is derived solely from a female parent.

Sexual reproduction has two essential qualities: it introduces new combinations of genes from different individuals (the parents), and it produces offspring that increase the number of individuals in the population. The special cells from the parents that combine to form the first cell of the offspring are called **gametes,** often also referred to as sex cells. The nuclei of gametes contain only one set of chromosomes and are said to be **haploid,** abbreviated **1$n$.** When two haploid gametes merge, they reconstitute the double, or **diploid ($2n$),** number of chromosome sets in the cells of the offspring. In humans, the gametes are the sperm and egg (oocyte) cells. These cells have half the number of chromosomes as a somatic (body) cell, such as a nerve, muscle, or leaf cell. Some species, including ciliates and certain fungi, produce gametic (haploid) nuclei as part of sexual reproduction but do not enclose them in separate cells.

### Reviewing Concepts

- Reproduction is a vital mechanism for maintaining a species.
- Single-celled organisms use binary fission and asexual reproduction most often.
- Sexual reproduction ensures genetic diversity in the offspring to maintain a species. Haploid gametes are combined to establish new offspring.
- Asexual reproduction is useful in stable environments where genetic changes might be detrimental.

## 10.2 Variations on the Sexual Reproduction Theme

Organisms reproduce in many ways. Some species exhibit distinctly different phases in their life cycles. Consider the yeast *Saccharomyces cerevisiae,* which is a single-celled eukaryote

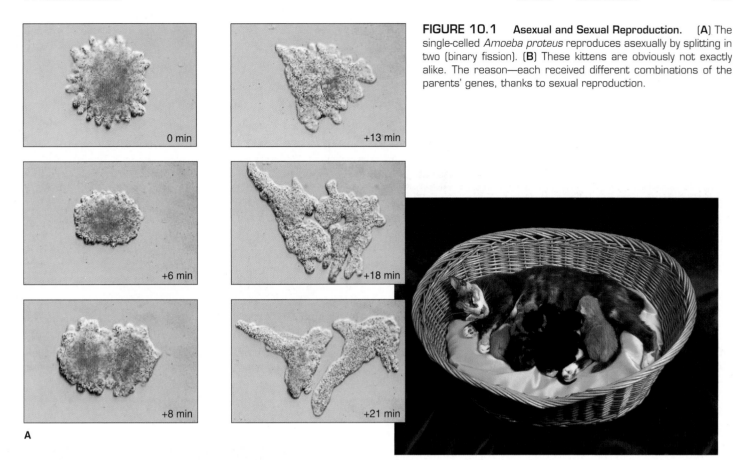

**FIGURE 10.1** **Asexual and Sexual Reproduction.** **(A)** The single-celled *Amoeba proteus* reproduces asexually by splitting in two (binary fission). **(B)** These kittens are obviously not exactly alike. The reason—each received different combinations of the parents' genes, thanks to sexual reproduction.

that reproduces both asexually and sexually. A single diploid cell of the yeast can replicate its genetic material and "bud," yielding genetically identical diploid daughter cells. This is asexual reproduction, because each offspring has one parent. Alternatively, yeast can form a structure called an ascus, which gives rise to specialized haploid cells. Two of these haploid cells can fuse, restoring the diploid chromosome number. This is sexual reproduction, because two individuals produce a third. The false spider mite *Brevipalpus phoenicis* is the only known haploid animal.

## Haploid Cells Arise from the Sexual Life Cycle

Sexual life cycles describe the proportion and timing of an individual's existence that is spent in the haploid and/or diploid state. The two main events of a sexual life cycle are meiosis and fertilization. **Meiosis** is a form of cell division that halves the genetic material. **Fertilization** is the joining of haploid nuclei that result from meiosis, which reconstitutes the diploid cell. The fertilized cell becomes the beginning of a new individual. In humans, meiosis leads to formation of the sperm or egg cells, and fertilization occurs following sexual intercourse.

Mitosis provides more cells as the body grows through the rest of the sexual life cycle. The timing and extent of mitosis varies from one sexually reproducing species to another. In green algae, cellular slime molds, and some fungi, mitotic growth occurs between meiosis and fertilization, resulting in a body that is haploid. Yet in ciliates, brown algae, water molds, and animals, the period of mitotic activity occurs between fertilization and meiosis, producing a diploid body.

## Alternation of Generations Involves Two Multicellular Life Stages

Land plants and some other organisms have a more complex life cycle that includes multicellular diploid and haploid stages. These stages are called generations, and because they alternate within a life cycle, sexually reproducing plants are said to undergo **alternation of generations** (figure 10.2). Some plants can also reproduce asexually when a "cutting" from a larger plant grows into a new organism.

In plants, the diploid generation, or **sporophyte,** produces haploid spores through meiosis. Haploid spores divide mitotically to produce a multicellular haploid individual called the **gametophyte.** Eventually, the gametophyte produces haploid

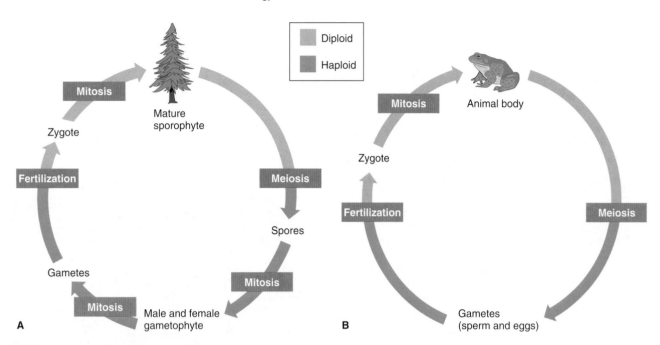

**FIGURE 10.2    Sexual Reproduction.**  Sexual life cycles of plants (**A**) and animals (**B**) alternate between haploid and diploid stages. The difference between the two life cycles pictured is that plants have a multicellular haploid stage, called the gametophyte generation, that animals almost never have.

gametes—eggs and sperm—which combine through fertilization to form a **zygote.** The zygote grows into a sporophyte, and the cycle begins anew. Chapter 29 discusses flowering plant reproduction.

## Conjugation May Provide Evolutionary Clues to Sexual Reproduction

Learning how diverse organisms reproduce and exchange genetic material today can provide clues to how sexual reproduction may have evolved. The earliest process that combines genes from two individuals appeared about 3.5 billion years ago, according to fossil evidence. This form of bacterial gene transfer, called **conjugation,** is still prevalent today (see chapter 21). In conjugation, one bacterial cell uses an outgrowth called a sex pilus to transfer genetic material to another bacterial cell (**figure 10.3**). Bacterial conjugation is not true reproduction because it alters the genetic makeup of an already existing organism rather than producing new individuals.

● **horizontal gene transfer, p. 395**

Like the bacteria, *Paramecium* is a unicellular eukaryote that can separate reproduction and the creation of genetic diversity. Paramecia reproduce asexually by binary fission, yet two individuals can conjugate by aligning their oral cavities and forming a bridge of cytoplasm between them. Nuclei are transferred across this bridge, resulting in new gene combinations but no net increase in individuals.

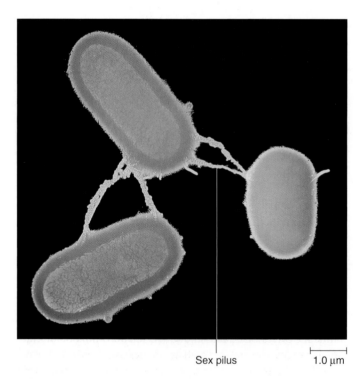

Sex pilus                    1.0 µm

**FIGURE 10.3    Conjugation.**  The bacterium *Escherichia coli* usually reproduces asexually by binary fission. But *E. coli* can also transfer genetic material to another bacterium using an appendage called a sex pilus to insert DNA into another cell. DNA transfer between bacterial cells, called conjugation, is similar to sexual reproduction in that it produces a new combination of genes. It is not reproduction, however, because a new individual does not form.

# Student Study Guide

## Key Terms

acrosome  *175*
allele  *168*
alternation of generations  *165*
asexual reproduction  *164*
autosome  *168*
binary fission  *164*
centromere  *168*
chromatid  *168*
conjugation  *166*
crossing over  *168*
diploid (2*n*)  *164*

epididymis  *175*
equational division  *168*
fertilization  *165*
gamete  *164*
gametogenesis  *168*
gametophyte  *165*
germ cell  *168*
haploid (1*n*)  *164*
homologous pairs  *168*
independent assortment  *169*
meiosis  *165*

oogenesis  *176*
oogonium  *176*
ovary  *176*
parthenogenesis  *164*
polar body  *176*
primary oocyte  *176*
primary spermatocyte  *174*
reduction division  *168*
secondary oocyte  *176*
secondary spermatocyte  *174*
seminiferous tubule  *175*

sex chromosome  *168*
sexual reproduction  *164*
spermatid  *174*
spermatogenesis  *174*
spermatogonium  *174*
spermatozoa  *175*
sporophyte  *165*
testis  *175*
vas deferens  *175*
zygote  *166*

## Chapter Summary

### 10.1  Reproduction: The Perpetuation of Species

1. Sexual reproduction uses haploid gametes to produce genetic diversity in offspring, which ensures species survival.

2. Asexual reproduction produces clones and is useful in stable environments where genetic changes might be detrimental.

### 10.2  Variations on the Sexual Reproduction Theme

3. Meiosis is cell division that forms haploid cells for use in sexual reproduction or for the stages of some species, such as yeast, that live as haploid organisms.

4. Plants alternate between haploid and diploid stages in the alternation of generations.

5. Some species use different mating types to mimic sexual reproduction in higher organisms.

6. Bacteria can exchange genetic material through conjugation, but this is not equivalent to sexual reproduction.

### 10.3  Meiosis: One Replication, Two Divisions

7. Meiosis produces gametes for sexual reproduction with half the number of parental chromosomes by using two rounds of division of chromosomes after a single event of DNA replication.

8. The phases of meiosis are similar to those of mitosis and ensure that offspring get a complete genetic complement from the parents.

9. Genetic variation for sexual reproduction is increased through independent assortment and crossing over, producing nonidentical homologs of the parental chromosomes.

### 10.4  The Sculpting of Human Gametes

10. The formation of gametes in humans produces four nonidentical cells from one parent cell under the influence of hormonal signals.

11. Following cyclic hormonal signals, human females produce a single ovum and use polar bodies to reduce the chromosome numbers.

12. Human males are under constant hormonal signals that produce millions of sperm each day. Each parent cell yields four nonidentical sperm.

13. Human ova complete meiosis only when fertilization has taken place.

## What Do I Remember?

1. Describe the differences between sexual and asexual reproduction and the circumstances where each would have an advantage.

2. Describe each of the steps of meiosis and indicate how many DNA molecules would be present at each phase.

3. What is the alternation of generations, and how is this useful?

4. How do the structures of the human male and female reproductive systems differ? Describe how these differences are used to form a new individual.

5. Describe the mechanisms in meiosis that form new genetic combinations.

6. How many genetically different offspring are possible due to meiosis?

**Fill-in-the-blanks**

1. The _____ are visible evidence of crossing over in chromosomes.

2. Human sperm are manufactured in the _____.

3. The _____ is the site of fertilization in humans.

4. A(n) _____ is one of a pair of the same type of chromosome.

5. To produce a functional human ovum, as many as three _____ are formed during meiosis.

**Multiple Choice**

1. A _____ is the haploid cell in humans used for sexual reproduction.
   a. gamete
   b. ovum
   c. zygote
   d. parthogenote

2. The immediate product of fertilization is called a(n)
   a. gamete.
   b. ovum.
   c. zygote.
   d. parthogenote.

3. There are _____ DNA molecules in each of the two cells produced by the first round of meiosis.
   a. 23
   b. 46
   c. 92
   d. 4

4. Sperm are stored and complete their maturation in the
   a. vasa deferentia.
   b. seminiferous tubules.
   c. epididymis.
   d. acrosomes.

5. _____ is the release of a mature human ovum once each month.
   a. Ovulation
   b. Fertilization
   c. Chiasmata
   d. Conjugation

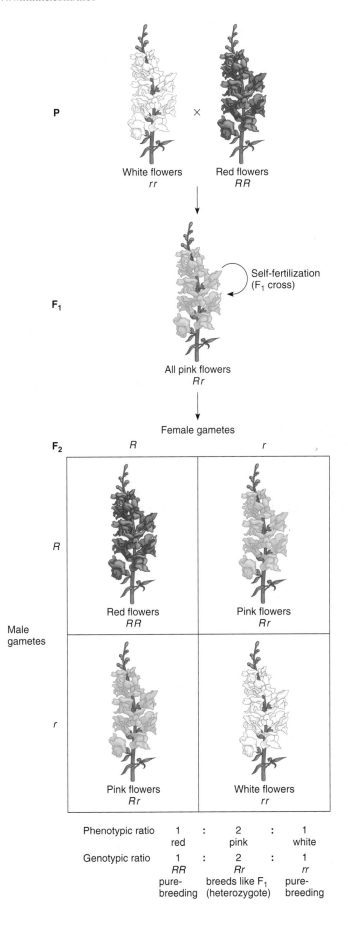

P

White flowers
*rr*

×

Red flowers
*RR*

Self-fertilization
(F₁ cross)

F₁

All pink flowers
*Rr*

Female gametes

F₂    R           r

R

Red flowers
*RR*

Pink flowers
*Rr*

Male
gametes

r

Pink flowers
*Rr*

White flowers
*rr*

| Phenotypic ratio | 1 | : | 2 | : | 1 |
|---|---|---|---|---|---|
| | red | | pink | | white |
| Genotypic ratio | 1 | : | 2 | : | 1 |
| | *RR* | | *Rr* | | *rr* |
| | pure-breeding | | breeds like F₁ (heterozygote) | | pure-breeding |

## Different Types of Dominance Can Cause a Mixed Phenotype

Mendel envisioned simple traits that are dominant or not, producing a complete trait or none at all. However, some genes show **incomplete dominance,** with the heterozygous phenotype intermediate between those of the two homozygotes. For example, the piebald trait in domestic cats confers varying numbers of white spots and shows incomplete dominance. Cats of genotype *SS* have many spots, *ss* cats have no spots, and *Ss* cats have an intermediate number of spots. Another example of incomplete dominance is the snapdragon plant (**figure 11.18**). A red-flowered plant of genotype *RR* crossed with a white-flowered *rr* plant gives rise to a pink-flowered *Rr*. An intermediate amount of pigment in the heterozygote confers the pink color.

Another variation in Mendel's conclusions results from equal dominant alleles. Different alleles that are both expressed in a heterozygote are termed **codominant.** The *A* and *B* alleles of the *I* gene for blood type are codominant. People of blood type AB have both antigens A and B on the surfaces of their red blood cells (**figure 11.19**).

### Reviewing Concepts

- Complex interactions between several genes, along with the influences of the environment, can explain apparent exceptions to Mendel's laws.
- Some allele combinations are lethal and distort Mendel's ratios. Some traits have more than two alleles that can be passed in various combinations.
- The environment can affect how severely a trait is expressed, or whether it is expressed at all, in individuals with the genes for that trait.
- Due to their biochemical role, some genes affect more than a single trait. Epistasis results from the masking of one trait by the expression of a separate set of genes.
- Dominant alleles can both be expressed, combining phenotypes.

**FIGURE 11.18** Incomplete Dominance in Snapdragon Flowers. A cross between a homozygous dominant plant with red flowers (*RR*) and a homozygous recessive plant with white flowers (*rr*) produces a heterozygous plant with pink flowers (*Rr*). When *Rr* pollen fertilizes *Rr* egg cells, one-quarter of the progeny are red-flowered (*RR*), one-half are pink-flowered (*Rr*), and one-quarter are white-flowered (*rr*). The phenotypic ratio of this monohybrid cross is 1:2:1 (instead of the 3:1 seen in cases of complete dominance) because the heterozygous class has a phenotype different from that of the homozygous dominant class.

| Genotypes | Phenotypes | |
| --- | --- | --- |
| | Antigens on surface | ABO blood type |
| $I^A I^A$ $I^A i$ | A A | Type A |
| $I^B I^B$ $I^B i$ | B B | Type B |
| $I^A I^B$ | AB | Type AB |
| $ii$ | None | Type O |

**Type A**

| | $I^A$ | $I^A$ |
| --- | --- | --- |
| Type B $I^B$ | $I^A I^B$ AB | $I^A I^B$ AB |
| $I^B$ | $I^A I^B$ AB | $I^A I^B$ AB |

**Type A**

| | $I^A$ | $i$ |
| --- | --- | --- |
| Type B $I^B$ | $I^A I^B$ AB | $I^B i$ B |
| $I^B$ | $I^A I^B$ AB | $I^B i$ B |

**Type A**

| | $I^A$ | $I^A$ |
| --- | --- | --- |
| Type B $I^B$ | $I^A I^B$ AB | $I^A I^B$ AB |
| $i$ | $I^A i$ A | $I^A i$ A |

**Type A**

| | $I^A$ | $i$ |
| --- | --- | --- |
| Type B $I^B$ | $I^A I^B$ AB | $I^B i$ B |
| $i$ | $I^A i$ A | $ii$ O |

**FIGURE 11.19 Codominance.** Even though the $I^A$ and $I^B$ alleles of the $I$ gene are codominant, they still follow Mendel's law of segregation. These Punnett squares follow the genotypes that could result by crossing a person with type A blood with a person with type B blood. Is it possible for parents with type A and type B blood to have a child who is type O?

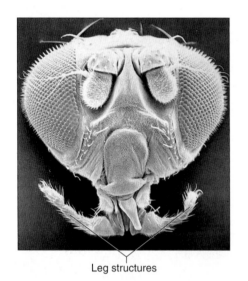

Leg structures

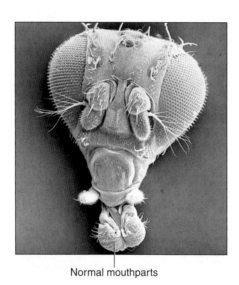

Normal mouthparts

Antennal structures

**FIGURE 11.20 A Conditional Mutant.** A temperature-sensitive mutant gene in the fruit fly *Drosophila melanogaster* transforms normal mouthparts (*center*) into leg structures at high temperatures (*left*) and into antennal structures at low temperatures (*right*).

## 11.5   Environmental Influences on Gene Expression

The expression of certain genes is exquisitely sensitive to the environment. Temperature influences gene expression in some familiar animals—Siamese cats and Himalayan rabbits have dark ears, noses, feet, and tails because these parts are colder than the animals' abdomens. Heat affects the abundance of pigment molecules that produce coat color.

Temperature-sensitive alleles also lead to striking phenotypes in the fruit fly (**figure 11.20**). In flies mutant for a gene called proboscipedia, mouthparts develop as antennae when the flies are raised at low temperatures and as legs when the flies are raised at high temperatures. When raised at room temperature, the flies' mouthparts are mixtures of leg and antennal tissue.

A phenotype that is expressed only under certain environmental conditions is termed conditional.

A trait that appears to be inherited but actually is caused by something in the environment is called a **phenocopy.** It may either coincidentally resemble a Mendelian disorder's symptoms or mimic inheritance because it occurs in more than one family member. For example, in the 1960s, the tranquilizing drug thalidomide caused grossly shortened limbs in children whose mothers took it while pregnant. This birth defect is a phenocopy of an inherited illness called phocomelia. Doctors realized it was a phenocopy and not an inherited birth defect because the incidence suddenly increased when the drug came into use. An infection can also appear to be a Mendelian disorder and therefore be a phenocopy. For example, children who have AIDS may contract the infection from HIV-positive parents. But these children acquire AIDS by viral infection, not by inheriting a gene.

Unlike an inherited disorder, an infection does not recur with a predictable frequency. Its recurrence depends upon exposure to the infective agent.

## Reviewing Concepts

- The environment, particularly disease, can mimic genetic traits.
- Environment also plays a role in whether certain genes are expressed and how the resulting trait appears.

## 11.6 Genetic Heterogeneity

Another complication of Mendel's laws is that several different genes can produce the same phenotype, a phenomenon called **genetic heterogeneity.** For example, 132 forms of deafness are transmitted as autosomal recessive traits in humans. If a man who is heterozygous for a deafness gene on one chromosome has a child with a woman who is heterozygous for a deafness gene on a different chromosome, then that child faces only the general population risk of inheriting either form of deafness—not the 25% risk that Mendel's law predicts for a monohybrid cross. This is because the parents are heterozygous for different genes.

Genetic heterogeneity can occur when genes encode different enzymes that participate in the same biochemical pathway. For example, 11 biochemical reactions lead to blood clot formation. Clotting disorders may result from abnormalities in genes specifying any of these enzymes, causing several types of bleeding disorders. The phenotypes are the same—poor blood clotting—but the genotypes differ. Researchers recently found that cystic fibrosis can be caused by more than one gene.

**Table 11.4** summarizes phenomena that appear to alter Mendelian inheritance.

## Reviewing Concepts

- Several different genes can produce the same trait, making it difficult to identify which genes might be responsible for any given trait.
- The complexity of biochemical pathways adds to the complexity of genetic heritability.

## TABLE 11.4

### Factors That Alter Mendelian Phenotypic Ratios

| Phenomenon | Effect on Phenotype | Examples |
|---|---|---|
| Lethal alleles | A phenotypic class stops developing | Spontaneous abortion |
| Multiple alleles | Produces many variants of a phenotype | Rabbit coat color |
| Penetrance | Some individuals inheriting a particular genotype do not have the associated phenotype | Polydactyly |
| Expressivity | A genotype is associated with a phenotype of varying intensity | Polydactyly |
| Pleiotropy | The phenotype includes many symptoms, with different subsets in different individuals | Porphyria variegata; Marfan syndrome |
| Epistasis | One gene masks another's phenotype | Bombay phenotype |
| Incomplete dominance | A heterozygote's phenotype is intermediate between those of two homozygotes | Snapdragon flower color; piebald trait |
| Codominance | A heterozygote's phenotype is distinct from and not intermediate between those of the two homozygotes | ABO blood types |
| Conditional mutations | An environmental condition affects a gene's expression | Temperature-sensitive coat colors |
| Phenocopy | An environmentally caused condition whose symptoms and recurrence in a family make it appear to be inherited | Infection; environmentally caused birth defect |
| Genetic heterogeneity | Mutations in different genes produce same phenotype | Deafness; clotting disorders |

# 11.7   Polygenic and Multifactorial Traits

Mendel's data were clear enough for him to infer laws of inheritance because he observed characteristics caused by single genes with two easily distinguished alternate forms. Inherited traits can also result from the actions of more than one gene.

A trait can be described as either Mendelian (single gene) or **polygenic.** A polygenic trait, as its name implies, reflects the activities of more than one gene, and the effect of these multiple inputs is often additive, although not necessarily equal. Both Mendelian and polygenic traits can also be **multifactorial,** which means they are influenced by the environment. (Multifactorial traits, whether Mendelian or polygenic, are also termed "complex.")

The combined actions of the genes that contribute to a polygenic trait produce a continuum, or continuously varying expression, as **figure 11.21** shows for skin color. The genes of a polygenic trait follow Mendel's laws individually (unless they are on the same chromosome), but they don't produce typical ratios because they are neither dominant nor recessive to each other. Skin color is also multifactorial because sun exposure affects the phenotype.

Eye color is a rare polygenic trait that is not multifactorial because the environment does not influence pigmentation of the iris (except if one wears colored contact lenses). Enzymes control production and distribution of the pigment melanin. Darker eyes

have more clumps of melanin, which absorb incoming light, than do lighter eye colors. In plants, continuously varying traits include flower color, petal length, blade length, and stomata density on the leaves. When the frequencies of all the phenotypes associated with a polygenic trait are plotted on a graph, they form a characteristic bell-shaped curve, as seen for human height in **figure 11.22**.

Most human diseases are multifactorial. Cystic fibrosis and sickle cell disease, for example, are single gene disorders, but since both include increased susceptibility to infection, the course of each illness depends upon which infectious agents a person encounters. Single gene disorders are usually quite rare. More common are conditions caused by interacting genes as well as the environment. For example, we know that multiple sclerosis (MS) has a genetic component, because siblings of an affected individual are 25 times as likely to develop MS as siblings of people who do not have MS. One model of MS origin is that five susceptibility genes have alleles, each of which increases the risk of developing the condition. Those risks add up, and in the presence of an appropriate (and unknown) environmental trigger, symptoms begin. Body weight and intelligence are other traits that are both polygenic and multifactorial.

## Gene Input and Environmental Effects Can Be Predicted

Information from population studies and from family relationships helps describe the inherited component of multifactorial

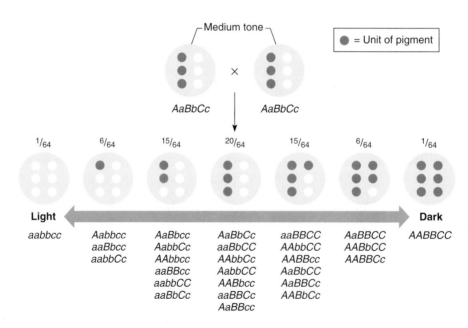

**FIGURE 11.21   Skin Color Follows a Polygenic Pattern of Inheritance.**    Multiple copies of genes, or different versions of genes, combine to increase the pigment in skin cells. Different numbers of the contributing genes produce the intermediate phenotypes.

**FIGURE 11.22**  **Height Is Polygenic and Multifactorial.**   Many genetics classes demonstrate polygenic (continuously varying) traits by having students line up by height. However, the fact that recent lineups have more tall students than lineups done early in the twentieth century reveals the influence of the environment as well. Still, all such lineups display the characteristic bell-shaped curve of a polygenic trait. This class from the University of Connecticut at Storrs lined up in 1997 to update a famed similar photo taken at the same institution, then called the Connecticut Agricultural College, around 1920. Back then, the tallest student was 5'9"; in 1997, the tallest was 6'5".

traits. **Empiric risk** is a prediction of recurrence based on a multifactorial trait's incidence in a specific population. In general, empiric risk increases with the severity of the disorder, the number of affected family members, and how closely related the person is to affected individuals.

Empiric risk helps, for example, to predict the likelihood that a neural tube defect (NTD) will recur in a family. An NTD is an opening or lesion in the brain or spinal cord in an embryo that occurs on about the 28th day of prenatal development in humans. In the United States, the overall population risk of carrying a fetus with an NTD is about 1 in 1,000. However, if a person has an NTD, the risk to a sibling is 3%, and if two siblings are affected, the risk to a third child is even greater. The statistics come from direct observations of the prevalence of NTDs. These birth defects have a variety of causes, including mutation and folic acid deficiency. The environmental risk of NTDs may be greatly reduced by supplementing a pregnant woman's diet with folic acid.

**Heritability** is another way to describe a multifactorial trait. It estimates the proportion of phenotypic variation in a group that can be attributed to genes. Because the environment can change, so too can heritability. For example, the heritability of human skin color in a Canadian population is greater in the winter, when sun exposure is less likely to darken skin. Heritability equals 1.0 for a trait that is completely the result of gene action and 0 if it is entirely caused by an environmental influence. Most traits lie in between. For example, verbal aptitude has a heritabil-

ity of 0.7, whereas mathematical aptitude has a heritability of 0.3. Given this information, which skill do you think could be improved more by taking a course to prepare for a standardized exam such as the SAT?

Heritability can be measured in several ways using statistical methods. One way to measure heritability is to consider pairs of individuals who are related in a certain way and compare the frequency with which Mendel's laws predict they should share a trait to the actual frequency. Such a calculation depends on the proportion of genes that two people related in a particular way share (**table 11.5**). For example, parents and children, as well as siblings to each other, theoretically share 50% of their genes because of the mechanism of meiosis. If the heritability of a trait is very high, then out of a group of 110 pairs of siblings, nearly 50 would be predicted to both have it. But for height, which has a greater environmental component, only 40 of a group of 110 sibling pairs are both the same height. Heritability for height among this group is 0.40/0.50 (observed/expected), or 0.80. That is, about 20% of their height is attributed to environmental factors, such as diet.

## Certain Studies Can Separate Genetic from Environmental Influences

Multifactorial inheritance analysis does not easily lend itself to the scientific method, which ideally examines one variable. Two

## TABLE 11.5

### Relatives Share Genes

| Relationship | Percent Shared Genes |
| --- | --- |
| Sibling to sibling (non-identical) | 50% (1/2) |
| Parent to child | 50% (1/2) |
| Uncle/aunt to niece/nephew | 25% (1/4) |
| First cousin to first cousin | 12.5% (1/8) |

types of people, however, have helped geneticists tease apart the genetic and environmental components of complex traits: adopted individuals and twins.

**Adopted Individuals**  A person adopted by nonrelatives shares environmental influences, but not genes, with his or her adoptive family. Conversely, adopted individuals share genes, but not the exact environment, with their biological parents. Therefore, biologists assume that similarities between adopted people and their adoptive parents reflect environmental influences, whereas similarities between adopted people and their biological parents mostly reflect genetic influences. Information on both sets of parents can reveal to what degree heredity and the environment contribute to a trait.

Many adoption studies use the Danish Adoption Register, a list of children adopted in Denmark from 1924 to 1947. One study examined causes of death among biological and adoptive parents and offspring. If a biological parent died of infection before age 50, the adopted child was five times more likely to die of infection at a young age than a similar person in the general population. This may be because inherited variants in immune system genes increase susceptibility to certain infections. In support of this hypothesis, the risk that adopted individuals would die young from infection did not correlate with adoptive parents' death from infection before age 50. The study also revealed that environment affects longevity. If adoptive parents died before age 50 of cardiovascular disease, their adopted children were three times as likely to die of heart and blood vessel disease as a person in the general population. Can you think of a factor that might explain this correlation?

**Twins**  Identical twins, or **monozygotic (MZ) twins,** are always of the same sex and have identical genes because they develop from one fertilized ovum. Fraternal twins, or **dizygotic (DZ) twins,** are no more similar genetically than any other two siblings, although they share the same prenatal environment because they develop at the same time from two fertilized ova.

A trait that occurs more frequently in both members of identical twin pairs than in both members of fraternal twin pairs is at least partly controlled by genes. The **concordance** of a trait is the degree to which it is inherited, and it is calculated as the percentage of twin pairs in which both members express the trait (**figure 11.23**). Twins can be used to calculate heritability, which

equals approximately double the difference between MZ and DZ concordance values for a trait.

Diseases caused by single genes, whether dominant or recessive, are always 100% concordant in MZ twins—that is, if one twin has it, so does the other. However, among DZ twins, concordance is 50% for a dominant trait and 25% for a recessive trait, the same Mendelian values that apply to any two siblings. For a trait determined by several genes, concordance values for MZ twins are significantly greater than for DZ twins. Finally, a trait molded mostly by the environment exhibits similar concordance values for both types of twins.

Using twins to study genetic influence on complex traits dates to 1924, when German dermatologist Hermann Siemens compared school transcripts of identical versus fraternal twins. Noticing that grades and teachers' comments were much more alike for identical twins than for fraternal twins, he concluded that genes contribute to intelligence. Siemens also suggested that a better test would be to study identical twins that were separated at birth and then raised in very different environments.

Twins separated at birth provide natural experiments for distinguishing nature from nurture. Much of what they have in common can be attributed to genetics, especially if their environments have been very different, which the famous cartoon in

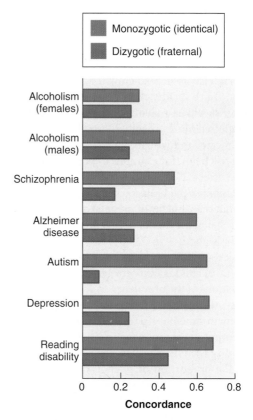

**FIGURE 11.23  Concordance.**  A trait that is more often present in both members of monozygotic twin pairs than it is in both members of dizygotic twin pairs has a significant inherited component.
Source: Robert Plomin, et al., "The genetic basis of complex human behaviors," *Science,* 17 June 1994, vol. 264, pp. 1733–1739. Copyright 1994 American Association for the Advancement of Science.

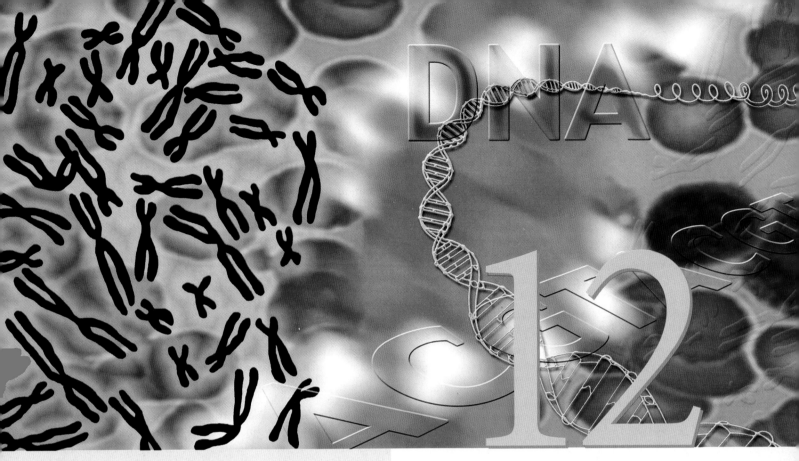

# Chromosomes

Human chromosomes.

## Displaying Chromosomes

Today, many couples expecting a child can see a photograph of the fetus's chromosomes displayed in a size-ordered chart called a karyotype.

Chromosomes are easiest to see if the cell is dividing, when they are the most condensed and likely to soak up a dye. From early on, geneticists used colchicine, extracted from the autumn crocus, to arrest cells in mitosis. In the 1970s, Swedish researchers developed specific stains that home in on DNA sequences that are rich in the base pairs A and T or C and G. These stains created banding patterns unique to each chromosome type.

A much more specific technique is FISH, which stands for "fluorescence *in situ* hybridization." FISH uses DNA probes, pieces of DNA attached to molecules of a dye that glows when hit with light. Probes for specific genes are added to chromosomes spread out on a microscope slide. A probe binds to the complementary DNA sequence, and light is applied. A flash of color results, revealing the location of a particular gene. By using many different probes and dyes, researchers can "paint" each chromosome a unique color.

Today, computerized karyotype devices scan cell preparations and select one in which the chromosomes are the most visible and well spread. Image-analysis software then recognizes each chromosome pair, size-orders pairs into a chart, and prints it. If an abnormal pattern is detected, similar karyotypes are supplied from a database along with clinical information.

# Chapter Preview

1. A chromosome is a continuous double-stranded molecule of DNA, with RNA and associated proteins that provide scaffolding or help carry out replication or transcription.

2. Each species has a characteristic number of chromosomes, which are distinguished by size, centromere position, and banding patterns of dark-staining heterochromatin and light-staining euchromatin.

3. Genes on the same chromosome are linked; rather than demonstrating independent assortment, they produce a large number of parental genotypes and a small number of recombinant genotypes.

4. Genotype predictions and linkage maps are derived from knowing allele configurations and crossover frequencies, which are directly proportional to distances between genes.

5. Sex determination mechanisms are diverse. Chromosomes carrying genes that determine maleness are combined in specific ways to produce males or females.

6. An X-linked trait passes from mother to hemizygous son because the male inherits his X chromosome from his mother and his Y chromosome from his father. Y-linked traits are very rare.

7. X inactivation shuts off one X chromosome in the cells of female mammals, equalizing the number of active X-linked genes in each sex.

8. Polyploid cells have extra full chromosome sets, and aneuploids have extra or missing individual chromosomes.

9. Chromosomal rearrangements disrupt meiotic pairing, which can delete or duplicate genes.

## 12.1 Chromosome Structure

Chromosome stability and integrity are essential to trait transmission. A **chromosome** is a long, continuous piece of DNA, plus RNA and several types of associated proteins. A species has a characteristic number of chromosomes. Each chromosome is distinguished by size, centromere position, and banding patterns when stained. Modern technology makes use of DNA probe patterns to identify chromosomes.

Several biochemicals associate with the DNA of a chromosome, including RNA and various proteins that help replicate the DNA or transcribe it into RNA. Other proteins serve as scaffolds around which DNA tightly entwines and coils particularly tightly

during mitosis (see figure 9.4). Chromosomes also include proteins specific to certain cell types.

## Microscopic Techniques Allow Chromosomes to be Distinguished

Although each species has a characteristic number of chromosomes, the number does not reflect the complexity of the organism nor necessarily how closely related two species are. A mosquito has 6 chromosomes; a grasshopper, rice plant, and pine tree each have 24; a dog has 78; a carp has 104; humans have 46.

To study chromosomes, somatic cells are fixed to slides while in metaphase and stained. The special stains applied to chromosomes create patterns of bands, which differ among the chromosome types. Most stains do this by binding preferentially to tightly wound DNA with many repetitive sequences, called **heterochromatin,** which stains darkly. In contrast is lighter-staining **euchromatin,** which harbors more unique sequences and is looser (**figure 12.1**). Heterochromatin comprises the telomeres (chromosome tips) and centromeres and may help maintain chromosomes' structural integrity. Euchromatin encodes proteins and may be more loosely wound so that its information is accessible. For ease of study, chromosomes stained in this way are often size-ordered into **karyotype** charts.

Scientists often study chromosomes as they appear in metaphase of mitosis when they are most highly condensed. The centromere position visible at this time is also used to distinguish chromosomes. Recall from figure 9.4 that a centromere is a characteristically located constriction, consisting of repetitive DNA sequences and special centromere-associated proteins, to which spindle fibers attach in mitosis. A chromosome is **telocentric** if the centromere is very close to a chromosome tip; **acrocentric** if the centromere pinches off only a small amount of material; **submetacentric** if the centromere establishes one long arm and one short arm; and

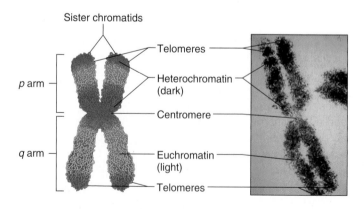

**FIGURE 12.1** **Anatomy of a Chromosome.** Dark-staining chromosomal material (heterochromatin) was once thought to be nonfunctional. Even though heterochromatin does not encode as many proteins as the lighter-staining euchromatin, it is important; it stabilizes the chromosome.

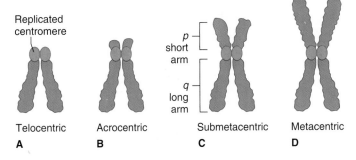

**FIGURE 12.2** **Centromere Position Is Used to Distinguish Chromosomes.** **(A)** A telocentric chromosome has the centromere very close to one end. **(B)** An acrocentric chromosome has the centromere near an end. **(C)** A submetacentric chromosome's centromere creates a long arm (*q*) and a short arm (*p*). **(D)** A metacentric chromosome's centromere creates equal-sized arms.

**metacentric** if the centromere divides it into two arms of approximately equal length (**figure 12.2**). The long arm of a submetacentric or acrocentric chromosome is termed *q*, and the short arm, *p*. Some chromosomes also have bloblike ends called satellites that extend from a stalklike bridge from the rest of the chromosome.

## Artificial Chromosomes Reveal Required Elements

We can sequence genes and analyze genomes, but the basis of a chromosome's integrity remains somewhat mysterious. What are the minimal building blocks necessary to form a chromosome that persists throughout the repeated rounds of DNA replication and reassortment that constitute the cell cycle?

To remain stable, a chromosome requires three basic parts:

- telomeres;
- origins of replication, which are sites where DNA replication begins;
- centromeres.

What would it take to construct a chromosome? There are two ways to tackle this question: (1) pare down an existing chromosome to see how small it can get and still hold together or (2) build up a new chromosome from DNA pieces.

To cut an existing chromosome down to size, researchers swap in a piece of DNA that includes telomere sequences. New telomeres form at the insertion site, a little like prematurely ending a sentence by adding a period. This technique forms small chromosomes, but they can't be easily isolated from cells for further study. The alternative approach, building a chromosome, was challenging because researchers did not know the sequences of origin of replication sites.

Huntington Willard and colleagues at Case Western Reserve University solved the problem. To create "human artificial chro-

mosomes," they sent separately, into cultured cells' telomere DNA, repetitive DNA known to form a crucial part of centromeres, and, because they didn't know the origin of replication sites, random pieces of DNA from the human genome. In the cells, these pieces associated in a correct orientation and formed functional and stable structures about 5 to 10 times smaller than the smallest natural human chromosome. The hardy human artificial chromosomes, or "HACs," withstand repeated rounds of cell division. Since creation of the first artificial chromosomes in 1997, researchers have made even smaller ones, revealing what it takes to be a chromosome—an autonomous, nucleic acid/protein partnership that can replicate.

### Reviewing Concepts

- Chromosomes are complex structures that contain the DNA for a cell.
- The position of the centromere and the pattern of stained bands in metaphase chromosomes are useful characteristics in identifying chromosomes.
- Scientists have produced artificial chromosomes in the quest for understanding the minimal components needed to maintain the genetic material.

## 12.2 Gene Linkage on Chromosomes

At the beginning of this century, genetic researchers began building artificial chromosomes to learn the minimal requirements for these bearers of genetic information. At the start of the last century, when chromosomes were first visualized shortly after the rediscovery of Mendel's laws, they were quite mysterious.

In 1902, German biologist Theodor Boveri and U.S. graduate student Walter Sutton independently realized that chromosomes transmit inherited traits. The association of particular chromosomes with particular traits, including abnormalities, constitutes the field of **cytogenetics.** It began with the study of corn chromosomes, but today is a medical specialty.

As biologists in the early decades of the twentieth century cataloged traits and the chromosomes that transmit them in several species, it soon became clear that the number of traits far exceeded the number of chromosomes. Fruit flies, for example, have four pairs of chromosomes, but dozens of different bristle patterns, body colors, eye colors, wing shapes, and other characteristics. How might a few chromosomes control so many traits? The answer: chromosomes contain many genes that are **linked,** or inherited together, and do not assort independently, just as the cars of a train arrive at the same destination at the same time, whereas automobiles headed for the same place do not arrive exactly together. The seven traits that Mendel followed in his pea plants were transmitted on different chromosomes. Had the same chromosome carried these genes near each other, Mendel would have generated markedly different results in his dihybrid crosses. • **tracing inheritance of two genes, p. 191**

The different inheritance pattern of linked genes was first noticed in the early 1900s, when William Bateson and R. C. Punnett observed offspring ratios in pea plants that were different from the ratios Mendel's laws predicted. They looked at different traits, crossing true-breeding plants with purple flowers and long pollen grains (genotype *PPLL*) with true-breeding plants with red flowers and round pollen grains (genotype *ppll*). Then they crossed the F₁ plants, of genotype *PpLl*, with each other. Surprisingly, the F₂ generation did not show the expected 9:3:3:1 phenotypic ratio for an independently assorting dihybrid cross (**figure 12.3**).

Two types of F₂ peas—those with the same phenotypes and genotypes as the parents, *P_L_* and *ppll*—were more abundant than predicted, while the other two progeny classes—*ppL_* and *P_ll*—were far less common. Bateson and Punnett hypothesized that the prevalent parental allele combinations reflected genes transmitted on the same chromosome and the genes therefore did not separate during meiosis.

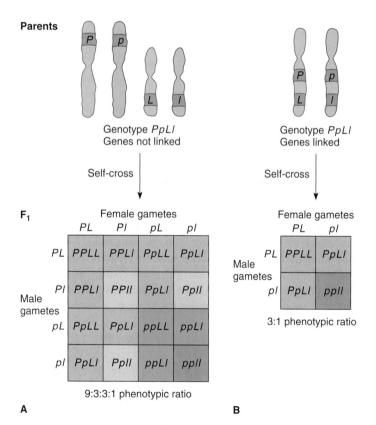

**FIGURE 12.3** **Expected Results of a Dihybrid Cross.** (**A**) When genes are not linked, they assort independently. The gametes then represent all possible allele combinations. The expected phenotypic ratio of a dihybrid cross would be 9:3:3:1. (**B**) If genes are linked on the same chromosome, two allele combinations are expected in the gametes. The expected phenotypic ratio would be 3:1, the same as for a monohybrid cross.

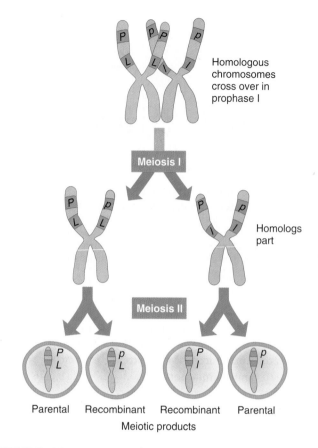

**FIGURE 12.4** **Crossing Over.** Genes linked closely on the same chromosome are usually inherited together. Linkage between two genes can be interrupted if the chromosome they are located on crosses over with its homolog at a point between the two genes. This packages recombinant arrangements of the genes into gametes.

If the two genes were on the same chromosome, why did the researchers see nonparental trait combinations at all? These offspring classes arise because of another meiotic event, crossing over. Recall that crossing over is an exchange between homologs that mixes up maternal and paternal gene combinations in the gametes (see figure 10.8). **Figure 12.4** follows the fate of alleles during crossing over.

**Recombinant chromosomes** result from the mixing of maternal and paternal alleles into new combinations in the meiotic products. **Parental** chromosomes retain the gene combinations from the parents. "Parental" and "recombinant" are relative terms, however, depending on the parents' allele combinations. Had the parents in Bateson and Punnett's crosses been of genotypes *ppL_* and *P_ll* (phenotypes red flowers, long pollen grains and purple flowers, round pollen grains), then *P_L_* and *ppll* would be the recombinant rather than the parental classes.

Two other terms describe the arrangement of linked genes in heterozygotes. Consider a pea plant with genotype *PpLl*. These alleles can be on the same chromosome in two different positions. If the two dominant alleles travel on one chromosome cell and the ability to yield an exact replica of itself to pass to daughter cells. But the recognition of DNA's vital role in life was not always as obvious as it is now.

# DNA, Not Protein, Carries Genetic Information

Swiss physician and biochemist Friedrich Miescher was the first investigator to chemically analyze the contents of a cell's nucleus. In 1869, he isolated the nuclei of white blood cells obtained from pus in soiled bandages. In the nuclei, he discovered an unusual acidic substance containing nitrogen and phosphorus. Miescher and others went on to find it in cells from a variety of sources. Because the material resided in cell nuclei, Miescher called it nuclein in his 1871 paper; subsequently, it was called a nucleic acid due to its acidic nature.

Miescher's discovery, like those of his contemporary Gregor Mendel, was not appreciated for years. Instead, most investigators researching inheritance focused on the association between inherited disease and proteins. Proteins were known to be different from species to species and held the potential for far more diversity.

In 1909, English physician Archibald Garrod was the first to associate inheritance and protein. Garrod noted that people with inherited "inborn errors of metabolism" lacked certain enzymes. Other researchers added supporting evidence: they linked abnormal or missing enzymes to unusual eye color in fruit flies and nutritional deficiencies in bread mold variants. But how do enzyme deficiencies produce traits? Experiments in

bacteria would answer the question and return, eventually, to Miescher's nuclein.

In 1928, while searching for a vaccine against pneumonia, English microbiologist Frederick Griffith inadvertently contributed the first step in identifying DNA as the genetic material. As a model, Griffith was studying pneumonia in mice caused by a bacterium, *Diplococcus pneumoniae.* He identified two types of bacteria—type S and type R. Type S bacteria form smooth colonies when grown in a Petri dish. The smooth appearance was the result of their excreting a polysaccharide capsule. When injected into mice, type S bacteria cause pneumonia. Type R bacteria form rough-shaped colonies in culture and, when injected into mice, do not cause pneumonia. Therefore, the smooth polysaccharide coat seemed to be necessary for infection. Griffith hypothesized that he could use the R strain to form a vaccine but recognized the need for the capsule from the S strain.

Griffith started by heating type S bacteria to kill them and then injected them into mice. When he used the heat-killed S bacteria, they no longer caused pneumonia. However, when he injected both the type R bacteria with a solution of heat-killed type S bacteria—neither able to cause pneumonia alone—the mice died of pneumonia (**figure 13.2**). Upon examination, Griffith was able to isolate live type S bacteria from the dead mice. Griffith made note of this interesting effect but did not attempt to explain how the R strain had been changed within the mice.

In the 1940s, U.S. physicians Oswald Avery, Colin MacLeod, and Maclyn McCarty offered an explanation as they repeated portions of Griffith's experiment. They hypothesized that something

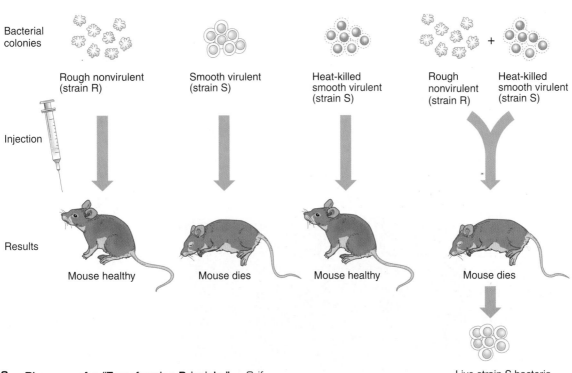

**FIGURE 13.2** **Discovery of a "Transforming Principle."** Griffith's experiments showed that a molecule in a lethal strain of bacteria can transform nonkilling bacteria into killers.

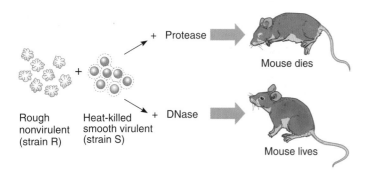

**FIGURE 13.3    DNA Is the "Transforming Principle."**    Avery, MacLeod, and McCarty identified Griffith's transforming principle as DNA. By adding enzymes that either destroy proteins (protease) or DNA (DNase) to the types of mixtures that Griffith used in his experiments, they demonstrated that DNA transforms bacteria—and that protein does not.

from the heat-killed type S bacteria entered and "transformed" the normally harmless type R strain into a killer. Was this "transforming principle" a protein? When they treated the heat-killed S strain solution with a protein-destroying enzyme (protease), the R strain was still transformed. Therefore, protein could not be responsible for transmitting the killing trait. Treating the solution from the heat-killed S bacteria with a DNA-destroying enzyme (DNase), however, did prevent the transformation. Could DNA transmit the killing trait?

Avery, MacLeod, and McCarty confirmed that DNA transformed the bacteria by isolating DNA from heat-killed type S bacteria and injecting it along with type R bacteria into mice (**figure 13.3**). The mice died, and their bodies contained active type S bacteria. The conclusion: DNA from type S bacteria altered the type R bacteria, enabling them to manufacture the smooth coat necessary to cause infection.

Biologists at first were rather hesitant in accepting DNA as the biochemical of heredity. More was known about proteins than about nucleic acids, and it was thought that protein, with its 20 building blocks, was more versatile and therefore more likely to be able to encode many more traits than DNA, with its four

types of building blocks. In 1950, U.S. microbiologists Alfred Hershey and Martha Chase showed conclusively that DNA—not protein—is the genetic material.

Hershey and Chase used a very simple system—*Escherichia coli*, a bacterium, and T4, a virus that infects bacteria (called a bacteriophage). T4 bacterial viruses consist of only a protein coat and a DNA core (**figure 13.4**). Hershey and Chase wanted to know which part of the virus controls its reproduction (referred to as replication for viruses)—the DNA or the protein coat. To track the two components, they grew one batch of viruses in the presence of radioactive sulfur. The sulfur becomes incorporated into the molecules of protein and is easy to detect. Protein contains sulfur, but DNA does not. To label the DNA, another batch of viruses was grown in the presence of radioactive phosphorus, an element found in DNA but not in protein (**figure 13.5**). (Recall that Miescher had identified phosphorus in nuclein nearly a century earlier.)

They used each type of labeled virus to infect a separate batch of bacteria and allowed several minutes for the virus particles to bind to the bacteria. Then they agitated each mixture in a blender, which knocked the remaining viruses and empty protein coats from the surfaces of the bacteria. Hershey and Chase examined the contents of the bacteria and the growth media. In the test tube containing sulfur-labeled virus, the virus-infected bacteria were not radioactive, but the fluid portion of the material in the tube was. In the other tube, where the virus contained radioactive phosphorus, the infected bacteria were radioactive, but the fluid was not. The "blender experiments" therefore showed that the only part of the virus that entered the bacteria was the DNA from the virus particles.

We now know that when the virus infects the bacterial cell, it injects its DNA, and the protein coat is left attached loosely to the bacterium (see figure 13.4). The viral DNA then takes over the bacterial cell's metabolic machinery to manufacture more virus particles. New virus particles are released when the cell bursts. By removing the protein coats from the bacteria with the blender, this team of scientists proved quite convincingly that DNA was all that was needed to direct the cell to manufacture entire virus progeny. The genetic material, therefore, was DNA

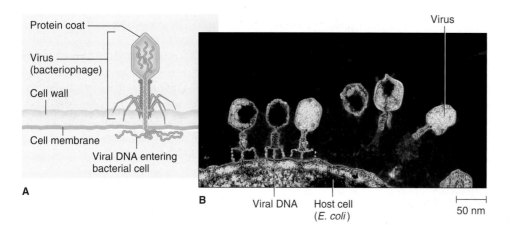

**FIGURE 13.4    A Virus (Bacteriophage) Infects a Bacterium.**    **[A]** A bacteriophage is a virus. It consists of a nucleic acid in a protein coat. The virus uses the protein coat to attach and inject its DNA into a bacterial cell. **(B)** This photo shows several bacteriophages infecting a bacterium.

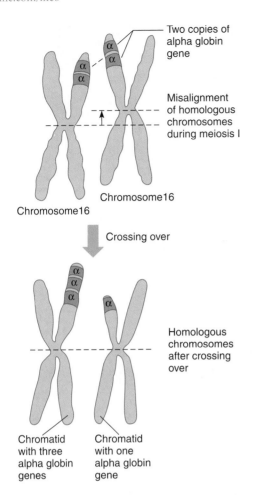

**FIGURE 13.33**   **Gene Duplication and Deletion.**    The repeated nature of the alpha globin genes makes them prone to mutation by mispairing during meiosis. A person missing one alpha globin gene can develop anemia.

two copies of the gene (**figure 13.33**). This type of mutation is responsible for color blindness.

The spontaneous mutation rate for most genes is quite low, occurring in about 1 in 100,000 bases. For studying mutants to learn how genes function, this rate is not high enough to efficiently provide the variants needed. So researchers use mutagens that can induce a mutation rate as high as 1 in 100 bases. For example, chemicals called alkylating agents alter bases in a way that causes the cell to incorporate the wrong base on the other strand. Other chemical mutagens can add or remove several nucleotides in a row. Other sources of mutations include ionizing radiation, which can break the DNA strands, and ultraviolet light, which induces thymine dimers.

Researchers test the ability of substances to induce mutation by exposing cells growing in culture. The ability of sodium nitrite in smoked meats and of certain pesticides to cause mutation was detected this way. A test called the Ames test routinely examines chemicals for the ability to induce mutation in bacteria. Because about 90% of mutagens cause cancer, this test is one

of many used to screen new drug candidates for their ability to cause cancer.

## Mutations Are Classified by Effect

From a molecular point of view, there are several types of mutations. A **point mutation** is a change in a single DNA base. Two types of point mutations are distinguished by their extent and consequences. A **missense mutation** changes a codon that normally specifies a particular amino acid into one that encodes a different amino acid. If the substituted amino acid is found at a critical location in the protein, such as the active site or a region involved in protein folding, the resulting protein will be very different in function. The phenotype will change. Sickle cell disease results from a missense mutation. A missense mutation can also greatly affect a gene's product if it alters a site controlling intron removal. The encoded protein has additional amino acids, even though the causative mutation affects only a single DNA base.

A **nonsense mutation** is a change in a single base that results in changing a codon specifying an amino acid into a "stop" codon. This shortens the protein product, which can profoundly influence the phenotype. If a missense mutation changes a normal stop codon into a codon that specifies an amino acid, the resulting protein has extra amino acids and may not function properly.

In genes, the number three is very important, because triplets of DNA bases specify amino acids. Adding or deleting bases by any number other than a multiple of three usually devastates a gene's function. It disrupts the reading frame, and, therefore, it also disrupts the sequence of amino acids. Such a change is called a **frameshift mutation.** Even adding or deleting multiples of three can alter a phenotype if an amino acid is added that alters the protein's function or a crucial amino acid is removed.

But the number three is also important because the genetic code has extra information—61 types of codons specify 20 types of amino acids. Different codons that encode the same amino acid are termed **synonymous codons,** and the genetic code is said to be degenerate because of this redundancy. This characteristic protects against the effects of mutation because many alterations of the third codon position do not alter the specified amino acid and are therefore "silent." For example, both CAA and CAG mRNA codons specify glutamine. Mutations in the second codon position often cause one amino acid to replace another with a similar conformation, which may not disrupt the protein's form drastically. For example, if a GCC codon mutates to GGC, glycine replaces alanine; both are very small amino acids.

Missing genetic material—even a single base—can greatly alter gene function. A DNA deletion can be so large that detectable sections of chromosomes are missing, or it can be so small that only a few genes or parts of a gene are lacking. Mutations can also be the result of the duplication of large sections of the DNA. In each case, the result can disrupt the formation of the correct protein. **Table 13.2** illustrates the effects of various

## TABLE 13.2

### Types of Mutation

A sentence of three-letter words can serve as an analogy to demonstrate the effects of mutations on gene sequence:

| | |
|---|---|
| Wild type | THE ONE BIG FLY HAD ONE RED EYE |
| Missense | TH**Q** ONE BIG FLY HAD ONE RED EYE |
| Nonsense | THE ONE BIG |
| Frameshift | THE ONE **Q**BI GFL YHA DON ERE DEY |
| Deletion of three letters | THE ONE BIG HAD ONE RED EYE |
| Duplication | THE ONE BIG FLY **FLY** HAD ONE RED EYE |
| Insertion | THE ONE BIG **WET** FLY HAD ONE RED EYE |
| Expanding mutation | P$_1$ THE ONE BIG FLY HAD ONE RED EYE |
| | F$_1$ THE ONE BIG FLY **FLY FLY** HAD ONE RED EYE |
| | F$_2$ THE ONE BIG FLY FLY **FLY FLY FLY FLY** HAD ONE RED EYE |

types of mutations on the final products, using an analogy of a sentence structure.

Mutations can affect somatic cells or sex cells. A **somatic mutation** usually affects only a single cell and its descendents. By contrast, a change in the DNA of a gamete, called a **germinal mutation**, is passed to every cell in the new body. A somatic mutation can result in the formation of a tumor but cannot affect every cell in the body. But the germinal mutations cause systemwide problems. Instead of a local group of cancerous cells, germinal mutations produce effects, including tumors, throughout the body.

### Reviewing Concepts

- Mutations are changes in the genetic instructions within a gene.
- Single changes in bases, called point mutations, can be invisible or produce tremendous changes in the resulting protein.
- Frameshift mutations often terminate the protein early, resulting in a loss of function for the cell.
- Other point mutations change portions of the protein and likely add to life's diversity.
- Scientists classify mutations by their cause and effect. Radiation and chemicals are the most common mutagens. But most mutations occur spontaneously through replication errors.

## Connections

The careful choreography of cell division preserves the integrity of genetic information. That information is used to manufacture proteins, which do everything for the cell, including carry out metabolism, manufacture other molecules, process signals, and so on. A change in any gene could produce disastrous results for the cell. The discovery of the relationship among genes, proteins, and traits has opened a tremendous future of possibilities. Although we still do not understand how a cell decides which genes to use, and where and when, we can do much with what we know. Chapter 14 explores some of the techniques that can alter genetic information in useful ways. As you read chapter 14, remember that DNA is just an information molecule for a cell. The functions of the whole organism arise from the functions of the cells, which arise from the genes they express.

# Student Study Guide

## Key Terms

anticodon  *243*

antiparallel  *232*

codon  *243*

complementary base
   pair  *232*

conservative  *235*

deoxyribose  *231*

dispersive  *235*

DNA polymerase  *237*

excision repair  *239*

exon  *244*

frameshift mutation  *251*

genetic code  *244*

germinal mutation  *252*

histone  *234*

intron  *243*

ligase  *237*

messenger RNA
   (mRNA)  *239*

mismatch repair  *239*

missense mutation  *251*

mutagen  *250*

mutation  *249*

nonsense mutation  *251*

nucleosome  *234*

operon  *241*

origin of replication site  *237*

photoreactivation  *238*

point mutation  *251*

promoter  *240*

purine  *232*

pyrimidine  *232*

replication fork  *237*

repressor  *240*

ribose  *231*

ribosomal RNA (rRNA)  *239*

RNA polymerase  *239*

RNA primer  *237*

semiconservative  *234*

somatic mutation  *252*

synonymous codon  *251*

transcription  *239*

transcription factor  *241*

transfer RNA (tRNA)  *239*

translation  *239*

## Chapter Summary

### 13.1  Identification of the Genetic Material

1. A historical series of elegant experiments provided the evidence needed to conclude that DNA was both necessary and sufficient to direct the functions of an entire organism.

2. DNA, not protein, is the genetic material, and any changes in the DNA will change the organism.

3. DNA has a relatively simple, yet elegant, structure that facilitates the functions of the molecule in storing genetic information.

### 13.2  The Double Helix

4. The DNA molecule is a double helix with sugar-phosphate rails and pyrimidine-purine pairs as rungs.

5. The two strands of DNA run antiparallel to one another, in what is referred to as the 5′ to 3′ direction.

6. DNA is tightly wrapped around histones so it fits within the nucleus.

### 13.3  DNA Replication: Passing on the Code

7. DNA replication makes use of the functions of several enzymes to accurately copy the information, including helicase, primase, DNA polymerase and ligase.

8. Replication begins at hundreds of origins of replication to rapidly duplicate the entire DNA in most cells.

### 13.4  DNA Repair: Keeping the Code Accurate

9. Cells contain different enzymes that repair errors such as thymine dimers or mismatched bases that arise during DNA replication.

10. Lack of repair systems can lead to serious problems for the cell and organism.

### 13.5  Gene Expression: Putting the Code into Action

11. The functions and unique characteristics of cells arise from the expression of different genes, which is controlled by regulating transcription.

12. Gene expression is controlled as proteins bind to promoters to activate or block transcription in response to cellular signals.

### 13.6  Translation: From Codons to Amino Acids

13. RNA is a multifunctional molecule that participates in protein synthesis by carrying information, matching codons to amino acids, and forming peptide bonds.

14. The introns of eukaryotic mRNA must be removed before translation to join together a set of exons as a functional gene sequence.

### 13.7  The Genetic Code: From Gene to Protein

15. The genetic code is the correspondence between a codon in the DNA or RNA and a single amino acid. It is redundant, nonoverlapping, and contains start and stop signals.

16. All organisms use the same genetic code, with a few rare exceptions in mitochondria.

17. Ribosomes use the genetic code to assemble proteins, following instructions encoded in mRNA.

18. Chaperones and other proteins help fold the protein into its final conformation, which is ultimately dictated by the amino acid sequence itself.

### 13.8  Mutation: Strange New Rabbits

19. Mutations are changes in the nucleotides or sequences within a gene resulting from altered or missing bases.

20. Mutations can be silent or cause tremendous changes in the resulting protein.

21. Frameshift mutations often terminate the protein early, resulting in a loss of function for the cell. Other point mutations change portions of the protein, and likely add to life's diversity.

22. Radiation and chemicals are the most common mutagens. But most mutations occur spontaneously through replication errors.

# What Do I Remember?

1. Describe the contributions of each of these researchers in our understanding of DNA:

   Miescher; Griffith; Hershey and Chase; Avery, MacLeod, and McCarty; Levene; Franklin and Wilkins

2. Why were the experiments of Watson and Crick so significant?

3. Describe what is meant by "base pairing" and why this is so important to the functions of DNA.

4. Translate the following mRNA sequence to its corresponding protein (remember to look for the start codon):

   CACUAACGUAAUGCCACGUUUCGGCAACAUACG-
   GUAGUAAACCG

5. Describe each of the steps that must occur during translation to produce a functional protein.

6. What would be the consequence for the cell if the anticodons were changed on half of the tRNA molecules?

7. Why are eukaryotic cells not immediately affected by a mutation in a gene?

8. Describe the functions of the lactose operon with and without lactose present.

## Fill-in-the-Blank

1. The two strands of a DNA molecule run _____ to each other.

2. The _____ are eukaryotic proteins used to package DNA.

3. The codons _____, _____, and _____ signal an end to translation.

4. DNA replication is _____ because the two daughter molecules contain half of the parent molecule.

5. A(n) _____ mutation results from the deletion of one or two bases.

6. Anything that alters DNA is a(n)_____.

## Multiple Choice

1. A 146-nucleotide piece of DNA wrapped around a set of eight organizing proteins is called a
   a. histone.
   b. nucleosome.
   c. centromere.
   d. chromosome.

2. What would result from the deletion of two bases from a codon in the middle of a gene and one base from the end of the next codon, assuming no stop codons are created?
   a. The majority of the protein would be produced correctly.
   b. One amino acid would be deleted; another would be altered.
   c. The protein might be completely unaffected.
   d. All of these would happen.
   e. None of these would happen.

3. Prior to the removal of any nucleotides, the 5′ end of each newly replicated piece of DNA contains
   a. introns.
   b. RNA.
   c. amino acids.
   d. a series of adenines.

4. A gene composed of 600 nucleotides (not counting the promoter or other control elements) would encode a protein of
   a. 300 amino acids.
   b. 600 amino acids.
   c. 200 amino acids.
   d. 1,800 amino acids.
   e. 1,200 amino acids.

5. For a protein to fold correctly, it must bind to other proteins called
   a. promoters.
   b. histones.
   c. chaperones.
   d. enhancers.

6. In the lactose operon, lactose has what effect?
   a. It binds to the repressor, activating transcription.
   b. It binds to the promoter, activating RNA polymerase.
   c. It binds to RNA polymerase, inactivating the repressor.
   d. It binds to the mRNA, blocking transcription.

Elephants can detect sounds far below the human hearing range.

## Early Warning Systems? Adaptation Has Produced Unique Abilities

The day after Christmas 2004 demonstrated to the world the deadly effect of the tsunami. Racing at nearly supersonic speeds, a wave only a foot high in the deep ocean headed toward the shores of Sri Lanka. There was no warning.

Or was there? Although hundreds of people were killed, there are few reports of equivalent numbers of animals being found dead. But dozens of reports of animals behaving strangely emerged from the devastated areas. Dogs refused to go to the beach; bats were seen flying in broad daylight; and elephants moved to higher ground, sometimes carrying terrified tourist riders with them!

Did these animals have psychic abilities? We can't be sure—but a more direct explanation may apply. It is possible that the wave produced a distinctive, low frequency sound as it moved across the ocean floor. Such sounds have been shown to travel great distances. In fact, the Savanna Elephant Vocalization Project in Kenya, Africa, has discovered that elephants can communicate over a distance of up to 2.5 km using frequencies nearly two octaves below human perception. Perhaps the elephants literally heard the wave approaching.

This may be the answer to the responses of other animals as well—but a more interesting question is how animals were able to interpret the sound (if that is what they did) as some kind of danger signal. Tsunamis might not occur within an individual animal's life span.

Did humans once have the ability to hear such sounds? In the modern world, bombarded by constant noise around us, we may no longer recognize some of nature's signals.

# 15

# The Evolution of Evolutionary Thought

# Chapter Preview

1. Biological evolution is change in allele frequencies in populations. Evolution has occurred in the past and is constant and ongoing.

2. Evolution consists of large-scale, species-level changes (macroevolution) as well as gene-by-gene changes (microevolution).

3. Before Darwin, attempts to explain life's diversity were human-centric and subjective. Lamarck was the first to propose a mechanism of evolution, but it was erroneously based on acquired rather than inherited traits.

4. Geology laid the groundwork for evolutionary thought. Some people explained the distribution of rock strata with the idea of catastrophism (abrupt changes due to natural disasters). The more gradual uniformitarianism (continual remolding of Earth's surface) became widely accepted. The principle of superposition states that lower rock strata are older than those above, suggesting a time frame for fossils within them.

5. During the voyage of the HMS *Beagle,* Darwin observed the distribution of organisms in diverse habitats (biogeography) and their relationships to geological formations. He noted that similar adaptations can lead to convergent evolution. After much thought, and considering input from other scientists, he synthesized his theory of the origin of species by means of natural selection.

6. The bountiful evidence in Darwin's treatise is organized as "one long argument." However, people who believed Earth to be young, humans to be unique, and nature to move toward perfection, had difficulty with his ideas.

7. Darwin's theory was based on the observations that populations include individuals that vary for inherited traits; that many more offspring are born than survive; and that life is a struggle for use of limited resources. According to the theory of natural selection, individuals least adapted to their environments are less likely to leave fertile offspring, and, therefore, their genes will diminish in the population over time. The genes of those better adapted to the particular environment will persist. Natural selection caused the diversification, or adaptive radiation, of finches on the Galápagos Islands.

8. Sexual selection is a form of natural selection in which certain inherited traits make an individual more likely to mate and produce viable offspring.

9. Epidemiology explores connections among genetics, evolution, and infectious diseases in populations.

10. Emerging and returning infectious illnesses and increasing resistance of bacteria to antibiotic drugs illustrate evolution in action.

11. Changes in the severity of illness or types of symptoms of viral infections can reflect evolution.

## 15.1 Evolution: Changing Forms Over Time

Evolution is genetic change in a population over time, and a population is a group of interbreeding organisms (members of a species) that live in the same area. Evolution occurs when the frequencies (percentages) of alleles change from one generation to the next. As chapters 15 through 19 will repeatedly demonstrate, evolution is a process that is ongoing and everywhere, and obvious in many ways. It serves as such a compelling conceptual framework for many observations about life that noted geneticist Theodosius Dobzhansky entitled a much-quoted article "Nothing in biology makes sense except in the light of evolution."

Evolution is a continuing process that explains the history of life on Earth, as well as the diversity of organisms today, in terms of the number of species and of variation within species. Evolution also explains the great unity of life—why organisms diverse in many ways nonetheless use the same genetic code, the same reactions to extract energy from nutrients, even the same or very similar enzymes and other proteins. Shared ancestry—that is, descent from a common ancestor—explains the similarities among species. **Natural selection**—the differential survival and/or reproductive success of individuals with particular genotypes in response to environmental challenges—accounts for much of the diversity of life. Natural selection operates on individuals, but evolution occurs in populations. The result is a planet packed with spectacular living diversity, of millions of variations of the same underlying biochemical theme.

Biological evolution includes large-scale events, such as new species appearing and existing ones dying out. Such large changes are called **macroevolution.** However, evolution also includes changes in individual allele frequencies within a population, termed **microevolution.** Often, macroevolutionary change is the consequence of accumulating microevolutionary changes. Macroevolutionary events tend to span very long time periods, whereas microevolutionary events happen so rapidly that we can sometimes observe them over periods of just a few years. Scientific investigations have shown how adaptations spread within a population of animals, demonstrating short-term microevolution. Chapters 16 and 17 explore microevolution and macroevolution in more detail. The discussion of bacterial antibiotic resistance at the chapter's end provides another compelling example of modern evolution.

## Evolution Explains the Diversity of Organisms

The term **evolution** has been around for a very long time (**figure 15.1**), although it has had many meanings. Up to the mid-1700s, most people used the term to describe the process of development from embryo to adult. Greek philosopher Aristotle (384 BC to 322 BC) applied the term to mean the progression from imperfection to perfection based upon innate potential. Although Aristotle recognized that all organisms are related in a hierarchy of simple to complex forms, he believed that all organisms were created with an infinite potential to become a perfect version of that particular type or "essence." By "essence," Aristotle meant that all members of a species were identical in form and capacity. This idea influenced scientific thinking for nearly 2,000 years.

Several ideas, not original with Aristotle, were also considered to be fundamental principles of science well into the 1800s. Among them was the concept of a "special creation," the sudden appearance of organisms on Earth. People believed that this creative event was planned and purposeful, that species were fixed and unchangeable, and that Earth was relatively young (some still believe this). The idea of a special creation also implied there could be no extinctions. Scientists studying nature believed they were studying the work of God as seen in the laws and diversity of life. Belief in a creator, however, is not the hypothesis-testing of science.

As scientists observed the many different varieties of organisms, they struggled to reconcile the concept of fixity of species with compelling evidence that species could in fact change. Fossils, discovered at least as early as 500 BC, were at first thought to be accidents of nature, oddly shaped crystals or faulty attempts at life that arose spontaneously in rocks. By the mid-1700s, the increasingly obvious connection between organisms and fossils argued against the idea that fossils were accidents or only coincidentally resembled life. To explain how fossils came to be, yet not deny the role of a creator, scientists suggested that fossils represented organisms killed during the biblical flood. A problem, though, was that some of the fossils depicted organisms not seen before. Since people believed firmly that species created by God could not become extinct, these fossils presented a paradox. The conflict between ideology and observation widened as geologists discovered that different rock layers revealed different groups of fossils, all now extinct. How could this be?

In 1749, French naturalist Georges Buffon (1707–1788) became one of the first to openly suggest that species were changing. According to his early writings, new species were degenerate forms of existing species. An ass was just a less perfect version of a horse, according to Buffon's reasoning. All other species could, therefore, have come from one "inner form." He was quite wrong, but he did suggest that species change, a radical idea at the time. By moving the discussion into the public, he made possible a new consideration of evolution and its causes from a scientific point of view. Buffon also suggested that Earth would be much older than a few thousand years based upon how long it would take to cool after moving away from the sun.

## The Earth's History Provides Clues About Ancient Life-Forms

In the early 1700s and 1800s, geology attracted more scientists than did biology, and much of the study of nature focused on geology. Most scientists of the time thought that rapid, catastrophic events, such as floods and earthquakes, were responsible for most geological formations. In 1785, physician James Hutton (1726–1797) proposed that the forces that formed the earth acted in a gradual, yet uniform, way, producing profound changes over time. For instance, rain has a small effect in a single day, but runoff created the Grand Canyon over millions of years. His ideas became known as **uniformitarianism,** but were not widely accepted at the time.

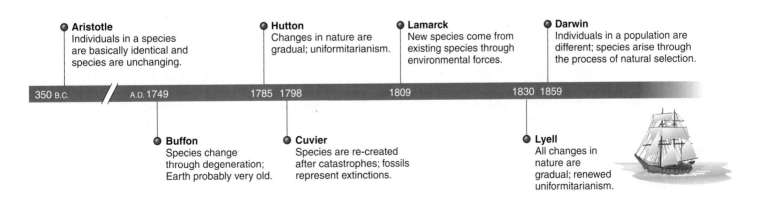

**Aristotle**
Individuals in a species are basically identical and species are unchanging.

**Hutton**
Changes in nature are gradual; uniformitarianism.

**Lamarck**
New species come from existing species through environmental forces.

**Darwin**
Individuals in a population are different; species arise through the process of natural selection.

350 B.C.     A.D. 1749     1785   1798      1809      1830   1859

**Buffon**
Species change through degeneration; Earth probably very old.

**Cuvier**
Species are re-created after catastrophes; fossils represent extinctions.

**Lyell**
All changes in nature are gradual; renewed uniformitarianism.

**FIGURE 15.1** **Science Evolves.** Significant contributions were made by many scientists, over many years, to develop the foundation Darwin used to describe natural selection as the mechanism for evolution.

**FIGURE 15.2   Rock Layers Reveal Earth History and Sometimes Life History.**    **(A)** Layers, or strata, of sedimentary rock formed from sand, mud, and gravel deposited in ancient seas. The rock layers on the bottom are older than those on top. Rock strata sometimes contain fossil evidence of organisms that lived (and died) when the layer was formed and provide clues about when the organism lived. **(B)** Sediment layers are visible along the Grand Canyon. Hiking there is like taking a journey through time. Although the rim now rises over 2,000 meters (6,500 feet) above sea level, it has repeatedly been submerged and uplifted.

The father of paleontology, Georges Cuvier (1769–1832), was an outspoken opponent of uniformitarianism. He challenged making predictions from limited data, which he believed Hutton had done. Instead, Cuvier's observations of the action of water on coastlines convinced him that a series of catastrophes had caused the geographic changes, an idea called **catastrophism.** Profound changes on a geologic scale must, Cuvier argued, occur abruptly through actions such as volcanic eruption. Cuvier also described the anatomical similarities among organisms. Because of his knowledge of anatomy, he was able to identify many fossils from just a few bones. He was also the first to recognize that older, simpler fossils appeared at the lower layers of rock (**figure 15.2**)—a concept known as the **principle of superposition.** Although he had to accept that certain species must have become extinct, he refused to believe that they were not originally formed through creation. Applying geology to biology, he argued that catastrophes would destroy most of the organisms in an area, but then new life would be created or arrive from surrounding areas. To Cuvier, each rock layer represented a separate destructive event, followed by influx of new life.

In 1809, French taxonomist Jean-Baptiste de Lamarck (1744–1829) proposed a radical new theory of the formation of new species. Once fossils were recognized as evidence of extinct life, it became clear that species could in fact change. Still, no one had proposed how this might happen. Lamarck hypothesized that the environment could exert a powerful influence on populations, guiding changes. Whereas Cuvier had worked mostly with fossils of complex animals, such as elephants, Lamarck studied invertebrates, such as clams and mussels. By comparing fossils with living examples of these animals, Lamarck made a breakthrough in thinking—he con-cluded that one species becomes extinct by becoming a new, different species.

Lamarck reasoned that organisms that used one part of their body repeatedly would increase their abilities, very much like weight lifters developing strong arms. He proposed that the resulting changes in individuals would give them the ability to get more food in a changing environment. Other individuals would die if they did not similarly change. But to explain how those traits were passed to the next generation, Lamarck applied the (then) accepted theory of the inheritance of acquired characteristics. To illustrate his point, he suggested that the sons of a blacksmith were born strong because the father worked so much with his arms. The mechanism is absurd in light of what we know today about genetics, but Lamarck was the first to propose a mechanism for evolution and to suggest that animals could change or even become extinct in response to their interactions with their environment.

Geologist Charles Lyell (1797–1875) renewed the argument for uniformitarianism in 1830 in a lengthy work on principles of geology, suggesting that natural processes are slow and steady. One obvious conclusion from his contribution is that gradual changes in some organisms could be represented in successive fossil layers. Although many scientists at the time held to the views of Cuvier, Lyell was so persuasive that some people began to support the idea of gradual geologic change. With these new theories and ideas, people were beginning to accept the concept of evolution but could not understand how it could result in the formation of new species. Charles Darwin was schooled in geology and became convinced that Lyell's explanations of geologic change were correct. Ultimately, Darwin recognized their application to the changing diversity of life on Earth.

## Dogs and Cats—Products of Artificial Selection

The pampered poodle and graceful greyhound may win in the show ring, but they are poor specimens in terms of genetics and evolution. Human notions of attractiveness can lead to bizarre breeds that may never have evolved naturally. Behind carefully bred traits lurk small gene pools and extensive inbreeding—all of which may harm the health of highly prized and highly priced show animals. Purebred dogs suffer from more than 300 types of inherited disorders.

The sad eyes of the basset hound make this dog a favorite in advertisements, but these runny eyes can be quite painful (**figure 16.A, top right**). Short legs make the dog prone to arthritis, the long abdomen encourages back injuries, and the characteristic floppy ears often hide ear infections. The eyeballs of the Pekingese protrude so much that a mild bump can pop them out of their sockets. The tiny jaws and massive teeth of pugs and bulldogs cause dental and breathing problems, as well as sinusitis, bad colds, and their notorious "dog breath" (**figure 16.A, top left**). Folds of skin on their abdomens easily become infected. Larger breeds, such as the Saint Bernard, have bone problems and short life spans. A Newfoundland or a Great Dane may suddenly die at a young age, its heart overworked from years of supporting a large body (**table 16.1**).

We artificially select natural oddities in cats, too. One of every 10 New England cats has six or seven toes on each paw, thanks to a multi-toed ancestor in colonial Boston (**figure 16.A, bottom left**). Elsewhere, these cats are quite rare. The sizes of the blotched tabby populations in New England, Canada, Australia, and New Zealand correlate with the time that has passed since cat-loving Britons colonized each region. The Vikings brought the orange tabby to the islands off the coast of Scotland, rural Iceland, and the Isle of Man, where these feline favorites flourish today.

A more modern breed appealing to cat fanciers is the American Curl cat, whose origin is traced to a stray female who wandered into the home of a cat-loving family in Lakewood, California, in 1981. This cat passed her unusual, curled-up ears to kittens in several litters (**figure 16.A, bottom right**). A dominant gene that makes extra cartilage grow along the outer ear causes the trait. Cat breeders attempting to fashion this natural peculiarity into an official show animal are hoping that the gene does not have other, less lovable effects. Cats with floppy ears, for example, are known to have large feet, stubbed tails, and lazy natures.

All these examples make one genetic truth clear: you may be able to breed desired characteristics into a dog or cat, but you can't always breed other traits out.

Bulldog

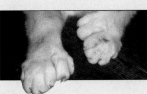
Polydactylous cat

Basset hound

American Curl cat

**FIGURE 16.A    Pets Exhibit a Variety of Unusual Traits Due to the Genetic Variations Selected Through Breeding.**    The bulldog (*top left*) was selected for its flattened face and fierce demeanor. Some traits are simply attractive or unusual, such as curled ears (*bottom right*). Multi-toed cats (*bottom left*) make better mousers, and the mournful expression of the basset hound (*top right*) accompanies its heightened sense of smell. All of these traits originally occurred as natural genetic variation.

### TABLE 16.1
### Purebred Plights

| Breed | Health Problems |
|---|---|
| Cocker Spaniel | Nervousness |
| | Ear infections |
| | Hernias |
| | Kidney problems |
| Collie | Blindness |
| | Bald spots |
| | Seizures |
| German shepherd | Hip dysplasia |
| Golden retriever | Lymphatic cancer |
| | Muscular dystrophy |
| | Skin allergies |
| | Hip dysplasia |
| | Absence of one testicle |
| Great Dane | Heart failure |
| | Bone cancer |
| Labrador retriever | Dwarfism |
| | Blindness |
| Shar-pei | Skin disorders |

bottleneck. The South African cheetahs are so genetically alike that even unrelated animals can accept skin grafts from each other. Researchers attribute the genetic uniformity of cheetahs to two bottlenecks—one that occurred at the end of the most recent ice age, when habitats changed drastically, and another when humans slaughtered many cheetahs during the nineteenth century. The loss of genetic diversity among the cheetahs presents a potential disaster: A single change in the environment might doom them all.

## Reviewing Concepts

- Genetic drift is the random variation in allele frequencies due to chance events.
- The founder effect occurs when a few individuals start a new population isolated from the original one, eliminating alleles not present in the founding group.
- The bottleneck effect eliminates alleles from a population without selection, but rather due to natural disasters.

# 16.5    Mutation and Variability

Recall from chapter 13 that a mutation is a change in an organism's DNA. Mutations are the raw material for evolution, because natural selection acts on phenotypes, and phenotypes arise from genes. Sexual reproduction ensures that a population includes individuals with different phenotypes. Should an environmental condition harm individuals of one particular phenotype, others may survive. Mutations help to provide this variability. The genetic makeup of populations, and ultimately species, changes as natural selection permits differential survival of variants that are adapted to a particular environment. For example, bacterial populations become resistant to antibiotics by a mutation that arises at random in one individual. As bacteria without the mutation die, the mutant thrives and reproduces. Soon the entire population consists of mutants.

Most mutations are neither beneficial nor useful, with no effect on phenotype; some are harmful (deleterious), resulting in defects in protein production that can lead to disease. Most of these harmful genetic traits are recessive, and, therefore, the alleles persist in a population through heterozygotes. In evolutionary terms, the deleterious alleles constitute a **genetic load** that may be the target of some future natural selection on a given population. The potential protection that genetic diversity offers is why inbreeding is so detrimental. As related individuals mate, heterozygosity is diminished. Some mutations are beneficial. For example, mutations in the genes that encode certain receptors on T cells in humans protect against HIV infection.

## Reviewing Concepts

- Deleterious alleles in a population constitute the genetic load and arise from mutation or are perpetuated in heterozygotes.
- Mutations provide the raw material for evolution by introducing new genetic variants.

# 16.6    Natural Selection as Evolutionary Force

Allele frequencies may change in response to environmental change. This is natural selection in action. Different types of natural selection are distinguished by their effects on phenotypes (**figure 16.4**).

## Selection Acts in Different Ways

In **directional selection,** a changing environment selects against one phenotype, allowing another to gradually become more prevalent. For example, populations of approximately 100 insect species have undergone color changes enabling them to blend into polluted backgrounds. This adaptive response of darkening is called **industrial melanism.** The rise of antibiotic resistance among infection-causing bacteria also reflects directional selection, as does the increase in pesticide-resistant plants.

In **disruptive selection** (sometimes called diversifying selection), two extreme expressions of a trait are the most fit, and both come to predominate. For example, in a population of marine snails that live among tan rocks encrusted with white barnacles, the animals near the barnacles are white and camouflaged, and those on the bare rock are tan and likewise blend in. The snails that are not white or tan or that lie against the opposite-colored background are more often seen and eaten by predatory shorebirds.

In a third form of natural selection, called **stabilizing selection,** extreme phenotypes are less adaptive, and an intermediate phenotype has greater survival and reproductive success. Human birth weight illustrates this tendency to stabilize. Newborns who are under 5 pounds (2.27 kilograms) or over 10 pounds (4.54 kilograms) are less likely to survive than babies weighing between 5 and 10 pounds.

## Balanced Polymorphism Reflects a Homeostatic State

Stabilizing selection can result in **balanced polymorphism,** the maintenance of a genetic disease in a population even though the illness usually diminishes the fitness of affected individuals. (Polymorphism means "multiple forms" and refers specifically to genetic variants.) The inherited disease persists because carriers (heterozygotes) have some health advantage over individuals who have two copies of the wild type allele. Balanced polymorphism explains some fascinating links between certain inherited and infectious diseases. **Figure 16.5** depicts the following example, and **table 16.2** lists some others.

**Sickle Cell Disease and Malaria** Sickle cell disease is an autosomal recessive disorder that causes anemia; joint pain; a swollen spleen; and frequent, severe infections. Homozygous individuals usually do not feel well enough, or live long enough, to reproduce. Sickle cell disease carriers, who do not have symp-

toms, are resistant to malaria, which is an infection by any of four species of the protistan genus *Plasmodium.* Mosquitoes transmit the parasite to humans, causing the agonizing cycle of severe chills and fever of malaria. • **sickle cell disease, p. 249**

Discovering the sickle cell-malaria link took clever medical sleuthing. In 1949, British geneticist Anthony Allison found that the frequency of sickle cell carriers in tropical Africa was unusually high in regions where malaria raged. Blood tests of children hospitalized with malaria revealed that nearly all of them were homozygous for the wild type allele, giving them normal blood cells. The few sickle cell carriers among them had the mildest

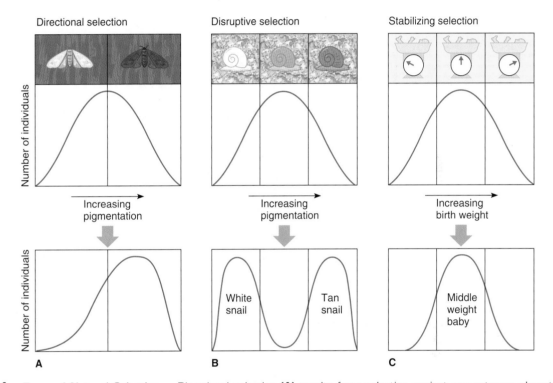

**FIGURE 16.4    Types of Natural Selection.**     Directional selection (**A**) results from selection against one extreme phenotype. In disruptive selection (**B**), two extreme phenotypes each have a selective advantage, and both persist. Stabilizing selection (**C**) maintains an intermediate expression of a trait by selection against extreme variants.

### TABLE 16.2

## Balanced Polymorphism

| Person Who Has or Carries | Is Protected From | Possibly Because |
|---|---|---|
| Cystic fibrosis | Diarrheal disease | Carriers have too few functional chloride channels in intestinal cells, blocking toxin |
| G6PD deficiency | Malaria | Red blood cells inhospitable to malaria parasite |
| Phenylketonuria (PKU) | Spontaneous abortion | Excess amino acid (phenylalanine) in carriers inactivates ochratoxin A, a fungal toxin that causes miscarriage |
| Sickle cell disease | Malaria | Red blood cells inhospitable to malaria parasite |
| Tay-Sachs disease | Tuberculosis | Unknown |
| Non-insulin-dependent diabetes mellitus | Starvation | Associated tendency to gain weight once protected against starvation during famine |

Distribution of malaria, 1920s

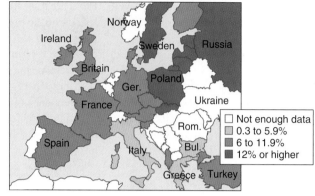

Distribution of AIDS-resistant gene

☐ Not enough data
☐ 0.3 to 5.9%
☐ 6 to 11.9%
■ 12% or higher

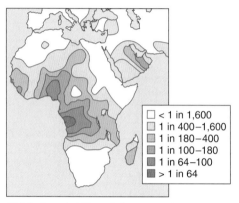

☐ < 1 in 1,600
☐ 1 in 400–1,600
☐ 1 in 180–400
☐ 1 in 100–180
☐ 1 in 64–100
■ > 1 in 64

Distribution of sickle cell disease carriers

**A**

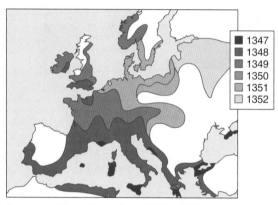

■ 1347
■ 1348
■ 1349
■ 1350
☐ 1351
☐ 1352

Progress of black plague from 1347 to 1352

**B**

**FIGURE 16.5 Balanced Polymorphism.** Being a carrier for a specific inherited illness can protect against another type of condition. Maps that compare frequency distributions of both disorders are evidence of this phenomenon. **(A)** The classic example of balanced polymorphism is sickle cell disease heterozygosity protecting against malaria. **(B)** Evidence is starting to accumulate for a similar relationship between people who have one or two alleles conferring resistance to HIV infection and protection against bubonic plague, an infectious disease that ravaged Europe in the Middle Ages.

cases of malaria. Was malaria somehow selecting against the wild type allele? In the United States, where malaria is very rare, sickle cell disease is also less common, suggesting a relationship between the two conditions.

Further historical evidence supports the hypothesis that being a sickle cell carrier in a malaria-ridden environment confers a selective advantage. The rise of sickle cell disease parallels cultivation of crops that provide breeding grounds for the malaria-carrying *Anopheles gambiae* mosquito. About 1000 BC, sailors from southeast Asia traveled in canoes to East Africa, bringing new crops of bananas, yams, taros, and coconuts. When the jungle was cleared to grow these crops, mosquitoes came, offering a habitat for the malaria parasite in the early part of its life cycle.

When an infected mosquito feeds on a human who does not have or carry sickle cell disease, the malaria parasite enters the red blood cells, eventually bursting them and releasing the parasite throughout the body. In all the red blood cells of a person with sickle cell disease, and in about half of the cells of a carrier, the abnormal beta globin chains adhere to one another to form aggregates that bend the cell into the characteristic sickle shape, which also thickens the blood (see figure 13.31). For a reason still unknown, the sickled cells are inhospitable to the parasite.

When malaria first invaded East Africa, sickle cell carriers, who remained healthier, had more children and passed the protective allele to approximately half of them. Gradually, over 35

generations, the frequency of the sickle cell allele in East Africa rose from 0.1% to 45%. However, whenever two carriers produced a child who suffered from sickle cell disease—a homozygote—the child paid the price for this genetic protection.

A cycle set in. Settlements with many sickle cell carriers escaped debilitating malaria. Their residents were therefore strong enough to clear even more land to grow food—and support the disease-bearing mosquitoes. Even today, sickle cell disease is more prevalent in agricultural societies than among people who hunt and gather their food.

Given the large number of genes in any organism, the changes in allele frequencies that constitute microevolution must occur nearly all the time. **Figure 16.6** summarizes these changes. Chapter 17 considers the easier-to-see macroevolutionary changes of speciation and extinction.

## Reviewing Concepts

- Natural selection may favor one phenotype, two extreme phenotypes, or an intermediate phenotype.
- Balanced polymorphism maintains a deleterious allele by giving the heterozygote some advantage.
- Usually, this advantage is resistance to a specific, usually infectious, illness.

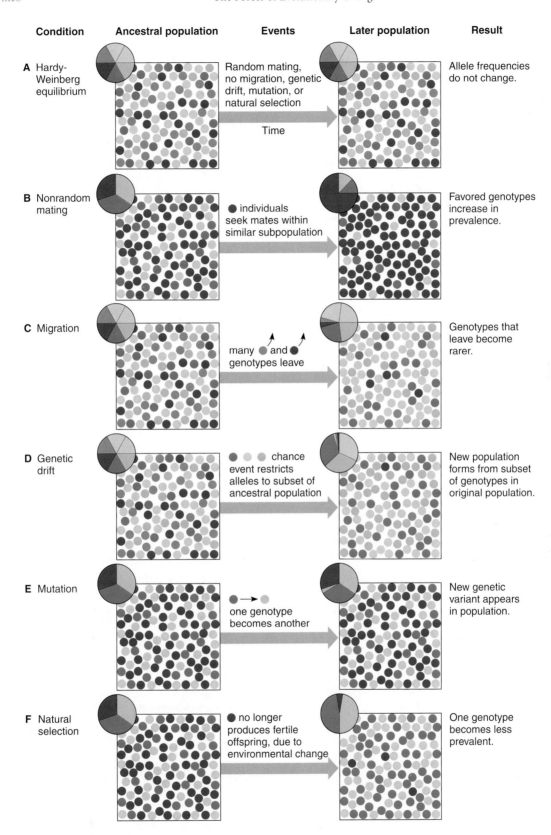

| Condition | Ancestral population | Events | Later population | Result |
|---|---|---|---|---|
| **A** Hardy-Weinberg equilibrium | | Random mating, no migration, genetic drift, mutation, or natural selection

Time | | Allele frequencies do not change. |
| **B** Nonrandom mating | | ● individuals seek mates within similar subpopulation | | Favored genotypes increase in prevalence. |
| **C** Migration | | many ● and ● genotypes leave | | Genotypes that leave become rarer. |
| **D** Genetic drift | | ● ● ● chance event restricts alleles to subset of ancestral population | | New population forms from subset of genotypes in original population. |
| **E** Mutation | | ● → ● one genotype becomes another | | New genetic variant appears in population. |
| **F** Natural selection | | ● no longer produces fertile offspring, due to environmental change | | One genotype becomes less prevalent. |

**FIGURE 16.6** **Factors That Alter Allele Frequencies and Thereby Contribute to Evolution.** **(A)** In Hardy-Weinberg equilibrium, allele frequencies stay constant. **(B)** Nonrandom mating increases some allele frequencies and decreases others because individuals with certain phenotypes are more attractive to the opposite sex. **(C)** Migration removes alleles from, or adds alleles to, populations. **(D)** Genetic drift samples a portion of a population, altering allele frequencies. **(E)** Mutation creates new alleles. **(F)** Natural selection operates when environmental conditions prevent individuals of certain genotypes from reproducing successfully.

# Connections

Mutation provides the raw material for evolution to act upon. But that is just the start. There are several different ways for those alleles to become more and more prevalent in the population. In fact, the Hardy-Weinberg equations, and conditions for equilibrium, provide an obvious conclusion that evolution is not only possible, but highly probable. The conditions are nearly impossible to meet to achieve equilibrium. This view helps us to see that nature is constantly in flux in fundamental ways. The diversity in each population becomes not just a nice thing to view, but a necessity for survival of the species and the formation of new ones.

# Student Study Guide
# Key Terms

| | | | |
|---|---|---|---|
| balanced polymorphism  *294* | founder effect  *291* | genetic load  *294* | population  *288* |
| cline  *291* | gene flow  *288* | Hardy-Weinberg | population bottleneck  *292* |
| directional selection  *294* | gene pool  *288* |   equilibrium  *288* | stabilizing selection  *294* |
| disruptive selection  *294* | genetic drift  *291* | industrial melanism  *294* | |

# Chapter Summary

## 16.1  The Inevitability of Evolution

1. Evolution occurs at the population level as gene (allele) frequencies change.

2. Algebra can be used to represent Hardy-Weinberg equilibrium, the unlikely situation in which microevolution does not occur.

## 16.2  Mate Choice and Evolution: Some Rabbits Reproduce More Than Others

3. Nonrandom mating is one way for evolution to occur through sexual selection.

4. Behavior and migration change allele frequencies.

5. Random changes in gene pools cause some alleles to become more common or less common in future generations.

## 16.3  The Effects of Migration on Population Allele Frequencies

6. Gene flow introduces new alleles into populations and disrupts genetic equilibrium by removing others.

7. Migration provides some of the raw material for evolution to act upon.

8. Geographical barriers create clines—genetic variation across a changing landscape.

## 16.4  Genetic Drift

9. Genetic drift is the random variation in allele frequencies due to chance events.

10. The founder effect occurs when a few individuals start a new population isolated from the original one, eliminating alleles not present in the founding group.

11. The bottleneck effect eliminates alleles from a population without selection, but rather due to natural disasters.

## 16.5  Mutation and Variability

12. Deleterious alleles in a population constitute the genetic load and arise from mutation or are perpetuated in heterozygotes.

13. Mutations provide the raw material for evolution by introducing new genetic variants.

## 16.6  Natural Selection as Evolutionary Force

14. Natural selection may favor one phenotype, two extreme phenotypes, or an intermediate phenotype.

15. Balanced polymorphism maintains a deleterious allele by giving the heterozygote some advantage, such as resistance to a specific, usually infectious, illness.

**FIGURE 18.1    A Gallery of Spectacular Fossils.    (A)** *Archaefructus* is one of the earliest fossils of flowers and fruits, from the Liaoning Province of northeast China about 138 million years ago. **(B)** This emu (*Genyornis newtoni*) eggshell from Lake Eyre, central Australia, is unchanged chemically from its original state, from about 14,000 years ago. **(C)** A ground sloth, *Nothrotheriops shastensis*, dropped this coprolite—fossilized excrement—about 20,000 years ago in a cave in Las Vegas.

A

B

C

More meaningful than an evolutionary tree diagram based on apparent similarities is a **cladogram,** which is another type of treelike diagram built using specific features common to only one group of organisms. The group is called a **clade,** and the distinguishing feature is a **derived character.** A clade indicates monophyletic evolution, or a single pathway. For example, the presence of feathers is a trait suitable as the basis of a clade because they are found only among birds. Not useful is a trait such as the flippers of a penguin and a porpoise, which arise from different structures (a wing and a leg, respectively) and are adaptations to swimming.

**Figure 18.2** depicts the subtle distinction between a traditional evolutionary tree diagram and a cladogram, highlighting vertebrates (animals with backbones). A traditional tree is based on the overall similarities among organisms, whether or not traits are derived. The traditional evolutionary tree clearly separates mammals and birds from reptiles. In contrast, the cladogram places them on a continuum that includes reptiles.

To construct a cladogram, a researcher begins by selecting traits that are of evolutionary import—that is, traits that probably reflect descent from a shared ancestor. As an analogy, consider a class of 24 students on a Monday morning. Six of them show up wearing identical T-shirts from a concert the night before. It is more likely that those students got their shirts from a recent shared experience—the concert—than that they all just happened to pick out the same shirts from vast collections to wear on the same day. Similarly, shared characters probably reflect a shared origin—the most logical explanation.

The next step is to make a chart listing which organisms under consideration have which traits. Then a tree is built, with those species sharing the most derived characters occupying the branches farthest from the root branch. The nodes indicate where two new groups arise from a common ancestor. **Figure 18.3** demonstrates how to construct a cladogram. It depicts key dinosaurs on the path to birds, using derived characters that are important in the acquisition of flight. We will return to this cladogram later in the chapter.

A problem with cladistics is that the diagrams can become enormously complicated when many species and derived characters are used. Mathematically, several trees can accommodate any one data set. Computers are used to derive trees, and the tree that

## Hippos and Whales—Closer Relatives Than We Thought?

Classifying cetaceans—a group that includes whales, porpoises, and dolphins—has always been a problem. The cetaceans have many adaptations to life in the water that complicate comparisons to terrestrial vertebrates, and according to fossil evidence, they went from land to water very rapidly, in just a few million years. About a century ago, biologists placed the cetaceans closest to the ruminants (hoofed, grazing mammals such as deer, cows, sheep, and giraffes), based on skeletal similarities. The ruminants are part of a larger group, the artiodactyls, or even-toed ungulates, that also includes hippos, pigs, peccaries, camels, and llamas. All of these mammals first appeared in modern form about 48 to 50 million years ago.

The most defining feature of the artiodactyls is a very mobile heel joint, something that cannot be observed in the legless cetaceans. The axis of symmetry that splits artiodactyl feet is also nonexistent in the footless whales, porpoises, and dolphins. Artiodactyls are also classified by a triple row of cusps (raised areas) on the back molars, a trait that may have disappeared in the cetaceans because of dietary differences from their land-dwelling relatives. The traditional family tree, based largely on these superficial resemblances, placed the cetaceans closest to the ruminants, and then most closely related to the hippos, which, in turn, were closest to pigs and peccaries (**figure 18.A**). In commonsense terms, a hippo looks more like a pig than it does a porpoise!

But molecules told a different story. Starting in 1994, evidence began to accumulate that showed striking biochemical similarities between hippos and cetaceans. DNA sequences from the nucleus and the mitochondria support a tree that places the cetaceans, ruminants, and hippos as one

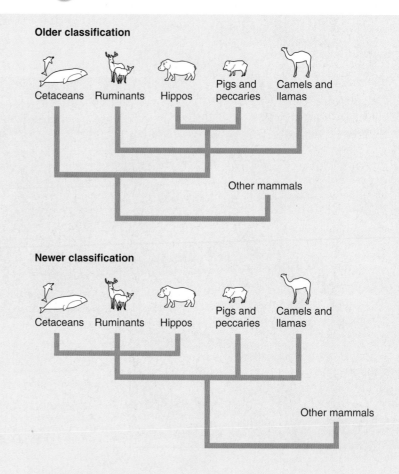

**FIGURE 18.A  Classification Based on Molecules.**  Molecular evidence led to reconsideration of the relationships among hippos, pigs, and whales. Several types of molecular measures indicate that cetaceans (whales, dolphins, and porpoises), ruminants (deer, cows, and sheep), and hippos form a monophyletic group, or clade.

clade, or monophyletic group. This means that they are set off together by at least one derived character. Very definitive molecular evidence came in 1997 in the form of highly distinct DNA sequences called "short interspersed elements," or SINES for short. These sequences are so species-specific that it is very unlikely for them to appear in different species by chance alone—inheritance from a shared ancestor is much more probable. Nearly identical SINES are in the

genomes of cetaceans, ruminants, and hippos—but notably absent in the genomes of pigs, peccaries, camels, and llamas.

As the molecular evidence continues to mount that hippos are more closely related to whales than to the similar-appearing pigs, paleontologists and evolutionary biologists are looking for clues—fossils of the animals that gave rise to the cetaceans—that could explain the apparent contradiction between the two types of trees.

that simple animals may have been abundant a billion years ago, as the sequence data suggest, but left few fossils from that time. The recent discoveries of never-before-seen microscopic algae and animal embryos from south-central China (see figure 18.7E) predating the Cambrian explosion suggest the molecular data were indeed on the right track.

As the stories of life told in the rocks and in the sequences of informational biomolecules accumulate, converge, and overlap, we will learn more about where and how life began, the subject of chapter 19.

### Reviewing Concepts

- There are disagreements between DNA data and fossil evidence.
- It is important to consider all types of evidence, and their limitations, when drawing conclusions.
- Taken together, they show how all life is related.

## Connections

We have learned much about the natural world since the days of Darwin and others who had little more than fossils and their imaginations to work with. The more data we acquire, the more we see the grand tapestry of life on this planet and how all species are intimately connected. We gain a great appreciation for the resilience, but also the delicacy, of life. While we cannot go back in time and visit the dinosaurs, we can use new molecular techniques to understand more and more what they were like and how they relate to life today. Concepts of physics and astronomy begin to apply to genetics, and genetics helps us to understand geology. The fields of science, once functioning in relative isolation, are becoming more and more dependent upon each other. In many ways, evolution has become a grand unifying theme of science.

## Student Study Guide

## Key Terms

| | | | |
|---|---|---|---|
| absolute dating *331* | derived character *323* | impression fossil *327* | relative dating *331* |
| analogous *333* | geological timescale *327* | molecular clock *339* | synteny *337* |
| clade *323* | half-life *331* | paleontology *322* | systematics *322* |
| cladogram *323* | highly conserved *336* | petrifaction *328* | vestigial *334* |
| compression fossil *327* | homologous *333* | radiometric dating *331* | |

## Chapter Summary

### 18.1 Reconstructing the Stories of Life

1. Evolutionary biologists assemble clues from fossils as well as from structures and informational molecules in modern species to paint portraits of past life.
2. Cladograms are diagrams that show the relationship of one species to another based on sets of shared characteristics.

### 18.2 Fossil Evidence

3. Fossils, formed in a variety of ways, can reveal the immediate environment of an ancient organism, clues to a prehistoric location, or even global change.

4. The age of a fossil can be estimated by comparison of its position in rock strata or more definitively by measuring the breakdown of incorporated radioactive isotopes.

### 18.3 Structural Evidence

5. Comparing bones in various dinosaurs and birds vividly illustrates evolutionary change.
6. Homologous structures are retained from a shared ancestor, whereas analogous structures reflect adaptation to a similar environment.
7. Vestigial organs and similarities among embryos are also evidence of evolution.

### 18.4 Molecular Evidence

8. Molecular evidence for evolution includes similarities at the gene, protein, chromosomal, and genome levels.

9. Gene sequence differences among species can be placed in a time frame derived from mutation rates.

10. Molecular data can be used as a "clock" to measure the differences among species and how long ago they diverged.

### 18.5 Reconciling the Evidence

11. Since there are disagreements between DNA data and fossil evidence, it is important to consider all types of evidence.

12. Taken together, there is significant evidence that all life is related.

# What Do I Remember?

1. What types of information supplement fossils in investigating evolutionary relationships?
2. How is a cladogram constructed?
3. What information can fossils reveal about life?
4. Distinguish between relative and absolute dating of fossils.
5. What is the best explanation for the presence of vestigial organs and homologous structures?

**Fill-in-the-Blank**

1. _____ is the study of the evolutionary relationship of organisms.
2. A(n)_____ is an anatomical remnant that does not seem to function any longer.
3. A(n)_____ is the common link present in all members of a clade.
4. _____ evolution contains no branching off, but a linear progression from one species to another.
5. The process of _____ produces a fossil as mineral replaces soft tissues of trees and plants.

**Multiple Choice**

1. Which of these statements is true concerning fossils?
   a. All fossils are made in the same fashion.
   b. All fossils represent actual living organisms.
   c. Fossils usually represent rare organisms.
   d. Soft body tissues are never captured in fossilization.
   e. Fossils do not necessarily mean that the organisms once lived on Earth.

2. The bones in the human arm contain the same anatomical arrangement as those in the wing of a bat. This is an example of
   a. a vestigial organ.
   b. analogous structures.
   c. homologous structures.
   d. allopatric speciation.
   e. convergent evolution.

3. Suppose you find a mummified cat in your backyard. An analysis of the mummy shows that there is 1/8 of the $^{14}C$ that would be found in a live cat. How long ago did your mummified cat die?
   a. 2,865 years
   b. 716 years
   c. 45,840 years
   d. 15,041 years
   e. impossible to tell from this data

4. Which of the following is an accurate description of what is meant by a "molecular clock"?
   a. using changes in molecules to determine how long ago two species diverged
   b. using DNA and protein sequences to predict what functions an organism has
   c. determining how long ago an organism died by changes in its DNA
   d. a clock constructed entirely of biological molecules to tell time
   e. none of these

5. If two organisms possess a high degree of synteny, we can conclude that
   a. they were the same species not so long ago.
   b. they were never related to one another.
   c. they share only one or two genes found in all organisms.
   d. they cannot successfully reproduce.

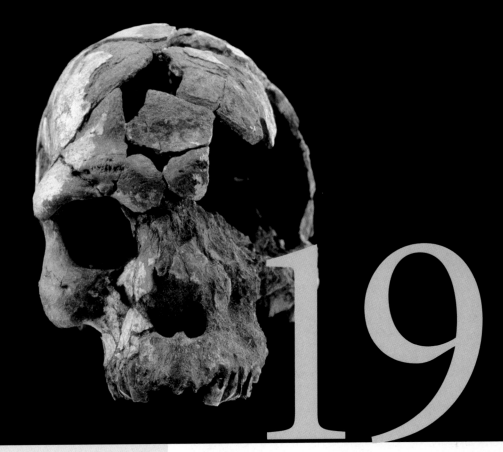

# The Origin and History of Life

The preserved skull of a *Homo sapiens idalltu* is remarkably similar to ours. This ancestor lived about 156,000 years ago in Ethiopia.

## On The Cusp of Humanity

One rainy autumn, paleoanthropologist Tim White, of the University of California, Berkeley, led a team through the village of Herto, Ethiopia, along a bend of the Awash River. In one area, White spotted a hippopotamus skull, near stone tools. Returning with helpers on a sunnier day, they discovered three humanlike skulls. Two were large and intact; the other was a shattered child's skull.

Analysis of surrounding volcanic rock established an approximate date of 160,000 years ago to the humanlike skulls. The researchers excitedly realized that the "Herto people," named *Homo sapiens idaltu,* lived when DNA evidence placed the dawn of humankind.

The ancient skulls were slightly larger than ours. Along the base of the child's skull were fine, parallel lines, which must have been made when the stony skull was still soft bone. No other body parts were found nearby, suggesting that the skull had been kept separately and handled.

The hippo and buffalo bones had been carefully removed from the skeletons and bore cut marks made with tools that had probably sliced off meat. The tools were much more sophisticated than the flaked tools of ancestors who dwelled in the area 2 million years ago. Overall, the clues paint a picture of a people who not only understood the concept of death, but who practiced mortuary rituals in keeping the skulls. Were these signs of the first inklings of human culture?

# Chapter Preview

1. The solar system formed about 4.6 billion years ago, and life left evidence on Earth 700 million years later.

2. When life began, Earth was geologically unstable, and the atmosphere was high in hydrogen and low in oxygen.

3. Prebiotic simulations have demonstrated how the fundamental biochemical units of life could have arisen from a prebiotic environment.

4. The RNA world theory proposes that RNA or a similar but more stable molecule preceded formation of the first cells because it could replicate, encode information, change, and catalyze reactions.

5. Phospholipid sheets that formed bubbles around proteins and nucleic acids may have formed cell precursors, or progenotes.

6. Metabolic pathways may have originated when progenotes mutated in ways that enabled them to use alternate or additional nutrients.

7. Life began in the seas. The earliest fossils are of cyanobacteria from 3.7 billion years ago. The oldest eukaryotic fossils, of algae, date from 1.9 billion years ago. Evidence of multicellularity dates to 1.2 billion years ago.

8. The Ediacarans were soft, flat organisms, completely unlike modern species. They lived in the late Precambrian and early Cambrian periods.

9. The Cambrian explosion introduced many species, notably those with hard parts.

10. Dinosaurs prevailed throughout the Mesozoic era, when forests were largely cycads, ginkgos, and conifers. In the middle of the era, flowering plants became prevalent.

11. Molecular evidence dates the origin of mammals to 100 million years ago, and fossil evidence indicates their adaptive radiation beginning 65 million years ago. Placental mammals eventually replaced many marsupial species.

12. *Aegyptopithecus* and other primates preceded the hominoids, which were ancestral to apes and humans. Fossil evidence provides a lineage of hominids, ancestors to humans, stretching back 7 million years.

## 19.1 Life's Origins

Reconstructing life's start is like reading all the chapters of a novel except the first. A reader can get an idea of the events and

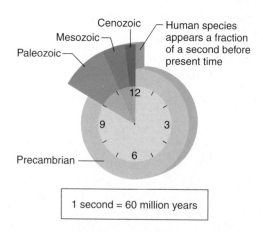

**FIGURE 19.8 Precambrian Time Accounts for Five-Sixths of Earth's History.** This clock face compresses the approximate time life has existed on Earth into 1 minute. The time when scientists think life on Earth has been abundant—the Paleozoic, Mesozoic, and Cenozoic eras—accounts for only one-sixth of the planet's history.

setting of the opening chapter from clues throughout the novel. Similarly, scattered clues from life through the ages reflect events that may have led to the origin of life. This evidence includes experiments that simulate chemical reactions that may have occurred on early Earth; fossilized microorganisms that might have been among Earth's first residents; exploring how clays or minerals might have molded biochemical building blocks into polymers (chains) important in life; and identifying ancient gene and protein sequences retained in modern organisms.

The history of the Earth is vast compared with the span of human existence. In fact, for billions of years, no life as we know it existed on Earth. As we continue with the study of life's origins, consider the scale shown in **Figure 19.1,** which illustrates key time periods in Earth's history.

## The Early Earth Was an Inhospitable Place

The study of the chemistry that led to life actually begins with astronomy and geology (**figure 19.2**). Earth and the other planets of the solar system formed about 4.6 billion years ago, as solid matter condensed out of a vast expanse of dust and gas. The red-hot ball had cooled enough to form a crust by about 4.2 to 4.1 billion years ago, when the temperature at the surface ranged from 500°C to 1,000°C (932°F–1,832°F) and atmospheric pressure was 10 times what it is now.

During the planet's first 500 to 600 million years, comets, meteors, and possibly asteroids bombarded the surface, repeatedly boiling off the seas and vaporizing rocks to carve the features of the fledgling world. The oldest rocks that remain today, and house the oldest known fossils, are from Greenland, in an area called the Isua geological formation. But most of the initial crust was torn down and built up again into sediments, heated and compressed, or dragged into Earth's interior at deep-sea trenches and possibly recycled to the surface. The oldest hints of life are organic deposits in quartz crystals that are rich in the carbon isotopes found in organisms, dating to about 3.85 billion years ago. Some of the old-

# Connections

The fossil record captures snapshots spanning the 4-billion-year history of Earth. From the earliest cell, we have representative fossils of a dizzying array of approaches to the problems of being alive. But we also see an orderly progression from simple to complex forms, illustrating how species adapt and improve as they meet life's challenges. There are lessons in the fossil record that humans should not view themselves as the most important species that has ever lived here. We can see evidences of the progression from simple to complex in the species that have survived to this day. As we explore the different kingdoms of life in chapters 20 through 26, we will use this as a model for organizing and classifying the many diverse species on Earth. We gain a greater respect for life as we realize that some species present today have been here for millions, if not billions of years. They have solved the problems of survival quite efficiently!

# Student Study Guide

## Key Terms

amniote egg　*357*　　　　hominid　*361*　　　　prebiotic simulation　*345*　　　reverse transcriptase　*348*

bipedalism　*361*　　　　hominoid　*361*　　　　progenote　*348*　　　　　　"RNA world"　*348*

## Chapter Summary

### 19.1　Life's Origins

1. Clues from geology, paleontology, and biochemistry provide evidence for the gradual formation of the first living cells from interacting collections of chemicals on long-ago Earth.

2. Modern experiments have shown that the prebiotic environment could have produced organic molecules necessary for life.

### 19.2　The Origins of Replication and Metabolism

3. RNA or a similar molecule that could encode information, replicate, and change may have provided the basis for the first stable, reproducible metabolism.

4. Lipids could have captured groups of molecules in a way that protected and reproduced them.

5. Complex pathways would have collected within these progenotes, leading to the first primitive cell.

### 19.3　The Emergence of Cells

6. Fossil evidence suggests the first prokaryotic cell arose 3.8 billion years ago.

7. The first photosynthetic cell followed at 3.5 billion years ago.

8. This was followed by the diversification of cells to form the precursors of today's different cell types.

### 19.4　The Ediacarans

9. The first multicellular life is represented by fossils of the Ediacarans, which were simple organisms without any obvious complex organs or systems, just an association of cells to form tissues.

### 19.5　The Paleozoic Era: Life "Explodes"

10. The Cambrian explosion introduced tremendous varieties of life.

11. Complex animals and plants became abundant and changed over the next several million years.

12. A succession of different periods came and went as organisms adapted to live in all kinds of environments.

### 19.6　The Mesozoic Era: Reptiles and Flowering Plants

13. The Mesozoic era was highlighted by the expansion of reptiles and the Age of Dinosaurs.

14. Flowering plants also appeared during this era.

### 19.7　The Cenozoic Era: Mammalian Radiation

15. Mammals first appeared in the Mesozoic but radiated extensively in the Cenozoic, as the climate changed.

## 19.8 The Evolution of Humans

16. The hominoids were ancestral to apes and humans; extinct hominids were animals that were ancestral to humans only.

17. Neanderthals have been shown to be a parallel species to *Homo sapiens*, rather than a direct ancestor of modern humans.

# What Do I Remember?

1. Describe the environment of primordial Earth and what chemical elements had to have been in primordial "soup" to generate nucleic acids and proteins.

2. Cite four reasons why RNA or a similar molecule is a likely candidate for the most important biochemical in the origin of life.

3. List the major life events of the five geological eras, and trace human evolution through time.

4. How might metabolic pathways have originated and evolved?

5. When and how did prokaryotic and eukaryotic cells arise?

**Fill-in-the-Blank**

1. The _____ were flat, strange organisms without complex organ systems that lived prior to the Cambrian period.

2. Reptiles and flowering plants were predominant during the _____ era.

3. The scientists _____ and _____ conducted the first definitive experiments that showed how biomolecules could have arisen from primordial soup.

4. The first photosynthetic organism appeared _____ years ago.

5. The _____ era was the "Age of Fishes."

**Multiple Choice**

1. The adaptive radiation of mammals occurred during the
   a. Mesozoic era.
   b. Cenozoic era.
   c. Cretaceous period.
   d. Cambrian period.
   e. Paleozoic era.

2. The first dinosaurs appeared during the
   a. Cambrian period.
   b. Cretaceous period.
   c. Triassic period.
   d. Jurassic period.
   e. Tertiary period.

3. The ancestors to apes and humans constitute only the
   a. hominids.
   b. hominoids.
   c. primates.
   d. Homo species.

4. Which of the following is an early version of our own species?
   a. *Homo erectus*
   b. *Homo habilis*
   c. Cro-Magnon
   d. Neanderthals
   e. *Aegyptopithecus*

5. The first fossil primate to show evidence of extensive tool use, giving rise to its name, was the species
   a. *Homo erectus*
   b. *Homo habilis*
   c. *Homo neanderthalensis*
   d. *Homo sapiens*
   e. *Australopithecus garhi*

replicate viral DNA and produce capsomer proteins, until, eventually, enough components accumulate so that hollow virions begin to be assembled. Then proteins insert viral DNA into the cores, forming complete **progeny virions,** the new viral particles. The host cell swells as hundreds of viruses assemble simultaneously. Finally, a viral protein breaks open the bacterial cell, releasing a flood of progeny virions and killing the cell. Within hours, all of the bacteria in the culture are dead. Considering the fact that just one virion infecting one cell starts this overwhelming infection process, it isn't surprising that a viral infection can kill cells, or cause symptoms, so quickly.

In eukaryotic cells, viruses follow a similar route to replication. However, they are able to use the organelles of these more complex host cells to manufacture larger and more intricate proteins, glycoproteins, and lipids. For example, some viruses surround themselves with a membrane envelope by using the secretory network formed by the endoplasmic reticulum, the Golgi apparatus, and associated vesicles (see figure 4.11) to manufacture certain viral proteins that lodge in the cell membrane. As progeny virions exit the cell, they bind to these embedded proteins, wrapping bits of the cell membrane around themselves, thereby creating envelopes. As more and more virions carry off segments of the cell membrane, the cell eventually loses its integrity and lyses. Viruses that lack envelopes use a variety of other pathways to leave the cell, including bursting out in the manner of bacteriophages. ● **secretion, p. 54**

## The Lysogenic Pathway Involves Integration into Host DNA

In **lysogeny,** viral DNA integrates into the DNA of the host and remains "hidden" for some time. A lysogenic virus cuts the host cell DNA with an enzyme, then uses other enzymes to join its DNA to the host DNA molecule. When the cell divides, it replicates the viral genes too. Lysogeny does not damage the host cell and, at the same time, ensures survival of the virus, because the cell cannot identify and remove the stowaway foreign DNA. Only a few viral proteins are transcribed and translated, and most of these proteins function as a switch to determine whether and when the virus should become lytic and exit the cell. At some signal, such as stress from DNA damage or lack of nutrients for the cell, these viral proteins cut the viral DNA out of the host genome and proceed to a lytic infection cycle. The resulting progeny virions then infect other cells, where they may enter a lytic or lysogenic state, depending upon the condition of the new host cells.

Some animal viruses can remain dormant as DNA molecules within a cell until conditions make it possible, or necessary, to replicate. Viruses in this state are **latent.** They may remain in cells for a long time without producing progeny virions. An example of a latent virus is the herpesvirus that causes cold sores. After the infection, the herpesvirus DNA remains in the infected cells indefinitely. When the cell becomes stressed or damaged, the DNA is transcribed, and viral proteins are made. New virions are assembled and leave the cell to infect other cells. Cold sores often

recur at the site of the original infection. Epstein-Barr virus (EBV)—the virus that causes mononucleosis—is another latent virus. EBV is so efficient at remaining undetected that more than 80% of the human population carries it. Because latent viruses persist by signaling their host cells to divide continuously, they may cause cancer.

HIV has a unique variation of lysis (**figure 20.9**). The HIV genome is incorporated into the host DNA as a part of its replication cycle. Shortly after infection, many HIV virions are produced and released. Eventually, an infection persists as infected cells produce small numbers of virus particles. Infected individuals have almost no symptoms, yet HIV is present in their bloodstreams. This persistent phase can last for many years. Finally, in response to some as yet unidentified stimulus, the virus enters a fully productive stage. So many viruses are produced that nearly all of the target cells in the body are killed. Vital immune system cells are destroyed, leaving the body unable to defend itself from infections or cancer. AIDS is the result.

### Reviewing Concepts

- A viral infection begins as a virus binds to a specific molecule on a cell surface and then injects its genetic material or is engulfed.
- The virus may set into motion its own reproduction, eventually bursting the cell to release viral progeny (lytic pathway), or it inserts into the host's DNA and persist for some time (lysogenic pathway).
- The steps of viral replication depend upon whether the genetic material is RNA or DNA, single-stranded or double-stranded.

## 20.4 Cell Death and Symptoms of Infection

The specific symptoms of a viral infection reflect the types of host cells that are destroyed and the host's responses. For example, rhinoviruses grow most efficiently in the mucous-producing cells in a person's nose, throat, and lungs, causing the symptoms of the common cold. Papillomaviruses infecting cells lining the reproductive tract cause growths called genital warts. The virus that causes hoof-and-mouth disease infects cells in the limbs and mouths of cattle.

Symptoms of viral infections felt at the whole-body level are generally caused by molecules that the immune system releases. The immune response renders the body inhospitable to the infectious agent, raising body temperature, causing the aches and pains of inflammation, and increasing secretion of mucus, among other actions. Unfortunately, these responses can also make the host rather miserable!

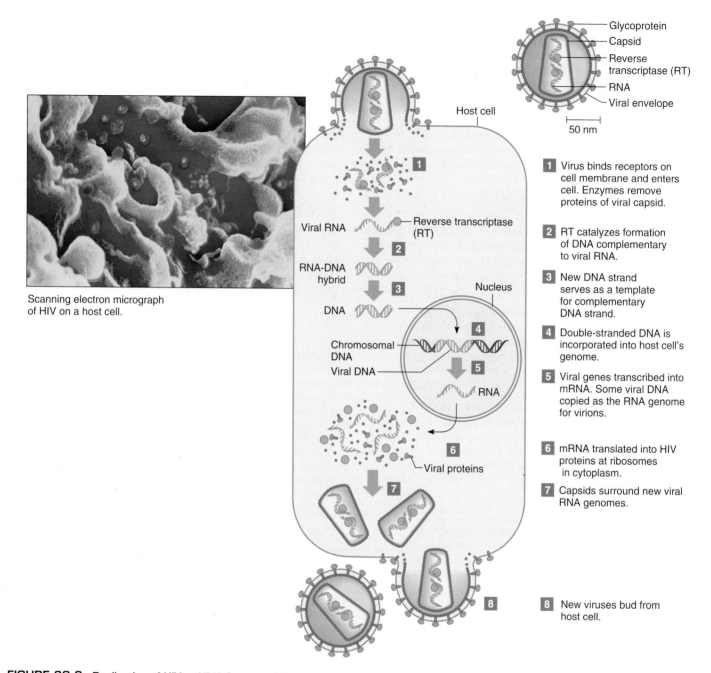

FIGURE 20.9  **Replication of HIV.**    HIV infects a cell by integrating first into the host chromosome, then producing more virions.

## Viruses Infect Particular Cells of Certain Species

Viruses are very specific about the types of cells they can invade. They infect only cells that have a specific target attachment molecule, or receptor, on their surfaces. Some target molecules are on a very small subset of cells in an organism, whereas others may be found in an entire group of related organisms. For example, the rabies virus can infect humans, skunks, and bats because they all share some common target molecules on their cells.

The kinds of organisms a virus can infect constitute its **host range.** The types of cells a virus can infect contain the enzymes and other molecules that the virus requires to effectively reproduce. However, a virus will enter any cell that contains the appropriate attachment molecule, whether it can efficiently replicate or not. Sometimes a virus can infect and replicate within a species

# Health

## From the Birds: Avian Influenza

Influenza has caused much human suffering and death over the centuries. The disease's origins trace to waterfowl, the natural viral reservoir.

Flu probably began about 4,500 years ago in China, where the virus moved from wild to domesticated ducks. In the seventeenth century, the Chinese brought domesticated ducks to live among rice paddies, where the birds ate insects and crabs, but not rice. In this way, flu-infested ducks came to live close to people, as well as near pigs and chickens. Occasionally, pigs would snort about in infected duck droppings, triggering a curious molecular mixing. The cells that line a pig's throat bear receptors for both avian (bird) and human flu viruses. In pig throats long ago, an avian flu virus mutated in a way that enabled it to infect humans. Avian and human flu viruses commonly intermingle in pigs and exchange segments of their RNA genomes, generating new viruses that can evade our immune systems, causing epidemics. But genetic exchange isn't the only contributing factor to global epidemics (pandemics). In wartime, the sickest people, in hospitals, spread the illness.

The flu pandemic of 1918 killed more people in the United States than both world wars, the Korean war, and the Vietnam war combined. Unlike more recent flu outbreaks that kill mostly the very young and very old, the 1918 flu felled those aged 20 to 40. The first reported case came from Camp Funston, Kansas, on March 4, 1918. It spread to several cities, and then U.S. troops brought it overseas. Over the summer, the virus became more virulent, and starting in August, it devastated Europe, where the death rate peaked in the fall (**figure 20.A**). The 1918 flu spread and killed fast. A young man might awaken in the morning with a stuffed nose and die by midnight, his lungs filled with fluid. Nurses would estimate how close a man was to death by the color of his feet, which would turn black as ravaged lungs failed to deliver oxygen. Researchers today are trying to determine why the 1918 flu was so deadly by studying viruses in lung tissue from several preserved victims.

Flu pandemics also occurred in 1957, 1968, and 1977. Two decades later, epidemiologists began to fear a possible repeat of pandemics past when a new flu variant appeared to jump from birds directly to humans. Infection from birds causes more severe symptoms. If such a flu can then spread from person to person, rather than just from bird to person, the stage would be set for another pandemic. Evidence is mounting that this may be happening.

In May 1997, three schoolchildren in Hong Kong died of a fierce flu caused by a virus identified previously only in birds. The children likely contracted the infection from pet chickens in the classroom. To avert an epidemic, the government killed every chicken in Hong Kong. Researchers realized by 2005, however, that such efforts are futile, both because birds are so intimately associated with humans in this part of the world and because the viruses lurk in other species, including ferrets, mice, and wild cats. Fortunately, the 1997 outbreak affected only 18 people, killing six of them.

A few dozen other human cases of bird flu have since occurred in China, Thailand, Vietnam, and Malaysia. Most were contracted directly from birds, and so the outbreaks petered out. Then in September 2004, in Thailand, a young mother died of bird flu after holding her 11-year-old dying daughter. The mother had been far away in a city, while the girl had been handling dead fowl in her village. The only way the mother could have acquired the virus was through her daughter. Researchers confirmed this by copying the RNA from viruses infecting the mother and child into DNA, sequencing it, and demonstrating a match. The case may have been the first detection of the long-feared human-to-human transmission of bird flu.

A flu virus, whose genetic material is RNA, has a remarkable ability to reinvent itself, mutating in great sweeps that require us to constantly change vaccine components, and at the same time, in a base-by-base manner. The bird flu, technically called H5N1 for the composition of its outer components, is changing. Besides infecting other species and persisting longer in the environment, the current bird flu virus can also cause different symptoms. Two boys who died of it in 2005 developed brain inflammation and not respiratory symptoms.

Cases of human-to-human transmission of bird flu continue, and epidemiologists are preparing for a possible pandemic. Vaccines for bird flu may be available by the time you read this—only time will tell if they can prevent a pandemic.

**FIGURE 20.A  Guarding Against the Flu.** The 1918 flu epidemic spread so quickly that cities took drastic steps. Policemen in London wore surgical masks to avoid infection.

without causing symptoms. Such carriers of a virus are called **reservoirs** because they act as a source of that virus that can infect other species and produce disease in them. Often, insects such as mosquitoes or leafhoppers are reservoirs (see figure 20.6) and spread disease.

Identifying the reservoir for a virus can help epidemiologists develop strategies for preventing spread of infection. Ebola virus is passed in body fluids and causes a rapidly fatal uncontrollable bleeding in humans. Attempts to limit outbreaks of Ebola hemorrhagic fever have been stymied because the reservoir—where the virus "hides" between surfacing in human settlements—is unknown. In contrast, smallpox virus vanished after successful global vaccination. With only one host—us—the virus had no other cells to infect. It exists today only in certain laboratories, but because at least 16 nations have developed it as a bioweapon, vaccination programs may resume.

## Viruses May Move to New Species

Another feature of viruses is their ability to change as they infect new species. This is rare, but can produce frightening results. For instance, HIV may descend from a monkey or chimpanzee virus that became able to infect humans.

A virus can "jump" species, widening its host range to include another species, only if the cells of that species have receptors to which the virus can bind. Then, the new host must have the enzymes necessary to replicate the viral genetic material. If a species' cells are not compatible with the virus in these ways, then the virus cannot infect it and replicate efficiently. For this reason, a pet cat cannot transmit the feline form of the AIDS virus to its owner, nor can a tomato plant pass bushy-stunt disease to a gardener. However, a person can get the flu from ducks, which serve as reservoirs. Influenza is an example of a **zoonosis,** which is an infection that jumps from another animal species to humans.

Influenza is perhaps the most familiar viral zoonosis, able to infect the cells that line the respiratory tracts of various species of birds and mammals. A well-documented infectious route is from human to pig to duck and back to humans. Along the way, because these viruses lack genetic repair systems, mutations accumulate, making the virus appear different to the human immune system from one year to the next. Epidemiologists must track the base sequences of influenza virus genetic material—which is single-stranded RNA—among people, pigs, and various fowl species the world over, continually, to predict the coming year's strains and develop vaccines. Scientists continue to take a closer look at the role of birds in flu epidemics and in encephalitis caused by the West Nile virus.

Viruses that acquire new types of hosts can cause new infectious diseases. Many zoonoses probably never affect more than a few people, because they often are transmitted only from originating animal to human, but not from human to human. For example, most outbreaks of monkeypox—a disease similar to smallpox—soon vanish because people must catch it from infected monkeys, not from other people.

**FIGURE 20.10** **Bats and Birds Can Transport Viruses to New Hosts.** In Australia, when people moved closer to bats, a new respiratory illness arose. The virus jumped from pregnant bats to horses to people.

Often birds or bats are the agents of a new zoonosis because they can travel long distances, spreading an illness that might disappear if left in just one host population. In Australia, for example, a new respiratory illness in humans came from horses, who, in turn, got it from fruit bats carrying Hendra virus (**figure 20.10**). The disease killed several thoroughbred horses and their trainer in a suburb of Melbourne in the fall of 1994. After similar cases were reported, epidemiologists identified what looked like a large version of a known virus, equine morbillivirus. Because the cases were in different geographic regions, virus-trackers suspected a bird or bat source. They focused on bats for two reasons—they are more closely related to horses and humans than are birds, and the expanding suburbs of Australian cities have been encroaching on bat habitats.

The bat link came from people who cared for injured bats they found in their backyards. Blood from the bat-rescuers contained antibodies to the newly described virus, indicating they

with a mysterious illness called kuru that affected the Foré people of New Guinea, until they abandoned a centuries-old cannibalism ritual. A young U.S. government physician, D. Carleton Gajdusek, began unraveling the story of kuru in 1957. He connected scrapie, kuru, and the very rare but similar Creutzfeldt-Jakob disease (CJD) in humans, which in 1972 caught the attention of a young physician, Stanley Prusiner. Prusiner has been investigating prions ever since and proposed the explanation for their mechanism of action shown in figure 20.12. Both men won the Nobel prize for their work on prion diseases, which are still not completely understood.

CJD was known to affect one in a million people, usually over age 60, either sporadically (no known cause) or from contamination by surgical instruments or transplanted tissue. In 1996, evidence began to mount that a "new variant" of CJD could be transmitted to humans who ate nerve tissue from cows suffering from the prion disease bovine spongiform encephalopathy (BSE), which quickly became known as "mad cow disease."

New-variant CJD, which became known simply as variant CJD (vCJD), seems to have been limited to about 120 cases (so far), mostly in the United Kingdom.

Prions may be more common than has been realized. Similar infectious proteins are known in yeast, where they have metabolic functions different from those associated with the prions of mammals. Prions may also illustrate a general disease mechanism of a cascade of proteins being changed into abnormal forms that cause them to clump.

## Reviewing Concepts

- Viroids are infectious RNA molecules that harm plants.
- Prion diseases result when abnormal prion proteins bend normal prion proteins into an alternate shape.

# Connections

Viruses challenge our definition of life. They are so prolific when reproducing that they can overwhelm a host in an incredibly short time period. As we learn more about viruses, we are finding they are responsible for a wide range of human ailments never before associated with germs or disease. It is a wonder that we have survived at all! We are learning much about how our own cells function through the study of viruses and their effects. But viruses are also being used to modify cells. They are so very good at invading cells that they have become an ideal tool for inserting genes into target cells. The future will likely reveal much we have yet to suspect about the role of viruses in life.

# Student Study Guide

## Key Terms

| | | | |
|---|---|---|---|
| bacteriophage  *370* | latent  *375* | prion  *380* | virion  *370* |
| capsid  *370* | lysogenic infection  *372* | progeny virion  *375* | viroid  *379* |
| capsomer  *370* | lysogeny  *375* | reservoir  *378* | virus  *368* |
| envelope  *370* | lytic infection  *372* | transmissible spongiform | zoonosis  *378* |
| host range  *376* | plaque  *368* | encephalopathy  *380* | |

## Chapter Summary

### 20.1  The Discovery and Study of Viruses

1. A virus is not alive and is much simpler than a cell, consisting of a nucleic acid wrapped in protein and perhaps a fatty envelope.

2. Viruses use living cells to reproduce by taking over the enzyme machinery that makes biomolecules.

3. The invaded cell is disrupted and often killed, resulting in disease.

4. Cultures of the target cells are grown in the laboratories to provide a means for growing, and thereby studying, viruses.

## 20.2 Viral Structure

5. Viruses are classified by the type of nucleic acid contained within the virus particle, or virion.
6. The protein package is composed of capsomers assembled into a capsid.
7. Some viruses also contain an outer layer of lipids, called an envelope.

## 20.3 Viral Reproduction Within Cells

8. A viral infection begins as a virus binds to a specific molecule on a cell surface and then injects its genetic material or is engulfed.
9. The virus may set into motion its own reproduction, eventually bursting the cell to release viral progeny (lytic pathway), or insert into the host's DNA and persist for some time (lysogenic pathway).
10. The steps of viral replication depend upon whether the genetic material is RNA or DNA, single-stranded or double-stranded.

## 20.4 Cell Death and Symptoms of Infection

11. The symptoms of a viral illness arise from direct effects of the virus on certain cells and from the immune system's response.
12. A virus can infect only specific cells of particular species.
13. Through mutation, some viruses "jump" from one species to another.
14. Sometimes viruses persist without causing harm in reservoir species.
15. Drugs can only rarely affect viruses, since they lack a unique metabolism.

## 20.5 Other Noncellular Infectious Agents

16. Viroids are infectious RNA molecules that harm plants.
17. Prion diseases result when abnormal prion proteins bend normal prion proteins into an alternate shape.

# What Do I Remember?

1. Compare and contrast viruses, bacteria, and eukaryotic cells.
2. What factors determine whether a virus infects a cell?
3. How do viruses infect bacterial, animal, and plant cells?
4. Describe a lytic and lysogenic "life cycle" of a virus.
5. Diagram and label the parts of a virus particle.
6. How do viruses cause symptoms, directly and indirectly, and why do dog viruses rarely infect humans?

**Fill-in-the-Blank**

1. The virus that causes AIDS is named _____.
2. A viral _____ is a species where a virus can "hide," then emerge and cause a disease outbreak.
3. A(n)_____ is an irregular protein that behaves like an infectious agent.
4. A(n)_____ is a virus that infects bacteria.
5. Viruses are classified by the type of _____ they use as genetic material.

**Multiple Choice**

1. The genetic material of a virus can be
   a. single-stranded RNA.
   b. single-stranded DNA.
   c. double-stranded RNA.
   d. double-stranded DNA.
   e. any of these.

2. A _____ virus is hidden within the genetic material of the host cell.
   a. latent
   b. lytic
   c. phage
   d. flu
   e. toxic

3. The _____ of a virus describes which species within which it can infect and replicate.
   a. nucleic acid
   b. host range
   c. lysogeny
   d. capsomer

4. The viral capsid is composed of individual proteins known as
   a. virions.
   b. capsomers.
   c. ligands.
   d. histones.
   e. plaques.

5. Why are antibiotics ineffective in treating a viral disease such as influenza?
   a. Viruses are able to destroy antibiotics with their enzymes.
   b. The virus uses the host cell's enzymes to reproduce.
   c. Antibiotics cannot penetrate the envelope of some viruses.
   d. Viral metabolism uses a different set of amino acids.

A confocal microscope and special computer software can be used to create high-resolution images of complex structures like this biofilm. The red dots are fluorescent tracer beads. The blue dots are bacterial cells, which are typically 1 micrometer in diameter.

## Bacterial Biofilms: "Mob Mentality"

Contrary to common conception, bacteria are not loners. Instead, they often build complex communities in which cells communicate with each other, protect each other, and even form differentiated structures. These organized aggregations of bacterial cells are called biofilms.

In contrast to free-floating life in the laboratory shaker flask, bacteria in many "real" habitats settle and reproduce on solid surfaces. Once a critical cell density is reached, the individual lifestyle gives way to community living. The cells begin to express genes that trigger secretion of a sticky polysaccharide slime. As cells continue to divide, three-dimensional mushroom-shaped structures form. Occasionally, cells are released from the biofilm and colonize new habitats.

Microbiologists are still learning about the molecular signals that trigger biofilm formation. For example, different bacteria use different signaling molecules to detect the density of cells around them. Thus, within the first 6 hours of biofilm formation, the bacteria turn off genes coding for the proteins that form the flagellum—motility is obviously not required for success in a sedentary lifestyle. In contrast, genes for proteins that build attachment structures called pili are activated.

As microbiologists learn more about biofilm formation, they may be able to develop new treatments to combat medically important bacteria. Conversely, learning how to trigger biofilm formation in beneficial soil bacteria may enhance efforts to clean up toxic wastes.

# Bacteria and Archaea

**21.1 Prokaryotes: A Success Story**
- Prokaryotes Changed the Earth
- Bacteria and Archaea Exhibit Vast Diversity and Exist in Huge Numbers

**21.2 Parts of a Prokaryotic Cell**
- Internal Structures Are Simple Compared to Eukaryotes
- External Structures Include Protective Layers and Extensions

**21.3 Classification of Prokaryotes**
- Traditional Methods Consider Morphology, Physiology, and Habitat
- Molecular Data Reveal Evolutionary Relationships
- Archaea Are Clearly Different from Bacteria and Eukaryotes

**21.4 Gene Transfer in Prokaryotes**
- Binary Fission Provides Vertical Gene Transfer
- Horizontal Gene Transfer Occurs in Three Ways

**21.5 The Human Impact of Prokaryotes**
- Bacteria Infect Humans and Cause Disease in Several Ways
- Bacteria Can Be Used as Bioweapons
- Industrial Microbiology Borrows Prokaryotes' Products and Processes

# Chapter Preview

1. Prokaryotes are important in many ways. They contribute gases to the atmosphere, form the bases of food webs, and fix nitrogen. More than a billion years ago, they gave rise to certain organelles that now characterize eukaryotic cells.

2. Prokaryotes are very abundant and diverse and occupy a great variety of habitats, and some bacteria survive harsh conditions protected in endospores.

3. A cell wall surrounds most prokaryotic cells. Gram-staining reveals differences in cell wall architecture to group bacteria.

4. Traditionally, microbiologists have classified prokaryotes based on cell morphology, metabolic capabilities, physiological tests, or habitat.

5. Bacteria are spherical cocci, rod-shaped bacilli, spiral-shaped spirilla, or variations of these. Archaea may have other shapes.

6. Prokaryotes can be classified based on how they acquire carbon and energy and their requirements for the presence of oxygen.

7. Bacteria and archaea differ in many key characteristics, including genetic sequences, antibiotic sensitivity, and cellular components.

8. Domain Archaea contains three taxa: Euryarchaeota, Crenarchaeota, and Korarchaeota. Like some bacteria, many archaea thrive in extreme environments.

9. Binary fission is division of a prokaryotic cell to yield two genetically identical daughter cells. Prokaryotic cells are susceptible to three routes of horizontal gene transfer.

10. Koch's postulates are used to demonstrate whether a type of microorganism causes a particular set of symptoms. Pathogenic bacteria are transmitted in air, by direct contact, via arthropod vectors, or in food and water.

11. Bioweapons are formulated from highly virulent bacteria, viruses, or fungi. A bioweapon attack would be difficult to detect, trace, and treat.

12. Prokaryotes are used in the manufacture of many foods, drugs, and other chemicals. They are also used for treating sewage and cleaning the environment.

# 21.1 Prokaryotes: A Success Story

If all of the animals and plants on Earth were to become extinct, microbial life would nevertheless continue. But if all the prokaryotes on the planet were to vanish, all the rest of life would cease too. The prokaryotes—single-celled organisms lacking nuclei and membrane bounded organelles—were the first life-forms and remain essential to life in many ways.

## Prokaryotes Changed the Earth

Prokaryotes have had a tremendous impact on Earth's natural history. The first cells were probably more like existing prokaryotes than any other type of organism. Along the road of evolution, prokaryotes were probably the precursors of the chloroplasts and mitochondria of eukaryotic cells (see figure 19.10). Ancient photosynthetic prokaryotes also contributed oxygen gas ($O_2$) to Earth's atmosphere, creating a protective ozone layer and paving the way for aerobic respiration.

Even now, the seemingly simple prokaryotes lie at the crux of the continual cycling of chemical elements between organisms and the nonliving environment (see chapter 44). Besides their role in decomposing dead organic matter, only bacteria can regularly convert atmospheric nitrogen gas ($N_2$) into biologically useful forms. **Nitrogen fixation** is a process in which bacteria reduce $N_2$ to ammonia ($NH_3$), which other organisms can incorporate into organic molecules. Some of these bacteria, such as those in the genus *Rhizobium*, induce the formation of nodules in the roots of certain host plants (**figure 21.1**). Inside the nodules, *Rhizobium* cells share the nitrogen that they fix with their hosts, offering the host plant a growth advantage in nutrient-poor soils. In exchange, the bacteria receive nutrients and protection from their hosts. Other nitrogen-fixing bacteria live independent of plant hosts.

Biologist Mark Wheelis summed up the importance of the **Bacteria** and **Archaea,** the two prokaryotic domains of life: "The earth is a microbial planet, on which macroorganisms are recent additions—highly interesting and extremely complex in ways that most microbes aren't, but in the final analysis relatively unimportant in a global context."

## Bacteria and Archaea Exhibit Vast Diversity and Exist in Huge Numbers

Bacteria and archaea were once thought to comprise half of all living matter. The percentage is probably much higher. We have likely underestimated the *diversity* of the prokaryotic domains because these organisms are not easy to observe. But when we try to survey the *abundance* of prokaryotes, their numbers are astounding. Three researchers did just that.

William Whitman, David Coleman, and William Wiebe from the University of Georgia divided Earth into types of habitats and consulted many studies that evaluated the number of prokaryotes in each place. This analysis led to the estimate that more than 5 million trillion trillion prokaryotes live on or in the planet—that's 5,000,000,000,000,000,000,000,000,000,000 prokaryotes.

Many people think of prokaryotes in terms of those that cause disease in humans. But prokaryotes live in an incredible diversity of places—within rocks and ice, in thermal vents (see figure 6.3), nuclear reactors and hot springs, and in extremely low pH conditions. They live as far up as 40 miles (69.4 kilome-

ters worse, many microorganisms are difficult, if not impossible, to culture in numbers sufficient to study them. In spite of these hurdles, microbiologists have devised taxonomic groupings based on characteristics they could observe. These traditional groupings do not necessarily reflect evolutionary relationships, and they are slowly being revised as more microbial genes and genomes are sequenced.

## Traditional Methods Consider Morphology, Physiology, and Habitat

Cell structure, or morphology, is one important criterion for classifying prokaryotes. Individual cells can take a variety of forms. Three of the most common are cocci (spherical), bacilli (rod-shaped), and spirilla (spiral) (**figure 21.6**). The arrangement of the cells in pairs, tetrads, grapelike clusters (*staphylo-*), or chains (*strepto-*) also is sometimes important. *Staphylococcus*, for example, causes infections in humans; its spherical cells form clusters. Other morphological characteristics include the presence or absence of certain structures such as flagella, pili, endospores, or a glycocalyx.

Besides direct microscopic observation of cells, staining is used to divide prokaryotes into groups based on differences in cell wall structure. For example, the Gram stain technique, described earlier, is one of the methods used to place bacteria into major subgroups.

Prokaryotes may be structurally uncomplicated, but they have great metabolic diversity. The methods by which prokaryotes acquire carbon and energy form another basis for their classification. The most fundamental distinction reflects carbon source. Recall from chapter 6 that **autotrophs** acquire carbon from inorganic sources such as carbon dioxide ($CO_2$), and **heterotrophs** use more reduced and complex organic molecules, typically from other organisms. An autotroph or heterotroph may also be a **phototroph** or **chemotroph,** which refers to the energy source. Phototrophs derive energy from the sun, and chemotrophs get energy by oxidizing inorganic or organic chemicals.

Combining these terms describes how specific microorganisms function and fit into the environment. Cyanobacteria, for example, are photoautotrophs—they use sunlight (photo) for energy and $CO_2$ (auto) for carbon. Many pathogenic bacteria are chemoheterotrophs, because they use organic molecules from their hosts to acquire both carbon and energy. Chemoautotrophic bacteria acquire carbon from $CO_2$ and energy by oxidizing inorganic nitrogen- or sulfur-containing molecules. The metabolic diversity of prokaryotes allows them to live in many environments that do not have eukaryotes.

Prokaryotes are also classified by physiological tests such as temperature and pH optima and oxygen requirements. For example, **obligate aerobes** require oxygen for generating ATP. For **obligate anaerobes** such as *Clostridium tetani* (the bacterium that causes tetanus), oxygen is toxic, and they live in habitats that lack it. **Facultative anaerobes,** which include *E. coli* and *Salmonella,* can either use oxygen or not.

Finally, prokaryotes may also be classified based on where they live: acidophiles live in acidic conditions; thermophiles live

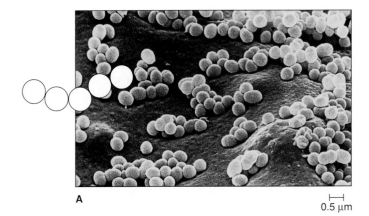

A    0.5 μm

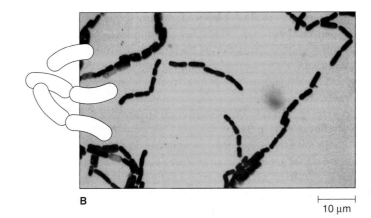

B    10 μm

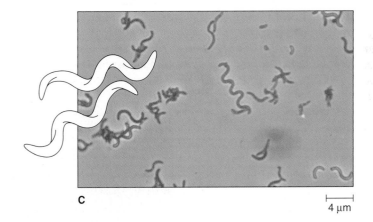

C    4 μm

**FIGURE 21.6 Prokaryotes Can Be Classified by Cell Shapes.** *Micrococcus* (**A**) are spherical (cocci). *Bacillus megaterium* (**B**) are rods (bacilli). *Rhodospirillum rubrum* (**C**) are spiral-shaped (spirilla).

in hot springs; halophiles live in highly salty lakes. But classification based on habitat can be confusing because distantly related organisms may share habitats. For example, both the bacterium *Thermus aquaticus* and the archaeon *Thermoplasma volcanium* are thermophiles, living in hot conditions. Another complication of this method is that one organism might fall into more than one category.

## Molecular Data Reveal Evolutionary Relationships

Traditional classification systems are still in widespread use because they are based on characteristics that are easily observed. One problem with describing microorganisms based on their appearance and methods of energy acquisition, however, is that members of the groups may be only distantly related to one another. Molecular data are leading scientists closer to a classification system that reflects evolutionary relationships among all organisms, including prokaryotes.

Starting in the late 1970s, microbiologists began to compare organisms by ribosomal RNA sequences, which tend to change little over evolutionary time and therefore might reveal common ancestries. Such studies compare **signature sequences,** which are short stretches of nucleotides that are unique to certain types of organisms. If two varieties of bacteria share sequences that no others have, then one explanation is that they descended from a recent shared ancestor. Such information is used to build phylogenetic trees. Recall from chapter 3 that RNA sequencing of prokaryotes was how Woese distinguished archaea from bacteria.

• **archaean cells, p. 51**

RNA analysis can also be valuable in studying microorganisms that are difficult to culture. This was the case for *Epulopiscium fishelsoni,* discovered in 1993 in the intestine of a large fish in the Red Sea. Microbiologists thought it was a eukaryote, because it was thousands of times the size of *E. coli* and even larger than the unicellular eukaryote *Paramecium* (**figure 21.7**). *E. fishelsoni* survives for only 20 minutes outside of the fish, so laboratory culture did not look promising. However, because of its large size, researchers were able to isolate the organism, retrieve its DNA, and use the polymerase chain reaction (see figure 14.4) to amplify the sequences that encode rRNA. The signature sequences were unmistakably those of bacteria, not eukaryotes. *Epulopiscium fishelsoni* is a giant bacterium.

By knowing the types of proteins that a prokaryote manufactures, researchers are gaining insights into how these organisms live. Consider *Treponema pallidum,* which causes syphilis. Microbiologists have long wondered why this bacterium is remarkably comfortable in the human body for decades, yet it dies within days in laboratory culture. The genome sequence revealed that *T. pallidum* cannot manufacture many nucleotides, fatty acids, and enzyme cofactors, but it does have transport proteins that obtain these essentials from its host. Laboratory culture simply can't provide what a warm human body can.

## Archaea Are Clearly Different from Bacteria and Eukaryotes

The tendency to lump together all organisms that lack nuclei and membrane-bounded organelles hid much of the diversity in the microbial world for many years. It was only when microbiologists began incorporating molecular sequence data that the distinction between domains Bacteria and Archaea was discovered. Although microbiologists haven't yet scrutinized many archaean species intensely and many distinctions probably have not yet been discovered, these cells do differ from bacterial cells in some key ways (see figure 4.4):

• Archaea exhibit different rRNA and tRNA sequences, ribosomal proteins, and antibiotic sensitivity patterns.

• Archaea have certain cell wall components and membrane lipids not seen in bacteria.

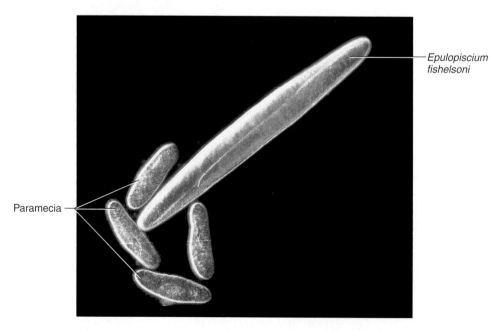

**FIGURE 21.7    Giant Bacterium.**    *Epulopiscium fishelsoni* is a million times larger than many other prokaryotes, and it is even larger than the single-celled eukaryote *Paramecium* 200).

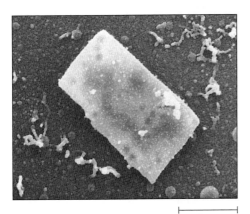

2 µm

**FIGURE 21.8  Rectangular Cell.** Archaea can take forms not seen in bacteria, indicating a different type of cell structure.

- An archaean genome contains more introns and repeated sequences than does a bacterial genome.
- Some archaea are thin rectangles (**figure 21.8**), a form not seen among bacteria, indicating that members of the two domains differ in their cellular organization and possibly composition.

Archaea are often collectively described as "extremophiles" because the first to be discovered were found in habitats with high temperatures, acidity, or salinity. They are therefore sometimes informally divided into three groups: methanogens, thermophiles, and halophiles. As more archaea are discovered in moderate environments such as soil or the open ocean, however, formal taxonomic descriptions become more important.

Microbiologists currently divide domain Archaea into three taxa: Euryarchaeota, Crenarchaeota, and Korarchaeota. The Euryarchaeota contains the methanogenic archaea that inhabit stagnant waters and the anaerobic intestinal tracts of many animals. These prokaryotes generate methane as a metabolic byproduct, contributing to the greenhouse effect (see chapter 45). This group also includes halophilic archaea that live in very salty habitats such as seawater, evaporating ponds and salt flats.  ● **global warming, p. 917**

The second group of archaea, the Crenarchaeota, includes thermophiles that thrive in hot springs or at hydrothermal vents. However, this group also contains a wide variety of soil and water microorganisms with moderate temperature optima. Other thermophiles are classified among the Korarchaeota, which may represent the oldest archaean lineage. This third group is known mostly from genes extracted from their habitats.

Considering the long history of microbiology and the short time molecular analyses have been in use, it is not surprising that microbial taxonomy is in flux. **Table 21.2** reviews selected

## TABLE 21.2

### Selected Groups of Prokaryotes

| Group | Features | Example |
|---|---|---|
| **Domain Bacteria (groups reflect traditional taxonomic criteria)** | | |
| Purple and green sulfur bacteria | Bacterial photosynthesis using $H_2S$ (not $H_2O$) as electron donor | *Chromatium vinosum, Chlorobium limicola* |
| Cyanobacteria | Photosynthesis releases $O_2$; some fix nitrogen; free-living or symbiotic with plants, fungi, or protists | *Nostoc, Anabaena* |
| Spirochetes | Spiral-shaped; some pathogens of animals | *Borrelia burgdorferi* (causes Lyme disease) *Treponema pallidum* (causes syphilis) |
| Enteric bacteria | Rod-shaped, facultative anaerobes in animal intestinal tracts | *Escherichia coli, Salmonella* spp. |
| Vibrios | Comma-shaped, facultative anaerobes common in aquatic environments | *Vibrio cholerae* (causes cholera) |
| Gram-positive endospore-forming bacteria | Aerobic or anaerobic; rods or cocci | *Bacillus anthracis* (causes anthrax) *Clostridium tetani* (causes tetanus) |
| **Domain Archaea (groups reflect phylogeny)** | | |
| Euryarchaeota | Many are methanogens, generating $CH_4$, a metabolic by-product; others are extreme halophiles | *Methanococcus, Halobacterium* |
| Crenarchaeota | Most are thermophiles | *Sulfolobus* |
| Korarchaeota | Little-known thermophiles, detected mostly by DNA | Unnamed |

groups from both prokaryotic domains. For convenience, and because of their long history in microbiology, most bacteria are still grouped according to traditional criteria. In contrast, the listed groups of archaea reflect current understanding of their evolutionary relationships.

## Reviewing Concepts

- Prokaryotes are distinguished first by their having one of three basic shapes—spherical, rod, or spiral.
- They are also classified by their method for acquiring food and their tolerance or need of oxygen.
- Comparisons of molecular sequences have provided more precise means for identification and have also revealed the existence of another unique type of cell—the archaea.
- Members of domain Archaea share features found in both prokaryotic and eukaryotic cells, suggesting evolutionary relationships for these organisms.

# 21.4   Gene Transfer in Prokaryotes

Genes are transferred from generation to generation but also from cell to cell. Gene transfer between distantly related cells may complicate our ability to discover what the earliest cells were like, as well as to understand how prokaryotes are related to each other today.

Genes pass from one cell to another in two general ways. **Vertical gene transfer** is the transmission of DNA from a parent cell to daughter cells as the original cell divides. **Horizontal gene transfer** occurs laterally, from one cell to another. That is, vertical gene transfer entails a generational difference (inheritance), but horizontal gene transfer does not.

## Binary Fission Provides Vertical Gene Transfer

Prokaryotes reproduce by **binary fission,** a process that replicates DNA and distributes it and other cellular constituents into two daughter cells. Binary fission therefore provides vertical gene transfer, from one generation to the next.

In prokaryotic cells, DNA is attached to a point on the inner face of the cell membrane (**figure 21.9**). In binary fission, the DNA begins to replicate, and the cell membrane grows between the two DNA molecules and separates them. Then the cell membrane dips inward, pinching off two daughter cells from the original one. Formation of cell walls completes the process.

Binary fission superficially resembles mitosis, but it is different because it lacks spindle fibers and many of the types of proteins that are associated with chromosomes in more complex

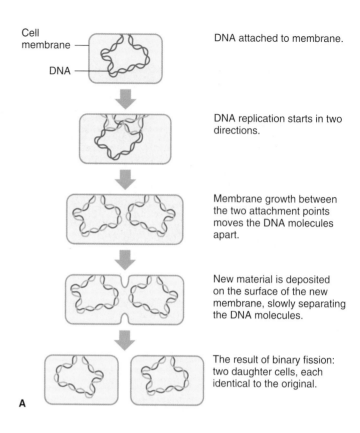

Cell membrane — 
DNA — 
DNA attached to membrane.

DNA replication starts in two directions.

Membrane growth between the two attachment points moves the DNA molecules apart.

New material is deposited on the surface of the new membrane, slowly separating the DNA molecules.

The result of binary fission: two daughter cells, each identical to the original.

A

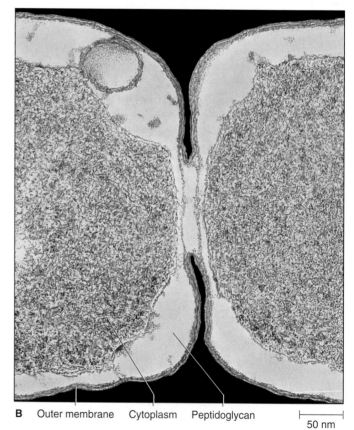

B   Outer membrane   Cytoplasm   Peptidoglycan
50 nm

**FIGURE 21.9 Prokaryotic Cell Division.** **(A)** In binary fission, the cell membrane grows and then indents as the DNA replicates, separating one cell into two. **(B)** This false-color image shows two new *E. coli* cells on the verge of separating from each other. The cytoplasm (green) has already divided, and the cell walls are nearly complete.

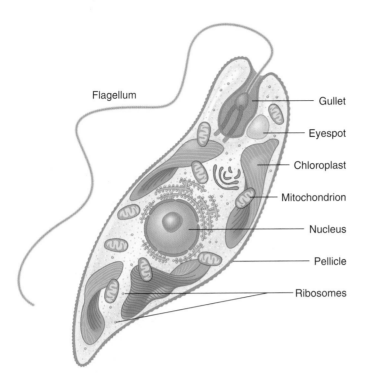

**FIGURE 22.8** *Euglena.*    The pond-dwelling *Euglena* has a flagellum and chloroplasts but can also ingest food particles. A typical *Euglena* cell is about 50 micrometers long.

## Reviewing Concepts

- The basal eukaryotes descend from the earliest lineages of eukaryotes.
- Some lack the mitochondria that characterize cells of most other eukaryotes.
- The slime molds live as both single- and multi-celled organisms.
- The mechanisms they use to aggregate can teach us how multicellular organisms might have evolved.

## 22.3    The Alveolates

The **alveolates** are crown eukaryotes that have in common a series of flattened sacs, or alveoli, just beneath the cell membrane. Many species have three membrane layers—the cell membrane and two membranes formed by the alveoli. These organisms also possess mitochondrial cristae that are tubular. (The cristae in other protista may be disc-shaped or flat.) Molecular sequence evidence from rRNA and from mitochondrial DNA also places these protista together. The alveolates include three major groups (dinoflagellates, apicomplexa, and ciliates) and several groups with fewer members. Among these smaller groups is the

**FIGURE 22.9 Trypanosomes.**    These protista share a unique organelle, the kinetoplast (**A**). Trypanosomes (**B**) cause Chagas disease, African sleeping sickness, and leishmaniasis. The mucocutaneous form of leishmaniasis has deformed this man's face (**C**).

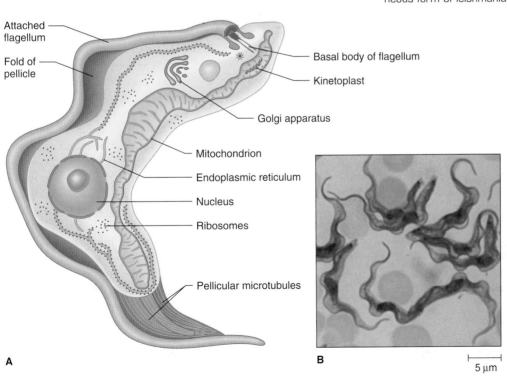

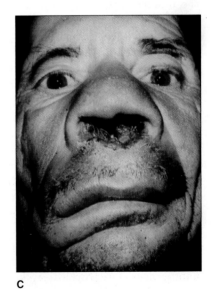

A

B            5 µm

C

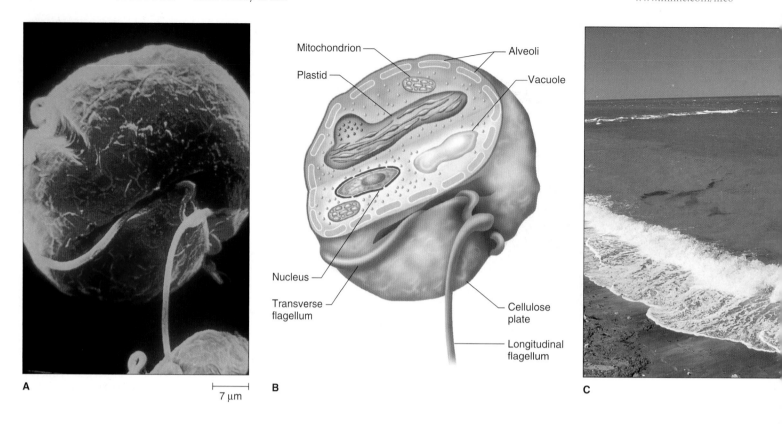

**FIGURE 22.10** **Dinoflagellate Features.**    *Gonyaulax tamarensis* exhibits classic dinoflagellate structure—note the characteristic transverse and longitudinal flagella (**A** and **B**). Blooming populations of dinoflagellates are responsible for red tides (**C**).

foraminifera, whose durable shells have left microscopic fossils over millions of years. These tiny remnants can provide important clues to the age and environment in which a rock formed.

## Dinoflagellates Are Whirling Protists

The **dinoflagellates** take their name from the Greek *dinein,* which means to whirl. They have two flagella, of different lengths, one of which is longitudinal and the other transverse in orientation (**figure 22.10**). The transverse flagellum beats in a way that propels the cell with a whirling motion. Another characteristic of many dinoflagellates is that their cell walls are divided into 2 to 100 overlapping cellulose plates. The shape and pattern of these plates is the main basis for classification of the dinoflagellates.

Dinoflagellates are important ecologically because they help form plankton, the microscopic food that supports the ocean's vast food webs. About half of the 2,000 or so species are photosynthetic; the remainder obtain nourishment from other organisms, either as predators or parasites. Some species are bioluminescent; others are symbionts of giant clams, jellyfish, sea anemones, mollusks, or corals.

Many dinoflagellates can enter a dormant cyst stage, with a cell wall and lowered metabolic demands. Cysts enable the dinoflagellate to survive a temporarily harsh environment or provide a means of transport from one host to another. When environmental conditions improve, food becomes available, or an appropriate host arrives, the dinoflagellate bursts from its protective cyst. Sometimes, populations of the dinoflagellates *Gymnodinium* and *Gonyaulax* "bloom," with many cysts bursting at once, producing a "red tide" effect (figure 22.10C). The dinoflagellates also may release a neurotoxin, which becomes concentrated as fish eat it and pass it up food chains (see figure 44.18). In humans, the neurotoxin inhibits sodium ion transport in the nerve cells (see chapter 32) and causes the numb mouth, lips, face, and limbs that result from paralytic shellfish poisoning.

Another dinoflagellate that produces a toxin and can make people ill is *Pfiesteria piscicida,* discovered in 1988 in the waters off the coast of North Carolina. *Pfiesteria* has 24 stages to its life cycle but assumes three basic forms—flagellated cells, amoeboid cells, and cysts (**figure 22.11**). Researchers suspect that these organisms normally exist in nontoxic forms in estuaries but that excrement and secretions from schooling fish such as Atlantic menhaden trigger a switch to the toxin-producing forms. Just 300 *Pfiesteria* cells per milliliter of water (about 20 drops) is enough to cause bloody sores on fish or kill them. In some areas of extensive "fish kills," the density of dinoflagellates reaches 250,000 cells per milliliter! The toxin release takes only a few hours, and then the poison breaks down rapidly, but the fish die over the next several days. The short-lived nature of the toxin has made it very difficult to study.

Ginkgo leaves, powder, and capsules.

# Plantae

## Nature's Botanical Medicine Cabinet

Plant-based medicines have been a staple of many cultures for millennia. In Europe, they are called phytomedicines; in the United States and Canada, herbal remedies or botanicals. Before chemists began synthesizing drugs in the 1930s, pharmaceuticals came entirely from nature—mostly plants. Today, 25% of prescription drugs in the United States are derived from plants, and the proportion is higher in Europe.

Many other plant products are sold as food supplements. In the United States, plant products need not undergo rigorous testing, and manufacturers are not permitted to claim that the products treat specific illnesses. Still, many people take *Ginkgo biloba* to improve memory, St. John's wort to fight depression, and *Echinacea* to bolster immunity. Clinical trials are testing whether scientific evidence supports anecdotal reports of benefits from these and other plant extracts.

Medicinal use of botanical substances dates from ancient times. Remains of prehistoric settlements include parts of poppy plants that once yielded opiates used to dull pain. Native peoples in East Africa chewed twigs from the Indian neem tree to prevent tooth decay. Other animals may use plant biochemicals, too. A Navajo Indian story tells of a bear teaching people to use the flowering plant *Ligusticum* to heal damaged skin. In an experiment, Kodiak bears in captivity chewed the roots of the plant and rubbed the mash on sore paws. Chimpanzees and certain birds eat plants that contain compounds which kill bacteria or parasitic worms.

# Chapter Preview

1. Plants are multicellular eukaryotes that have cellulose cell walls and use starch as a carbohydrate reserve. Most photosynthesize.

2. Plants have alternation of a sporophyte (diploid) phase and a gametophyte (haploid) phase.

3. Plants are classified by presence or absence of vascular tissue, seeds, and flowers and fruits. Bryophytes lack vascular tissue. Vascular plants have vascular tissue and may be seedless or produce seeds. Gymnosperms have naked seeds, and angiosperms have seeds enclosed in fruits.

4. Plants originated about 480 million years ago from a type of green alga. By 360 million years ago, plants thrived on dry land. Adaptations that made this possible include a vascular tissue, roots and leaves, a waterproof cuticle, and stomata. By 200 million years ago, the angiosperms appeared and by 90 million years ago had diversified greatly.

5. Bryophytes are small green plants lacking vascular tissue, supportive tissue, and true leaves and stems. Lignin hardens bryophytes, and rhizoids anchor them to the ground, where they absorb water and nutrients.

6. In bryophytes, the gametophyte is dominant, and sexual reproduction requires water for sperm to travel through. Sperm from an antheridium travel to eggs in an archegonium.

7. The three divisions of bryophytes are liverworts, hornworts, and mosses.

8. Seedless vascular plants have the same pigments, reproductive cycles, and starch storage mechanisms as bryophytes, but they also have vascular tissue and a dominant sporophyte generation.

9. The seed-producing vascular plants are the gymnosperms and angiosperms. Neither requires water for sperm to meet eggs.

10. Angiosperms are mostly terrestrial. The two major groups are monocotyledons and dicotyledons. The success of the angiosperms may reflect coevolution with animals and dinosaur foraging habits.

## 23.1 Plant Characteristics and Origins

The planet would not be what it is today were it not for plants. When plants and fungi settled the land together hundreds of millions of years ago, they set into motion a complex series of changes that would profoundly affect both the living and nonliving worlds. The explosion of photosynthetic activity from plants altered the atmosphere, lowering carbon dioxide levels and raising oxygen content. Because plants are autotrophs, they came to form the bases of the intricate food webs that sustain life. Tall trees provided diverse and nutrient-packed habitats for many types of animals, while flowers fed insects and birds. Leaf litter accumulating on forest floors created a rich soil for countless microorganisms, insects, and worms, and when washed into streams and rivers, fueled a spectacular diversification of fishes and other aquatic animals.

In our lives, plants provide food, fuel, clothing, shelter, and medicines, as the chapter opening essay describes. Yet the plants that we cultivate are only a tiny percentage of the vastly diverse kingdom Plantae. This chapter introduces the diversity of plants, and chapters 27 through 30 expand upon this information.

## Plants Have Cell Walls and Are Largely Autotrophic

Plants are multicellular eukaryotes that have cellulose-rich cell walls and use starch as a nutrient reserve. The vast majority of plants are photoautotrophic, using photosynthesis to convert solar energy into chemical energy in glucose. The cells of photosynthetic plants have chloroplasts that contain chlorophyll *a* and *b* and other pigments. Unlike the green algae from which they arose, most plants live in terrestrial habitats.

Plants and certain algae have a characteristic type of life cycle termed **alternation of generations. Figure 23.1** repeats figure 10.2, which depicts how a plant alternates between a diploid stage called a **sporophyte** ("spore-making body") and a haploid stage called a **gametophyte** ("gamete-making body"). The sporophyte develops from a zygote that forms when gametes come together at fertilization. In turn, certain sporophyte cells undergo meiosis and produce haploid **spores,** which divide mitotically to form the gametophyte. It is the gametophyte that produces gametes that fuse at fertilization, starting the cycle anew.

Complex plants, distinguished by specialized tissues that transport water and nutrients, have a sporophyte that is larger and more visible than the gametophyte. In the flowering plants, for example, the gametophyte is microscopic, and among the conifers, it is visible but small. A blade of grass, a maple tree, and a rose bush are all sporophytes. In contrast, simpler plants such as mosses have a more prominent gametophyte stage and a reduced sporophyte. However, there are exceptions. Botanists thought the Killarney fern (*Trichomanes speciosum*) was extinct because changing weather conditions proved challenging for the sporophyte, which became rare. The gametophyte, however, persisted in many parts of Europe because it was better adapted to the drier environment that has prevailed since the 1700s.

Classification of plants is fairly straightforward compared with some other groups of organisms, although scientists disagree on some aspects (**figure 23.2**). Plants can be classified by whether or not they have vascular tissue to transport water and nutrients. The simplest plants, which lack vascular tissue and

A

C

B

D

**FIGURE 23.14** **Gymnosperm Diversity.** **(A)** *Zamia furfuracea* is a cycad that produces large cones. **(B)** *Ginkgo biloba* is also called the maidenhair tree. Plants are either male or female. Male trees are preferred for cultivation because seeds on female trees emit an odor of rot. **(C)** The bristlecone pine (*Pinus longaeva*) is one of the most ancient plants known. It is a conifer. **(D)** *Ephedra* is a gnetophyte that is the source of the stimulant and decongestant ephedrine.

coastal redwood. The Pacific yew is a source of paclitaxel, the drug used to treat certain cancers.

**Gnetophytes** (division Gnetophyta) include some of the most distinctive (if not bizarre) of all seed plants: *Ephedra*, *Gnetum*, and *Welwitschia*. Botanists have struggled with the classification of these plants. Originally, scientists placed them with the conifers, but certain life history details led to speculation that gnetophytes were closely related to the flowering plants (see figure 23.13). Molecular evidence, however, may yet relocate these puzzling plants back to the conifers.

*Ephedra,* known as Mormon tea, or Ma Huang, grows in desert regions worldwide. Some varieties of this plant contain a powerful decongestant, ephedrine, that also stimulates the central nervous system and is sometimes abused as the drug "herbal ecstasy" (figure 23.14D). Some diet pills formerly contained ephedrine, even

though this drug increases the risk of heart attack, stroke, and sudden death. Another compound from this plant, pseudoephedrine, has the decongestant effect without affecting the nervous system and is used in many preparations that treat the common cold.

*Welwitschia* is a slow-growing plant that lives in African deserts, where it gets most of its water from fog. Mature plants have a single pair of large, strap-shaped leaves that persist throughout the life of the plant.

## The Pine Exemplifies the Gymnosperm Life Cycle

Like other plants, pines have an alternation of generations. Cones are the reproductive structures of pines (**figure 23.15**). Large female cones bear two ovules (megasporangia) on the upper

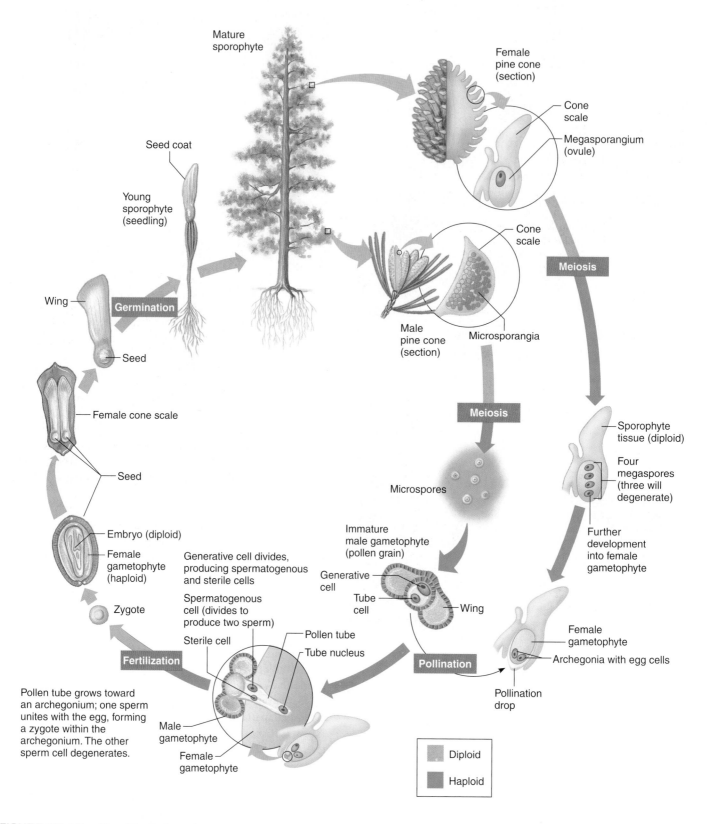

**FIGURE 23.15    Pine Life Cycle.**    Alternation of generations occurs in all plants, but in gymnosperms (and angiosperms), the gametophyte generation consists of just a few cells (*lower right* part of cycle). As in all plants, male and female gametophytes produce sperm and egg cells. Each female cone scale has two ovules (only one is visible in the figure), each of which produces a gametophyte with two or three archegonia, but only one egg per ovule is fertilized. Pollen, the male gametophyte, delivers a sperm cell to an egg via a pollen tube. More than a year after pollination, each fertilized egg (zygote) completes its development into a seed, which is the young sporophyte. Once the sporophyte is a mature tree, it produces male and female cones to complete the life cycle.

surface of each scale. Through meiosis, each ovule produces four haploid structures called megaspores, three that degenerate and one that continues to develop into a female gametophyte. Over many months, the female gametophyte undergoes mitosis. Finally, two to six archegonia form, each of which houses an egg ready to be fertilized.

Small male cones bear pairs of microsporangia on thin, delicate scales. Through meiosis, these microsporangia produce microspores, which eventually become pollen grains (immature microgametophytes). The microsporangia burst, releasing millions of winged pollen grains. Tapping a male cone in the spring releases a cloud of pollen. Most of these pollen grains, however, never reach a female cone.

Pollination occurs when airborne pollen grains drift between the scales of female cones and adhere to drops of a sticky secretion. Female cones are above male cones on most species, which makes it unlikely that pollen will drift upwards to land on female cones on the same tree. This is an adaptation that encourages outcrossing (mating between different individuals), which fosters new combinations of genes in the next generation.

After pollination occurs, the cone scales grow together, and a structure called a **pollen tube** begins growing through the ovule toward the egg. But the pollen is not yet mature—before the pollen tube reaches the egg, the pollen grain must divide twice more to become a mature microgametophyte. Two of the cells become active sperm cells, and one of them fertilizes the egg cell. The whole process is so slow that fertilization occurs about 15 months after pollination.

Within the ovule, the haploid tissue of the megagametophyte nourishes the developing embryo. Following a period of metabolic activity, the embryo becomes dormant, and the ovule develops a tough, protective seed coat. It may remain in this state for another year. Eventually, the seed is shed, to be dispersed by wind or animals. If conditions are favorable, the seed germinates, giving rise to a new tree.

---

### Reviewing Concepts

- Evolution of seeds continued the trend of adapting to life on land.
- The gymnosperms have "naked" seeds.
- Angiosperms enclose their seeds in fruits.
- Both groups are very diverse.

---

## 23.5   Angiosperms: Flowering Vascular Plants

Angiosperms (division Anthophyta or Magnoliophyta) are plants that produce flowers (**figure 23.16**). The angiosperm life cycle is complex, with alternation of generations and double fer-

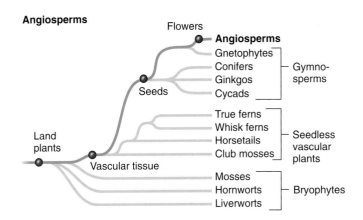

**FIGURE 23.16**   **Angiosperms.**   The angiosperms are distinguished by flowers and by seeds that are enclosed in fruits.

tilization. Chapter 29 considers angiosperm reproduction in detail, and figures 29.4 and 29.5 depict the life cycle.

### Angiosperms Are a Huge, Diverse Group

More than 95% of modern plants are angiosperms (**figure 23.17**), and about half of the more than 230,000 species live in tropical rain forests. Except for conifer forests and moss-lichen tundras, angiosperms dominate major terrestrial zones of vegetation. Angiosperms, especially grain crops such as wheat, rice, and corn, are staples in the human diet. There are even rare types of flowering plants that "eat" other types of organisms.

The two major classes of angiosperms are **monocotyledons** (monocots) and **dicotyledons** (dicots), which are distinguished by whether there are one or two embryonic seed leaves, or cotyledons (**figure 23.18**). The groups also differ by the other characteristics described in **Table 23.1** (many of these structural differences are further described in chapter 27). However, these differences are far from absolute, and molecular evidence indicates that several types of plants thought to be dicots based on their seed leaves are actually more closely related to monocots. The monocots apparently are monophyletic (forming a clade), but the dicots are a more complex group. • **systematics, p. 322**

### Why Have Angiosperms Been So Successful?

Flowering plants dominate the contemporary landscape, yet they are relatively recent arrivals. What accounts for the success of the angiosperms?

Coevolution with animal pollinators may have played a part in the angiosperm success story. Insects, reptiles, birds, and mammals transport pollen more efficiently than wind, which most gymnosperms rely on. Enclosing seeds within fruits also encourages widespread dispersal by animals.

Another hypothesis to explain the success of the flowering plants is that the dietary habits of dinosaurs benefited

# Investigating Life

## Carnivorous Plants

To a human observer, the sundew plant is a magnificent member of a swamp community. To an insect, however, the plant's beckoning club-shaped leaves, bearing tiny nectar-covered tentacles, are treacherous. Once the insect alights on the plant and begins to enjoy its sweet, sticky meal, the surrounding hairlike tentacles begin to move. Gradually, they fold inward, entrapping the helpless visitor and forcing it down toward the leaf's center. Here, powerful digestive enzymes dismantle the insect's body and release its component nutrients. After 18 hours, the leaves open. All that remains of the previous day's six-legged guest is a few bits of indigestible matter.

The Venus's flytrap (*Dionaea muscipula*) is another insect-eating plant, native only to North and South Carolina. Its two-sided leaves have highly sensitive trigger hairs that capture insects, which are then digested. (These plants, which are endangered in the wild because of habitat loss, are commercially produced by cloning.) The pitcher plant (*Sarracenia*) entices insects to explore a hornlike structure that collects rainwater. When the hapless insect falls in, digestive enzymes go to work. By summer's end, pitcher plants contain many leftover insect parts. **Figure 23.A** shows the sundew, Venus's flytrap, and pitcher plant.

Only about 450 plants display a carnivorous lifestyle, which is an adaptation that permits survival in nutrient-poor soils. All carnivorous plants were thought to attract and attack insects, until the recent discovery of a small flowering plant from South Amer-

**FIGURE 23.A    Carnivorous Plants.    (A)** The sundew plant grows in the swamps of upstate New York and the Pine Barrens of New Jersey. Its sticky leaves are insect traps. **(B)** The leaves of the Venus's flytrap snap shut to catch prey. This plant is native only to North and South Carolina. **(C)** Leaves of the pitcher plant *Sarracenia* trap and kill insects, which the plant digests.

A

B

C

# 23.1

ica and tropical Africa whose victims are ciliates and other protista. ● **ciliates, p. 413**

*Genlisea aurea* is an unusual plant among unusual plants (**figure 23.B**). It lives in white sands and on outcroppings of moistened rock, with a small rosette of leaves close to the ground and yellow or purple flowers extending up a foot or so from the leaves. Underground are extensive outgrowths that look like roots, but are actually highly modified leaves. Rather than photosynthesizing, as leaves typically do, these

structures have long hairs that ensure that ciliates swimming toward the plant cannot retreat.

Botanists had known for some time that *Genlisea aurea* was a carnivorous plant, but no one had observed its prey, assumed to be an insect. But the size of the opening between the traps and the absence of insect evidence suggested another type of meal. So, researchers grew the plant in a greenhouse and fed it something that seemed the right size—ciliates. The results were striking. The

ciliates headed right for the underground leaves, which are booby-trapped with glands that secrete digestive enzymes. The ciliates did not approach roots of other plants, sending researchers looking for a chemical attractant, which they indeed found. Further experiments showed that when ciliates were fed radioactive food, the radioactivity appeared 2 days later in the rosette of leaves. Finally, sampling from the field identified several types of ciliates in the vicinity of *Genlisea*'s unusual underground traps.

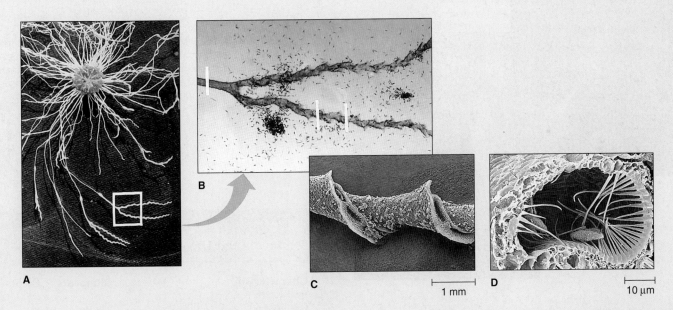

**FIGURE 23.B    A Ciliate-Eating Plant.**    (**A**) The rootlike structures in this *Genlisea* plant are actually underground leaves, highly modified to trap ciliated protista. (**B**) Ciliates swarm around a *Genlisea* leaf. (**C**) A close-up view of the leaf reveals the tiny ciliate traps and the long hairs (**D**) that prevent escape.

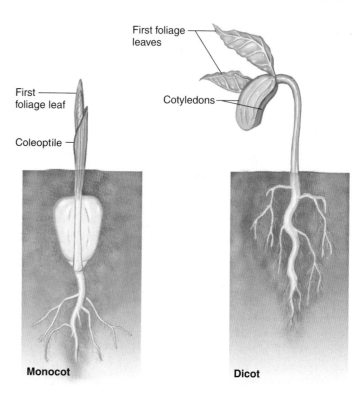

**FIGURE 23.18   Monocots and Dicots.** Monocots have a single seed leaf (not visible in this figure), whereas dicots have two. Other characteristics distinguish these groups. Molecular evidence indicates the monocots share a common ancestor, whereas the dicots form a more complex group.

**FIGURE 23.17   Gallery of Angiosperms.** The angiosperms exhibit an astonishing variety of flowers and fruit. **(A)** In bananas (genus *Musa*), flowers occur in clusters. Yellow flowers and green developing fruits are visible in this photograph. **(B)** Red maple (*Acer rubrum*) produces small, bright red flowers that develop into winged fruits after fertilization. **(C)** The red passion vine (*Passiflora coccinea*) is a tropical plant with very showy flowers. **(D)** Cattails (*Typha latifolia*) are familiar wetland plants. The brown cylindrical "tails" are actually spikes of tiny brown flowers.

## TABLE 23.1

### Monocots and Dicots

| Characteristic | Monocots | Dicots |
|---|---|---|
| Number of seed leaves (cotyledons) | 1 | 2 |
| Number of flower parts is a multiple of | 3 | 4 or 5 |
| Presence of wood (secondary growth) | No | Yes |
| Configuration of leaf veins | Parallel | Reticulate (meshlike) |
| Pattern of vascular bundles in stem | Throughout | Ring |
| Root system | Fibrous | Taproot |
| Examples | Grasses | Roses |
|  | Orchids | Oaks |
|  | Lilies | Beans |

angiosperms while harming gymnosperms. The enormous sauropods, such as *Apatosaurus* and *Brachiosaurus,* roamed in great herds, browsing on the tops of conifers and other gymnosperms. This nibbling would not have killed large trees or harmed the smaller seedlings, which the giants could not reach. But by the late Cretaceous, smaller dinosaurs had begun to replace the large herbivores (**figure 23.19**). The low browsers probably devastated the gymnosperms by eating small seedlings before the plants could reach maturity and produce seeds. Because the first angiosperms were smaller and herbaceous (nonwoody), they probably grew and reproduced more rapidly than the woody gymnosperms and therefore were more likely to produce seeds before they were eaten. Furthermore, small dinosaurs eating gymnosperms opened up new habitats for angiosperm invasion and evolution. When the dinosaurs became extinct in the Tertiary period, angiosperms exploded in abundance and diversity.

Perhaps dinosaurs' dining habits and plants' coevolution with other types of animals influenced the rapid evolution of angiosperms. Whatever the precise triggers, the spurt of angiosperm diversity in the Cretaceous period set the stage for their spectacular global diversification.

## Reviewing Concepts

- Angiosperms are plants with flowers and fruits.
- They represent the most complex form of plant life, using the flowers and fruits to disseminate the species.
- Most of the colorful and edible plants we use are members of this group.

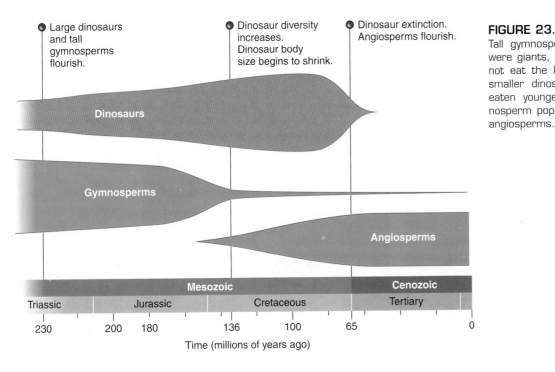

**FIGURE 23.19** **The Rise of Angiosperms.** Tall gymnosperms flourished when dinosaurs were giants, perhaps because the reptiles did not eat the lower portions of the plants. As smaller dinosaurs appeared, they may have eaten younger gymnosperms, reducing gymnosperm populations and opening habitats for angiosperms.

## Connections

Plants represent a very successful life-form. They vary in complexity and approaches to reproduction. Their evolution provided the means for various plants to invade and thrive on land, and it is represented in the features of living plants today. These differences show how life finds a way to survive in nearly every imaginable environment. As we explore the fungi, notice how they take a different approach to reproducing and maintaining life. The photosynthetic abilities of plants, combined with their existence in a variety of environments, provide the energy needed for nearly all life. As we learn more about the different ways that other organisms use to solve the problems of life, we learn more about ourselves.

# Student Study Guide

## Key Terms

alternation of
   generations *422*
angiosperm *423*
antheridium *425*
archegonium *425*
bryophyte *423*
charophyte *423*
club moss *430*
conifer *432*

cycad *432*
dicotyledon *435*
fronds *431*
gametophyte *422*
gemma *425*
ginkgo *432*
gnetophyte *433*
gymnosperm *423*
hornwort *427*

horsetail *430*
lignin *423*
liverwort *425*
monocotyledon *435*
moss *427*
phloem *423*
pollen grain *431*
pollen tube *435*
rhizoid *425*

rhizome *430*
seedless vascular plant *423*
sporangium *425*
spore *422*
sporophyte *422*
true fern *431*
whisk fern *430*
xylem *423*

## Chapter Summary

### 23.1  Plant Characteristics and Origins

1. Plants are diverse multicellular organisms that photosynthesize, have cellulose cell walls, and exhibit alternation of generations.

2. Plants evolved from green algae and are classified by the presence or absence of transport tissues, seeds, flowers and fruits, and by molecular evidence.

### 23.2  The Bryophytes: Liverworts, Hornworts, and Mosses

3. The liverworts, hornworts, and mosses are simple plants that lack transport vessels and roots, so they must absorb nutrients and water directly from their surroundings.

4. With lignin, these species were able to form more elaborate structures than algae could.

### 23.3  Seedless Vascular Plants: Club Mosses, Horsetails, Whisk Ferns, and True Ferns

5. The club mosses, horsetails, whisk ferns, and true ferns retain features that initially enabled plants to live on dry land—a protective cuticle and stomata.

6. They do not have seeds, but they do have true leaves, stems, and roots.

### 23.4  Gymnosperms: Vascular Plants with "Naked Seeds"

7. Evolution of seeds continued the trend of adapting to life on land.

8. The gymnosperms have "naked" seeds, whereas the angiosperms enclose their seeds in fruits.

### 23.5  Angiosperms: Flowering Vascular Plants

9. Angiosperms are plants with flowers and fruits.

10. They represent the most complex form of plant life, using the flowers and fruits to disseminate the species.

11. Most of the colorful and edible plants we use are members of this group.

## What Do I Remember?

1. Describe the alternation of generations in three of the major groups of plants.
2. What factors enabled plants to live on land?
3. What do all plant life cycles have in common?
4. Describe the importance of lignin to plant evolution.
5. List the major characteristics that distinguish the major groups of plants.
6. Compare and contrast the gymnosperms and angiosperms.

**Fill-in-the-Blank**

1. The _____ were the green algae that gave rise to modern plants.
2. Hairlike extensions called _____ help the bryophytes absorb water.
3. A(n) _____ is usually the diploid stage of a plant that we recognize.
4. The word "gymnosperm" literally means _____ _____.

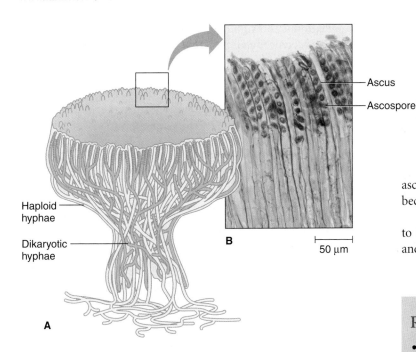

**A**

Haploid hyphae

Dikaryotic hyphae

**B**    50 μm

— Ascus

— Ascospore

**FIGURE 24.11  Ascomycete Reproduction.** (A) Cross section of a typical cup-shaped ascomycete fruiting body, with developing asci. (B) Asci often contain eight ascospores, which are the products of meiosis and one subsequent mitotic division.

ascomycetes. They were formerly considered deuteromycetes because sexual structures were not observed.

Ascomycetes and basidiomycetes are sister groups. In addition to sharing DNA sequences, these fungi also have septate hyphae, and before sexual reproduction, they enter a dikaryotic stage.

### Reviewing Concepts

- The sac fungi include the delectable morels and truffles but also include species that cause disease.
- Their cells can have two nuclei, but most are haploid for most of the life cycle.
- Ascomycetes reproduce asexually by budding and sexually with ascospores.

## 24.5 Basidiomycetes: The Familiar Club Fungi

The 25,000 or so species of basidiomycetes account for about a third of all recognized fungi. They are named club fungi because during sexual reproduction, their hyphae tips form club-shaped, spore-bearing swellings called **basidia** (singular: basidium). Basidia often grow in pores or along strips of tissue called gills that are on the underside of the mushroom. Basidiomycete species differ from each other by the form of the fruiting body, the shape of the basidia, how many spores a basidium produces, and the organization of the gills or pores.

Familiar basidiomycetes include mushrooms, toadstools, puffballs, stinkhorns, shelf fungi, rusts, bird's nest fungi, and smuts (**figure 24.12**). In extreme cases, these fungi may grow quite large—a puffball from Canada is 9 feet (2.75 meters) in circumference, a mushroom in England is 100 pounds (45.4 kilograms), and the underground mycelium of *Armillaria gallica* growing in northern Michigan covers 37 acres (15 hectares; see figure 24.1B). Some mushrooms are edible, some are deadly, and others are hallucinogenic. Smuts and rusts are pathogens of cereal crops. The extensive mycelia of some basidiomycetes grow outward from their food source in a circle beneath the ground, depleting the soil of nutrients. Mushrooms poke up above these spreading mycelia, creating the "fairy rings" of folklore.

delicacies out of the ground. Squirrels and chipmunks disperse the spores of uprooted truffles. The French have attempted to farm truffles. Natural or farmed, they sell for $500 a pound! Morels grow aboveground and are easy to spot, although care must be taken to eat only nontoxic species. Morels grow in shaded areas that are moist and full of nutrients, such as orchards, swamps, flower beds, and forests, especially after a fire.

Reproduction among ascomycetes is both asexual and sexual. For example, the well-studied yeast *Saccharomyces cerevisiae* reproduces asexually by budding (see figure 24.4) but also reproduces sexually. In multicellular ascomycetes, asexual spores (conidia) pinch off from the ends of exposed hyphae. In sexual reproduction, compatible mating types attract each other with pheromones. Their hyphae fuse, but unlike in zygomycetes, the individual nuclei from the two parents do not immediately merge. The result is a dikaryotic cell, which occurs only in ascomycetes and basidiomycetes. When dikaryotic cells divide, the two nuclei undergo mitosis separately, each retaining its genetic identity.

In ascomycetes, the sexual fruiting body consists of haploid tissue (often in a cuplike or bottlelike shape) that houses a fertile layer of dikaryotic cells. Eventually, the two nuclei in each dikaryotic cell fuse to form a diploid zygote, which immediately undergoes meiosis to form four haploid nuclei. Usually each haploid nucleus divides once mitotically to yield a total of eight haploid **ascospores,** so named because they form in saclike structures called **asci** (singular: ascus), as shown in **figure 24.11**. After dispersal, ascospores either persist in the environment or germinate to yield new haploid fungi.

The 30,000 ascomycete species include 500 that are single-celled yeasts, many that are multicellular, and some that are both. Most ascomycetes are haploid for much of their life cycles. Molecular data have placed *Penicillium* and *Candida* within the

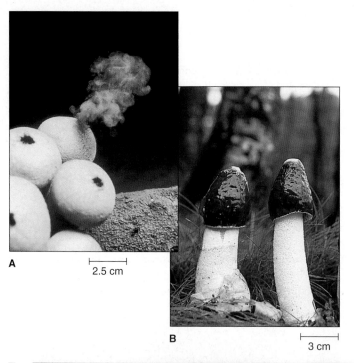

**FIGURE 24.12** A Gallery of Club Fungi (Basidiomycetes). (**A**) Puffballs (genus *Lycoperdon*), (**B**) stinkhorns (*Phallus impudicus*), (**C**) turkey tail bracket fungi (*Trametes versicolor*), and (**D**) bird's nest fungi (order Nidulariales).

Reproduction among the basidiomycetes is asexual and sexual. Asexual reproduction occurs by budding, fragmentation, and production of asexual spores. **Figure 24.13** shows how a mushroom is a key player in the sexual reproductive cycle. Lining the mushroom's gills (or pores) are dikaryotic basidia, each housing two nuclei of compatible mating types. The haploid nuclei in each basidium fuse, giving rise to a diploid zygote, which immediately undergoes meiosis to yield four haploid nuclei. Each nucleus migrates to a **basidiospore,** which, after being shed, germinates to produce a haploid hypha. Two hyphae of compatible mating types fuse to create a secondary mycelium whose cells are once again dikaryotic. This mycelium grows, producing a mushroom, and the cycle begins anew.

### Reviewing Concepts

- The club fungi include many familiar species and may grow quite large.
- They have two genetically different nuclei per cell for most of the life cycle.
- Spore-bearing structures on mushrooms carry out sexual reproduction.

## 24.6 Fungal Symbiotic Relationships

Although relatively few fungi cause disease in humans, the natural histories of fungi and plants are intimately intertwined. Fungi accompanied plants when they moved onto the land. Today, they cause thousands of plant diseases, yet they also remain an integral part of the underground root systems that sustain plant communities. Fungi can also associate with certain types of bacteria and algae to form a kind of compound organism called a **lichen.** Yet another fascinating fungal interaction is their cultivation by certain species of ants.

### Mycorrhizal Fungi Live in Association with Plant Roots

Fungal hyphae in association with roots constitute **mycorrhizae** (literally, fungus roots). Nearly all land plants have mycorrhizae, and some, such as orchids, cannot live without their mycorrhizal associates. The symbiotic relationship benefits both partners—the plant obtains water and minerals that the hyphae absorb from the surroundings, and the fungus gains carbohydrates from the plant's photosynthetic activity. The relationships among root fungi and plants can become complex, with a single plant attracting several types of fungi, and the extensive hyphae from different fungi interconnecting the roots of many plants.

gen, proteoglycans, glycoproteins, and integrins. Depending upon its exact composition, this matrix enables some animal cells to move, others to assemble into sheets, and yet others to embed in supportive surroundings, such as bone or shell.

## Animals Are Heterotrophic

Animals are heterotrophs: they obtain carbon and energy from nutrients that are part of or released by other organisms, instead of manufacturing nutrients through photosynthesis. In an ecological context, animals are consumers because they eat others. An animal may be a parasite; a detritivore that eats dead organisms or nutrients in soil; or a predator. A carnivore is a predator that eats other animals, and an herbivore eats plants. We will return to these designations in chapter 44.

## Animals May Be Categorized by Symmetry and Cephalization

Recall from chapter 1 that multicellular organisms consist of cells, which associate with others of the same or similar type to form tissues, which, in turn, build organs, which may be linked into organ systems (see chapter 31). These levels of organization are reflected in the levels of complexity of animal phyla.

One way to divide the animal kingdom is on the basis of cellular organization. Animals in subkingdom Parazoa have individual cells that do not work together in a coordinated fashion (that is, their cells do not form tissues; see chapter 31). Subkingdom Parazoa contains only one extant phylum (Porifera, or sponges). In sponges, cells aggregate to form a body, and different cells have distinct structures and functions, but they do not interact as they would in a true tissue. In contrast, all other animals (subkingdom Eumetazoa) have true tissues. For example, some cnidarians

(jellyfishes, sea anemones, and corals) have cells linked into "nerve nets" that coordinate movement. In more complex eumetazoans, tissues form organs linked into systems that carry out specific functions. The simplest animals to show this level of complexity are the flatworms (phylum Platyhelminthes), which, among others, demonstrate organ systems in their reproductive, excretory, and nervous structures.

Biologists also describe animal bodies by symmetry, which is the organization of body parts around an axis (**figure 25.2**). Many sponges are bloblike and therefore lack symmetry—they are asymmetrical. Animals with radial or bilateral symmetry are classified as Radiata or Bilateria. In an animal with **radial symmetry,** any plane passing from the oral end (the mouth) to the aboral end (opposite the mouth) divides the body into mirror images. Hydras have radial symmetry. In an animal with **bilateral symmetry,** such as a crayfish, only one plane divides the animal into mirror images. Bilaterally symmetrical animals have anterior (head) and posterior (tail) ends. These animals move through their environment in a headfirst manner. This behavior is linked evolutionarily with **cephalization,** the tendency to concentrate neural elements such as sensory organs and a brain at the anterior end of the organism. Bilaterally symmetrical body forms also have back sides (dorsal) and belly or undersides (ventral).

## Reviewing Concepts

- Animals eat other organisms, move, have cells that lack walls, and secrete extracellular matrix.
- Biologists classify animals by body symmetry, presence or absence of a body cavity, mechanisms of reproduction and development, and molecular sequences.

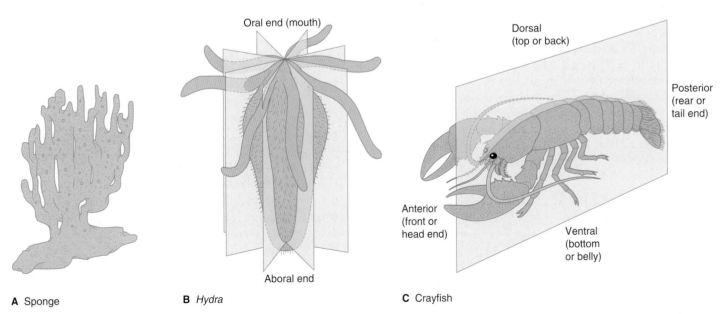

**A** Sponge  **B** *Hydra*  **C** Crayfish

**FIGURE 25.2  Animal Body Forms.** (**A**) Some sponges have an asymmetrical body form. (**B**) *Hydra* has a radially symmetrical body form. Any plane passing through the oral–aboral axis divides the animal into mirror images. (**C**) A crayfish has a bilaterally symmetrical body form. Only one plane divides the animal into mirror images.

# 25.3 Animal Reproduction and Development

Animals reproduce sexually, although there are variations on this theme. Some species can reproduce asexually, such as the budding of a hydra. Earthworms are **hermaphroditic,** producing sperm and eggs in the same individual. Among the social insects, such as bees and wasps, certain members are haploid. But for the most part, an animal's somatic cells are diploid, and the gametes (sperm and oocytes) are haploid. A sperm cell fertilizes an oocyte, restoring the diploid state in the zygote that results. The zygote—the first cell of the new organism—undergoes a series of rapid mitoses called cleavage divisions, forming the developing organism, or embryo. Chapter 41 discusses prenatal animal development in more detail.

## Embryonic Development Involves Generation of Tissue Layers

From the zygote, the embryo develops as a solid ball of cells that hollows out to form a **blastula.** The space inside is called the blastocoel. Soon, complex foldings generate a structure called a **gastrula,** composed of tissue layers called primary germ layers. Some animal species have two germ layers and are termed **diploblastic.** The immature form of a jellyfish, for example, is diploblastic and consists of two layers—**ectoderm** and **endoderm**—that sandwich a layer of noncellular material. Other species have a third layer, **mesoderm,** and are termed **triploblastic (figure 25.3).** A human embryo is triploblastic with ectoderm and endoderm sandwiching the cellular mesoderm.

● **human prenatal development, p. 806**

Development from fertilized egg to adult happens in either of two ways. Animals that undergo **direct development** hatch from eggs or are born resembling adults of the species. A newborn elephant, for example, looks like an adult elephant. Animals that directly develop have no larval stage. In contrast, an animal that develops indirectly may spend part of its life as a **larva,** which is an immature stage that does not resemble the adult of the species. Larvae undergo a developmental process called **metamorphosis,** in which they change greatly. Tadpoles and caterpillars are familiar larvae. Many aquatic animals metamorphose, some developing through several larval stages.

## The Two Major Lineages Reflect Different Development Paths

Among the bilaterally symmetrical animals (Bilateria), anatomical, developmental, and more recently, molecular evidence identifies two major lineages, called **protostomes** and **deuterostomes.** Protostomes include flatworms (tapeworms, flukes, and others), mollusks (snails, clams, slugs, octopuses, and others), annelids (earthworms, leeches, and others), roundworms (nematodes and others), and arthropods (insects, spiders, crabs, and others).

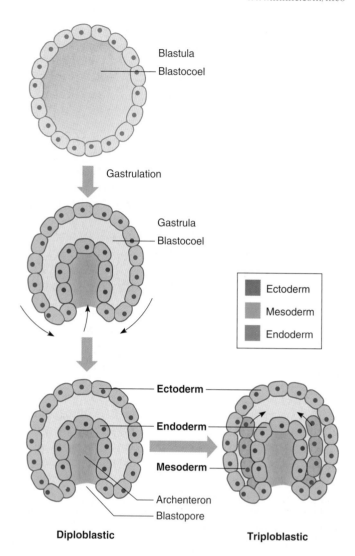

**FIGURE 25.3   Gastrulation Produces Embryonic Tissue Layers.** Early during development of many animal embryos, a hollow ball of cells called a blastula undergoes gastrulation, which generates primary germ layers. In diploblastic animals, these are the ectoderm (outer—blue) and endoderm (inner—green) layers. In triploblastic animals, a third layer, the mesoderm (red), forms between the other two.

Deuterostomes include echinoderms (sea stars, sea urchins, and other spiny-skinned animals) and chordates, the group to which vertebrates belong.

Protostomes and deuterostomes show three major distinctions very early in development (**figure 25.4**). The first major difference between protostomes and deuterostomes is the developmental potential of the first cells (blastomeres) resulting from early cleavage divisions. In protostomes, if an early cell is removed, development is severely altered. Both the removed and remaining cells go on to form disorganized masses that lack tissues derived from the missing pieces. This phenomenon, called **determinate cleavage,** occurs because the fate of each cell of the embryo is predetermined. In contrast, in deuterostomes, if one cell is removed from the four-celled stage, both the remaining stage called a trochophore, which may develop into a larva with a tiny mantle, shell, foot, and swimming organ. The trochophore larva settles to the bottom of the sea, where it develops into an adult. Freshwater mussel larvae are parasites, clamping onto gills or fins of fishes. When the larvae are developed, they detach, settle to the bottom, and mature into free-living adults. In cephalopods, terrestrial gastropods, and some other mollusks, all larval development occurs inside the egg. The hatchlings resemble adults.

Mollusks feed in diverse ways. Bivalves such as oysters, clams, scallops, and mussels are suspension feeders, eating organic particles and small organisms they strain out of the water. Mollusks concentrate pollutants or toxins produced by dinoflagellates and may become poisonous to human seafood eaters. Slugs consume vegetation so voraciously that they can rapidly defoliate certain garden plants. Some marine snails drill holes in bivalve shells to eat them. Cephalopods are active predators. Their keen eyes, closed circulatory systems, and ability to move by "jet propulsion," squirting water out of modified feet, allow them to detect and catch fast-moving prey such as fishes. Other cephalopods, such as the octopus and chambered nautilus, scavenge dead fishes.

## Reviewing Concepts

- Although mollusks appear quite different from each other, they all share certain basic body parts, such as a mantle, muscular foot, and visceral mass.
- They have complete digestive tracts, open circulatory systems, and well-developed nervous systems.
- The mantle and radula are unique to this phylum.

## 25.9   The Annelids (Annelida)

Segmented worms belong to the phylum Annelida. **Figure 25.17** shows the place of annelids on the protostome branch of the animal family tree. The annelids' main identifying characteristic is metamerism, having a body composed of repeated segments.

Phylum Annelida includes three easily recognized classes. Oligochaetes, such as earthworms, have a few small bristles called setae on their sides (**figure 25.18**). The clitellum, a saddlelike thickening around a few body segments, secretes mucus when the worms copulate and a cocoon that protects the eggs and embryos early in development. The most diverse marine segmented worms, the polychaetes, have pairs of fleshy, paddlelike appendages called parapodia that they use for locomotion. Parapodia are located on the sides of segments and have many long setae embedded in them. Setae anchor annelids and keep them from slipping backwards when they move. The third class of annelids, the Hirudinea, includes the leeches. They lack parapodia and setae and have suckers and superficial rings called annuli within each segment.

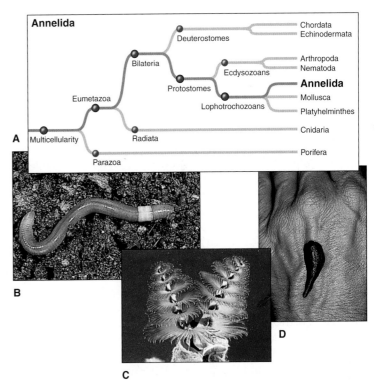

**FIGURE 25.17   Annelid Diversity.** (**A**) Annelids are segmented protostomes (lophotrochozoans). (**B**) Oligochaetes, such as this earthworm (*Lumbricus terrestris*), show the metamerism (repeated body parts) characteristic of phylum Annelida. (**C**) *Spirobranchus giganteus*, the Christmas-tree worm, is a sedentary, tube-dwelling polychaete that uses its double crown of radioles to filter organic particles from seawater. (**D**) The third class of annelids, Hirudinea, includes the leeches, which have suckers and lack parapodia and setae.

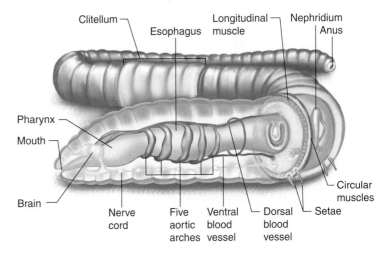

**FIGURE 25.18   Anatomy of an Earthworm.** Earthworms show the body segmentation, both internal and external, that is typical of annelids. Groups of cell bodies (ganglia) above the pharynx fuse to form a primitive brain that connects by nerve rings to the ventral nerve cord. The blood system is closed, and the anterior end of the dorsal blood vessel contracts rhythmically, forcing blood through five aortic arches that connect with the ventral blood vessel. Setae are bristles that provide traction for locomotion. Pairs of nephridia in each segment remove waste from the coelomic fluid and blood.

Annelids feed in several ways. Earthworms and several other annelids are deposit feeders, which take in large quantities of soil or aquatic sediment and strain out organic material for food. Many polychaetes are suspension feeders, using their crowns of ciliated tentacles, called radioles, to filter small organisms and organic particles from seawater. Some polychaetes are predators with formidable jaws. Many leeches suck blood from vertebrates, but most eat small organisms such as arthropods, snails, or other annelids. In medicine, a blood-thinning chemical (anticoagulant) from leeches is used to promote free circulation to surgically reattached digits and ears.

Annelids have complex organ systems. Polychaetes have various external structures that carry out gas exchange. Oligochaetes and leeches respire more simply, by diffusion through the body wall. Annelids have a **closed circulatory system,** meaning the blood is always confined to blood vessels. The coelom serves as a hydrostatic skeleton, which provides support with a constrained fluid; muscles work against it as the worm crawls, burrows, or swims. The nervous system consists of a brain connected around the digestive tract to a ventral nerve cord, with lateral nerves running through each segment. • **hydrostatic skeletons, p. 679**

Leeches and oligochaetes reproduce differently than polychaetes. Leeches and oligochaetes are hermaphroditic and also cross-fertilize. Two individuals copulate, each discharging sperm that its partner temporarily stores. Eggs and sperm then are shed into a cocoon that each worm's clitellum secretes. In earthworms the clitellum is visible at all times, but in leeches it is visible only during breeding season. Juvenile leeches and earthworms, resembling adults, hatch from the fertilized eggs and develop within the cocoon.

In contrast to leeches and oligochaetes, polychaetes have separate sexes. Eggs and sperm are usually shed into the ocean, where fertilization occurs. The animals develop indirectly, passing through a trochophore larval stage before undergoing metamorphosis to the adult form. The Samoan palolo worm has a particularly interesting method of reproduction, using specialized posterior body segments. On one night with a full moon in October or November, these segments detach, swarm to the surface of the ocean, and break open, releasing many gametes. The anterior part of the body survives and can regenerate the posterior portion the following year.

One type of polychaete, the Pompeii worm *Alvinella pompejana,* tolerates the largest known environmental temperature gradient. Adhering to the sides of deep-sea hydrothermal vents off the coast of Mexico, the anterior end of the worm might be 81°C (178°F), while its posterior end is 21°C (70°F)! The temperature gradient occurs because at the vent sites, 300°C (572°F) water spews from Earth's interior and hits the seawater, which is 2°C (36°F). The worms, which are about 6 centimeters (2.4 inches) long, spend most of their time in papery tubes that they secrete, poking their heads out, but they can travel as far as a meter to eat bacteria (**figure 25.19**).

Other polychaetes thrive in cold. The "ice worm" *Hesiocaeca methanicola* lives on the bottom of the Gulf of Mexico in deposits of methane that are so cold they form solid blocks. The worms consume bacteria that produce nutrients by metabolizing the methane. If the temperature rises slightly, the ice (methane hydrate) turns into gas. But while it is solid, it swarms with polychaete worms.

**FIGURE 25.19    The Pompeii Worm.**    The polychaete Pompeii worm *Alvinella pompejana* lives along deep-sea hydrothermal vents, where it tolerates extreme temperature gradients. Biologists don't know how the worms do it! A typical Pompeii worm is about 10 centimeters long. Masses of bacteria make up the furlike "coat" on its back.

## Reviewing Concepts

- Annelids are segmented worms such as leeches and earthworms.
- These animals have bilateral symmetry and complex organ systems, and some thrive in extreme environments.

# 25.10    The Roundworms (Nematoda)

Roundworms of the phylum Nematoda are cylindrical worms without segments. Biologists have recently reclassified nematodes from the pseudocoelomate lineage (see figure 25.6, inset) to the ecdysozoan branch of the protostome clade (**figure 25.20**).

## TABLE 25.1

### Nematode Parasites of Humans and Domestic Animals

| Worm (Common Name) | Host | Mode of Infection | Distribution |
|---|---|---|---|
| *Enterobius vermicularis* (pinworm) | Humans | Inhaling eggs; fingers to anus | Worldwide;* most common worm parasite in United States |
| *Necator americanus* (New World hookworm) | Humans | Larvae burrow into feet from contaminated soil; can cause anemia | New World;† southeastern United States |
| *Ascaris lumbricoides* (giant human intestinal roundworm) | Humans | Eating infective-stage eggs in contaminated food | Worldwide;* Appalachia and southeastern United States |
| *Trichinella spiralis* (trichina worm) | Humans, pigs, dogs, cats, rats | Eating undercooked meat with encysted larvae; can be fatal | Worldwide;* throughout United States |
| *Wuchereria bancrofti* (filarial worm) | Humans | Mosquito; can cause elephantiasis (severe swelling) | Tropics |
| *Dirofilaria immitis* (heartworm) | Dogs | Mosquito; can be fatal | Worldwide |

*Distribution is scattered throughout the world; appears to varying degrees in different places.

†Distribution is scattered throughout the New World western hemisphere; appears to varying degrees in different places.

They are tapered at both ends and have a surrounding nonliving cuticle. Most are barely visible to the unaided eye and are free-living in soil or in sediments of aquatic ecosystems. Nematodes are extremely abundant. One sample of sediment from Loch Ness, the largest lake in Great Britain, yielded 27 species of nematodes, some never seen before! Some roundworms parasitize plants or animals and are acquired in a variety of ways (**table 25.1**). Many developmental biologists use the nematode *Caenorhabditis elegans* as a model organism to trace cell lineages. As a result, it is among the best studied of all animals.

Roundworms have only longitudinal (lengthwise) muscles, which limits their range of movement. They have complete digestive systems (one-way flow from mouth to anus) and are thus more complex than flatworms, but they still have poorly developed anterior ends and lack circulatory or respiratory organs. A roundworm's hydrostatic skeleton consists of a fluid-filled pseudocoelom. An external layer of tissue, called the hypodermis, secretes the cuticle.

Most species of nematodes have separate sexes. Females of parasitic species produce large numbers of eggs that are extremely resistant to harsh environmental conditions such as dryness and exposure to damaging chemicals. Most nematodes undergo direct development, with most parasitic forms living in only one host. Mosquitoes can transmit certain roundworm infections.

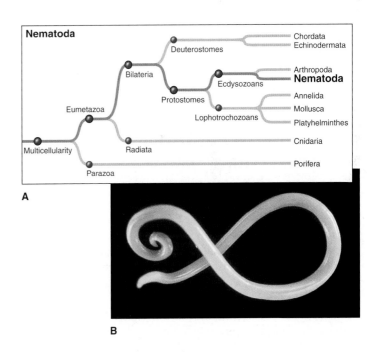

**FIGURE 25.20** **Roundworms.** **(A)** Roundworms are pseudocoelomate protostomes (ecdysozoans) and are among the most abundant animals. **(B)** Millions of humans harbor the giant intestinal roundworm *Ascaris lumbricoides*. The parasite is acquired through contact with human feces containing eggs, which hatch in the intestine. This mature male is about 25 centimeters long.

## Reviewing Concepts

- A chunk of soil houses hundreds, maybe thousands, of roundworms.
- Nematodes are unsegmented, lack respiratory and circulatory organs, and are distinguished from the simpler flatworms by their complete digestive systems.

## 25.11    The Arthropods (Arthropoda)

If biological success is judged in terms of diversity, perseverance, and sheer numbers, then the phylum Arthropoda certainly qualifies as the most successful group of animals. Biologists hypothesize that more than 75% of all animal species are arthropods, the group that includes insects, crustaceans, and arachnids. More than a million species have been recorded already, and biologists speculate this number could double.

Arthropods intersect with human society in about every way imaginable, both to our benefit and detriment. Mosquitoes, flies, fleas, and ticks, for example, spread infectious diseases when they consume human blood (**see table 25.2**, figure 22.13). Bees sting, termites chew wood, and many moth larvae destroy crops. Yet, entire industries exploit certain arthropod activities. Bees produce wax and honey; silkworms contribute their secretions to our finest garments; and shrimp, crabs, and lobsters are delicacies. Less apparent to the casual observer are parasitic wasps that deposit their eggs in caterpillars or in fly pupae (cocoons). The larvae hatch inside the pest, using its nutrients to mature new wasps, which are harmless to humans. In fact, some farmers exploit these insects, called parasitoids, as biocontrol agents. The larvae of some mosquito species prey on the larvae of other mosquito species. Wolf spiders eat many crop pests, and some farmers place small hay refuges in their fields to attract these beneficial arachnids. Finally, in our homes, dust mites process billions of human skin cells. • **communities, p. 866**

### Arthropods Have Exoskeletons and Complex Organ Systems

Part of the arthropod success story traces to their lightweight **exoskeleton**, which protects from the outside. The versatile exoskeleton is made of chitin (a nitrogen-containing polysaccharide, see figure 2.14B) bound with protein. Chitin toughens the exoskeleton. In crustaceans, such as crabs and lobsters, calcium salts harden the exoskeleton. This outer covering can be thick and rigid in some places and thin and flexible in others, providing for moveable joints between body segments and within appendages. (Arthropoda means "jointed foot.") Internal muscles span the joints, creating hollow lever systems for movement (see figure 34.9). The exoskeleton can also form a variety of mouthparts, sensory structures, copulatory organs, ornaments, and weapons. An exoskeleton has a drawback, though—to grow, an animal must crawl out of it (molt) and secrete a bigger one. • **exoskeletons, p. 679**

Arthropods have **open circulatory systems**. A dorsal heart propels the blood, called **hemolymph**, which circulates freely through the body cavity, or hemocoel. Unlike in humans, in many terrestrial arthropods the circulatory system plays only a limited role in gas exchange. Instead, most land arthropods have a respiratory system that consists of body wall holes called spiracles that open into a series of branching tubes—the tracheae and smaller tracheoles. The smallest tracheoles serve individual cells. The tracheal system efficiently transports oxygen and carbon dioxide to and from tissues. Other arthropod respiratory systems include gills in aquatic arthropods and stacked plates called book lungs in spiders and scorpions. Unique organs called **Malpighian tubules**, working along with specialized cells in the rectum to regulate water and ion balance, carry out excretion in insects and spiders (see figure 38.12).

An arthropod's nervous system is similar to an annelid's, with a dorsal brain and a ventral nerve cord. Their sensory systems detect many forms of energy, including chemical, mechanical, and electromagnetic. The antennae of insects and crustaceans are packed with microscopic hairs, each containing nerve cells sensitive to airborne or waterborne chemicals. Most

## TABLE 25.2

### Arthropods of Medical Importance in the United States

| Arthropod | Effect on Human Health |
| --- | --- |
| **Spiders** | |
| *Latrodectus* species (black widows) | Venomous bite |
| *Loxosceles reclusa* (brown recluse, violin spider, or fiddleback) | Venomous bite |
| **Mites** | |
| *Trombiculid* mites (chiggers) | Dermatitis |
| *Sarcoptes scabiei* (itch mite or scabies) | Dermatitis |
| **Ticks** | |
| *Ixodes dammini* (deer tick) | Transmits Lyme disease |
| *Dermacentor* species (dog tick, wood tick) | Transmits Rocky Mountain spotted fever |
| **Scorpions** | Venomous sting (not dangerous in most U.S. species) |
| **Centipedes** | Venomous bite (not dangerous in U.S. species) |
| **Insects** | |
| Mosquitoes | Female transmits disease (encephalitis, filarial worms, malaria, dengue fever) |
| Horseflies, deerflies | Female has painful bite |
| Houseflies and relatives | Many transmit bacteria, viruses, trypanosomes, and worms to food or water |
| Fleas | Dermatitis; transmit plague, tapeworms |
| Bees, wasps, ants | Venomous stings (single sting not dangerous unless person is allergic) |

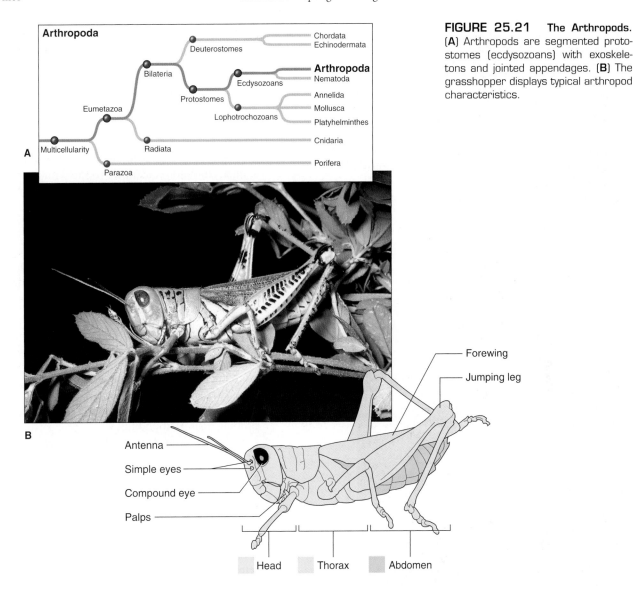

**FIGURE 25.21    The Arthropods.**
(**A**) Arthropods are segmented proto-stomes (ecdysozoans) with exoskele-tons and jointed appendages. (**B**) The grasshopper displays typical arthropod characteristics.

arthropods have taste hairs on their legs and other body parts, but scorpions and another group of arachnids called solpugids have special mid-body appendages that drag across the ground as they walk. These organs—pectines in scorpions, maleoli in solpugids—contain dense concentrations of taste cells and are like a mid-body tongue.

Spiders and scorpions also use exquisitely sensitive organs on their legs to track the vibrations of their prey. Spiders use these organs to detect the struggling insects in their webs, and sand scorpions track seismic waves that spread from a cricket's footsteps. Crickets themselves use sound to communicate. Male crickets court females by scraping the modified edges of their forewings together to produce their familiar "chirps." Females receive and interpret the songs, using special hearing organs on their forelegs.

Many arthropods have large **compound (faceted) eyes** (see figure 32.7B) that receive and process light from many angles. Honeybees use visual cues to identify and learn the route to nectar-rich flowers many kilometers away from the hive (see figure 41.19). Fireflies generate and use their own light (biolumines-cence) in a kind of visual Morse code. • **firefly bioluminescence, p. 97**

Moths, butterflies, houseflies, and some other arthropods undergo indirect development, metamorphosing from larva to pupa (cocoon) to adult (see figure 33.6).The extensive body remodeling that occurs during metamorphosis is adaptive. Juve-niles and adults of the same species usually have very different eating habits, reducing competition for food and space.

## Arthropods Exhibit Vast Diversity

Arthropods are protostomes, classified with nematodes as ecdysozoans (**figure 25.21A**). Different types of arthropods vary in the number of major body regions and appendages; the shape of mouthparts and genitalia; and whether certain appendages, such as antennae or legs, are branched. A **uniramous** appendage

**FIGURE 25.22    Chelicerate Arthropods.**    Subphylum Chelicerata includes horseshoe crabs (**A**) and arachnids, such as this spider
(**B**) and this scorpion (**C**), fluorescing green under ultraviolet light.

of subphyla and classes within the phylum is controversial. Older
references say that modern arthropods descended from the trilobites that dominated the oceans of the early Paleozoic era, but
were extinct before the Cenozoic era began. Trilobites had biramous appendages and a three-lobed body. Today, however, biologists hypothesize that trilobites formed a dead-end subphylum.
The remaining three subphyla are divided into two groups based
on the shape of their mouth parts—the chelicerates and the
mandibulates. The chelicerates have clawlike mouthparts (called
chelicerae) and constitute subphylum Chelicerata (spiders,
ticks, scorpions). The mandibulates have jawlike mouthparts
(mandibles) and contain the subphyla Crustacea (crabs, lobsters, shrimp) and Uniramia (insects, centipedes, millipedes).

**Subphylum Chelicerata** The chelicerates (**figure 25.22**)
include marine horseshoe crabs and the mainly terrestrial **arachnids,** such as mites, ticks, scorpions, and spiders. They have six
pairs of appendages. From front to back these are a pair of chelicerae, a pair of pedipalps that have sensory and feeding functions, and four pairs of walking legs (horseshoe crabs have one
pair of chelicerae and five pairs of walking legs). Arachnids lack
the antennae that are so prominent in the mandibulates,
although many species have appendages that function like antennae. Large, fierce-looking arachnids called vinegaroons and
whipscorpions wave their long, thin front legs in all directions,
probing their surroundings with dense patches of taste hairs.

Unlike insects, most arachnids and horseshoe crabs have two
major body regions: a cephalothorax (fused head and chest) and
an abdomen. In scorpions, however, the abdomen is further subdivided into the mesosoma (mid-body region) and tail, which
ends in a neurotoxic stinger.

Horseshoe crabs are common on the Atlantic coast of the
United States. These animals aren't true crabs, which are crustaceans. They are primitive-looking animals whose form has
remained mostly unchanged since the Triassic period. Their
name refers to their hard, horseshoe-shaped dorsal shield, called
a carapace, which covers a wide abdomen and a long tailpiece.
They walk on legs on land, but swim using abdominal plates. To
mate, the crabs clamber to shore at high tide. The females lay eggs
in holes they dig in the sand. Attentive males add their sperm
before the eggs are covered. With the next high tide, the hatched
larvae climb from the sand and go to the sea.

Arachnids are outstanding hunters, and they have incredibly
powerful sensory organs that pick up signs of prey. Some use
their pedipalps to subdue prey; others use venom. Chelicerae
hold the victim close to the pharynx (beginning of the digestive
tract). The arachnid then releases strong digestive enzymes into
the prey and sucks up the liquefied product.

Spiders and some other arachnids make silk from spigots in
abdominal structures called spinnerets. The silk is released as a
liquid but quickly hardens as it is drawn out. The uses of silk are
varied, including prey-snaring webs, silk tunnels for hiding,
silken egg cases, and nurseries for spiderlings. Some spiderlings
use silk to "balloon." They climb to a perch and draw off several
lines of attached silk. The wind catches the lines, and the spiderlings sail off to a new area.

is unbranched, whereas a **biramous** appendage branches in two.
Many arthropods have their body segments grouped into three
major body regions: head, thorax (chest), and abdomen (figure
25.21B).

Because of the great diversity of its members, phylum
Arthropoda is divided into four subphyla, one of which, containing animals called trilobites, is extinct. Sorting out the evolution

## T A B L E   2 5 . 3

### Some Animal Phyla

| Phylum | Examples | No. of Known Species | Body Symmetry | Body Form | Development | Specific Structures/ Characteristics (not necessarily in all species) |
|---|---|---|---|---|---|---|
| Porifera | Sponges | 9,000+ | Asymmetrical Radial | | | Archeocytes, pinacocytes, amoebocytes<br>Mesohyl<br>Skeleton of spicules and spongin |
| Cnidaria | Jellyfishes<br>Hydras<br>Corals<br>Sea anemones | 9,000+ | Radial | | Diploblastic | Polyp and medusa stages<br>Mesoglea<br>Gastrovascular cavity<br>Nematocysts |
| Platyhelminthes | Planarians<br>Flukes<br>Tapeworms | 20,000 | Bilateral | Acoelomate<br>No segments | Triploblastic<br>Protostome<br>(lophotrochozoan) | Intestinal ceca<br>Tegument<br>Protonephridia (flame cells) |
| Mollusca | Clams<br>Squids<br>Snails<br>Octopuses | 100,000 | Bilateral | Coelomate<br>No segments | Triploblastic<br>Protostome<br>(lophotrochozoan) | Mantle<br>Muscular foot<br>Visceral mass<br>Radula |
| Annelida | Earthworms<br>Leeches<br>Polychaetes | 15,000 | Bilateral | Coelomate<br>Segments | Triploblastic<br>Protostome<br>(lophotrochozoan) | Setae<br>Clitellum<br>Parapodia<br>Closed circulatory system |
| Nematoda | *C. elegans*<br>Pinworms<br>Hookworms | 12,000+ | Bilateral | Pseudocoelomate<br>No segments | Triploblastic<br>Protostome<br>(ecdysozoan) | Pseudocoel<br>Complete digestive system |
| Arthropoda | Insects<br>Arachnids<br>Crustaceans<br>Myriapods | 1,000,000 | Bilateral | Coelomate<br>Segments | Triploblastic<br>Protostome<br>(ecdysozoan) | Hemolymph in hemocoel<br>Spiracles<br>Jointed appendages<br>Exoskeleton |
| Echinodermata | Sea stars<br>Brittle stars<br>Sea urchins<br>Sand dollars<br>Sea lilies<br>Sea cucumbers | 6,500 | Bilateral larvae<br>Pentaradial adults | Coelomate<br>No segments | Triploblastic<br>Deuterostome | Ossicles<br>Pedicellariae<br>Tube feet<br>Water vascular system |

# Connections

The animals contain tremendous diversity in resolving the problems of life. Humans are not the major species, nor do they represent the best approach to being alive. We gain a great appreciation for the different ways that life can exist by understanding these various approaches. Squid are, by all ways of measuring, fairly intelligent, yet their brains are not protected by a hard skeleton. Nevertheless, they survive and thrive in their environments. Humans can learn from the unique adaptations of these other animals. The phyla in this chapter are organized in a way that illustrates the increasing complexity of animal life and that represents a possible route of evolution. Chapter 26 covers the chordates, which includes humans.

# Study Guide

## Key Terms

| | | | |
|---|---|---|---|
| acoelomate  465 | crustacean  481 | hermaphroditic  464 | protonephridium  472 |
| arachnid  480 | determinate cleavage  464 | indeterminate cleavage  465 | protostome  464 |
| bilaterally symmetrical  463 | deuterostome  464 | invertebrate  462 | pseudocoelomate  465 |
| biramous  480 | diploblastic  464 | larva  464 | radial cleavage  465 |
| bivalve  473 | direct development  464 | lophotrochozoan  465 | radially symmetrical  463 |
| blastopore  465 | ecdysozoan  465 | Malpighian tubule  478 | radula  473 |
| blastula  464 | ectoderm  464 | mantle  473 | spicule  467 |
| cephalization  463 | endoderm  464 | medusa  470 | spiral cleavage  465 |
| cephalopod  474 | endoskeleton  482 | mesoderm  464 | triploblastic  464 |
| closed circulatory system  476 | exoskeleton  478 | metamorphosis  464 | tube feet  483 |
| cnidocyte  468 | gastropod  473 | muscular foot  473 | uniramous  479 |
| coelom  465 | gastrovascular cavity  470 | nematocyst  468 | vertebrate  462 |
| coelomate  465 | gastrula  464 | open circulatory system  478 | visceral mass  473 |
| compound eye  479 | hemolymph  478 | polyp  470 | water vascular system  483 |

# Chapter Summary

## 25.1  Introduction to the Animals.

1. The animal kingdom includes more than a million very diverse species that live in nearly every habitat found on Earth.

2. Animals probably evolved from choanoflagellates, and their first fossils date from 580 to 570 million years ago.

## 25.2  Animal Characteristics

3. Animals eat other organisms, move, have cells that lack walls, and secrete extracellular matrix.

4. Biologists classify animals by body symmetry, presence or absence of a body cavity, mechanisms of reproduction and development, and molecular sequences.

## 25.3  Animal Reproduction and Development

5. Differences in embryonic development provide another characteristic whereby animals may be divided into two large groups—protostomes and deuterostomes.

6. Animals develop from up to three tissue layers, which are folded in precise ways during development and give rise to specific tissues and structures in the adult.

7. Early in development, the tissues form a ball of cells, the gastrula, whose opening, the blastopore, becomes either the mouth or the anus.

8. The pattern of cell division that produces the gastrula may be spiral or radial.

## 25.4  Animal Classification

9. Animals are also classified by the presence or absence of a coelom (body cavity), surrounded by one or two tissue layers.

10. Molecular data such as protein or DNA sequence comparisons have been used to further identify divisions and relationships.

## 25.5  The Sponges (Porifera)

11. Sponges are the simplest organism with the characteristics of an animal.

12. Sponges move, albeit slowly; distinguish self from nonself; have cell types but not true tissues; strain food particles from water; and have skeletons.

## 25.6  The Cnidarians (Cnidaria)

13. Cnidarians have radial symmetry and incomplete digestive tracts.

14. They have sessile (polyp) and motile (medusa) body forms, and many can sting, using a unique feature that includes cnidocytes and nematocysts.

15. Coral skeletons form huge reefs that host diverse living communities.

## 25.7  The Flatworms (Platyhelminthes)

16. Planarians, flukes, and tapeworms belong to phylum Platyhelminthes, whose members are adapted to a parasitic or free-living lifestyle.

17. Flatworms lack circulatory and respiratory systems, but they have features that were likely precursors to more complex systems in higher animals: protonephridia (excretory structures) and bundles of nerve cells.

## 25.8  The Mollusks (Mollusca)

18. Although mollusks appear quite different from each other, they all share certain basic body parts, such as a mantle, muscular foot, and visceral mass.

19. Mollusks have complete digestive tracts, open circulatory systems, and well-developed nervous systems.

20. The mantle and radula are unique to this phylum.

## 25.9  The Annelids (Annelida)

21. Annelids are segmented worms such as leeches and earthworms.

22. These animals have bilateral symmetry and complex organ systems, and some thrive in extreme environments.

## 25.10  The Roundworms (Nematoda)

23. Nematodes (roundworms) are one of the most abundant of animal species, second only to the arthropods.

24. Nematodes are unsegmented, lack respiratory and circulatory organs, and are distinguished from the simpler flatworms by their complete digestive systems.

## 25.11  The Arthropods (Arthropoda)

25. The majority of animal species are arthropods (they are everywhere.

26. Arthropods are segmented animals with jointed appendages and external skeletons of chitin.

27. They have complex organ systems, including sensitive sensory structures, and are classified as chelicerates or mandibulates.

## 25.12  The Echinoderms (Echinodermata)

28. With their spiny skins and five-part body plans, adult echinoderms look very different from all other animals.

29. Their embryos and larvae reveal an evolutionary connection to the chordates, the organisms discussed in chapter 26.

# What Do I Remember?

1. Describe the features that distinguish animals from the organisms in the other kingdoms.

2. List the eight major phyla of animals and provide a distinguishing characteristic for each.

3. In what three ways do protostomes and deuterostomes differ?

4. Distinguish between the three types of body cavities.

5. Describe three different approaches to reproduction found in the kingdom Animalia.

**Fill-in-the-Blank**

1. A jellyfish possesses a body plan with _____ symmetry.

2. The first body cavity in a(n) _____ becomes the mouth of the animal.

3. The _____ are stinging capsules found in cnidarians.

4. The _____ of cnidarians is most often associated with coral reefs.

5. A structure used for eating, called a(n) _____, is a structure found only in mollusca.

**Multiple Choice**

1. The middle tissue layer in a triploblastic organism is called the
   a. ectoderm.
   b. mesoderm.
   c. endoderm.
   d. myoderm.

2. The _____ of a flatworm is(are) the primitive precursor to excretory systems using kidneys.
   a. Malpighian tubules
   b. water vascular system
   c. nephridium
   d. lophophore
   e. protonephridia

3. The _____ of arthropods serve(s) to regulate water balance and function(s) as an excretory system.
    a. Malpighian tubules
    b. water vascular system
    c. nephridium
    d. lophophore
    e. protonephridia

4. Annelids are able to move by manipulating their
    a. water vascular system.
    b. hydrostatic skeleton.
    c. Malpighian tubules.
    d. lophophores.

5. Clawlike mouthparts found in some arthropods are called
    a. mandibles.
    b. chelicerae.
    c. lophophore.
    d. tube feet.
    e. nematocytes.

6. The only phylum we have discussed in this chapter that is comprised of deuterostomes is
    a. annelida.
    b. nematoda.
    c. mollusca.
    d. platyhelminthes.
    e. echinodermata.

Tunicate larvae have all four chordate characteristics, but after a period of metamorphosis, most adults retain only gill slits. The notochord disappears, and the nerve cord shrinks to nearly nothing. The free-swimming larvae, which resemble tadpoles, move vigorously, dispersing sediments and thereby affecting nutrient distribution in aquatic ecosystems. Many adults are sessile and attach to boat bottoms, dock pilings, mollusk shells, algae, or corals. Tunicates range in size from a few millimeters to 30 centimeters in length.

Tunicates vary considerably in appearance and behavior. They may be brilliantly colored, transparent, or bioluminescent. Members of the class Larvacea have an unusual way of feeding. A larvacean builds a "house" around its mouth, a hollow sphere made of mucus and fibers. A filterlike structure within the house strains out food particles. About six times a day, the filter becomes clogged, and the animal moves away from it and secretes another. Some sea squirts have an excretory organ that accumulates stones, and they are being studied as possible models for human kidneys.

## Lancelets Retain Chordate Characteristics Throughout Life

The **lancelets** resemble small, eyeless fishes with translucent bodies 2 to 3 or so inches (5 to 8 centimeters) long (**figure 26.4**). They live in shallow tropical and temperate seas all over the world, assuming a characteristic positioning with their tails buried in the sand or sediments and their anterior ends extending into the water. Some parts of the oceans contain more than 5,000 lancelets per square meter. The fossil record is mostly silent on the lancelets much as it is with the tunicates because of their lack of hard parts. Today, biologists recognize 25 species of lancelets.

Like tunicates, the lancelets filter the water for nutrients, capturing suspended food particles by using ciliated structures extending from their mouths. Cilia on the gills coupled with the mucus secreted by the pharynx effectively trap and move food particles into the digestive tract for processing and absorption.

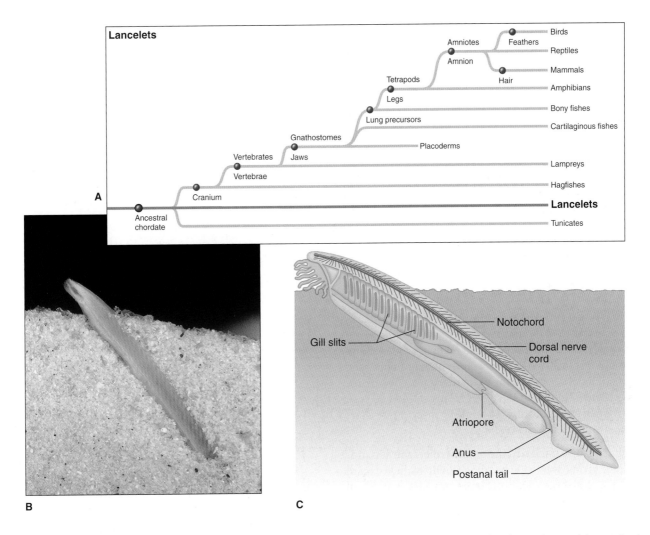

**FIGURE 26.4    Lancelets.    (A)** Lancelets, members of subphylum Cephalochordata, are invertebrate chordates whose adults retain chordate-defining characteristics. **(B)** This lancelet is in its suspension-feeding posture, with its head sticking up out of the substrate and its tail buried. The internal anatomy of the lancelet reveals the chordate characteristics **(C)**. *Amphioxus* is a well-studied lancelet.

The filtered water passes out through the gill slits, leaving the body through the atriopore, which is the counterpart to the sea squirt's excurrent siphon.

Lancelets have a circulatory system that lacks a heart but is closed—that is, the vessels are all connected. Pulsing of the vessels moves the blood. A lancelet's blood distributes nutrients but is not complex enough to handle gas exchange. The nervous system is a nerve cord with a slight swelling at the anterior end, plus sensory receptors on the body. The lancelet "brain" appears to share many of the same divisions as the more elaborate vertebrate brain. Sexes are separate, with gametes released through the atriopore. Fertilization occurs in the water, resulting in embryos that develop into ciliated, free-swimming larvae.

Lancelets were long considered the model chordates because they very clearly display all four major chordate characteristics, as well as inklings of the organ systems that appear in the vertebrates. Lancelets were once hailed as the direct ancestor of vertebrates, but because they lack the sophisticated sensory organs, brain, and mobility of vertebrates, it is more likely they branched from an early shared ancestor. Today, lancelets are the modern organisms that probably most closely resemble the ancestor to the vertebrates.

## Reviewing Concepts

- The invertebrate chordates—tunicates and lancelets—are small, cylindrical, aquatic animals that extract oxygen and trap nutrients from water using siphons and a ciliated pharynx.
- Although they may be missing in the adults, larval tunicates and adult lancelets clearly exhibit the four defining chordate features.

# 26.3    Introduction to the Vertebrates

The **vertebrates** are distinguished by a vertebral column (backbone) and a combination of other characteristics. For most vertebrates, these characteristics include paired limbs, an endoskeleton (on the inside) that includes a cranium surrounding a brain, and more complex organ systems. Many of these features are explored in detail in chapters 31 through 41, so they are just introduced here.

If the vertebrates are compared with the invertebrate chordates, three evolutionary trends emerge:

1. With increasing complexity, the notochord became less pronounced, giving way to a backbone built of linked hard parts, the **vertebrae.** The backbone provides a point of attachment for muscles and gives the animal a greater range of movement.

2. Jaws developed from gill supports. Having jaws greatly expanded the ways that the animals can feed, opening up new behaviors that led to the use of new habitats.

3. The evolution of lungs and limbs enabled some vertebrates—the tetrapods—to live on land.

Because of their endoskeletons, vertebrates have left a rich fossil record. A recent analysis of nearly 700 highly conserved genes used to estimate the time of origin of the different types of vertebrates supports this fossil evidence. **Figure 26.5** depicts an interpretation of this molecular dating information.

The 50,000+ modern (extant) species of vertebrates vary in several ways. Vertebrates occupy a diversity of areas, ranging from high mountains to great ocean depths, from dry hot deserts to wet tropical rain forests and icy polar regions. They locomote in diverse ways—many swim; others walk, jump, run, crawl, or fly. Some vertebrates are **oviparous** (egg laying), some are **ovoviviparous** (retaining eggs and giving birth to live offspring), and others are **viviparous** (bearing live young directly).

The eight classes of vertebrates include four classes of fishes, plus amphibians, reptiles, birds, and mammals. The four fish classes include an extinct group, jawless fishes, cartilaginous fishes, and bony fishes. Reptiles, birds, and mammals are called **amniotes** because they have membranes (the amnion, chorion, and allantois) that protect a developing fetus.

## Reviewing Concepts

- In addition to their namesake, the vertebrae of the backbone, most vertebrates share paired appendages, an endoskeleton that can include jaws, and organ systems.
- The evolution of lungs and limbs within this group made life on land possible for chordates.

# 26.4    The Fishes

Fishes are the most diverse and abundant of the vertebrates, with nearly 25,000 known living species. They vary greatly in size, shape, and color. Fishes exploit and occupy fresh, estuarine (salty and fresh), and marine waters worldwide. Most of these animals are **ectotherms,** meaning that the environment controls the body temperature. While fishes cannot tolerate hot springs, they can live in frigid waters by synthesizing antifreeze compounds that maintain their circulation.

In these diverse habitats, fishes eat in a variety of ways. Most sharks are voracious carnivores, while others like the tilapia are herbivores that feed on aquatic plants. Parrotfish nibble on coral reefs, some lampreys are parasitic, and hagfishes are scavengers. Very large fishes, such as the whale shark, can consume the smallest of prey, microscopic plankton.

The organs and organ systems of fishes are well developed. Fishes use structures called gills to extract oxygen from the water (see figure 37.4). Oxygen diffuses into the blood through the membranes that form the gills. A two-chambered heart pumps blood through an elaborate labyrinth of arteries and veins. Fish

vision, excellent coordination, and ability to grasp branches are essential for obtaining food and, in some species, using tools. Characteristics that taken together distinguish primates from other mammals include

- a large brain;
- color vision;
- binocular vision;
- five digits on each limb;
- nails instead of claws;
- an "opposable" thumb, meaning that it folds against the other fingers, enabling the animal to grasp.

**Figure 26.22** shows some members of order Primates. Suborder Prosimii includes the more primitive primates, such as lemurs, bush babies, tarsiers, and lorises. These small tree-dwellers look somewhat like squirrels. Suborder Anthropoidea ("resembling man") includes three subgroups that are monophyletic (each is a clade). The New World monkeys have a tail described as prehensile, which means it can wrap around objects, almost like a third arm. They also have flattened noses and nonopposable thumbs, and they lack cheek pouches and calloused rear ends (ischial callosities). The New World monkeys include the organ grinder, spider, and howler monkeys. Old World monkeys, in contrast, have close-set nostrils, cheek pouches, ischial callosities, and opposable thumbs, but lack prehensile tails. The group includes the mandrill and the rhesus and proboscis monkeys. The third group of Anthropoidea includes gibbons, orangutans, chimpanzees, gorillas, and humans. These primates lack cheek pouches and have tails reduced to a coccyx.

## Reviewing Concepts

- Mammals are distinguished by hair, mammary glands, red blood cells that lack nuclei, sweat and oil glands, two sets of teeth, and complex brains.
- We share the class with a few egg layers and pouched mammals and many other types of placental mammals.

**Primates**

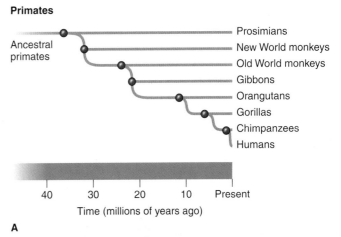

A

B            C            D

**FIGURE 26.22**    **Primate Diversity.**    **(A)** This phylogenetic tree shows proposed relationships among the primates and their approximate times of divergence. Representative primates include **(B)** white-faced capuchins (a New World monkey), **(C)** gibbons, and **(D)** chimpanzees.

## Connections

We have a tendency to judge the diversity of the living world according to what we can see. Chapters 20 through 26 should demonstrate that the slices of life with which we are familiar are very small, indeed. We humans are, biologically speaking, organisms like any other. Perhaps we can alter our environment to a greater extent. But we are a species, descended from ancestors with which we share many characteristics, yet with our own specific set of features. It is intriguing to think about where the human species is headed, which species will vanish, and how life will continue to diversify.

# Student Study Guide

## Key Terms

altricial *506*
amphibian *500*
amniote *494*
amniote egg *503*
anapsid *503*
bird *505*
bony fish *498*
caecilian *501*
cartilaginous jawed fish *497*
chordate *490*
chromatophore *501*
diapsid *503*

dorsal, hollow nerve
  cord *490*
ectotherm *494*
endothermy *505*
gill *490*
hagfish *496*
lamprey *496*
lancelet *493*
lateral line system *497*
lobe-finned fish *499*
lungfish *499*
mammal *508*

mammary gland *509*
marsupial *508*
marsupium *509*
monotreme *508*
notochord *490*
ostracoderm *497*
oviparous *494*
ovoviviparous *494*
paedomorphosis *501*
pharyngeal pouch *490*
pharynx *490*
placenta *509*

placental mammal *508*
placoderm *497*
postanal tail *490*
precocial *506*
ray-finned fish *499*
reptile *502*
swim bladder *498*
synapsid *503*
tetrapod *490*
tunicate *492*
vertebra *494*
vertebrate *494*
viviparous *494*

## Chapter Summary

### 26.1 Chordate Characteristics and Origins

1. Many familiar animals—fishes, amphibians, reptiles, birds, and mammals—belong to phylum Chordata.

2. Chordates all possess, at least at some point in their life, four key characteristics: a notochord; a dorsal, hollow nerve chord; pharyngeal gill slits; and a postanal tail.

### 26.2 Tunicates and Lancelets: The Protochordates

3. The invertebrate chordates—tunicates and lancelets—are small, aquatic animals that extract oxygen and trap nutrients from water using siphons and a ciliated pharynx.

4. Although they may be missing in the adults, larval tunicates and adult lancelets clearly exhibit the four defining chordate features.

### 26.3 Introduction to the Vertebrates

5. In addition to presence of a backbone, most vertebrates also possess paired appendages; an endoskeleton that can include jaws; and organ systems.

6. The evolution within this group of lungs, limbs, and different approaches to reproduction made life on land possible for chordates.

### 26.4 The Fishes

7. Fishes account for much of the diversity among vertebrate classes and are considered as two groups—jawed and jawless.

8. Fishes have skeletons of cartilage or bone, well-developed organ systems, and are exquisitely adapted to an aquatic lifestyle.

### 26.5 The Amphibians: Vertebrates with a Dual Lifestyle

9. Salamanders, newts, frogs, toads, and the wormlike caecilians are doubly adapted to life in water and on land.

10. In the amphibians, gills gave way to lungs, circulatory systems grew more efficient, the skeleton became denser, and senses attuned to terrestrial existence sharpened.

11. Amphibians reproduce in several ways, and some care for young.

### 26.6 The Reptiles: Adapted to Fully Terrestrial Life

12. The introduction of the leathery amniote egg, with its built-in food supplies and protective membranes to seal in moisture, provided a mechanism for reproduction without the need for a pond.

13. Reptiles with their better brains and more efficient respiratory, circulatory, and excretory systems became fully adapted to dry land.

### 26.7 The Birds: Adapted for Flight

14. Although other species can fly, birds are currently the only organisms that have feathers.

15. Just as the fish is a living machine adapted to water, a bird's body is a collection of adaptations to a life spent partially in flight.

### 26.8 The Mammals: Animals That Suckle Their Young

16. Mammals are distinguished by hair, mammary glands, red blood cells that lack nuclei, sweat and oil glands, two sets of teeth, and complex brains.

17. In addition to humans, mammals are a taxonomic class with a few egg layers and pouched mammals and many other types of placental mammals.

# What Do I Remember?

1. What are the four defining characteristics of chordates?
2. What are the major groups of vertebrates?
3. How do diapsids, anapsids, and synapsids differ?
4. What characteristics allowed reptiles to live exclusively on land?
5. Why are birds believed to be descended from reptiles?
6. How are birds adapted for flight?

**Fill-in-the-Blank**

1. A(n)_____ is an invertebrate member of the phylum Chordata.
2. Animals that are _____ bear live young.
3. Cartilaginous jawed fishes belong to the class _____.
4. _____ found in the skin of amphibians allows them to change the color and patterns of their skin.
5. The _____ system of a shark allows it to detect fish in distress from some distance away.

**Multiple Choice**

1. Which of the following is a reason for considering tunicates and lancelets to be chordates?
   a. They are found in oceans.
   b. They possess a notochord.
   c. They siphon water for food.
   d. They reproduce sexually.
   e. all of these

2. Bony fishes possess a(n)_____ that provides buoyancy without the need for constant movement.
   a. postanal tail
   b. lateral line system
   c. swim bladder
   d. fleshy lobe for fins
   e. amniote egg

3. Birds and humans possess a _____, which provides a more efficient way of oxygenating blood than that used by reptiles.
   a. notochord
   b. four-chambered heart
   c. pharynx
   d. covering of hair or feathers
   e. liver

4. Mammals that lay eggs belong to the group known as
   a. marsupials.
   b. monotremes.
   c. placentals.
   d. xenarthra.

5. Species belonging to the group known as _____ all possess five digits per limb, opposable thumbs, and binocular vision.
   a. monotremes
   b. primates
   c. marsupials
   d. carnivores

The paper of U.S. currency is made of cotton and linen fibers, not wood. The Chinese invented papermaking about 2,000 years ago.

# Plant Form and Function

## Versatile Paper

Manufacturing paper today demands a constant supply of wood pulp from millions of trees. Over the centuries, however, paper has been made from cotton and linen rags, hemp, jute, bamboo, sugarcane, wheat, and rice straw.

To make paper, plant material is gently broken apart in water, forming a slurry of fibers. The material is compressed into a sheet, then dried. Several additional chemical steps improve the transformation of wood pulp into paper. For example, to make white paper, manufacturers treat wood fibers with strong chemicals to dissolve lignin, leaving cellulose behind. Currently, people in the United States use almost 200,000 tons of paper each day, despite increasing use of electronic communication.

To conserve wood resources, magazines and books are printed on recycled paper, and napkins, towels, and paperboard come from recycled paper. Old newspapers, magazines, and junk mail are placed into huge tanks called pulpers, to which solvents and detergents are added to remove inks. The fibers are then reassembled into new paper.

One large American publishing company took a clever approach to saving paper—it trimmed 2.5 centimeters from the width of toilet paper rolls used in its building. The employees still used the same number of rolls each month. Making narrower rolls of toilet paper in the United States would save a million trees a year. Other inventive ideas could undoubtedly save millions more.

# Chapter Preview

1. The tissues of a flowering plant are meristems, ground tissue, dermal tissue, and vascular tissue, which includes phloem and xylem. The plant body consists of a shoot and a root.

2. Meristems are localized collections of cells that divide throughout the life of the plant. Apical meristems located at the plant's tips provide primary growth, and lateral meristems add girth, or secondary growth.

3. Most of the primary plant body is ground tissue. A chlorenchyma cell is a type of parenchyma cell that photosynthesizes. Collenchyma supports growing shoots, and sclerenchyma supports plant parts that are no longer growing.

4. Dermal tissue includes the epidermis, a single cell layer covering the plant. The epidermis secretes a waxy cuticle that coats aerial plant parts.

5. Vascular tissue is specialized conducting tissue. Xylem transports water and dissolved minerals from roots upwards.

6. Phloem is a living tissue that transports dissolved carbohydrates and other substances throughout a plant.

7. Simple leaves have undivided blades, and compound leaves form leaflets, which may be pinnate (with a central axis) or palmate (extend from a common point). Leaves are the main sites of photosynthesis.

8. Leaf modifications include tendrils, spines, bracts, storage leaves, insect-trapping leaves, and cotyledons.

9. Roots absorb water and dissolved minerals. Taproot systems have a large, persistent major root, whereas fibrous root systems are shallow, branched, and shorter-lived.

10. Two lateral meristems, the vascular cambium and cork cambium, produce outward growth.

## 27.1 Plant Tissue and Cell Types

A cactus, an elm tree, and a dandelion look very different from one another, yet each consists of the same basic parts. Most people are familiar with the vegetative, or nonreproductive, plant organs—the roots, stems, and leaves that support each other. Through photosynthesis, the aboveground part of a plant, or **shoot,** produces carbohydrates. A portion of this sugar supply nourishes the **roots,** which are usually belowground. In turn, roots absorb water and minerals that are transported to the shoots.

The plant organs are composed of specialized tissues that make and store food, acquire and transport water and dissolved nutrients, grow, provide support, and protect the plant from predators. Specialized cells give these tissues their unique properties. This chapter begins by exploring the tissue and cell types that perform the functions of plant life, then describes how these cell and tissue types work together to produce the diversity of stems, leaves, and roots found in herbaceous (nonwoody) and woody plants. The emphasis is on flowering plants—angiosperms—the most diverse and abundant plants. All the plant's organs consist of the same four basic tissue types (**figure 27.1 and table 27.1**). Meristematic tissue adds new cells that enable a plant to grow and specialize. Ground tissue makes up the bulk of the living plant tissue and stores nutrients or photosynthesizes. Dermal tissue covers and protects the plant, controls gas exchange with the environment, and, in roots, absorbs water. Vascular tissues, including xylem and phloem, conduct water, minerals, and photosynthate (sugars) throughout the plant.

### Meristems Contain Cells Capable of Rapid Division

**Meristems** are localized regions in a plant that undergo mitotic cell division and are the ultimate source of all the cells in a plant. These regions of active cell division account for the elongation of root and stem tips, the growth of buds, and the thickening of some stems and roots. Meristematic tissues function throughout a plant's life; because of them, some plants never stop growing.

**Apical meristems** are near the tips of roots and shoots in all plants. **Figure 27.2** depicts the location of the apical meristem in a shoot, and figure 27.15 in a root. Cells in the apical meristems are small and unspecialized. When the meristematic cells divide, the root or shoot tip lengthens in what is called **primary growth.** Apical meristems give rise to three other types of meristems—the ground meristem, protoderm, and procambium—which produce ground tissue, epidermal tissue, and vascular tissue.

In contrast to apical meristems, which lengthen a plant, **lateral meristems** (also called cambia) grow outward to thicken the plant. This process, called **secondary growth,** does

## TABLE 27.1

### Tissue Types in Angiosperms

| Tissue | Function |
|---|---|
| Meristem | Cell division and growth |
| Ground | Bulk of interiors of roots, stems, and leaves |
| Dermal | Protects plant; controls gas exchange; absorbs water |
| Vascular | Conducts water, dissolved minerals, and photosynthate |

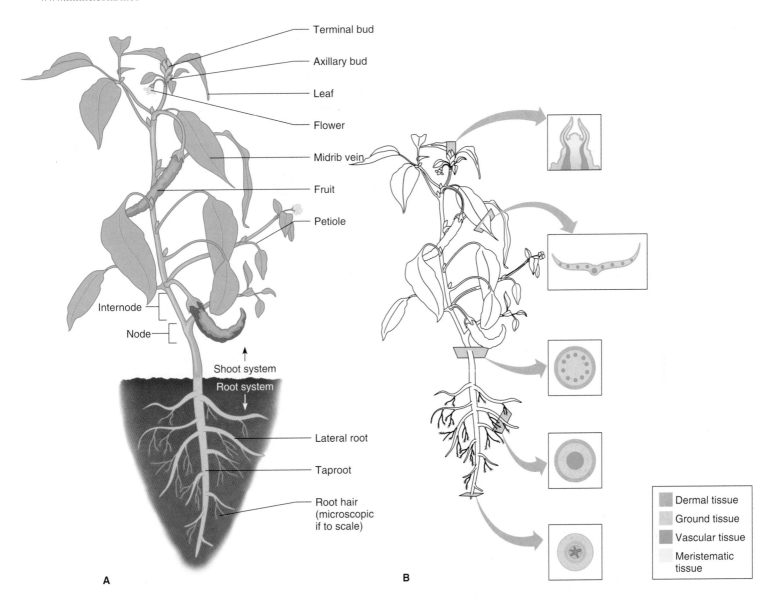

**FIGURE 27.1**    **Parts of a Flowering Plant.**    **(A)** A plant consists of a root system and a shoot system. Roots, stems, and leaves are vegetative organs; flowers and fruits are reproductive structures. **(B)** Four tissue types build plant organs.

not occur in all plants. Wood forms from secondary growth, which is discussed in section 27.5.

In some plants, intercalary meristems occur between areas that are more developed. Grasses, for example, tolerate grazing (and mowing) because the bases of their leaves have intercalary meristems that divide to regrow the leaf when it is munched off. **Table 27.2** summarizes meristem types.

## Ground Tissue Cells Have Many Functions

**Ground tissue** makes up most of the primary body of a flowering plant, filling much of the interior of roots, stems, and leaves.

| **T A B L E  27.2** | |
|---|---|
| **Meristem Types** | |
| **Type** | **Function** |
| Apical meristem | Growth at root and shoot tips |
| Lateral meristem | Growth outward, thickening plant |
| Intercalary meristem | In grass, allows rapid regrowth of mature leaves |

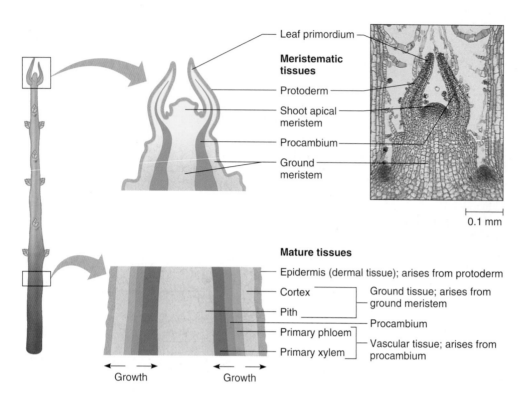

**FIGURE 27.2    Primary Growth of a Dicot's Shoot.** Looking at the tissue layers that comprise the growing tip of a shoot reveals how apical meristems give rise to primary meristems, which form mature, specialized tissues visible in older parts of the shoot.

These cells have many functions, including storage, support, and basic metabolism. Ground tissue consists of three cell types: parenchyma, collenchyma, and sclerenchyma.

**Parenchyma** cells are the most abundant cells in the primary plant body. They are structurally relatively unspecialized, although they can divide, which enables the tissue to become specialized in response to injury or a changing environment. These living cells typically have thin primary (outer) cell walls.

Parenchyma cells store the edible biochemicals in plants, such as the starch in a potato or a kernel of corn. These cells may also store fragrant oils, salts, pigments, and organic acids. Parenchyma cells of oranges and lemons, for example, store citric acid, which gives them their tart taste. These cells also conduct vital functions, such as photosynthesis, cellular respiration, and protein synthesis. *Chlorenchyma* cells are parenchyma cells that photosynthesize. Their chloroplasts impart the green color to leaves (**figure 27.3A**). Other parenchyma cells, such as the "rays" in wood, are associated with vascular tissues.

**Collenchyma** cells are elongated living cells that differentiate from parenchyma and support the growing regions of shoots. Collenchyma cells have unevenly thickened primary cell walls that can stretch and elongate with the cells. As a result, collenchyma provides support without interfering with the growth of young stems or expanding leaves (figure 27.3B).

• cell walls, p. 80

**Sclerenchyma** cells have thick, rigid secondary cell walls (a trilayered structure inside the outer, or primary, cell wall). Lignin strengthens the walls of these cells, which are usually dead at maturity, supporting parts of plants that are no longer

growing. Two types of sclerenchyma form from parenchyma: sclereids and fibers.

**Sclereids** have many shapes, and they occur singly or in groups. Small groups of sclereids create a pear's gritty texture (figure 27.3C). Sclereids may form hard layers, such as in the hulls of peanuts. **Fibers** are elongated cells that usually occur in strands that vary from a few to a few hundred millimeters long (figure 27.3D). Paper often includes wood fibers, and many textiles are also sclerenchyma fibers. Humans now cultivate more than 40 families of plants for fibers and have fashioned cords from fibers since 8000 BC. The hard leaf fibers of *Agave sisalana*, commonly known as sisal, or the century plant, are used to make brooms, brushes, and twines. Linen comes from soft fibers from the stems of *Linum usitatissimum*, or flax.

## Dermal Tissue Is the Plant's "Skin"

**Dermal tissue** covers the plant. The **epidermis,** usually only one cell layer thick, covers the primary plant body. Epidermal cells are flat, transparent, and tightly packed. Special features of the epidermis provide a variety of functions.

The **cuticle** is an extracellular covering over all the aerial epidermis of a plant; it protects the plant and conserves water (**figure 27.4A**). The cuticle consists primarily of cutin, a waxy material that epidermal cells produce. This covering retains water and prevents desiccation. As a result, plants can maintain a watery internal environment—a prerequisite to survival on dry land. The cuticle and underlying epidermal layer also are a first line of defense against predators and infectious agents. In many

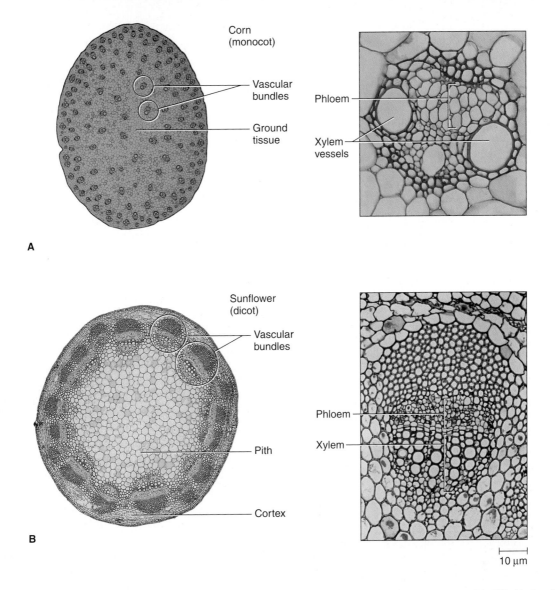

**FIGURE 27.8  Stem Anatomy.**  Cross sections of (**A**) a monocot stem (corn) and (**B**) a dicot stem (sunflower) (×10). Notice the scattered vascular bundles in the monocot stem and the ring of vascular bundles in the dicot stem.

figure 27.1). The angle between the stem and leaf stalk (petiole) is the leaf axil. Axillary buds are undeveloped shoots that form in leaf axils. Although axillary buds can elongate to form a branch or flower, many remain small and dormant.

Stems grow and differentiate at their tips, with new cells originating at the shoot's apical meristem (see figure 27.2). The shoot elongates as cells divide, grow, and become specialized into ground tissue, vascular tissue, or dermal tissue.

In most plants, stems also elongate in the internodal regions, and separate nodes may be easily distinguished. However, some plants have stems called rosettes that elongate very little. Rosettes have short internodes and overlapping leaves. A cabbage head is a rosette made of large, tightly packed leaves.

The epidermis surrounding a stem is a transparent layer only one cell thick. It contains stomata, but fewer than are in a leaf's epidermis. The epidermis of a stem also may have protective trichomes.

Vascular tissues in the stems of nonwoody flowering plants are organized into groups called **vascular bundles** that branch into leaves at the nodes. Phloem forms on the outer portions of a bundle, whereas xylem forms to the inside. Often, thick-walled sclerenchyma fibers associate with vascular bundles and strengthen the vascular tissue. The flax fibers used to make linen are one example.

Vascular bundles are arranged differently in two main groups of flowering plants: **monocotyledons** (monocots for short), which have one first, or "seed," leaf; and **dicotyledons** (dicots), which have two seed leaves (see figure 23.18). Monocots such as corn have vascular bundles scattered throughout their ground tissue, whereas dicots such as sunflowers have a single ring of vascular bundles (**figure 27.8**).

The ground tissue that fills the area between the epidermis and vascular tissue in a stem is called the **cortex.** This area is mostly parenchyma but may include a few supportive collenchyma strands. Some cortical cells are photosynthetic and store starch. The centrally located ground tissue in dicots is called **pith.**

Many plant stems are modified for special functions such as reproduction, climbing, protection, and storage (**figure 27.9**). Some specialized stems are described in the following list:

- Stolons, or runners, are stems that grow along the soil surface. New plants form from their nodes. Strawberry plants develop stolons after they flower, and several plants can arise from the original one.

- Thorns often are stems (branches) modified for protection, such as on hawthorn plants.

- Succulent stems of plants such as cacti are fleshy and store large volumes of water.

- Tendrils support plants by coiling around objects, sometimes attaching by their adhesive tips. Tendrils enable a plant to maximize sun exposure. The stem tendrils of grape plants readily entwine around anything they can touch. (Leaves may also be modified into tendrils.)

- Tubers are swollen regions of underground stems that store nutrients. Potatoes are tubers produced on stolons or underground stems called rhizomes.

- Rhizomes are underground stems that produce roots and new shoots. Ginger is a spice that is derived from a rhizome.

## Reviewing Concepts

- A plant's tissues are organized into stems, leaves, and roots, each of which can be modified.
- These tissues and organs associate to produce a great variety of specialized body forms that help support the plant, store water and nutrients, and aid in reproduction.

# 27.3   Leaves: Organs That Capture Energy

In addition to the stem, a shoot system has **leaves.** Like stems, leaves consist of epidermal, vascular, and ground tissues. Leaves are the primary photosynthetic organs of most plants. They provide an enormous surface area for the plant to capture solar energy. For example, a large maple tree has approximately 100,000 leaves, with a total surface area that would cover the area of six basketball courts (about 2,500 square meters).

## Leaf Morphology Is Adapted to Function and Environment

Leaves are extremely diverse in form. The leaves of some tropical palms may be 65 feet (20 meters) long, whereas the leaves of

*Azolla,* an aquatic fern, are only millimeters long. A mature American elm may have several million leaves, whereas the desert gymnosperm *Welwitschia mirabilis* produces only two leaves during its entire lifetime. Leaves may be needlelike, feathery, waxy, or smooth.

Botanists categorize leaves according to their basic forms (**figure 27.10**). Most leaves consist of a flattened **blade** and a supporting, stalklike **petiole.** The large vein down the center is called the midrib. Leaves may be simple or compound. Simple leaves have flat, undivided blades. Elm, maple, and zinnia have simple leaves. Compound leaves are divided into leaflets and are further distinguished by leaflet position. Pinnate (featherlike) compound leaflets are paired along a central line and include ash, rose, walnut, and *Mimosa*. Palmate compound leaflets all attach to one point at the top of the petiole, like fingers on a hand, and include lupine, horse chestnut, and shamrock.

Leaf epidermis covers the leaf and consists of tightly packed transparent cells. With the exception of guard cells, the epidermis is usually nonphotosynthetic. It may contain many stomata—more than 11 million in a cabbage leaf, for example. Water loss is minimized in many species because stomata are most abundant on the shaded undersides of horizontal leaves. The floating leaves of lily pads are unusual in having stomata on the upper surfaces only. Vertical leaves, such as those of grasses, have equal densities of stomata on both sides.

Xylem and phloem in leaves are connected to the stem's vascular tissue at the stem's nodes. Inside the leaves, the vascular tissue branches into an intricate network of veins. Sclerenchyma fibers and parenchyma cells support leaf veins. Leaf veins may be of two types: (1) netted, with minor veins branching off from larger, prominent midveins, or (2) parallel, with several major parallel veins connected by smaller minor veins (**figure 27.11**). Most dicots have netted veins, and many monocots have parallel veins. Vein endings are the blind ends of minor veins, where water and solutes move in and out of cells.

Leaf ground tissue, which is called **mesophyll,** is made up largely of parenchyma cells. Most of these cells are chlorenchyma and, therefore, photosynthesize and produce sugars. Horizontally oriented leaves have two types of chlorenchyma (**figure 27.12**). The long, columnar cells along the upper side of a leaf, called **palisade mesophyll cells,** are specialized for light absorption. Below the palisade layer are spongy mesophyll cells, which are irregularly shaped chlorenchyma cells separated by large air spaces. These cells are specialized for gas exchange and can also photosynthesize.

## Leaves Provide a Variety of Services

In addition to photosynthesizing, leaves may provide support, protection, and nutrient procurement and storage, with the following specializations.

- Tendrils are modified leaves that wrap around nearby objects, supporting climbing plants. Pea plants growing in a garden will "grab" a fence with leaf tendrils. (Both leaves and stems can be modified into tendrils.)

A                                    B                                    C

**FIGURE 27.18    Root Modifications.    (A)** The banyan tree has aerial roots growing out of its branches. **(B)** This tropical fig tree has buttress roots so enormous they resemble a trunk. **(C)** Prop roots on corn arise from the stem and support the plant.

- Storage is a familiar root specialization. Beet, carrot, and sweet potato roots store carbohydrates, and desert plant roots may store water.
- Pneumatophores are specialized roots that form on plants growing in oxygen-poor environments, such as swamps. Black mangrove trees have pneumatophores. These roots form underground and grow up into the air, allowing oxygen to diffuse in.
- Aerial roots are adventitious roots that form from stems and grow in the air. Orchids have aerial roots.
- Thick, enormous buttress roots at the base of a tree provide support, as do prop roots that arise from the stem, as seen in corn.

A plant's roots may interact with other organisms. Recall from chapter 24 that many roots form mycorrhizae with beneficial fungi. The fungi absorb water and minerals from soil, while the plants provide carbohydrates to the fungi. Roots of legume plants, such as peas, are often infected with bacteria of genus *Rhizobium*. The roots form nodules in response to the infection. The bacteria function as built-in fertilizer, providing the plant with nitrogen "fixed" into compounds it can use.

### Reviewing Concepts

- Roots provide the main source of water and mineral nutrients for the entire plant.
- They also are modified for special functions such as anchoring the plant and storing energy.
- The specializations within the tissues make them ideal for these functions.

## 27.5    Secondary Plant Growth

The tallest plants can intercept the most light. However, continued elongation poses a problem, because primary tissues cannot adequately support tall plants. Lateral meristems, which increase the girth of stems and roots by secondary growth, address this problem. These meristems are called the vascular cambium and cork cambium.

Secondary growth can be impressive. A 2,000-year-old tule tree in Oaxaca, Mexico, is 148 feet (45 meters) in circumference and only 131 feet (40 meters) tall. A 328-foot-tall (100-meter) giant sequoia in California is more than 23 feet (7 meters) in diameter. To support all of this extra tissue, a plant's transport systems must also become more complex and powerful.

### Vascular Cambium Produces New Xylem and Phloem

The **vascular cambium** is a ring of meristematic tissue that produces most of the diameter of a woody root or stem. Generally, it forms only in plants that exhibit secondary growth—primarily woody dicots and gymnosperms.

In roots and stems that undergo secondary growth, the vascular cambium forms a thin layer between the primary xylem and phloem. In stems with discrete vascular bundles, the vascular cambium extends between the bundles to form a ring (**figure 27.19**). The cells of the vascular cambium produce secondary xylem toward the inside of the cambium and secondary phloem on the outer side. Overall, the vascular cambium produces much more secondary xylem than secondary phloem.

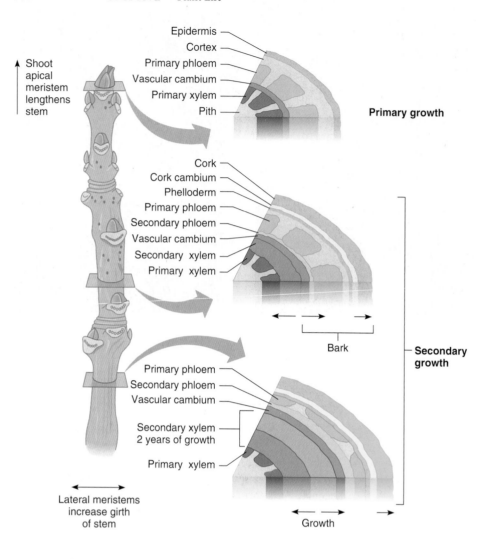

Shoot apical meristem lengthens stem

Epidermis
Cortex
Primary phloem
Vascular cambium
Primary xylem
Pith

**Primary growth**

Cork
Cork cambium
Phelloderm
Primary phloem
Secondary phloem
Vascular cambium
Secondary xylem
Primary xylem

Bark

Primary phloem
Secondary phloem
Vascular cambium
Secondary xylem 2 years of growth
Primary xylem

**Secondary growth**

Lateral meristems increase girth of stem

Growth

**FIGURE 27.19 Secondary Growth Produces Wood.** The secondary growth of a woody stem involves the activities of two types of lateral meristem—vascular cambium and cork cambium. In the top cross section, the microscopic vascular cambium has formed but not yet started producing secondary vascular tissue. The vascular cambium produces secondary xylem toward the inside of the stem and secondary phloem toward the outside, eventually crushing the primary vascular tissue, cortex, and epidermis. Although the two lower diagrams depict only one cork cambium, many trees have more than one.

Secondary xylem is more commonly known as wood. In temperate climates, cells in the vascular cambium divide to produce wood during the spring and summer. During the moist days of spring, wood is made of large cells and is specialized for conduction of water. During the drier days of summer, the vascular cambium produces summer wood that has small cells and is specialized for support. These seasonal differences in wood cell sizes generate visible demarcations called growth rings (**figure 27.20**). Secondary xylem can be used to measure the passage of time because the larger spring wood cells appear light colored and summer wood cells are smaller and darker colored. The contrast between the summer wood of one year and the spring wood of the next creates the characteristic annual tree ring. The most recently formed ring is next to the microscopic vascular cambium. Tree ring data can be aligned with known events and provide markers and confirmation of natural history. But without changes in seasons, no growth rings will be present. Tropical species, which have secondary growth all year long, do not have regular tree rings.

The older, innermost increments of secondary xylem gradually become unable to conduct water; this nonfunctioning wood

is called **heartwood.** The **sapwood,** located nearest the vascular cambium, transports water and dissolved minerals.

Woods differ in hardness. Dicots such as oak, maple, and ash are often called hardwood trees, and gymnosperms such as pine, spruce, and fir are called softwood trees. Dicot wood contains tracheids, vessels, and supportive fibers, whereas wood from softwood trees is more homogeneous, consisting mainly of tracheids. As a result, dicot wood is usually stronger and denser than wood from gymnosperms (the soft, light wood from the balsa tree, a dicot, is a notable exception). The terms "hardwood" and "softwood" are not scientific terms. They are used mainly in industry to reflect the observation that dicot wood is usually harder than wood from gymnosperms.

The tissues to the outside of the vascular cambium are collectively called **bark** (see figure 27.19). Secondary phloem forms the live, innermost layer of bark and transports phloem sap within the tree (the primary phloem is crushed as the stem or root continues to grow outward). The outer layer of bark includes the **periderm,** a protective layer of tissue that replaces the epidermis as the girth of a stem or root increases. The outer bark also includes dead tissues outside the periderm.

Water carried in fog condenses on the surface of a coastal redwood tree and drips to the forest floor. The tree's shallow roots absorb some of the water, which moves up the massive trunk and returns to the atmosphere through the tree's needles.

## Fog Drip

The coastal redwood *Sequoia sempervirens* lives in a narrow band of forest along the coast of southwest Oregon and northern and central California. It is at the center of an ecosystem that was once much larger, warmer, and wetter, until encroaching glaciers invaded its boundaries during the Cretaceous period.

The coastal redwood requires vast amounts of moisture, some thousand tons for every ton of its weight. Much of this water comes from an unusual source—fog. Thanks to these trees, the ecosystem captures up to 50 inches (127 centimeters) of water annually that it would otherwise not receive.

Fog consists of suspended particles of liquid water. It forms as moisture-rich air from the north falls over the west coast, and some of the water condenses on the many needle-covered branches and limbs of the redwoods. It then drips to the forest floor unimpeded because the lower third of the tree lacks branches.

The tree absorbs about 35% of the fog drip. The redwood's ability to tap so much of the moisture in fog may explain its great height. By keeping the immediate environment near-saturated, less water leaves the plant, so less water is pulled up by the roots.

Coastal redwoods, as suppliers of water to coastal soils, are critical components of these forest ecosystems. Ferns, for example, could not survive without the redwoods' fog drip. It isn't surprising that areas cleared of these trees often experience drought.

# 28

# Plant Nutrition and Transport

# Chapter Preview

1. In plants, xylem transports water and dissolved minerals, and phloem distributes photosynthate. Both tissue types are bundled together in a cylinder in roots and as veins in stems and leaves.

2. Root branches, root hairs, mycorrhizae, and root cortical cells provide abundant surface area to absorb water and dissolved minerals. Some plants secrete proteins to enhance mineral uptake.

3. Several types of bacteria fix nitrogen into forms that plants can use.

4. Water and dissolved minerals (xylem sap) are pulled up through xylem to replace water lost through transpiration in leaves. This is called the cohesion-tension theory.

5. The endodermis with its impermeable Casparian strip controls which minerals enter the nearby xylem. Water and dissolved minerals move through the root's cortex by the apoplastic pathway and by the symplastic pathway through cells.

6. Photosynthate may be used or stored in its cell of origin or moved in phloem to nonphotosynthetic cells in roots, flowers, or fruits. Phloem sap includes photosynthate and water and minerals from xylem.

7. Phloem sap flows through sieve tubes from a source to a sink, with pressure generated by continual influx of water from xylem. This mechanism is called the pressure flow theory.

## 28.1 Evolution of Complex Transport Systems

All organisms engage in constant exchange with the environment, acquiring energy and chemical building blocks and releasing the chemical by-products that accumulate as part of being alive. The single-celled organisms that were the first forms of life were small enough to handle this back and forth with their surroundings. The first multicellular forms, too, were probably sufficiently simple that each cell had easy access to the environment.

The earliest plants were probably descendants of green algae, perhaps resembling liverworts, as chapter 23 discusses. These organisms were aquatic, likely bathed in chemicals released from dead and decaying organisms, and diffusion was sufficient to transport materials into and out of each cell. When plants first colonized land, they formed carpetlike growths that could soak up water and dissolved nutrients from soil, dust, or rain. Over time, leaves greatly increased the photosynthetic capabilities of plants, and root systems grew more complex, expanding the range of nutrient and water sources available to plants. Long taproots plunged toward the water table, while other roots extended laterally, soaking up the nutrients released into the soil from weathered rocks and decaying organisms.

As organisms grew larger and more complex, bodies folded and contorted and developed extensions that increased surface area, which otherwise could not keep pace with increasing volume. Today, the members of the three multicellular kingdoms of life exhibit distinctive ways to maximize surface area and maintain contact with the environment. Fungi grow many long, slender strands of hyphae that permeate the soil and absorb its bounty of nutrients. Animals and plants have internal networks of tubules and tubes that deliver water and nutrients.

In chapter 27, you learned that xylem transports water and dissolved minerals, and phloem distributes photosynthate, the products of photosynthesis. These specialized tissues form a continuous distribution system connecting the plant's roots, stems, leaves, flowers, and fruits (**figure 28.1**). In roots, both types of vascular tissue are bundled together within a layer of parenchyma; together, these tissues are called the stele. In many plants, the xylem is at the center of the stele, and phloem is in strands toward the outside. In stems, the stele diverges to form discrete vascular bundles and associated ground tissue such as pith (see chapter 27).

### Reviewing Concepts

- The first forms of life on Earth were unicellular, simple enough that all cells could maintain contact with the environment.
- But as life evolved into more complex forms, mechanisms were needed to transport nutrients to cells separated from the environment.
- Among the plants, these adaptations included development of transport systems.

## 28.2 Plant Essential Nutrients

Like any other organism, a plant requires water as a medium for, and participant in, many of its metabolic reactions. A plant gets most of its water from the soil. The waterproof cuticle and stomata are adaptations for conserving water (see figure 27.4). Plants also need essential nutrients, which are chemicals that are vital for metabolism, growth, and reproduction.

In the mid-1800s, botanists discovered the essential plant nutrients by growing plants in water that contained known amounts of particular elements, a technique called hydroponics. In this way, they identified 16 elements vital to all plants (**figure 28.2**). Nine of these are **macronutrients**—elements that are required in fairly large amounts, at least 0.1% of the dry weight of the plant. The macronutrients are carbon, hydrogen, oxygen, nitrogen, potassium, calcium, magnesium, phosphorus, and sulfur. The **micronutrients,** required in much smaller amounts, are chlorine, iron, boron, manganese, zinc, copper, and molybdenum.

**FIGURE 28.1    Plant Transport Systems.**    In vascular plants, xylem transports water and dissolved minerals absorbed from soil, and phloem distributes the products of photosynthesis to fruits, roots, stems, and other plant parts.

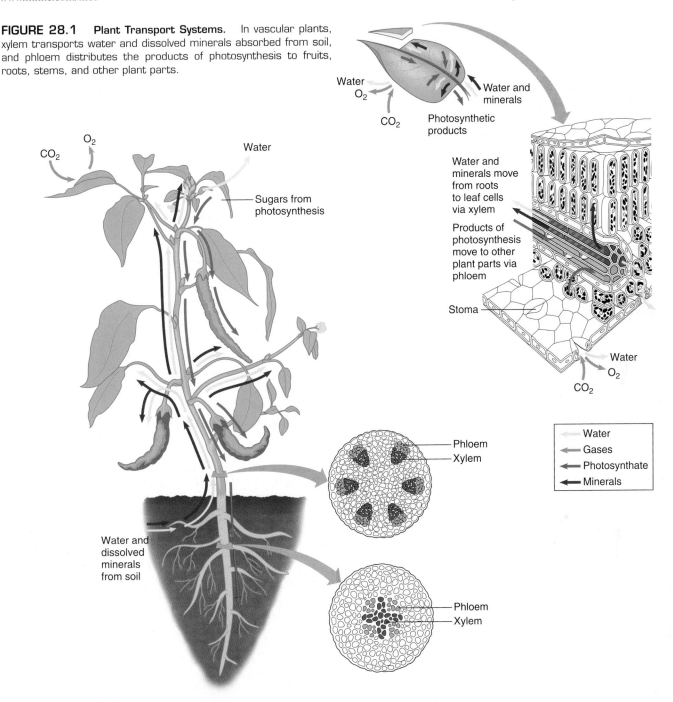

Figure 28.2 lists some of their functions. In general, macronutrients are required for production of structural and storage carbohydrates such as cellulose and starch, for enzyme synthesis and control, for altering osmotic potential, and as participants in metabolic reactions. Just as is the case for all organisms, nutrient deficiencies can harm plants, as **figure 28.3** shows.

Certain plants take up elements other than the 16 that all plants use. Horsetails and certain grasses, for example, accumulate silicon in their cells, and soybeans use nickel. Some other plants accumulate lead, cadmium, copper, zinc, or cobalt. Most interesting is the locoweed, *Astragalus*—more than 1% of its dry

weight is selenium. Selenium accumulation reduces the toxic effect of phosphorus, which is often present in high levels in selenium-rich soil. Cattle that eat the selenium-laden plants stagger about and seem intoxicated, and ranchers usually have to destroy them.

Accumulating minerals that may be toxic to other organisms could be an adaptation to avoid predation, but humans have also taken advantage of that capability. A biotechnology called phytoremediation uses plants with an affinity for toxic wastes to naturally remove heavy metals and other dangerous chemicals from the environment. Transgenic technology (see chapter 14) is being

| Macronutrients | Form taken up by plants | Percent dry weight | Selected functions |
|---|---|---|---|
| Carbon (C) | $CO_2$ | 45 | Part of organic compounds |
| Oxygen (O) | $H_2O$, $O_2$ | 45 | Part of organic compounds |
| Hydrogen (H) | $H_2O$ | 6 | Part of organic compounds |
| Nitrogen (N) | $NO_3^-$, $NH_4^+$ | 1.5 | Part of nucleic acids, amino acids, coenzymes, chlorophyll, ATP |
| Potassium (K) | $K^+$ | 1.0 | Controls opening and closing of stomata, activates enzymes |
| Calcium (Ca) | $Ca^{2+}$ | 0.5 | Cell wall component, activates enzymes, second messenger in signal transduction, maintains membranes |
| Magnesium (Mg) | $Mg^{2+}$ | 0.2 | Part of chlorophyll, activates enzymes, participates in protein synthesis |
| Phosphorus (P) | $H_2PO_4^-$, $HPO_4^{2-}$ | 0.2 | Part of nucleic acids, sugar phosphates, ATP, coenzymes, phospholipids |
| Sulfur (S) | $SO_4^{2-}$ | 0.1 | Part of cysteine and methionine (amino acids), coenzyme A |

| Micronutrients | Form taken up by plants | Percent dry weight | Selected functions |
|---|---|---|---|
| Chlorine (Cl) | $Cl^-$ | 0.01 | Water balance |
| Iron (Fe) | $Fe^{3+}$, $Fe^{2+}$ | 0.01 | Chlorophyll synthesis, cofactor for enzymes, part of electron carriers |
| Boron (B) | $BO_3^-$, $B_4O_7^{2-}$ | 0.002 | Growth of pollen tubes, sugar transport, regulates certain enzymes |
| Zinc (Zn) | $Zn^{2+}$ | 0.002 | Hormone synthesis, activates enzymes, stabilizes ribosomes |
| Manganese (Mn) | $Mn^{2+}$ | 0.005 | Activates enzymes, electron transfer, photosynthesis |
| Copper (Cu) | $Cu^{2+}$ | 0.0006 | Part of plastid pigments, lignin synthesis, activates enzymes |
| Molybdenum (Mo) | $MoO_4^{2-}$ | 0.00001 | Nitrate reduction |

Carbon, oxygen, and hydrogen (most abundant macronutrients) (96% of dry weight)

Other macronutrients (~3.5%)

Micronutrients (~0.5%)

**FIGURE 28.2   Essential Nutrients for Plants.**   The nine most abundant elements in plants are called macronutrients. The micronutrients occur in much lower concentrations but are also essential for plant survival. The table lists the nutrients present in all plants; some plants require additional micronutrients not listed here.

A

B

**FIGURE 28.3   Nutrient Deficiencies Produce Distinctive Symptoms.**   (**A**) Phosphorus deficiency causes dark green or purple leaves in seedlings. (**B**) Iron deficiency causes chlorotic (yellowed) leaves, but the veins remain green. A deficiency of manganese produces similar symptoms.

used to endow plants with bacterial "detox" genes, while using the plant's role in the environment to clean up polluted areas. For example, tobacco plants given a bacterial gene to produce an enzyme that detoxifies trinitrotoluene (TNT), an explosive, are being planted at sites of former weapon production. It is easier to cover such areas with transgenic plants than to permeate the soil with bacteria to get rid of the TNT.

Some plants assimilate elements, especially metals, that happen to be abundant in their particular environments. For example, plants growing near gold mines have gold in their tissues (which prospectors use as clues to promising deposits), plants drenched in cadmium-containing fertilizers contain that metal, and plants growing along superhighways take up lead accumulated from decades of leaded gasoline exhaust. Near nuclear test sites, plants absorb radioactive strontium—as do people who drink the milk from cows that graze on grasses exposed to nuclear fallout.

**FIGURE 28.4**    **Soil.**    Soil consists of layers called horizons, designated A through C. The top layer is litter, which covers the rich topsoil. Water passing through the topsoil deposits clay and minerals in the next layer, and below this is a layer of partially weathered rock.

### Reviewing Concepts

- All plants require at least 16 essential minerals, obtained from the atmosphere or soil.
- The ability of plants to extract minerals from the environment makes them ideal for phytoremediation efforts to remove toxic wastes from soils.

## 28.3   Nutrient Sources: Atmosphere and Soil

Plants and other autotrophs play a crucial role in ecosystem nutrient cycles, taking up inorganic molecules and using energy from the sun to manufacture organic compounds. The biogeochemical cycles for carbon, nitrogen, and phosphorus all rely on autotrophs. Both soil and the atmosphere are crucial links as well. • **biogeochemical cycles, p. 878**

Plants obtain carbon, oxygen, and hydrogen atoms from carbon dioxide gas ($CO_2$) and soil water ($H_2O$). $CO_2$ comprises 0.037% of the atmosphere. This gas enters the leaf through open stomata, dissolves in the water film surrounding the cells, and diffuses into the cytoplasm. As $CO_2$ is used in photosynthesis, its concentration drops in the cell relative to its surroundings, and more diffuses in from air spaces between the chlorenchyma cells. The gas continues to diffuse into the cells as long as its concentration in the air spaces exceeds its concentration in the cytoplasm. For $CO_2$ uptake to continue, the air in the intercellular spaces must be continuous with the air outside the leaf. That is, the stomata must be open.

Water and nutrients other than carbon come from soil, which consists of small particles of rocks and clay minerals mixed with decaying organisms and organic molecules. Soil is full of living organisms other than plants as well, which may compete with plant roots for nutrients. The characteristics of soil

change with depth (**figure 28.4**), and the composition and texture vary from place to place.

The mineral and rock particles in soil form from weathering of rocks on the Earth's surface. The texture of a soil depends upon the size of these particles. Sandy soils have coarse particles; silty soils have finer particles. Clay particles are the smallest. Many soils are a mixture of sand, silt, and clay, and soil scientists consider the relative amounts of each to classify the soil. The size of the particles also determines a soil's water-holding capacity. The finer the particles, the more surface area per unit of soil volume and the higher the soil's water-holding capacity.

Some of the minerals that plants extract from the soil come from rocks weathered into soil particles. Others are released during the decomposition of dead organisms. Most of the decaying material in soil comes from plants and forms a layer of pieces, called litter, on the soil surface. Most of the carbon in decaying litter is released as $CO_2$, but some complex organic chemicals that are hard to digest form a material called **humus** in the upper layer of soil, the **topsoil**. This layer, which is also called the A horizon, is a major source of water and nutrients for most plants growing in the soil. Below topsoil is the B horizon, which has less organic matter, although roots extend to this depth. Still lower is the C horizon, which consists almost entirely of partially weathered pieces of rocks and minerals. It is an interface between the bedrock below and the soil above.

To boost soil productivity, farmers and gardeners may apply nutrient-rich organic matter (such as manure) or inorganic fertilizer. Commercial fertilizer labels always include numbers that indicate the content of nitrogen, phosphorus, and potassium. For example, a product marked 10-20-10 contains 10% N, 20% $P_2O_5$, and 10% $K_2O$.

Chemically, nutrients from inorganic fertilizer are equivalent to those from manure. Adding organic matter to soil, however, has additional benefits, such as increasing the soil's water-holding capacity and providing food for soil-dwelling animals.

## Roots Tap Water and Dissolved Minerals

Unlike animals that can easily move in search of water, plants must take up water and the minerals in it with their roots, which is why these plant organs are usually very extensive. Roots offer a spectacular example of biological maximization of surface area. Not only do roots often extend well beyond the area covered by the aboveground plant body, but they also form millions of root hairs that increase surface area for water absorption (see figure 27.16). Fungi associated with roots, called **mycorrhizae,** increase this absorptive surface even more (see figures 24.1 and 24.14). Mycorrhizae boost the growth rate of onions 30-fold!

Inside the roots, cells of the cortex and endodermis provide yet more surface area because water has to pass through cell membranes to reach the xylem (see figure 27.17). These cell membranes also provide the selectivity that makes it possible for plants to have different concentrations of elements than are in the soil. The selectivity comes from membrane proteins that admit only certain ions. For example, a soil might be very rich in aluminum, but the plants that grow in it have little of the element because the number of carrier proteins limits uptake. On the other hand, plants often have higher concentrations of essential micronutrients than are present in the soil.

Minerals may enter a root cell by facilitated diffusion or active transport, depending on the relative concentrations inside and outside the cell. Specific minerals traverse specific membrane channels with the assistance of carrier proteins. Some plants secrete proteins to help them acquire scarce or insoluble elements. For example, monocots such as grains secrete small proteins called siderophores that bind to iron in soil, rendering it better able to enter root cells. A new biotechnology called **rhizosecretion** exploits the ability of roots to secrete proteins to manufacture proteins of commercial interest. • **how substances cross membranes, p. 68**

## Plants Obtain Nitrogen as Nitrates or Ammonium Ions

Plants combine nitrogen with Krebs cycle intermediates to manufacture certain amino acids, which, in turn, are used to build proteins, nucleic acids, hormones, chlorophyll, secondary metabolites, and other biochemicals. The nitrogen that is abundant in the atmosphere, however, is in the diatomic form—$N_2$. It is held together by three covalent bonds, which

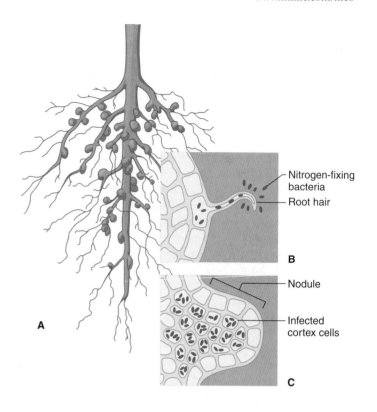

**FIGURE 28.5** **Bacteria Provide Usable Nitrogen to Plants.** In response to signals from leguminous plants such as beans, *Rhizobium* triggers the development of root nodules (**A**). The bacteria enter through root hairs (**B**) and live symbiotically within the plant's cells (**C**).

require more energy to break than a plant can muster. Instead, plants must acquire nitrogen as nitrate ($NO_3^-$) or ammonium ion ($NH_4^+$).

Several types of microorganisms "fix" atmospheric nitrogen into these more accessible forms. Certain types of bacteria, including those in the genus *Rhizobium,* are stimulated by signals from plants of the legume family (beans, peas, peanuts, soybeans, and alfalfa) to associate in growths called **nodules** on the plants' roots, where they use **nitrogen-fixation** enzymes to convert $N_2$ to usable forms (**figure 28.5**; see also figure 21.1). Cyanobacteria and *Azotobacter* are other types of bacteria that provide nitrogen to plants; some of these microorganisms form symbiotic relationships, while others are free-living.

*Rhizobium* reduces atmospheric $N_2$ to ammonium ion ($NH_4^+$), which also comes from fertilizer and from other types of bacteria that live on decaying organisms. Certain autotrophic bacteria, including *Nitrosomonas* and *Nitrobacter,* oxidize $NH_4^+$ into nitrites and nitrates (see figure 44.15). The form of nitrogen that plants actually take up, usually $NH_4^+$ or $NO_3^-$, depends on the plant species and soil conditions, such as pH. A few plants acquire nitrogen from more unusual sources. For example, carnivorous plants, which often live in nitrogen-poor soils, obtain nitrogen compounds from the insects they consume.

## Plants in Space

Plants will play a key part in space colonization, providing food and oxygen. But how will organisms that have evolved under constant gravity function in its near absence beyond Earth?

Researchers are studying the effects of microgravity on a variety of species by sending them on space shuttle voyages. These experiments can more realistically assess plant growth and development in space than simulations conducted on Earth, which use a rotating device called a clinostat to prevent plants from detecting the direction of gravity. So far, it appears that lack of gravity greatly affects plants, in everything from subcellular structural organization to the functioning of the organism as a whole.

### Subcellular Responses to Microgravity

Plant cells grown in space have fewer starch grains and more abundant lipid-containing bodies than their earthly counterparts, indicating a change in energy balance. Organelle organization is also grossly altered; endoplasmic reticula occur in randomly spaced bunches, and mitochondria swell. Nuclei enlarge, and chromosomes break. Chloroplasts have enlarged thylakoid membranes and small grana.

Interesting effects occur in the amyloplasts (starch-containing granules) of certain root cells. On Earth, amyloplasts aggregate at the bottoms of these cells, telling roots which way is down. But in space, amyloplasts are uniformly distributed in cells. For many years, botanists hypothesized that sinking starch granules in these cells are gravity receptors. An alternative hypothesis holds that cells sense gravity by detecting the difference in pressure on the cell membrane at the top versus the bottom of the cell. This difference results from protoplasm sinking in response to gravity, so there is less of a difference as gravitational force falls. The actual "sensing" may be carried out by a cell membrane protein called an integrin, and the starch granules may amplify the difference in pressure by sinking in the presence of gravity. Whatever the precise mechanism of gravity sensation, a root tip cannot elongate normally in microgravity.

### Cell Division

Microgravity halts mitosis, usually at telophase, which produces cells with more than one nucleus. Oat seedlings germinated in space have only one-tenth as many dividing cells as seedlings germinated on Earth. Microgravity also disrupts the spindle apparatus that pulls chromosome sets apart during mitosis.

Cell walls formed in space are considerably thinner than their terrestrial counterparts, with less cellulose and lignin. Microgravity also inhibits regeneration and alters cell distribution (**figure 30.A**). A decapped root will regenerate in 2 to 3 days on Earth but not at all in space. Lettuce roots have a shortened elongating zone when grown in space.

### Growth and Development

Germination is less likely to occur in space than on Earth because of chromosome damage, but it does happen. Early growth seems to depend upon the species—bean, oat, and pine seedlings grow more slowly than on Earth, and lettuce, garden cress, and cucumbers grow faster. Many species, including wheat and peas, cease growing and die before they flower. In 1982, however, *Arabidopsis* successfully completed a life cycle in space—indicating that human space colonies containing plant companions may indeed be possible.

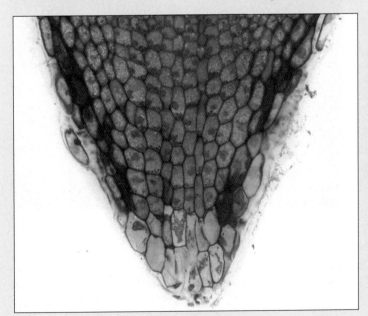

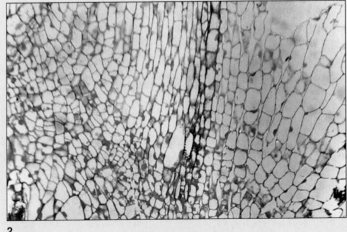

2

**FIGURE 30.A** *Plant Growth in Space.* (1) On Earth, a root whose cap has been removed regenerates an organized, functional root cap. (2) In the microgravity of space, however, regrowth of a decapped root is disorganized. Clearly, gravitational cues help direct normal regeneration.

1

of the motor cells by osmosis. The water loss shrinks the motor cells and causes thigmonastic movement. This response is considerably faster than a hormone-induced action. Reversal of the process unfolds the leaves in approximately 15 to 30 minutes.

Thigmonastic movements are protective. In some plants, closing leaflets expose sharp prickles and stimulate cells at the leaflet bases to secrete unpalatable substances called tannins, which discourage hungry animals. The Venus flytrap is famous for its dramatic thigmonastic response. Unlike *Mimosa,* in which thigmonastic movements result from reversible changes in turgor pressure, the Venus flytrap's movements result from increased cell size, which begins when the cell walls are acidified to pH 4.5 and below. The leafy traps each consist of two lobes, and each lobe has three sensitive "trigger" hairs overlying motor cells. When a meandering animal touches two of these hairs, it signals the plant's motor cells, which then initiate $H^+$ transport to epidermal cell walls along the trap's outer surface. Acidification promotes softening of the cell wall and osmotic swelling of the outer epidermal cells along the central portion of the leaf. Since epidermal cells along the inner surface of the leaf do not change volume, the flytrap shuts. Empty traps usually open after 8 to 12 hours.

**Nyctinasty,** the nastic response to daily rhythms of light and dark, is also known as "sleep movement" ("nyct-" means night). This is an example of a circadian (day-based) rhythm, discussed further in section 30.4. The prayer plant, *Maranta,* is an ornamental houseplant that exhibits such a response. Prayer plant leaves lie horizontally during the day, which maximizes their interception of sunlight. At night, the leaves fold vertically into a configuration resembling a pair of hands in prayer (**figure 30.11**).

The prayer plant's leaves move in response to light and dark as the turgor pressure changes in motor cells at the base of each leaf. In the dark, $K^+$ moves out of cells along the upper side and into cells along the lower side of a leaf base. This moves water, via osmosis, into cells along the lower side of the leaf base, swelling them, as cells along the upper side lose water and shrink. As the cellular volume changes, the leaf stands vertically. At sunrise, the process reverses, and the leaf again lies horizontally. Changes in leaf position can conserve water.

In addition to leaves, the flowers of some plants have similar sleep movements that occur at the same time each day. Carolus Linnaeus, the Swedish botanist who proposed a widely used taxonomic scheme in the eighteenth century, made clever use of these regular movements. He filled wedge-shaped portions of a circular garden with plants that had sleep movements at different times. By checking to see which plants in his so-called *horologium florae* (flower clock) were "asleep," Linnaeus could tell the time of day.

## Reviewing Concepts

- A plant can grow toward light, with or against gravity, or encircle an object in response to touch.
- Such tropisms are directed movements.

**FIGURE 30.10  Thigmonasty.**  *Mimosa pudica* is called the "sensitive plant" because of its thigmonastic movements. Leaves are erect in undisturbed plants (**A**), but touching a leaf (**B**) causes the leaflets to fold and the petiole to droop (**C**).

**FIGURE 30.11** **Nyctinasty.** The prayer plant, *Maranta*, exhibits nyctinasty. When the sun goes down, the prayer plant's leaves fold inward.

# 30.3 Plant Responses to Seasonal Changes

Seasonal changes affect plant responses in many ways. For instance, autumn in temperate regions brings cooler nights and shorter days, resulting in beautifully colored leaves, dormant buds, and decreased growth. In the spring, buds resume growth and rapidly transform a barren forest into a dynamic, photosynthetic community. These seasonal changes illustrate the complex interactions among environmental signals, hormones and other biochemicals, and the plant's genes.

## Flowering Is a Response to the Photoperiod

Flowering reflects seasonal changes. Many plants flower only during certain times of the year. Clover and iris flower during the long days of late spring and summer, whereas poinsettias and asters bloom in the short days of early spring or fall.

Studies of how seasonal changes influence flowering began in the early 1900s. W. W. Garner and H. A. Allard at a U.S. Department of Agriculture research center in Maryland were studying tobacco, which flowers during late summer in Maryland. One group of tobacco mutants did not flower but continued to grow vegetatively into autumn. These mutants became large, leading Garner and Allard to name them Maryland Mammoth. Since these oversized mutants had the potential for increasing tobacco crop yields, Garner and Allard moved their Mammoth plants into the greenhouse to protect them from winter cold and continued to observe their growth. To their surprise, the mutants finally flowered in December!

Could the plants somehow measure day length? To test this hypothesis, Garner and Allard set up several experimental plots of the tobacco, each planted approximately a week apart. All of the plants flowered at the same time, despite the fact that the staggered planting times resulted in plants of different ages and sizes. Garner and Allard suggested that the plants were exhibiting **photoperiodism,** the ability to measure seasonal changes by day length.

Photoperiodism attunes plant responses to a changing environment. At latitudes far from the equator, plants measure and respond to a critical day length rather than other climatic factors, such as rainfall or temperature, because weather is unpredictable from year to year. Day length is consistent due to the position of Earth as it moves around the sun.

Botanists classify plants into groups, depending upon response to photoperiod (duration of daylight) (**figure 30.12**). **Long-day plants** flower when light periods are longer than a critical length, usually 9 to 16 hours. These plants typically bloom in the spring or early summer and include lettuce, spinach, beets, clover, corn, and iris. **Day-neutral plants** do not rely on photoperiod to stimulate flowering. These include roses, snapdragons, cotton, carnations, dandelions, sunflowers, tomatoes, cucumbers, and many weeds. **Short-day plants** require light periods shorter than some critical length. These plants usually flower in late summer or fall. For example, ragweed plants flower only when exposed to 14 hours or fewer of light per day. Asters, strawberries, poinsettias, potatoes, soybeans, and goldenrods are short-day plants. Many short-day plants do not occur naturally in the tropics, where days are always too long to induce flowering.

The measuring system in plants is in the leaves: remove a plant's leaves, and it does not respond to changes in photoperiod. Response to photoperiod is also remarkably sensitive. Henbane, a long-day plant, flowers when exposed to light periods of 10.3 hours, but not when the light period is 10.0 hours. These flowering responses may ensure that flowers of different sexes open within a few days of each other.

## Plants Track Darkness, Not Light

Plant physiologists Karl Hamner and James Bonner continued Garner's and Allard's work by studying the photoperiodism of

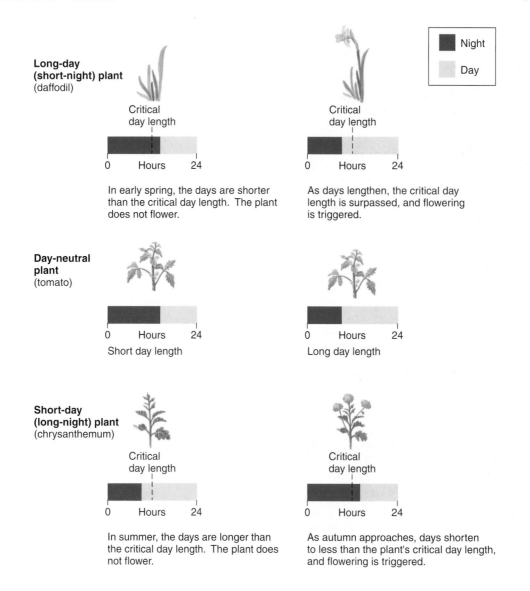

**Long-day (short-night) plant** (daffodil)

Critical day length

0    Hours    24

In early spring, the days are shorter than the critical day length. The plant does not flower.

Critical day length

0    Hours    24

As days lengthen, the critical day length is surpassed, and flowering is triggered.

Night
Day

**Day-neutral plant** (tomato)

0    Hours    24

Short day length

0    Hours    24

Long day length

**Short-day (long-night) plant** (chrysanthemum)

Critical day length

0    Hours    24

In summer, the days are longer than the critical day length. The plant does not flower.

Critical day length

0    Hours    24

As autumn approaches, days shorten to less than the plant's critical day length, and flowering is triggered.

**FIGURE 30.12    Flowering Responses to Day Length.**    Long-day plants flower in response to light periods longer than a critical period. Flowering in day-neutral plants does not rely on response to day length. Short-day plants respond to a light period shorter than a critical length.

the cocklebur, a short-day plant requiring 15 or fewer hours of light to flower. Hamner and Bonner used environmentally controlled growth chambers to manipulate photoperiods. They were startled to discover that plants responded to the length of the dark period rather than the light period. The cocklebur plants flowered only when the dark period exceeded 9 hours.

Hamner and Bonner also discovered that flowering did not occur if a 1-minute flash of light interrupted the dark period, even if darkness exceeded the required 9 hours. In the reverse experiment, darkness interrupting the light period had no effect on flowering. Furthermore, a long-day plant flowering on a photoperiod of 16 hours light to 8 hours dark also flowered on a photoperiod of 8 hours light to 16 hours dark if a 1-minute light flash interrupted the dark period. Other experiments with long- and short-day plants confirmed that flowering requires a specific

period of uninterrupted darkness, rather than uninterrupted light (**figure 30.13**). Thus, short-day plants are really long-night plants, because they flower only if their uninterrupted dark period exceeds a critical length. Similarly, long-day plants are really short-night plants.

## Phytochrome, a Blue Pigment, Controls Photoperiodism

How do plants measure the lengths of night and day? The "clock" is a blue pigment molecule called **phytochrome.**

Phytochrome exists in two interconvertible forms that have very similar structures (**figure 30.14**). The $P_r$ form absorbs red wavelengths of light (660 nanometers). The $P_{fr}$ form absorbs in the far-red portion of the electromagnetic spectrum (730 nm). $P_r$

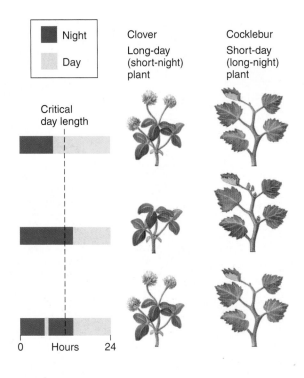

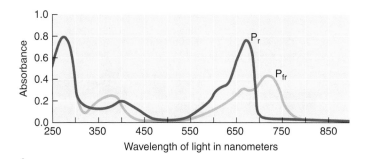

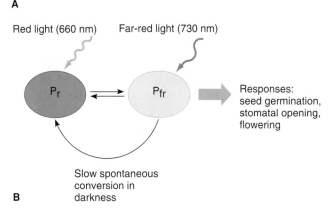

**FIGURE 30.13**   **Night-time Light Flashes Inhibit Flowering in Short-Day Plants.**   Length of day and night influence flowering in long-day plants such as clover and short-day plants such as cockle-bur. Long-day plants require a dark period shorter than a critical length. Short-day plants require an uninterrupted dark period longer than a critical period. Interrupting the dark period of a short-day plant inhibits flowering.

**FIGURE 30.14**   **The Blue Pigment Phytochrome Comes in Two Forms.**   (**A**) $P_r$ and $P_{fr}$ absorb slightly different wavelengths of red light; as a result, both appear blue. (**B**) $P_r$ absorbs red light and is rapidly converted into $P_{fr}$; the reverse occurs rapidly when $P_{fr}$ absorbs far-red light. $P_{fr}$ is also converted slowly to $P_r$ in the absence of light. $P_{fr}$ has a variety of biological effects.

is converted to $P_{fr}$ nearly instantaneously in the presence of red light, but the reverse transformation, $P_{fr}$ to $P_r$, is slow (unless plants are exposed to far-red light). It is the different amounts of red and far-red wavelengths in sunlight throughout a 24-hour period that the plant senses with its phytochrome system. Because sunlight has more red than far-red, in the daytime, $P_{fr}$ production is rapid, which somehow tells the plant that it is day. As night falls, the $P_{fr}$ form of phytochrome is slowly converted back to $P_r$.

Often in science, the existence of a structure or phenomenon is predicted based on indirect evidence and later confirmed by discovery or experiment. With phytochrome, researchers hypothesized that a pigment controlled flowering because light (or its absence) was clearly part of the response. Then, a series of experiments showed that a flash of red light during the dark period could inhibit flowering, but that a subsequent flash of far-red light could cancel this effect (**figure 30.15**). The presumed pigment was therefore sensitive to red and far-red wavelengths. In 1959, researchers isolated and identified phytochrome.

## Phytochrome Affects Other Plant Responses

Phytochrome also affects seed germination. In seeds of lettuce and many weeds, red light stimulates germination, and far-red light

inhibits it. Like the flowering response shown in figure 30.15, seeds alternately exposed to red and far-red light are affected only by the last exposure. Therefore, the phytochrome system can "inform" a seed whether sunlight is available for photosynthesis and thus promote germination under favorable conditions or inhibit germination until good conditions prevail. If seeds are buried too deeply in the soil, $P_{fr}$ is absent (due to lack of sunlight needed to convert $P_r$ to $P_{fr}$), and germination does not occur.

Phytochrome also controls early seedling growth. Seedlings grown in the dark are etiolated—they have abnormally elongated stems, small roots and leaves, a pale color, and a spindly appearance (**figure 30.16**). Bean sprouts used in Chinese cooking are etiolated. Etiolated plants rapidly elongate before exhausting their food reserves. Normal growth replaces etiolated growth once the plants are exposed to red light, even if for just 1 minute.

Phytochrome may also help direct shoot phototropism, in addition to the influence of auxin. Light coming from only one direction would presumably create a gradient of $P_r$ and $P_{fr}$ across the stem. $P_{fr}$ would be most abundant on the illuminated side of the stem, because in daylight, $P_r$ is rapidly converted to $P_{fr}$. $P_r$ would be most abundant on the shaded side of the stem. This phytochrome gradient could bend a shoot as $P_r$ promotes stem elongation and $P_{fr}$ inhibits it.

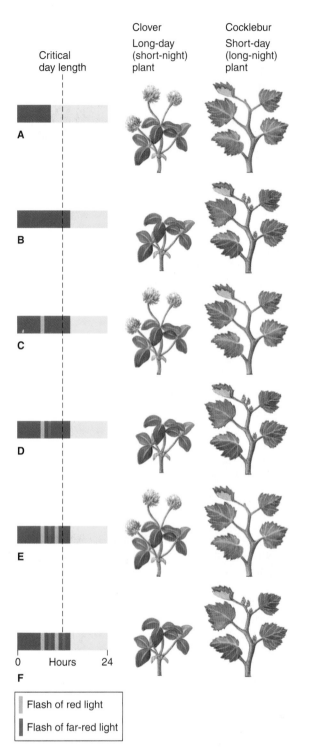

Critical day length

Clover
Long-day
(short-night)
plant

Cocklebur
Short-day
(long-night)
plant

A

B

C

D

E

0    Hours    24

F

| Flash of red light
| Flash of far-red light

**FIGURE 30.15    The Last Flash Matters.    (A)** A long-day plant flowers when daylight exceeds a critical length. **(B)** A short-day plant flowers when daylight is less than a critical length. **(C)** Interrupting night with a flash of red light shortens continuous darkness, and a long-day plant flowers, but a short-day plant does not because the time of uninterrupted darkness is too short. **(D)** A flash of far-red light closely following a flash of red cancels the effect of the red, so the results are the same as **(B)**. Three flashes **(E)**—red, far-red, red—has the same effect as one flash of red. Four flashes **(F)**—red, far-red, red, far-red—has the same effect as **(D)**. The conclusion: the last flash matters because it determines the prevalent form of phytochrome, which plants use to detect the length of the night.

**FIGURE 30.16    Growing Up in the Dark.**    An overlying log was lifted to reveal these etiolated seedlings, which germinated and grew in the absence of light. Note the pale, elongated stems and tiny leaves.

## Plants Undergo Seasonal Cycles of Aging and Dormancy

**Senescence,** or aging, is also a seasonal response of plants. Aging occurs at different rates in different species. The flowers of plants such as wood sorrel and heron's bill shrivel and die only a few hours after blooming. Slower senescence is seen in the colorful changes in leaves in autumn.

Whatever its duration, senescence is not merely a gradual cessation of growth but an energy-requiring process brought about by new metabolic activities. Leaf senescence begins during the shortening days of summer as nutrients are mobilized. By the time a leaf is shed, most of its nutrients have long since been transported to the roots for storage. Fallen leaves are little more than cell walls and remnants of nutrient-depleted protoplasm.

Destruction of chlorophyll in leaves is part of senescence. In autumn, the yellow, orange, and red carotenoid pigments, previ-

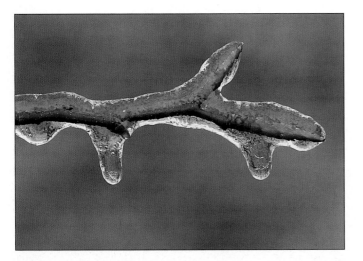

**FIGURE 30.17  Dormancy.**  Plants enter a seasonal state of dormancy, which enables them to survive harsh weather. The protective scales give the buds of this species a light brown color.

ously masked by chlorophyll, become visible. Senescing cells also produce pigments called anthocyanins. Loss of chlorophyll, visibility of carotenoids, and production of anthocyanins combine to create the spectacular colors of autumn leaves.

Before the onset of harsh environmental conditions such as cold or drought, plants often become **dormant** and enter a state of decreased metabolism (**figure 30.17**). Like leaf senescence, dormancy entails structural and chemical changes. Cells synthesize sugars and amino acids, which function as antifreeze, preventing or minimizing cold damage. Growth inhibitors accumulate in buds, transforming them into winter buds covered by thick, protective scales. These changes in preparation for winter are called acclimation.

Growth resumes in the spring as a response to changes in photoperiod and/or temperature. Lengthening spring days awaken birch and red oak trees from dormancy. Fruit trees such as apple and cherry have a cold requirement to resume growth. This is why a warm period in December, before temperatures have really plummeted, will not stimulate apple and cherry trees to bud, but a similar warm-up in late February will induce growth. The exact mechanism by which photoperiod or cold breaks dormancy is unknown, although hormonal changes are probably involved.

In some plants, factors other than photoperiod or temperature trigger dormancy. In many desert plants, for example, rainfall alone releases the plant from dormancy. In contrast, potatoes require a dry period before renewing growth.

### Reviewing Concepts

- Flowering, seed germination, early seedling growth, senescence, and dormancy are keyed to day length, which reflects seasonal changes.
- The ratio of two forms of a photopigment provides critical cues to day length.

## 30.4  Plant Responses to the Daily Cycle

The photoperiodic flowering response occurs in response to seasonal stimuli such as days becoming longer or shorter. Other rhythmic responses are not seasonal but daily, and these are called **circadian rhythms** (from the Latin *circa* = approximately and *dies* = day). Consider the common four-o'clock, which opens its flowers only in late afternoon, or the yellow flowers of evening primrose, which open only at nightfall. Similarly, the nyctinastic movements of prayer plants occur at the same time every day. Other circadian rhythms in plants include stomatal opening and nectar secretion. These daily rhythms allow plants to synchronize their activities. Flowers of a particular species can open when pollinators are most likely to visit; for example, bat-pollinated flowers of genus *Carnegiea* must open at night when bats are active (see figure 29.6). Such daily rhythms also occur in many protista, fungi, and animals.

### Control of Circadian Rhythms Is Largely a Mystery

The biological clocks that regulate circadian rhythms are controlled both internally (by genes that encode "clock" proteins) and externally, by environmental factors, but the precise mechanism (or mechanisms) remains unknown. Evidence for internal control is that in many species, circadian rhythms do not exactly coincide with a 24-hour day, but may be a few hours longer or shorter. In addition, circadian rhythms often continue under laboratory conditions of constant light or dark. They are ingrained.

Environmental factors, such as a change in photoperiod, can affect or reset a plant's circadian rhythms. This environmentally controlled resynchronization of the biological clock is called entrainment. However, entrainment to a new environment is limited. If the new photoperiod differs too much from a plant's biological clock, the plant reverts to its internal rhythms. Also, a plant maintained in a modified photoperiod over a long period of time reverts to its natural rhythms when placed in constant light.

### Solar Tracking Is a Circadian Rhythm

The sight of a field of sunflowers with their glorious heads turned to face the sun is so striking that anyone lucky enough to see it simply stares (**figure 30.18**). The response of flowers turning to face the sun, called **heliotropism**, is circadian because it is the sun that defines day and night. Heliotropism, more commonly known as solar tracking, is seen among flowers that grow in Arctic and alpine environments, where days are short. Snow buttercups (*Ranunculus adoneus*) and mountain avens (*Dryas integrifolia*) also display heliotropism.

By turning toward the sun, heliotropic plants maximize sun exposure, which is used not only for photosynthesis, but also to

**FIGURE 30.18**    **Heliotropism.**    Sunflowers turn to face the sun in a circadian response called heliotropism.

warm pollinators, which become more active. Charles Darwin described solar tracking, noting that it is both fast and reversible.

A bit of folklore inspired a series of experiments that traced heliotropism to auxin activity. In the nineteenth century, the French noticed that plants shielded by bottles of red wine did not point toward the sun, as others did that were not blocked. Reasoning that the red color blocked blue wavelengths of light, Candace Galen, a biologist at the University of Missouri at Columbia, devised a laboratory version of the wine phenomenon, depicted in **figure 30.19**.

Galen grew buttercups under three conditions: shielded with a red acrylic filter that blocked blue light; with no filter; and with a filter that allowed blue wavelengths through. The flowers not exposed to blue light did not orient toward the sun. To determine how the flowers sensed blue light, Galen followed a simple approach—she lopped off different parts of the plant, to see which one had to be missing for the plant to be unable to

detect blue light. The implicated area turned out to be the upper tip of the main stem (figure 30.19B). Finally, she used an opaque substance to cover different parts of the main stem and so discovered that blocking the tip eliminated heliotropism (figure 30.19C). By examining the plants under a microscope, she saw that the cells on the side of the middle of the stem away from the sun were larger than the cells facing the sun, suggesting that auxins are involved in the response. The conclusion: the top of the stem detects blue light and somehow passes this information to cells in the middle of the stem, where auxins accumulate on the shaded side, causing the plant to bend toward the sun.

This chapter has focused on plants' responses to environmental stimuli. These responses are easy to describe—leaves turn to intercept the sun, or flowers bloom. More difficult is untangling the many complex interactions that produce the responses. Plant hormones, which help control such fundamental functions

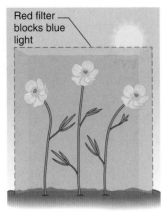

A Experimental         Control 1         Control 2

**Experiment 1**
Conclusion: Blue light is required for solar tracking.

B Decapitated        Control

**Experiment 2**
Conclusion: Solar tracking ability resides in stem.

C Block tip of stem     Block middle of stem     Block bottom of stem

**Experiment 3**
Conclusion: Solar tracking ability resides in upper stem.

**FIGURE 30.19 Experiments Reveal That Blue Light Controls Heliotropism Through Auxin.** (**A**) Blocking blue light destroys the ability of buttercups to track the sun. (**B**) The tip of the main stem just below the flower is the plant part that detects blue light. Even decapitated plants turn toward the sun, as long as their stems are intact. (**C**) By applying an opaque coating (Liquid Paper) to the tops, middles, and bottoms of stems, researcher Candace Galen showed that the tip detects the blue light. Microscopy revealed that an auxin response occurs in the middle of the stem. The tip must therefore communicate with the cells in the middle section.

as cell division and elongation, surely play some part in flowering, which in many plants occurs in response to photoperiod. But the precise interaction between phytochrome and plant hormones remains unknown. Furthermore, the hormones described in this chapter are by no means the only influences on plant growth and development. Many other mechanisms play roles in growth, cell division, and plant responses to the environment.

## Reviewing Concepts

- Biological clocks control activities that cycle according to regular schedules.
- Circadian rhythms, such as solar tracking, are based on the length of a day.
- Biological clocks are internally controlled but refined by the environment.

## Connections

Plants cannot move quickly in response to any kind of stimulus. Therefore, they have developed different mechanisms for responding to changes in their environments. We are fascinated when some of these appear to be similar to animal motion—like the snapping shut of a Venus flytrap. But all such responses in a plant are limited, as is all life, to the mechanisms available to its cells. The flytrap creates springlike tension through filling special cells with water. A trigger causes changes in proteins at the cell membrane, which moves water from place to place, simulating muscular motion. We are likewise fascinated by a plant's ability to grow toward the sun and to direct roots toward the earth. These illustrate the only other mechanism available to a plant—that of cell growth or death. In fact, the ability of a plant to shed its leaves and regrow them each year has suggested that similar mechanisms might be triggered in other species to regrow lost limbs. Finally, we have learned to use the growth signals to boost the production of plants by providing them with plant hormones. As we learn more about signaling in plants, we can apply those ideas to other species and gain a greater understanding of one of life's most fascinating attributes—the ability to respond to stimuli.

## Student Study Guide

## Key Terms

| | | | |
|---|---|---|---|
| abscisic acid (ABA) *574* | dormant *583* | long-day plant *579* | phytochrome *580* |
| apical dominance *573* | ethylene *573* | nastic movement *576* | senescence *582* |
| auxin *570* | gibberellin *571* | nyctinasty *578* | short-day plant *579* |
| circadian rhythm *583* | gravitropism *575* | $P_{fr}$, $P_r$ *580* | statolith *575* |
| cytokinin *572* | heliotropism *583* | photoperiodism *579* | thigmonasty *576* |
| day-neutral plant *579* | hormone *570* | phototropism *574* | thigmotropism *576* |
| | | | tropism *574* |

# Chapter Summary

## 30.1  The Influence of Hormones on Plant Growth

1. Plants use hormones to coordinate responses to environmental changes.
2. Plants have five major classes of hormones: auxins, gibberellins, cytokinins, ethylene, and abscisic acid.
3. Each hormone plays a role in growth or slowing growth and, ultimately, in dormancy.

## 30.2  Plant Movements: Motion Without Muscles

4. A plant can grow toward light, with or against gravity, or encircle an object in response to touch.
5. Such tropisms are directed movements.

## 30.3  Plant Responses to Seasonal Changes

6. Flowering, seed germination, early seedling growth, senescence, and dormancy are keyed to day length, which reflects seasonal changes.
7. The ratio of two forms of a photopigment provides critical cues to day length.

## 30.4  Plant Responses to the Daily Cycle

8. Biological clocks control activities that cycle according to regular schedules.
9. Circadian rhythms, such as solar tracking, are based on the length of a day.
10. Biological clocks are internally controlled but refined by the environment.

# What Do I Remember?

1. What is the definition of a hormone?
2. Describe the five major classes of plant hormones and include functions for each.
3. How is gravitropism in plants an adaptive trait?
4. Explain how phytochrome controls photoperiodism and seed germination.
5. What three factors can release a plant from dormancy?
6. What is the difference between photoperiodism and a circadian rhythm such as photonasty?

### Fill-in-the-Blank

1. Plant movements are _____ if they are not oriented to a stimulus.
2. _____ is an example of a circadian rhythm.
3. _____ is the aging of plants due to seasonal changes.
4. The pigment _____ is responsible for controlling plant responses to light.
5. A plant can measure the length of day and night through _____.

### Multiple Choice

1. _____ are hormones that delay leaf senescence and stimulate cell division.
   a. Auxins
   b. Gibberellins
   c. Cytokinins
   d. Abscisic acids
   e. Ethylenes

2. Tomatoes are often harvested while yet green and shipped to market, where they are sprayed with _____ to "ripen" them.
   a. auxins
   b. gibberellins
   c. cytokinins
   d. abscisic acids
   e. ethylenes

3. In order to exhibit gravitropism, a plant must have structures that function as
   a. statoliths.
   b. abscisic acid.
   c. plastids.
   d. thigmonasty.

4. Plants track the sun through the action of auxins in a process called
   a. heliotropism.
   b. photoperiodism.
   c. photonasty.
   d. gravitropism.
   e. thigmotropism.

5. Plants respond to touch in what is known as
   a. heliotropism.
   b. sensory periodicity.
   c. sensotropism.
   d. gravitropism.
   e. thigmotropism.

**Communication**  The nervous and endocrine systems integrate and coordinate the activities of all organ systems.

The **nervous system** in a human is a vast interconnected network of trillions of neurons, surrounded, nourished, and supported by several types of neuroglia. Organization within the nervous system ranges from simple sequences of neurons, to the central relay center that is the spinal cord, to the great complexity of the brain. The sense organs that provide information about the outside world are also part of the nervous system. Some neurons are specialized sensory receptors that detect changes from the environment or inside the body, while others receive the impulses transmitted from sensory receptors and relay, interpret, or act on them. Yet other neurons carry impulses from the brain or spinal cord to muscles or glands, which contract or secrete products in response. Nervous tissue infiltrates many organs. The digestive tract, for example, has its own mini-nervous system, which explains how emotions can affect appetite and digestion.

The **endocrine system** includes glands that secrete hormones, which are a type of communication biochemical. Most hormones travel within the circulatory system to target tissues, where they produce a characteristic response. For example, the pituitary gland in the brain produces the hormone prolactin, which signals the breasts of a woman who has just given birth to secrete milk. Hormones affect development, reproduction, mental health, metabolism, and many other functions.

**Support and Movement**  The skeletal and muscular systems enable us to stand upright and to move. Muscles also power movements within the body.

The **skeletal system** consists of bones, sheets of connective tissue called ligaments, and cartilage. Bones provide frameworks and protective shields for softer tissues and serve as attachments for muscles. The marrow within certain bones is also a vital tissue, producing the components of blood and storing inorganic salts.

Individual muscles are the organs that comprise the **muscular system.** Muscles contract when the protein filaments that comprise them slide past one another. This contraction provides the forces that move body parts. Muscle contraction also helps maintain posture, keeps food moving through the digestive tract, enables the lungs and heart to work, and is a major source of body heat.

**Getting Energy**  The **respiratory system** acquires oxygen from the atmosphere and releases carbon dioxide ($CO_2$); the **cardiovascular system** delivers $CO_2$ to the lungs for exhalation and carries oxygen throughout the body; and the **digestive system** provides nutrients. The oxygen the respiratory system collects and the cardiovascular system delivers is used to extract maximal energy from the chemical bonds of nutrient molecules.

All three of these systems maximize surface area through a generalized organization of parts being folded and subdivided into smaller and smaller pieces. The respiratory system is a series of linked tubes that collect, filter, and then deliver air to the lungs, where oxygen enters capillaries (tiny blood vessels) in millions of microscopic air sacs. Carbon dioxide leaves the capillaries here and is ultimately exhaled. The freshly oxygenated blood travels to the heart and is distributed through a vast system of blood vessels throughout the tissues, providing them with the oxygen they require to obtain energy to stay alive. Nerve and muscle tissues, with high energy requirements, are packed with capillaries, whereas tissues with low metabolic demands, such as cartilage and the outer skin layer, lack capillaries.

The digestive system is a series of tubes and chambers that conduct food into the body, breaking it apart mechanically and chemically along the way. Finally, at the small intestine, the molecules are small enough to be absorbed into capillaries, which then distribute the nutrients throughout the body.

**Protection**  The body protects itself in several ways. For example, the **integumentary system,** which consists of the skin and its outgrowths, is a barrier between an animal and its environment. In addition, the compositions of body fluids must remain within certain ranges for life to continue in the face of a changeable environment. The body must also fight infection, injury, and cancer.

The **urinary system** filters the blood, removing toxins and wastes, reabsorbing valuable substances, and maintaining the concentrations of a variety of ions, keeping the electrolytes in body fluids in balance. The system consists of the paired kidneys, which are each packed with tubular units called nephrons, along which filtered blood travels as the contents are adjusted; two tubes, called ureters, into which the kidney tubules drain; a urinary bladder to store urine; and the urethra, the tube leading to the outside.

The **immune system** is a huge army of several types of highly specialized cells, the vessels that transport them, and organs where these cells are produced and collect. This system is remarkably diverse—it can launch an attack against nearly any virus or invading organism, and it "remembers," which is how the body develops immunity to certain infections. The immune system recognizes cells of the body as "self," and launches an attack against anything that is foreign, or "nonself," such as certain viruses, microorganisms, and cancer cells.

**Biological Continuity**  The **reproductive system** is not vital to the functioning of an individual, but it is to the perpetuation of the species. The reproductive systems of male and female consist of organs where gametes form; tubules that transport gametes to parts of the body where the opposite types can meet; and glands whose secretions enable the gametes to mature and function. The female body also has a built-in system to nurture a developing offspring.

The reproductive system illustrates how the different organ systems are, in a sense, not separate at all. Consider the uterus, the pear-shaped sac that houses the embryo and fetus. It is innervated, which is why a woman feels cramps when it contracts; hormones stimulate these contractions. The majority of the uterus is muscle. The entire system is richly supplied with blood vessels, which also deliver cells and biochemicals of the immune system.

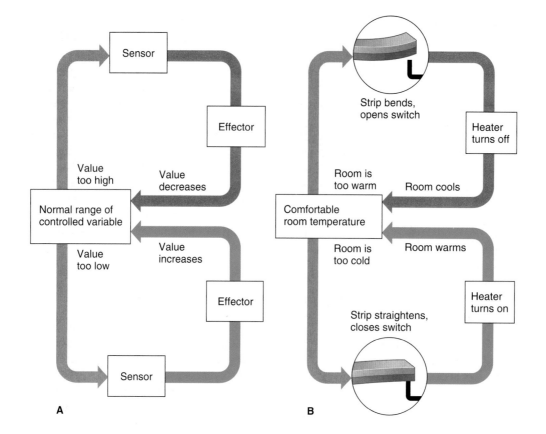

**FIGURE 31.6  Negative Feedback Systems Promote Homeostasis.** **(A)** In negative feedback systems, some variable is controlled within limits. Sensors monitor changes in the controlled variable and activate an effector if the value drifts too high or too low. The effector's response counteracts the original change in the controlled variable (negative feedback). **(B)** Some thermostats contain a bimetal strip that senses temperature change. As the room warms, the strip bends and eventually turns off the heater. As the room cools, the strip straightens, turning on the heater.

# 31.7  Organ Systems and Homeostasis

The environment changes. Light and dark cycle; temperatures rise and fall; food and water may be available, or not. In order to function in a changing environment, an animal's body must maintain a stable internal environment. Concentrations of nutrients and oxygen in body fluids must be within a certain range, and internal temperature and blood pressure must remain within certain parameters too. This state of internal constancy is termed **homeostasis.** Organ systems interact in ways that promote homeostasis. One of the great challenges in transplanting an organ is for the new body part to integrate and function with the rest of the body to maintain homeostasis.

Maintenance of body temperature illustrates the coordination and control that is necessary to maintain homeostasis. A part of the brain functions as a thermostat of sorts, sensing when body temperature deviates from the normal 98.6°F (about 37°C). If body temperature drops, the brain triggers activities that conserve or generate heat. Small groups of muscles contract— the body shivers. Simultaneously, blood vessels beneath the skin constrict, keeping warm blood away from the body's surface to minimize heat loss. In the opposite situation, a rising temperature, homeostatic changes promote loss of body heat. Sweat pours out, cooling the skin as it evaporates. Blood vessels in the skin dilate, releasing heat from deeper tissues to the environment. Heart rate increases, sending more blood to the surface vessels to release heat.

Maintenance of blood pressure is another bodily function controlled homeostatically. Rising blood pressure triggers receptors in the walls of blood vessels leading away from the heart to signal the brain to signal the heart to slow its contractions, sending less blood into the already stressed vessels. The pressure drops. If blood pressure falls too low, the brain center signals the heart to speed up, sending out more blood.

Most homeostatic responses demonstrate **negative feedback,** which is an action that counters an existing condition (see figure 6.17). When temperature or blood pressure is low, homeostasis raises it, and vice versa. **Figure 31.6** summarizes the negative feedback in a familiar situation—maintaining room temperature. Only a few biological functions demonstrate **positive feedback,** in

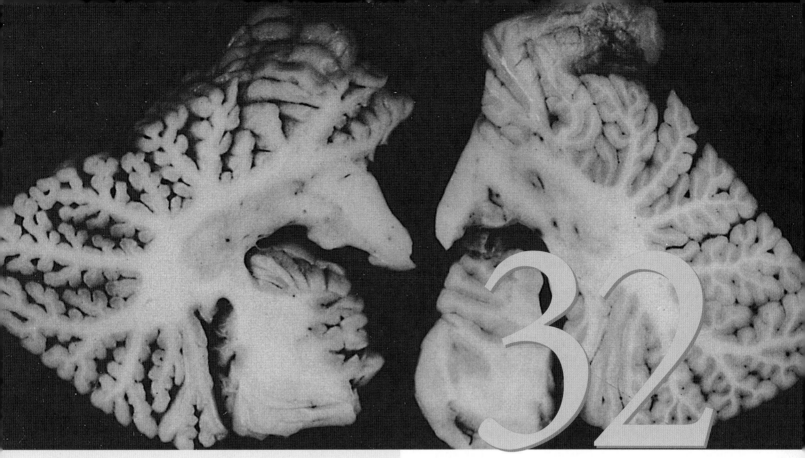

On autopsy, Dr. Karen Wetterhahn's cerebellum (*left*) was missing much tissue compared with a healthy cerebellum (*right*).

## Mercury: A Toxic Legacy

On August 14, 1996, Dartmouth College chemist Karen Wetterhahn was working with dimethylmercury when a few drops spilled onto her latex glove. She cleaned up the spill, then took off the glove. Although she recorded the accident in her notebook, she didn't think too much about it—until January, when symptoms of mercury poisoning began.

At first, Dr. Wetterhahn felt tingles in her fingers and toes. She then developed poor balance and slurred speech. Soon her visual field narrowed, she saw flashes of light, and her hearing acuity diminished. By the end of January, laboratory tests confirmed mercury poisoning.

On February 6th, she slipped into a coma. By the end of February, she no longer responded to light, sound, or gentle touch. She had bouts of crying, but never regained consciousness. She died on June 8, 1997. On autopsy, Karen Wetterhahn's brain was a mere vestige of normal.

Dimethylmercury is a "supertoxin" that rapidly crosses the blood-brain barrier and binds to sulfur-containing amino acids of neurons, slowly destroying these vital nerve cells. This compound is used to standardize nuclear magnetic resonance devices, and Dr. Wetterhahn was preparing such a standard when the deadly drops fell onto her glove. No one knew that latex gloves provided no protection.

In the short time before her brain function was destroyed, Karen Wetterhahn made heroic efforts to alert her colleagues to the dangers of dimethylmercury. As a result, improved safety standards have been instituted, including the wearing of plastic laminate gloves inside neoprene gloves when working with the compound.

# The Nervous System

## Chapter Preview

1. A nervous system receives information, integrates it, and initiates responses. Neurons and neuroglial cells are the foundation of nervous systems.

2. A neuron has a cell body; dendrites, which receive impulses and transmit them toward the cell body; and an axon, which conducts impulses away from the cell body.

3. A sensory neuron carries information toward the brain and spinal cord (central nervous system, or CNS). A motor neuron carries information from the CNS and stimulates an effector (a muscle or gland). An interneuron conducts information between two neurons and coordinates responses.

4. A neural impulse, measured as an action potential, is a change in electrical charge comprised of differences in concentration of $K^+$ and $Na^+$ across the cell membrane

5. Myelination increases the speed of neural impulse transmission as the neural impulse "jumps" from one node to the next, an action called saltatory conduction.

6. The several types of neurotransmitters, which are chemicals that transmit a neural impulse, include modified and unmodified amino acids.

7. Nervous systems are groups of interacting cells that help coordinate the activities of animals. Nervous systems range from simple nerve nets, through groups of cooperating neurons, up to complex brains.

8. The vertebrate nervous system includes the CNS and the peripheral nervous system (PNS), which includes all neural tissue outside of the CNS. The PNS consists of the somatic (voluntary) nervous system and the autonomic (involuntary) nervous system.

9. The brain has three regions that separate functions that maintain life, sensory information, and higher reasoning.

10. The cerebrum's two hemispheres contain an inner layer of white matter and an outer layer of convoluted gray matter, which comprises the cerebral cortex where information is processed and integrated.

11. Short-term memory may depend on temporal electrical activity in neuronal circuits. Long-term memories depend on permanent chemical or structural changes in neurons.

## 32.1 Communication and Homeostasis

The lynx is one of the few inhabitants of the snowy midwinter northern Montana plain. The cat is acutely attuned to its frigid environment, able to hear the quiet swish of a snowshoe hare on the icy cover before it sees its well-camouflaged prey (**figure 32.1**). The lynx watches. The hare doesn't realize it is being observed until the lynx bursts from behind a tree. The cat's snowshoelike feet enable it to quickly overtake and capture its fleeing prey.

The interaction between the lynx and the hare is orchestrated by each animal's nervous system, a vast collection of interconnected **neurons** and their associated **neuroglia** cells. The cat's adaptations as a predator—inch-long claws, knifelike fangs, compact powerful body, and snowshoe feet—require a nervous system that enables it to spot and attack prey. Similarly, the hare's nervous system triggers its swift flight—although too late in this case.

Overall, the function of the nervous system, and of all animal organ systems, is to maintain a range of internal constancy, or homeostasis, by detecting and reacting to changes in the environment. The nervous system maintains homeostasis by regulating virtually all other organ systems. The lynx's nervous system controls its respiratory system, which brings oxygen to the circulatory system, which delivers the oxygen plus nutrients to all cells. The nervous system also controls the excretory system, as cellular wastes move through the circulatory system to the kidneys. The nervous system enables muscles to act on bones so that the lynx can move; keeps the digestive system providing nutrients to cells for energy conversion and work; causes the endocrine system to secrete adrenaline, which sharpens the animal's awareness and coordinates the many systems for rapid action; and keeps the immune system on alert against infection.

The lynx's pursuit of the hare demonstrates three major roles of the nervous system—sensory input, sensory integration, and motor response. Sensory information arrives as odor, sound, and visual images. The lynx compares this information to past hunting experiences and decides how to act. The motor systems coordinate muscles to move the lynx into position to catch the hare. These events are initiated at the conscious level of the brain. Meanwhile, the heart beats, the lungs inhale and exhale, and the cat maintains its balance through sensory input, decision making, and motor output at the subconscious level.

Neurons make possible many sensations, actions, emotions, and experiences. Networks of interacting nerve cells control mood, appetite, blood pressure, coordination, and perception of pain and pleasure. The unique ability of neurons to communicate rapidly enables animals to be aware of and react to the environment and to screen out unimportant stimuli, to learn, and to remember. Yet despite their diverse functions, all neurons com-

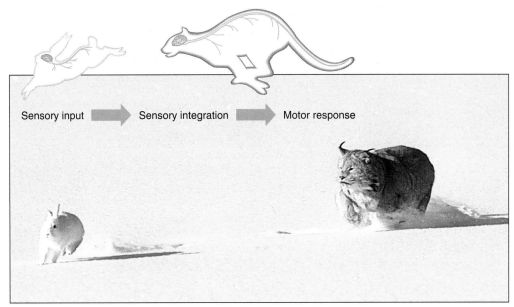

Sensory input ➡ Sensory integration ➡ Motor response

**FIGURE 32.1** **Nervous Systems in Action.** Networks of nerve cells control this lynx's attack—and the hare's attempt to escape. Sensory organs such as eyes and ears receive sensory input and relay it to the brain and spinal cord, which integrate the information and produce appropriate motor responses.

municate in a similar manner. This chapter first explores how neurons function and then considers the evolution and functioning of nervous systems.

> ### Reviewing Concepts
>
> - A nervous system enables an organism to detect, react to, and survive threats in the environment.
> - Networks of neurons and neuroglia form the foundation of nervous systems.

## 32.2  Neurons

All neurons have the same basic parts. However, these cells vary considerably in shape and size.

### Neuron Anatomy Includes Extensions for Distance

A typical neuron consists of a rounded central portion with many emanating long, fine extensions, a form well adapted to receiving, integrating, and conducting messages over long distances (**figure 32.2**). The central portion of the neuron, the **cell body,** does most of the neuron's metabolic work. It contains the usual organelles: a nucleus, extensive endoplasmic reticulum, mitochondria to supply energy, and ribosomes to manufacture proteins necessary to convey messages.

A neuron's extensions are of two types. The shorter, branched, and more numerous extensions are **dendrites.** They receive information from other neurons and transmit it toward the cell body. The many branching dendrites can receive input from many other neurons. The second type of extension from the cell body is the **axon,** which in many neurons conducts the message away from the cell body and toward another cell. Because a neuron's message may have to reach a cell quite far away, an axon is usually longer than a dendrite—sometimes surprisingly so. An axon that permits a person to wiggle a big toe, for example, extends from the base of the spinal cord to the toe. An axon is usually thicker than a dendrite, and a neuron usually has only one axon. Bundles of axons or dendrites from several cells are called **nerves.**

To picture the relative sizes of a typical neuron's parts, imagine the cell body is the size of a tennis ball. The axon might then be 1 mile (1.6 kilometers) long and half an inch (1.27 centimeters) thick. The dendrites would fill an average-size living room.

### Neurons Exhibit Different Functions

A neuron may have one of three general functions (**figure 32.3**). A **sensory** (or **afferent**) **neuron** brings information about the internal or external environment toward the **central nervous system** (CNS) (the brain and spinal cord). A touch of the skin, for example, stimulates a sensory neuron's dendrites just under the skin surface. The dendrites, in turn, converge on the sensory neuron's axon, which relays the information to another neuron whose dendrites are located within the spinal cord.

A **motor** (or **efferent**) **neuron** conducts its message outward, from the CNS toward muscle or gland cells. A motor neuron's

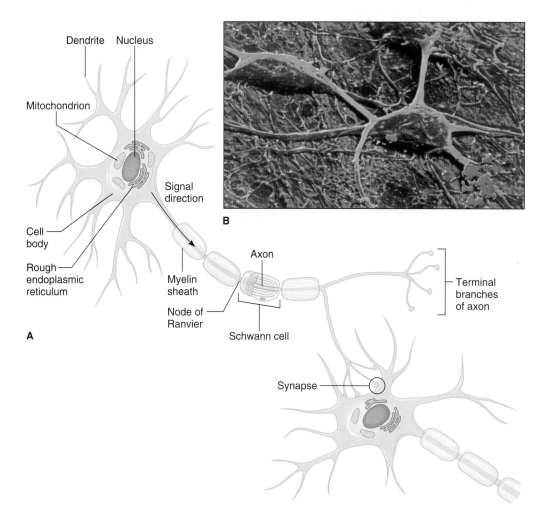

**FIGURE 32.2  Parts of a Neuron.**  (**A**) A neuron consists of a rounded cell body, "receiving" branches called dendrites, and a "sending" branch called an axon. The junction of one neuron and an adjacent neuron is called a synapse. Many axons are encased in fatty myelin sheaths. Unmyelinated regions between adjacent myelin sheath cells are called nodes of Ranvier. (**B**) These unmyelinated neurons from the cortex of a human brain are magnified 500 times. Note their entangled axons and dendrites.

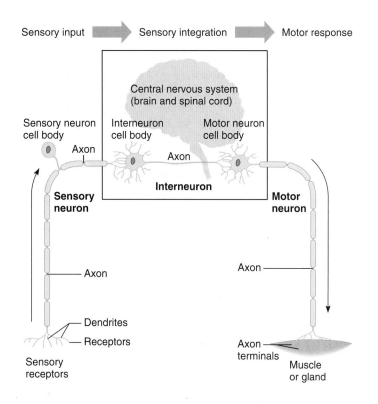

dendrites and cell body reside in the CNS. A cell body passes information from its dendrites to the axon, which extends from the CNS to the effector (muscle or gland). Thus, motor neurons stimulate muscle cells to contract and glands to secrete.

A third type of neuron, an **interneuron,** connects one neuron to another within the CNS to integrate information from many sources and coordinate responses. Large, complex networks of interneurons receive information from sensory neurons, process this information, and generate the messages that the motor neurons carry to effector organs.

**FIGURE 32.3  Categories of Neurons.**  Sensory neurons transmit information from sensory receptors in contact with the environment to the central nervous system. Motor neurons send information from the central nervous system to muscles or glands. Interneurons connect other neurons. Note that neuron shape varies considerably. The cell body of the sensory neuron is near one end of the cell and has a single branch that exits the cell body. The motor neuron's cell body is at one end of the cell and has several dendrites and a single axon.

# 32.3 The Action Potential and Nerve Impulses

The message a neuron conducts is called a **neural impulse**. This is an electrochemical change that occurs when ions move across the cell membrane. A measurement called an **action potential** describes the ionic changes that are the neural impulse. A neural impulse is the spread of electrochemical change (action potentials) along an axon. To understand how and why ions move in an action potential, it helps to be familiar with the resting potential, the state a neuron is in when it is not conducting an impulse.

## A Neuron at Rest Has a Negative Charge

The membrane of a resting neuron is polarized. That is, the inside carries a slightly negative electrical charge relative to the outside. This separation of charge creates an electrical "potential" that measures around –70 millivolts. (A volt measures the difference in electrical charge between two points. The minus sign indicates that the inside of the cell is negative when compared with the outside of the cell.) The charge difference across the membrane results from the unequal distribution of ions (**figure 32.4**).

How is this unequal distribution of ions established and maintained? First, the cell membrane is selectively permeable; it admits some ions but not others. Ions move through the membrane through small channels. Some channels are always open, but others open or close like gates, depending on proteins that change shape. A few of these gates are voltage regulated—whether a gate opens or closes depends upon the electrical charge of the membrane. Other gates open and close in response to certain chemicals. Some membrane channels are specific for sodium ions ($Na^+$), and others are specific for potassium ions ($K^+$).

Another property of the membrane that establishes and maintains ion distribution is the **sodium-potassium pump**. Recall from chapter 5 that this pump is a mechanism that uses cellular energy (ATP) to transport $Na^+$ out of, and $K^+$ into, the cell. The sodium-potassium pump actively transports $Na^+$ and

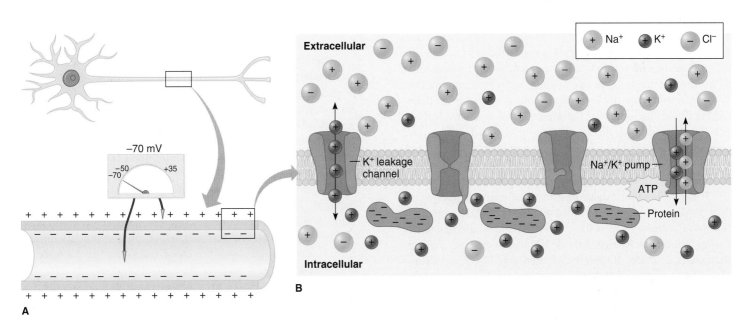

**FIGURE 32.4**   **The Resting Potential.**   (**A**) A voltmeter measures the difference in electrical potential between two electrodes. When one electrode is placed inside an axon at rest and the other is placed outside, the electrical potential inside the cell is approximately –70 millivolts (mV) relative to the outside due to the separation of positive (+) and negative (–) charges along the membrane. (**B**) At rest, the concentration of $Na^+$ is greatest outside the cell while the concentration of $K^+$ is greatest inside the cell. Large, negatively charged proteins are confined to the cell interior and help contribute to the net negative cell interior. The $Na^+/K^+$ pump uses ATP to continuously move three $Na^+$ out of and two $K^+$ into the cell. One class of $K^+$ channel is open at rest, and the diffusion of $K^+$ out of the cell is balanced by the electrical flow of $K^+$ back into the negative cell interior. Note that this figure shows $Cl^-$, the primary extracellular negative ion, but its role in the electrical behavior of the neuron is quite complex.

K⁺ against their concentration gradients, resulting in a concentration of K⁺ 30 times greater inside the cell than outside and a concentration of Na⁺ 10 times greater outside than inside.

● transport proteins, p. 71

Ions distribute themselves in response to two forces. First, ions follow an electrical gradient. Like charges (negative and negative; positive and positive) tend to repel one another. Opposite charges (negative and positive) attract. Second, ions follow a concentration gradient and diffuse from an area in which they are highly concentrated toward an area of lower concentration. Therefore, a particular ion enters or exits the cell at a rate determined by its permeability and concentration gradient.

Three mechanisms establish and maintain the **resting potential**. First, the sodium-potassium pump concentrates K⁺ inside the cell and Na⁺ outside. The pump ejects three Na⁺ for every two K⁺ it pumps in. Second, large negatively charged proteins (and other negative ions) are trapped inside the cell because the cell membrane is not permeable to them. Third, the membrane in the resting state is 40 times more permeable to K⁺ than to Na⁺.

## Resting Potential Is a Balance Between Gradients

Because of the concentration gradient and high permeability, K⁺ is able to diffuse out of the cell. As K⁺ moves through the membrane to the outside of the cell, it carries a positive charge, leaving behind large negatively charged molecules. A difference in charge, or potential, is therefore established across the membrane: positive on the outside and negative on the inside. The magnitude of the potential is determined by the balance of opposing forces acting on K⁺. The concentration gradient drives K⁺ outward, and the negative charge inside the cell holds K⁺ in. When these two opposing forces are equal, no net movement of K⁺ occurs, and the cell is in equilibrium, at the resting potential.

The importance of the sodium-potassium pump in maintaining the resting potential becomes evident when a metabolic poison such as cyanide disables the pump. K⁺ slowly diffuses out and Na⁺ in, destroying the concentration gradients. Nerve transmission is then impossible because a charge difference no longer exists across the membrane. Death occurs in minutes.

It is curious that a neuron uses more energy while resting than it does conducting an impulse. Presumably, expending energy to maintain the resting potential allows the neuron to respond more quickly than it could if it had to generate a potential difference across the membrane each time it received a stimulus. This is analogous to holding back the string on a bow to be continuously ready to shoot an arrow.

## A Nerve Impulse Is a Wave of Depolarization

During an action potential, Na⁺ and K⁺ quickly redistribute across a small patch of the cell membrane, creating an electrochemical change that moves like a wave along the nerve fiber. An action potential begins when a stimulus (a change in pH, a touch, or a signal from another neuron) changes the permeability of the membrane so that some Na⁺ begins to leak into the cell (**figure 32.5**). This permeability change occurs as some sodium channel activation gates open. As Na⁺ enters the neuron, the interior becomes less negative, or depolarizes, because of the influx of Na⁺. When enough Na⁺ enters to depolarize the membrane to a certain point, called the **threshold potential**, additional sodium gates sensitive to charge in that area of the membrane open, further increasing permeability to Na⁺. The threshold potential is about −50 mV. Driven by both the electrical gradient and the concentration gradient, enough Na⁺ enters the cell to positively charge the cell interior near the membrane. Na⁺ influx continues until the positive charge peaks.

Near this peak of the action potential, membrane permeability changes again. Permeability to Na⁺ halts as sodium channel inactivation gates close, but permeability to K⁺ suddenly increases as delayed potassium gates open. Now Na⁺ cannot enter in large numbers; however, exit of K⁺ begins, driven by both electrical and concentration gradients. K⁺ flows outward because it is more concentrated inside than outside and because the inside of the membrane is now positively charged due to the influx of Na⁺.

The loss of positively charged K⁺ restores the negative charge to the interior of the cell, repolarizing the cell membrane. Meanwhile, sodium channel activation gates close, and inactivation gates open, readying the membrane for a subsequent action potential. Figure 32.5 further describes the opening and closing of these ion channels.

While the Na⁺ and then the K⁺ gates are open, a second action potential cannot begin. Still, an action potential takes only 1 to 5 milliseconds, and the electrical change spreads rapidly to adjacent areas of membrane, away from the stimulus. This capacity to rapidly transmit neural impulses makes the nervous system an effective communication network.

The characteristic changes in membrane permeability that constitute the action potential travel along the neuron, usually from a trigger zone at the cell body-axon junction and down the axon toward the axon terminal. The neural impulse spreads because some of the Na⁺ rushing into the cell at a particular point moves to the neighboring part of the neuron and causes it to reach threshold. This triggers an influx of Na⁺ there, carrying the impulse forward.

A neural impulse is much like a line of falling dominoes. Each fall, like an action potential, is an all-or-none phenomenon—it either happens or it doesn't. When the first domino falls, it triggers the next, and so forth. While no domino has moved from its original position, the chain reaction creates a wave of energy change that spreads to the end of the series.

Interneurons discern the intensity of a stimulus from the frequency of action potentials from sensory neurons. Whereas a light touch to nerve endings in the skin might produce 10 impulses in a given time period, a hard hit might generate 100 impulses, intensifying the sensation. Although all action potentials are the same, neurons also distinguish the type of stimulation. We can tell light from sound because the neurons that light stimulates transmit impulses to a different place in the brain than sound-generated impulses reach.

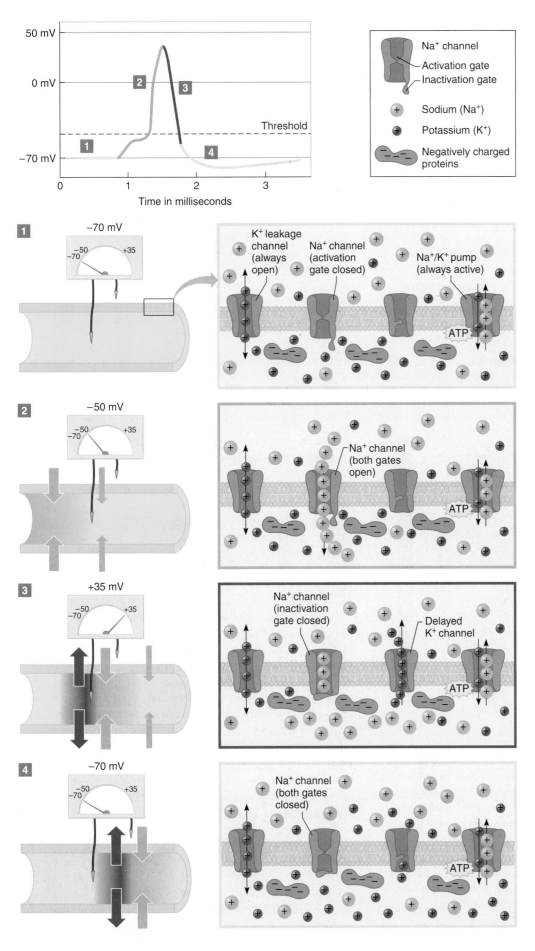

**FIGURE 32.5  The Generation of an Action Potential.** [1] Shown is a diagram of the neuronal membrane at rest. The K⁺ leakage channel is open, and K⁺ is in equilibrium between its diffusion force outward and its electrical force inward. The Na⁺ channel activation gate is closed, restricting the movement of Na⁺ into the cell; the membrane potential is about −70 mV. [2] Na⁺ channel activation gates open, and Na⁺ rushes into the cell, flowing down its electrochemical gradient; the membrane depolarizes. This local change in potential will trigger the opening of additional voltage-sensitive Na⁺ channels. If enough Na⁺ channels open, the cell reaches its threshold potential, an all-or-none action potential occurs, and the membrane reverses its polarity. [3] Na⁺ channel inactivation gates close after a split second, and delayed K⁺ gates open. K⁺ leaves the axon, restoring the polarized condition of the resting potential. [4] Na⁺ channel activation gates close, and the membrane briefly hyperpolarizes as the resting membrane is reestablished.

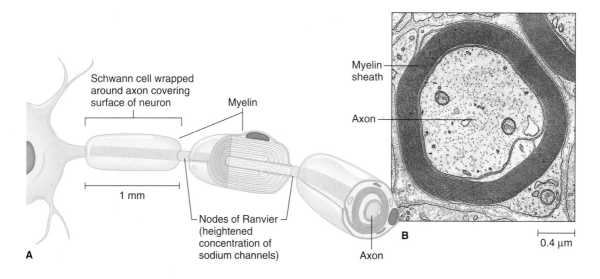

**FIGURE 32.6    Myelin Sheaths Encase Many Axons.**    A myelin sheath forms when a Schwann cell (**A**) winds around an axon so that several layers of its lipid-rich cell membrane surround the axon. The Schwann cell's cytoplasm and nucleus are often squeezed into the periphery of the sheath. The unmyelinated sections of axon between Schwann cells are called nodes of Ranvier. (**B**) TEM of a myelinated axon in cross section.

## Reviewing Concepts

- A resting neuron is negatively charged inside compared with outside due to the distribution of sodium and potassium ions, and proteins.

- During an action potential, which indicates a neural impulse, a neuron's interior near the cell membrane fleetingly becomes positively charged, a condition that spreads down the axon.

- Pumps restore and maintain the K$^+$ and Na$^+$ concentrations that form the basis for the action potential.

# 32.4   The Role of Myelin in Impulse Conduction

Not all nerve fibers conduct impulses at the same speed. Conduction speed depends on certain characteristics of the fiber. The greater the diameter of the fiber, the faster it conducts an impulse; however, thin vertebrate nerve fibers can conduct impulses very rapidly when they are coated with a fatty material called a **myelin sheath.**

Outside the brain and spinal cord, neuroglial cells called **Schwann cells,** which contain enormous amounts of lipid, form myelin sheaths. Each Schwann cell wraps around a small segment of an axon many times, forming a whitish coating (**figure 32.6**). Between Schwann cells is a short region of exposed axon called a **node of Ranvier.** Some neurons in the

brain and spinal cord are wrapped in myelin that neuroglial cells called **oligodendrocytes** produce.

In a myelinated axon, sodium channels are concentrated at the nodes of Ranvier. When a neural impulse travels along the axon, it "jumps" from node to node in a type of transmission called **saltatory conduction (figure 32.7).** The impulse appears to leap from node to node because the myelin insulation prevents ion flow across the membrane, but a small electrical current spreads instantly between nodes. Because a neural impulse moves faster when it jumps from node to node, saltatory conduction speeds impulse transmission. Myelinated axons may conduct impulses 100 times faster than unmyelinated axons, at speeds of up to 394 feet (120 meters) per second. This equals 270 miles (435 kilometers) an hour! A sensory message travels from the toe to the spinal cord in less than 1/100 of a second.

Myelinated fibers are found in neural pathways that transmit impulses over long distances. They make up the white matter of the nervous system. Cell bodies and interneurons that lack myelin usually specialize in interpreting multiple messages. These unmyelinated fibers, which make up the gray matter of the nervous system, form much of the nerve tissue in the brain and spinal cord.

## Reviewing Concepts

- Schwann cells, loaded with myelin, insulate nerve fibers, speeding conduction of nerve impulses via saltatory conduction.

- Gaps in the myelin sheath expose the neuron membrane at nodes of Ranvier, allowing action potentials to jump rapidly along the nerve fiber.

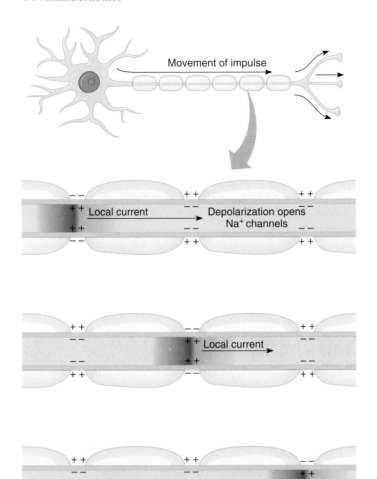

**FIGURE 32.7** **Saltatory Conduction.** In myelinated axons, the inward diffusion of Na⁺ can occur only at the nodes of Ranvier. Thus, action potentials appear to "jump" from one node to the next, which speeds up impulse transmission along the axon.

# 32.5 Neurotransmitters: Communication Between Cells

To form a communication network, a neuron must convey an impulse to another neuron or to a muscle or gland cell. Most neurons do not touch each other, so the impulse cannot travel directly from cell to cell. Instead, the impulse causes release of a chemical signal, called a **neurotransmitter,** that travels from a "sending" cell to a "receiving" cell across a tiny space. Once across the space, the neurotransmitter molecule binds to a receptor on the receiving cell's membrane, altering the permeability in a way that either provokes or prevents an action potential from occurring. The combination of the axonal terminal of the sending cell, the space, and the membrane of the receiving cell is called a **synapse.**

## Many Substances Act as Neurotransmitters

An individual neuron may produce one or several varieties of neurotransmitters. The **peripheral nervous system (PNS;** the part outside the brain and spinal cord) uses three neurotransmitters: acetylcholine, norepinephrine (also called noradrenaline), and epinephrine (adrenaline). The central nervous system, by comparison, uses more than 40 neurotransmitters, including serotonin, dopamine, and gamma aminobutyric acid (GABA).

Neurotransmitters can be classified chemically. Epinephrine, norepinephrine, dopamine, and serotonin are called monoamines, and they are modified amino acids. Unmodified amino acids can also function as neurotransmitters, and these include glycine, glutamic acid, aspartic acid, and GABA (see appendix C). Certain peptides (short chains of amino acids) function as **neuromodulators,** which means that they alter a neuron's response to a neurotransmitter or block the release of a neurotransmitter. For example, the enkephalins are five-amino-acid-long peptides that are produced in the CNS and relieve acute pain. Endorphins are longer molecules that provide more extended pain relief. Yet another neuromodulator is substance P, which is released by sensory neurons and transmits information about pain to interneurons. Endorphins relieve pain by blocking release of substance P from these sensory neurons.

Neurotransmitter function is enormously complex because the nervous system is enormously complex. A neurotransmitter's effect depends upon its concentration and the types and numbers of receptors and ion channels on the receiving cells' membranes. Plus, neurotransmitters affect each other's levels.

The same neurotransmitter can even have opposite effects on different types of cells. Consider acetylcholine. In the PNS, acetylcholine controls muscle contraction (see figure 35.16). In the brain's cerebral cortex, however, acetylcholine's role is not as straightforward. Here, this neurotransmitter coordinates function by enhancing nerve transmission in a horizontal direction (parallel to the brain surface), while at the same time inhibiting transmission in a vertical direction (perpendicular to the brain surface). The cortex is organized into columns, and horizontal transmission is necessary to coordinate functioning. Acetylcholine can both inhibit and stimulate because it can bind two types of receptors. One type, found on certain brain cells called basket cells, binds acetylcholine and then decreases levels of GABA production, which otherwise inhibits activity. The result is a stimulation of horizontal nerve transmission because GABA's inhibitory effect is lifted. The other type of acetylcholine receptor, located on brain cells called bipolar cells, increases GABA production. This strengthens the inhibition of vertical transmission.

Neurotransmission is very much a matter of balance, and too much or too little of a neurotransmitter can cause disease. Excess acetylcholine, for example, causes seizures. In contrast, death of acetylcholine-producing cells causes the forgetfulness and clouded thinking of Alzheimer disease.

## TABLE 32.1

**Mechanisms of Drug Action**

| Drug | Neurotransmitter Affected | Mechanism of Action | Effect |
|---|---|---|---|
| Cocaine | Dopamine | Binds dopamine transporters | Euphoria |
| Curare | Acetylcholine | Decreases neurotransmitter in synaptic cleft | Muscle paralysis |
| Monoamine oxidase inhibitors | Norepinephrine | Blocks enzymatic degradation of neurotransmitter in presynaptic cell | Mood elevation |
| Nicotine | Dopamine | Stimulates release of neurotransmitter | Increases alertness |
| Reserpine | Norepinephrine | Packages neurotransmitter into vesicles | Limb tremors |
| Ritalin | Dopamine | Binds dopamine transporters | Treats attention deficit disorder |
| Tricyclic antidepressants | Norepinephrine | Blocks reuptake | Mood elevation |
| Valium | GABA | Enhances receptor binding | Decreases anxiety |

## TABLE 32.2

**Disorders Associated with Neurotransmitter Imbalances**

| Condition | Imbalance of Neurotransmitter in Brain | Symptoms |
|---|---|---|
| Alzheimer disease | Deficient acetylcholine | Memory loss, depression, disorientation, dementia, hallucinations, death |
| Depression | Deficient norepinephrine and/or serotonin | Debilitating, inexplicable sadness |
| Epilepsy | Excess GABA leads to excess norepinephrine and dopamine | Seizures, loss of consciousness |
| Huntington disease | Deficient GABA | Personality changes, loss of coordination, uncontrollable movement, death |
| Hypersomnia | Excess serotonin | Excessive sleeping |
| Insomnia | Deficient serotonin | Inability to sleep |
| Myasthenia gravis | Deficient acetylcholine at neuromuscular junctions | Progressive muscular weakness |
| Parkinson disease | Deficient dopamine | Tremors of hands, slowed movements, muscle rigidity |
| Schizophrenia | Deficient GABA leads to excess dopamine | Inappropriate emotional responses, hallucinations |

**Table 32.1** lists drugs that alter specific neurotransmitter levels, and **table 32.2** describes illnesses that result from neurotransmitter imbalances.

## Neurotransmitters Act at Synapses

The end of an axon has tiny branches that enlarge at the tips to form **synaptic knobs.** These knobs contain many synaptic vesicles, which are small sacs that hold neurotransmitter molecules. A neural impulse passes down the axon of the **presynaptic neuron,** which is the cell sending the message. When the action potential reaches the membrane near the space, or **synaptic cleft,**

the permeability of the membrane changes, allowing calcium ions to enter the cell. Calcium ions cause the loaded vesicles to move toward the synaptic membrane, fuse with it, and dump their neurotransmitter contents into the synaptic cleft by exocytosis (**figure 32.8**).  • **exocytosis, p. 73**

Neurotransmitter molecules diffuse across the synaptic cleft and attach to protein receptors on the membrane of the receiving neuron, the **postsynaptic neuron.** A particular neurotransmitter fits only into a specific receptor type, as a key fits only a certain lock. When the neurotransmitter contacts the receptor, the shape of the receptor changes. This binding opens channels in the postsynaptic membrane, admitting specific ions and changing the probability that an action potential will occur.

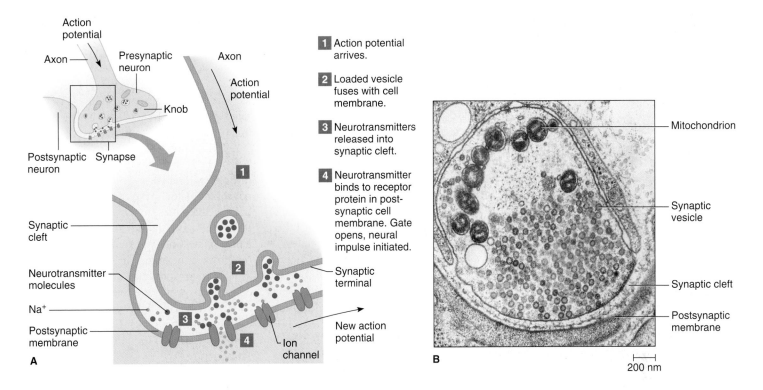

**FIGURE 32.8    Transmission Across a Synapse.    (A)** Arrival of a wave of depolarization—the action potential—at an axon of a presynaptic neuron triggers release of neurotransmitter from vesicles. The neurotransmitter crosses the synapse and binds with receptors in the postsynaptic neuron's cell membrane. This action opens ion channels in a way that increases or decreases the likelihood that an action potential will be generated. **(B)** This transmission electron micrograph of a synapse shows synaptic vesicles and a synaptic cleft.

## Neurotransmitters Must Be Removed from the Synapse

If a neurotransmitter stayed in the synapse, its effect on the receiving cell would be continuous, perhaps causing it to fire unceasingly and bombard the nervous system with stimuli. Three mechanisms control the amount of neurotransmitter in a synapse: (1) it can diffuse away from the synaptic cleft; (2) it can be destroyed by an enzyme; or (3) it can be taken back into the presynaptic axon soon after its release, an event called **reuptake.** Poisonous nerve gases and certain insecticides work by blocking the enzymatic breakdown of the neurotransmitter acetylcholine in the synaptic cleft. The twitching legs of a cockroach sprayed with certain insecticides demonstrate the effects. The excess acetylcholine activity overstimulates skeletal muscles, causing them to contract continuously.

Several classes of drugs work by blocking reuptake of serotonin. For example, the "selective serotonin reuptake inhibitors" (SSRIs) are antidepressants that block reuptake of serotonin, enabling the neurotransmitter to accumulate in the synapse, thereby offsetting a deficit that presumably causes the symptoms (**figure 32.9**). The reverse situation, excess serotonin, causes sleepiness. This is why people become sleepy after eating turkey. The meat contains abundant tryptophan, an amino acid required for the body to synthesize serotonin.

Two recent experiments examining serotonin reuptake demonstrated the similarity of nervous systems among vertebrates and invertebrates. Knowing that people who raise clams feed them serotonin to get them to release sperm and eggs at the same time, a biologist fed clams Prozac, the first SSRI to be used as a drug in humans. A tiny dose of the drug triggered sperm and egg release, giving clam farmers a much cheaper way to increase reproduction rates at their facilities. In another experiment, a researcher gave antidepressants to barnacles and mussels. The excess serotonin sped the development of the larvae, causing them to detach from surfaces. The drugs may be used as chemical agents to keep barnacles off ships! Yet another study is examining the effects of increasing serotonin on aggression in lobsters.

### Reviewing Concepts

- Neurotransmitters relay nerve signals across the synapses.
- Each neurotransmitter is responsible for a particular kind of signal.
- The neurons rapidly remove the signaling molecules by way of active transport.
- Psychoactive drugs act by interfering with binding or recycling of the transmitters or their receptors.

 Health

# 32.1

## Addiction!

Drug abuse and addiction are ancient as well as contemporary problems. A 3,500-year-old Egyptian document decries society's reliance on opium. In the 1600s, a smokable form of opium enslaved many Chinese, and the Japanese and Europeans discovered the addictive nature of nicotine. During the Civil War, morphine was a widely used painkiller; cocaine was introduced a short time later to relieve veterans addicted to morphine. Today, we continue to abuse drugs intended for medical use. LSD was originally used in psychotherapy, but was abused in the 1960s as a hallucinogen. PCP was an anesthetic before being abused in the 1980s. Why people become addicted to certain drugs lies in the complex interactions of neurons, drugs, and individual behaviors.

### The Role of Receptors

Eating hot fudge sundaes is highly enjoyable, but we usually don't feel driven to consume them repeatedly. Why do certain drugs compel a person to repeatedly use them in steadily increasing amounts—the definition of addiction? The biology of neurotransmission helps to explain how we, and other animals, become addicted to certain drugs.

Understanding how neurotransmitters fit receptors can explain the actions of certain drugs. When a drug alters the activity of a neurotransmitter on a postsynaptic neuron, it either halts or enhances synaptic

transmission. A drug that binds to a receptor, blocking a neurotransmitter from binding there, is called an antagonist. A drug that activates the receptor, triggering an action potential, or that helps a neurotransmitter to bind, is called an agonist. The effect of a drug depends upon whether it is an antagonist or an agonist and on the particular behaviors the affected neurotransmitter normally regulates.

Neural pathways that use the neurotransmitter norepinephrine control arousal, dreaming, and mood. Amphetamine drugs enhance norepinephrine activity, heightening alertness and mood. Amphetamine's structure is so similar to that of norepinephrine that it binds to norepinephrine receptors and triggers the same changes in the postsynaptic membrane.

Cocaine has a complex mechanism of action, both blocking reuptake of norepinephrine and binding to molecules that transport dopamine to postsynaptic cells. Cocaine's rapid and short-lived "high" reflects its rapid stay in the brain—its uptake takes just 4 to 6 minutes, and within 20 minutes, the drug loses half its activity.

### Opiates in the Human Body

Opiate drugs, such as morphine, heroin, codeine, and opium, are potent painkillers derived from the poppy plant. These drugs alter pain perception, making pain easier to tolerate, and they also elevate mood. When taken repeatedly by healthy individuals, opi-

ate drugs are addictive. When taken to relieve intense pain, such as that of cancer, opiates are usually not addictive.

The human body produces its own opiates, called endorphins (for "endogenous morphine"). These peptides are a type of neuromodulator. Like the poppy-derived opiates they structurally resemble, endorphins influence mood and perception of pain. Opiates and endorphins bind the same receptors in the human brain.

Humans produce several types of endorphins, which are released in response to stress or pain. Endorphins are released during a "runner's high," as a mother and child experience childbirth, and during acupuncture.

Endorphins help explain why some people addicted to an opiate drug such as heroin experience withdrawal pain when they stop taking the drug. Initially, the body interprets the frequent binding of heroin to its endorphin receptors as excess endorphins. To bring the binding down, the body slows its own production of endorphins. Then, when the person stops taking heroin, the body is caught short of opiates (heroin as well as endorphins). The result is pain.

Drug addiction is a powerful force, its causes rooted in both biology and psychology, and its effects an economic, sociological, and political problem. Fighting an addiction is far more difficult than "just saying no" to temptation, as one might to a hot fudge sundae.

## 32.6 Integration of Incoming Messages

The nervous system has two types of synapses—excitatory and inhibitory. The combination of excitatory and inhibitory synapses provides finer control over a neuron's activities.

Excitatory synapses depolarize the postsynaptic membrane, and inhibitory synapses increase the polarization, or hyperpolarize it. A neurotransmitter that acts at an excitatory synapse increases the probability that an action potential will

be generated in the second neuron by slightly depolarizing it. For example, when acetylcholine binds to the receptors at an excitatory synapse, channels open that admit $Na^+$ into the postsynaptic cell. Within a millisecond, half a million sodium ions flow into the cleft. If enough $Na^+$ enters to reach a threshold level of depolarization, it triggers an action potential in the postsynaptic cell.

On the other hand, a neurotransmitter may inhibit an action potential in the postsynaptic cell by making the cell's interior more negative than the usual resting potential. In this case, extra $Na^+$ must enter before the membrane depolarizes enough to generate an action potential.

620

| Nondepressed individual | Depressed individual, untreated | Depressed individual, treated with SSRI |
|---|---|---|

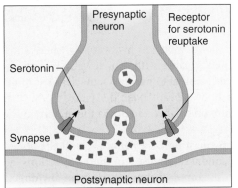
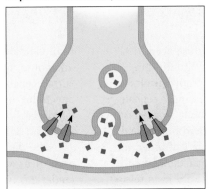
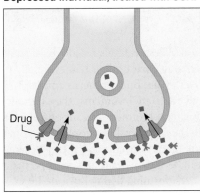

**FIGURE 32.9**   **Anatomy of an Antidepressant.**   Selective serotonin reuptake inhibitors (SSRIs) are antidepressant drugs that block the reuptake of serotonin, making more of the neurotransmitter available in the synaptic cleft. This restores a neurotransmitter deficit that presumably causes the symptoms. Overactive or overabundant reuptake receptors can cause the deficit. The precise mechanism of SSRIs is not well understood. Newer antidepressants block reuptake of both serotonin and norepinephrine.

A single neuron in the nervous system may receive input from tens of thousands of other neurons, some excitatory and others inhibitory. Nearly half of a neuron's receiving surface adjoins synaptic clefts. Whether that neuron transmits an action potential depends on the sum of the excitatory and inhibitory impulses it receives (**figure 32.10**). If it receives more excitatory impulses, the postsynaptic cell is stimulated; if inhibitory messages predominate, it is not. A neuron's evaluation of impinging nerve messages, which determines whether an action potential is "fired," is termed neural integration, or **synaptic integration.**

Synapses markedly increase the informational content of the nervous system. The human brain has a trillion neurons, each of which can be viewed as carrying bits of information. But if a synapse is also considered a unit of information, then the informational capacity of the brain increases a thousandfold, because a typical brain neuron has synaptic connections to a thousand other neurons, each sending or receiving messages hundreds of times per second.

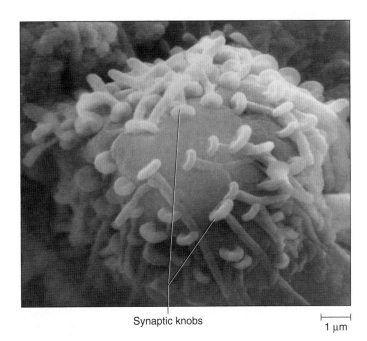

**FIGURE 32.10**   **Integration of Neural Impulses.**   The synaptic terminals from many neurons converge on the cell body of a single postsynaptic neuron, as shown in this scanning electron micrograph. Some of these terminals are inhibitory, while others are excitatory—the postsynaptic cell integrates the contributions from all active inputs and produces an action potential, or not, based on this calculation.

**Reviewing Concepts**

- The nervous system balances information flow by synaptic integration.
- This determines the kind of signal that passes to other neurons and when those signals are passed on.
- Each neuron integrates with thousands of others, increasing the potential informational content of the brain.

## 32.7   Evolutionary Trends in Nervous Systems

Multicellularity posed a profound new challenge to organisms—how would cells communicate with each other? In a single-celled organism, the cell membrane serves as a selective gateway to the outside environment. In a multicellular organism, cells maintain this gatekeeper function; they react to messages that coordinate cells' functions. These messages are electrical and chemical changes that transmit information by stimulating or inhibiting cell surface receptor molecules.

All nervous systems consist of organizations of interacting cells. Comparing nervous systems among diverse species reveals that nervous systems probably increased in complexity as new species evolved. A nervous system might be a loose network of relatively few cells or a structure as complex as the human brain.

Different organizations of nervous systems are adaptive in particular environments. Radially symmetrical animals such as sea anemones respond to stimuli that can come from any direction. These animals usually have simple, diffuse nerve nets that make all parts of their bodies equally receptive to stimuli.

Over evolutionary time, as animals began to move headfirst through their environments, they became increasingly cephalized. Bilateral symmetry meant one end of the body (the head) took in the most sensory information from the environment, and nervous tissue concentrated there. Nervous tissue continued to accumulate into a brain and sensory structures at the animals' head ends, forming highly intricate interconnections. Even relatively simple animals such as flatworms and roundworms have small brains, but the trend is most evident in the skull-encased brain of the vertebrate CNS. Vertebrates also have other concentrations of nervous tissue as part of the peripheral nervous system, which transmits information between the CNS and receptors, effectors, muscles, and glands. • **body form, p. 463**

## Invertebrates Have Nerve Nets, Ladders, or Simple Brains

The simplest nervous systems are diffuse networks of neurons, called **nerve nets.** They are a characteristic of members of phylum Cnidaria. In body walls of hydras, jellyfishes, and sea anemones, nerve nets synapse with muscle cells near the body surface, enabling each animal to move its tentacles. A stimulus at any point on the body spreads over the entire body surface. Cnidarians can maintain balance and detect touch, certain chemicals, and light.

The diffuse organization of a nerve net and the ability of each neuron to conduct impulses in both directions allow cnidarians to react to stimuli that approach from any direction, which is adaptive to an animal in the ocean where danger and food may approach from any direction. **Figure 32.11** illustrates some nervous systems of invertebrates.

Flatworms have rudimentary brains that consist of two clusters of nerve cell bodies, called **ganglia,** and two nerve cords that extend down the body and connect to each other to form a "ladder" type of nervous system. Motor structures and the neurons that control them are paired, which allows the worm to move in a coordinated forward motion. Paired symmetrical sense organs at the head end allow the flatworm to detect a stimulus and crawl toward it. Bilaterally paired receptors enable the animal to determine which direction stimulation is coming from by comparing

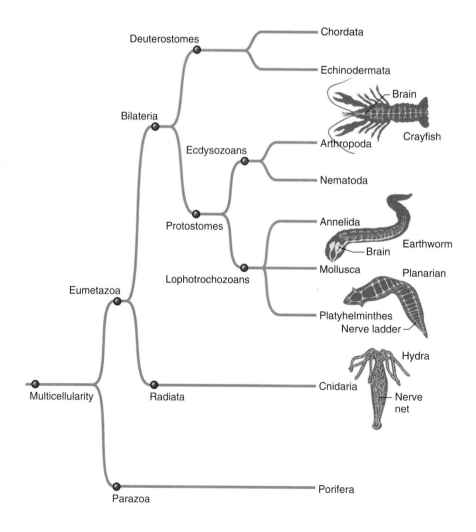

**FIGURE 32.11    Invertebrate Nervous Systems.**   As invertebrate bodies increased in complexity, so did their nervous systems.

receives a copy of motor input from the cerebral cortex (the outer portion of the cerebrum) and sensory feedback from the PNS. It then compares the action the cerebrum intended with the actual movement and makes corrections so that the two agree. For example, the cerebellum acts when we try to bring the tips of our two index fingers together with our eyes closed. Even if we miss initially, the cerebellum recognizes the error and makes corrections.

Many of our conscious activities have subconscious components that the cerebellum governs. If you had to consciously plan every movement in catching a ball, for instance, you would probably not be quick enough to prevent it from striking your head. Training and practice result in an automatic program of movements, many of which the cerebellum controls at a subconscious level.

The area above the medulla is the pons, which means "bridge." This is a suitable name for this oval mass, because white matter tracts in the pons form a two-way conduction system that connects higher brain centers with the spinal cord and connects the pons to the cerebellum. In addition, gray matter in the pons controls some aspects of breathing and urination.

The medulla and pons have changed very little through vertebrate evolution. The cerebellum, however, is greatly enlarged in more complex vertebrate species. In birds, the cerebellum is especially large because of the need for quick motor responses during flight.

A narrow region above the pons, the midbrain, is also part of the brain stem. The midbrain receives sensory information from touch, sound, visual, and other receptors and passes it to the forebrain. Much of this integrating activity occurs in a thickened area of gray matter at the roof of the midbrain called the tectum.

**The Forebrain**  The part of the brain that has changed the most through vertebrate evolution is the forebrain. It includes two major regions: (1) the telencephalon, which includes the cerebrum, and (2) the diencephalon, which lies in front of the midbrain and includes the thalamus, hypothalamus, pineal gland, and pituitary gland (see figure 34.8). Forebrain structures are important in complex behaviors such as learning, memory, language, motivation, and emotion.

The **hypothalamus** weighs less than 4 grams and occupies less than 1% of the brain volume, but it regulates many vital functions, including body temperature, heartbeat, water balance, and blood pressure. Groups of nerve cell bodies in the hypothalamus control hunger, thirst, sexual arousal, and feelings of pain, pleasure, anger, and fear. The hypothalamus also regulates hormone secretion from the pituitary gland at the base of the brain. Thus, the hypothalamus links the nervous and endocrine systems, the body's two communication systems. • **hormones in humans, p. 664**

The **thalamus** is a gray, tight package of nerve cell bodies and their associated neuroglial cells located beneath the cerebrum. It acts as a relay station for sensory input, processing incoming information and sending it to the appropriate part of the cerebrum.

The **reticular formation** is not really localized to the hindbrain, midbrain, or forebrain but is a diffuse network of cell bodies and nerve tracts that extends through the brain stem and into the thalamus. The name *reticular* ("little net") alludes to the formation's role in screening sensory information so that only certain impulses reach the cerebrum. If the reticular formation did not do so, the senses would be overwhelmed—you would be acutely aware of every cough, sneeze, and rustle of paper from those around you; every scent in the classroom; and even the touch of your clothing against your skin.

The reticular formation is also called the reticular activating system because it is important in overall activation and arousal. When certain neurons within the reticular formation are active, you are awake. When other neurons inhibit them, you sleep. The thalamus is a gateway between the reticular formation and regions of the cerebrum. On a cellular level, sleep is a state of synchrony, when many neurons in all of these regions are inhibited from firing action potentials at the same time.

The electrical activity of the brain—the brain waves detectable in an electroencephalogram (EEG) tracing—differs during wakefulness and sleep. We experience two types of sleep—REM sleep, named for the rapid eye movements that occur under closed lids, and nonREM sleep. During REM sleep, the nervous system is quite active, although movement is suppressed. The autonomic nervous system is alert, and heart rate and respiration are elevated. Brain temperature rises as its blood flow and oxygen consumption increase. Bursts of action potentials zip along certain brain pathways more frequently than during nonREM sleep or even during wakefulness. REM sleep is a time of vivid dreaming. NonREM sleep proceeds through four continuous stages, from light sleep to deep sleep, and the corresponding EEG brain waves become less frequent but greater in amplitude (size).

Sleep follows a predictable pattern—70 to 90 minutes of nonREM followed by 5 to 15 minutes of REM, repeated many times a night. If a person is deprived of REM sleep and therefore doesn't dream, the next night's pattern compensates to deliver more REM sleep.

We know little about sleep, but we do know that it is essential. Prions, the infectious proteins discussed in chapter 20 (see figure 20.12), can cause a condition called fatal insomnia. It begins with inability to fall asleep. As the months progress, the person suffers hallucinations and delusions. After about a year, he or she cannot distinguish reality from the dream state. Death soon follows.

## Reviewing Concepts

- The spinal cord receives impulses from the rest of the body, conducts reflexes, and communicates with the brain.
- Structures in the hindbrain control vital functions and connect to other regions; neurons in the midbrain process and integrate certain sensory input; in the forebrain, the cerebrum receives sensory input and directs motor responses, and the hypothalamus regulates secretion of certain hormones.

## Spinal Cord Injuries

On a bright May morning in 1995, actor Christopher Reeve sustained a devastating spinal cord injury when the horse he was riding in a competition failed to clear a hurdle. Reeve rocketed forward, striking his head on a fence. He landed on the grass, unconscious, and not moving or breathing.

Reeve had broken the first and second cervical vertebrae, between the neck and the brain stem. An onlooker kept him alive with CPR, until emergency medical technicians arrived and inserted a breathing tube and then stabilized him on a board. At a nearby hospital, Reeve received methylprednisolone, a drug that can save a fifth of the damaged neurons by reducing inflammation in the 8 hours after the injury. Reeve was then flown to a larger medical center for further treatment.

Reeve's rehabilitation was slow but inspiring. Despite discouraging words from his physicians, he persisted in trying to exercise, inspiring many others to follow his example and regain limited functions. Suspended from a harness, he moved his feet over a treadmill. He moved other muscles in a swimming pool and rode a special recumbent bicycle, with electrical stimulation to his legs enabling him to pedal an hour a day. Five years after the accident, he gradually started to move his fingers, and then his hips and legs, although he still required a wheelchair and could breathe without a respira-

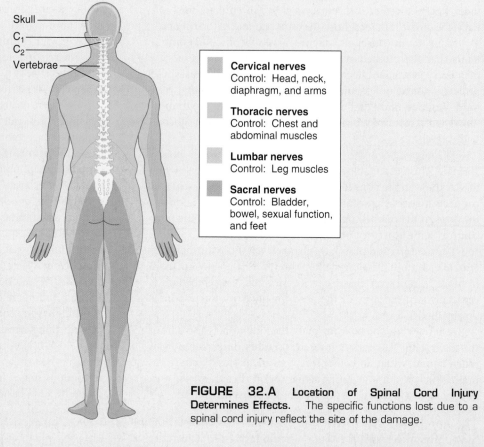

Skull

C₁

C₂

Vertebrae

**Cervical nerves**
Control: Head, neck, diaphragm, and arms

**Thoracic nerves**
Control: Chest and abdominal muscles

**Lumbar nerves**
Control: Leg muscles

**Sacral nerves**
Control: Bladder, bowel, sexual function, and feet

**FIGURE 32.A Location of Spinal Cord Injury Determines Effects.** The specific functions lost due to a spinal cord injury reflect the site of the damage.

tory for only short periods. Reeve died in October 2004 from heart failure that followed infection of a bedsore.

Thousands of spinal cord injuries occur each year. **Table 32.3** lists the top causes. A survivable spinal cord injury does not actually sever the cord, but crushes it, triggering a chain reaction of devastation. At the injured site, some neurons immediately die as they set off action potentials. The dying

## 32.9 The Cerebrum

The human brain has a large, highly developed cerebrum that controls the qualities of "mind," including intelligence, learning, perception, and emotion. The cerebrum is divided into two **cerebral hemispheres,** which make up 80% of total human brain volume.

The outer layer of the cerebrum, the **cerebral cortex,** consists of gray matter that integrates incoming information. In mammals, the cerebral cortex increases in size and complexity as species become more complex. Through evolutionary time, the cortex expanded so rapidly that it folded back on itself to fit into the skull, forming characteristic convolutions.

The cerebral cortex contains sensory, motor, and association areas (**figure 32.17**). Sensory areas receive and interpret messages from sense organs about temperature, body movement, pain, touch, taste, smell, sight, and sound. Motor areas send impulses to skeletal muscles. Association areas do not appear to be either sensory or motor, but they are the seats of learning and creativity.

A band of cerebral cortex extending from ear to ear across the top of the head, called the **primary motor cortex,** controls voluntary muscles. Just behind it is the **primary somatosensory cortex,** which receives sensory input from the skin, muscles,

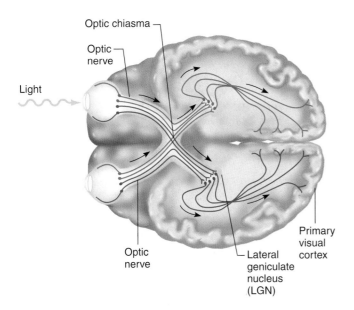

Optic chiasma

Optic nerve

Light

Optic nerve

Lateral geniculate nucleus (LGN)

Primary visual cortex

**FIGURE 33.11   From the Eyes to the Brain.** The optic nerve passes the light stimulus first to the lateral geniculate nucleus, then to the primary visual cortex, where the information is processed and integrated.

**Reviewing Concepts**

- Animals detect light with special light-sensitive cells called photoreceptors.
- Photoreceptors are grouped into several different types of eyes in invertebrates.
- In the vertebrate eye, light focuses onto the retina, where it stimulates rod photoreceptor cells to provide black-and-white vision and cone cells to provide color vision.

## 33.5   Mechanoreception

Mechanoreceptors enable an organism to respond to physical stimuli (touch) and sound waves (hearing) and to maintain balance and equilibrium.

### Invertebrates Exhibit Well-Developed Mechanoreceptor Systems

The sense of hearing operates when pressure waves hit and move receptors. In insects, hairlike structures called **setae** at the bases of the antennae vibrate in response to sound. The movement stimulates sensory cells that transmit the information to the brain.

Some insects have drumlike **tympanal organs,** which are thin membranes stretched over large air sacs that form resonating chambers for sound. Sensory cells beneath the membrane detect sound pressure waves and relay this information to the brain. Crickets and katydids have tympanal organs on their legs, while grasshoppers and some moths have them on their abdomens. These organs are usually on both sides of an insect's body, which enables it to sense not only the presence but also the direction of sound.

Setae and tympanal organs are highly sensitive. Insects can detect much higher pitches than the human ear can hear. The ability of certain moths to escape predatory bats vividly displays the sensitivity of insect hearing. Moths can detect the high-pitched cries of bats and dive down when the sound approaches. This confuses the bat's **echolocation,** which is the ability to locate airborne insects by bouncing ultrasound waves off them. When a moth flies down near the ground, the bat's echolocation bounces off of the moth and the ground at about the same time. Thus, the bat can no longer pick out the moth.

Mechanoreception also includes a sense of balance and equilibrium. In many animals, structures called **statocysts** provide this sense. A statocyst is a fluid-filled cavity containing sensory hairs and bits of mineral called **statoliths.** When the animal's body tilts, the statoliths touch the sensory hairs, stimulating the nervous system. In response, the animal can orient itself with respect to gravity. Such diverse invertebrates as jellyfishes, snails, squids, crayfishes, and earthworms use statocysts for balance. Recall that statoliths also function in a plant root's

Overall, vision is not simply a linear pathway (light to retina to brain) but an enormously intricate network of amplified and interacting signals, which the brain interprets with astonishing speed and accuracy. For example, much as a cook assembles a meal from specific ingredients, the brain assembles the idea of, say, a dog from its specific combination of features. The brain routes raw data about the dog's color, texture, size, shape, and motion through a hierarchy of increasingly comprehensive groupings. The retina perceives points of light. In the LGN, contiguous points are grouped into small edges. The visual cortex further groups these fragments into combinations such as corners or arcs. At still higher levels, these separate pieces of information eventually converge into the concept of "dog."

In many mammals, the brain also integrates the overlapping visual fields coming from each eye, creating the three-dimensional images necessary for depth perception. This is possible when an animal's eyes are located on the front of its head. Animals whose eyes are on the side of the head, such as fishes and rabbits, do not experience nearly the same degree of depth perception.

The visual cortex also fuses images formed at discrete moments of time into a fluid perception of motion. Inability to do this is like watching a movie in extremely slow motion, so each scene change is painfully obvious. It's hard to appreciate the importance of this ability, but one woman did when a stroke damaged part of her visual cortex, destroying her ability to perceive motion. She saw movement as a series of separate, static images. Her deficit had profound effects on her life. She could not pour a drink, because she could not tell when the cup would overflow. She could not cross a street because she could not detect cars moving toward her. Many animals can detect prey only if it moves. If she were a member of a different species, the inability to detect motion could be fatal.

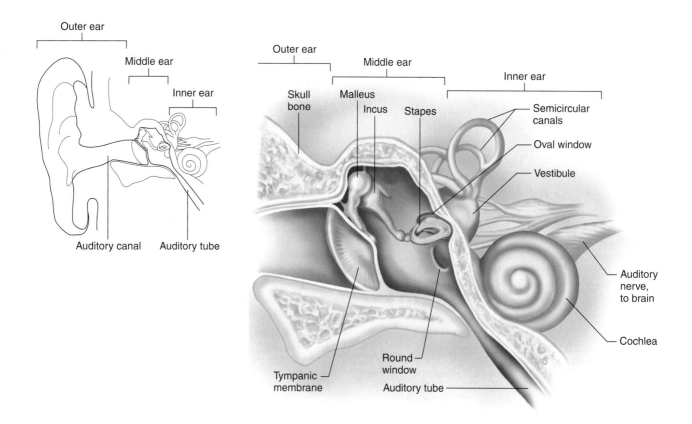

**FIGURE 33.12** **Anatomy of the Ear.** Sound enters the outer ear and, upon reaching the middle ear, impinges upon the tympanic membrane, which vibrates three bones (malleus, incus, and stapes). The inner ear houses the organs of hearing and balance. The vibrating stapes hits the oval window, and hair cells in the cochlea convert the vibrations into action potentials. The resulting neural impulses travel along the auditory nerve to the brain. Hair cells in the semicircular canals and in the vestibule sense balance. The auditory tube connects the middle ear to the rear of the nasal cavity, equalizing air pressure.

ability to detect gravity. Invertebrate statocysts are similar to structures in humans that carry out the same functions, discussed in a later section. • **gravitropism, p. 575**

Many invertebrates are startled by touch, which they sense using specialized mechanoreceptors. For example, an insect's body is laden with contact-sensitive hairs and with sensors that detect strain on its exoskeleton and joints. Some insect antennae function as both "feelers" and chemoreceptors. Spiders have modified strain gauges in their legs that feel the quivering of their webs when an insect becomes entangled. Scorpions have similar structures that detect the small ground-surface vibrations from nearby prey.

## Vertebrate Organs of Hearing Respond to Vibration in Air or Water

The clatter of a train, the sounds of a symphony, a child's wail—what do they have in common? All sounds, regardless of the source, originate when something vibrates. The vibrating object creates repeating pressure waves in the surrounding medium, such as air, water, or even the ground. The size and energy of these pressure waves determine the intensity or loudness of the sound, while the number of waves (cycles) per second determines the sound's frequency or pitch. The more cycles per second, the higher the pitch.

Vertebrates display an interesting variety of hearing mechanisms. Panamanian golden frogs lack outer and middle ears, yet they can detect and precisely localize the calls from other frogs. Perhaps they can hear because sound waves that hit the sides of the frog's body move the lungs, which then transmit vibrations to the intact inner ears for processing. Barn owls can pinpoint the height a sound emanates from because one ear is higher than the other and one points up, while the other points down. Bats' ears can swivel, and their receptor cells are sensitive to their own high-pitched calls that reflect off of nearby objects, such as their flying insect meals. Dolphins also use echolocation. High-frequency clicks and trills, generated in nasal sacs in their heads, are focused by an "acoustical lens" called the melon, located in their foreheads. In dolphins, the auditory canal typical of land-dwelling mammals is closed, so sound waves that bounce off of objects pass to the brain through the lower jaw and the middle ear. Using this sophisticated sense, dolphins can discriminate nearly identical objects, such as fishes.

**Anatomy of the ear** In humans, sound transduction and perception begins with the fleshy outer part of the ear (the pinna) that traps sound waves and funnels them down the **auditory canal** (ear canal) to the **tympanic membrane,** or eardrum (**figure 33.12**). The sound pressure waves set up vibrations in the eardrum, which moves three small bones, the **maleus, incus,** and **stapes,** located in the middle ear. These bones transmit the incoming sound and amplify it 20 times. From the middle ear, the vibrations of the stapes are transmitted through the **oval window,** a membrane that opens into the inner ear. Running between the middle ear and the air passageways in the back of the nose is a canal called the **eustachian** (or **auditory**) **tube.** This tube allows us to adjust pressure differences on the inside of the tympanic membrane if the outside pressure should change. "Ear popping" occurs when the internal and external pressures re-equilibrate and the tympanic membrane "pops" back into shape, such as when we yawn during an airplane takeoff.

The inner ear is a fluid-filled chamber that houses both balance and hearing structures. Two parts of the inner ear, the semicircular canals and the vestibule, control balance and are discussed in the next section. The remaining portion of the inner ear is the snail-shaped **cochlea.** The spirals of the cochlea are three fluid-filled canals called the vestibular, cochlear, and tympanic canals. The vestibular and tympanic canals actually form a continuous U-shaped tube. Between them lies the cochlear canal. When the last bone of the middle ear, the stapes, moves, it pushes on the oval window, transferring the vibration to the fluid in the vestibular canal. The pressure of each vibration is dissipated by the movement of the round window at the other end of the U. The vibration of the fluid also moves the **basilar membrane,** which forms the lower wall of the cochlear space, and initiates the change of mechanical energy to receptor potentials.

**Translating sound into nerve impulses**　Sound is translated into the universal action potential language of the nervous system within the cochlea. Here, specialized **hair cells** (the mechanoreceptors) lie between the basilar membrane below and another sheet of tissue, the **tectorial membrane,** above (**figure 33.13**). The basilar membrane is narrow and rigid at the base of the cochlea

and widens and becomes more flexible closer to the tip. Because of this variation in width and flexibility, different areas of the basilar membrane vibrate more intensely when exposed to different frequencies. The high-pitched tinkle of a bell stimulates the narrow region of the basilar membrane at the base of the cochlea, while the low-pitched tones of a tugboat whistle stimulate the wide end.

When a region of the basilar membrane vibrates, the hair cells are displaced relative to the tectorial membrane, which initiates action potentials in fibers of the **auditory nerve.** This nerve carries the impulses to the brain, which interprets the input from different regions as sounds of different pitches (**figure 33.14**). Louder sounds produce greater vibrations of the basilar membrane. As a result, individual hair cells fire more action potentials, and more cells are provoked to fire. The brain interprets the resulting increase in the rate and number of neurons firing as an increase in amplitude, or loudness.

Hearing loss may be temporary or permanent. It can result from conductive deafness (blocked transmission of sound waves through the middle ear) or from sensory deafness (damage to the nervous system). Many cases of hearing loss result from exposure to loud noise. Sound intensity is measured in decibels, with each 10-decibel increase representing a 10-fold increase in sound intensity. Damage to the inner ear's hair cells begins to occur at about 80 decibels, which is as loud as heavy traffic. The degree of damage depends both upon decibel level and duration of exposure to the sound.

Many rock stars, and their fans, of the 1960s have hearing loss today. The damage begins as the hair cells develop blisterlike bulges that eventually pop. The tissue beneath the hair cells swells and softens until the hair cells and sometimes the neurons leaving the cochlea become blanketed with scar tissue and degenerate.

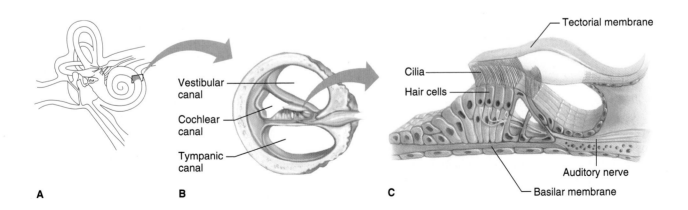

**FIGURE 33.13**　**Hearing.**　Hearing is the transduction of vibrations to neural impulses. The cochlea, shown in anatomical context in (**A**) and in cross section in (**B**), is a spiral-shaped structure consisting of three fluid-filled canals. When the stapes moves against the oval window, the fluid in the vestibular canal vibrates. The vibrations press on the hair cells that lie between the basilar and tectorial membranes (**C**). The hair cells push against the tectorial membrane, and this triggers action potentials in the auditory nerve. Cilia fringe the tops of the hair cells.

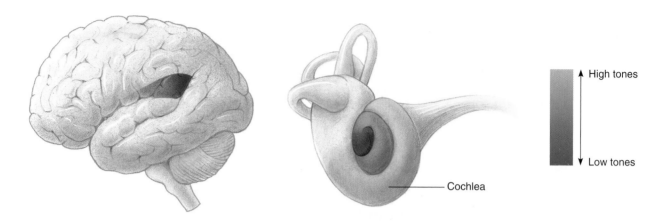

High tones

Low tones

Cochlea

**FIGURE 33.14    Correspondence Between Cochlea and Cortex.**    Sounds of different frequencies (pitches) excite different sensory neurons in the cochlea. These neurons, in turn, send their input to different regions of the auditory cortex.

## Equilibrium and Balance Are Detected by the Inner Ear

The **semicircular canals** and the **vestibule** of the inner ear regulate the sense of balance (**figure 33.15**). The semicircular canals tell us when the head is rotating and help us maintain the position of the head in response to sudden movement. The enlarged bases of the semicircular canals, the ampullae, are lined with small, ciliated hair cells. A caplike structure called a cupula covers the hair cells. Because the semicircular canals are perpendicular to each other, the fluid that fills a canal may or may not flow back and forth in response to a movement, depending on its direction. When the fluid in a canal moves, it bends the cilia on the hair cells in that ampulla, which stimulates action potentials in a nearby cranial nerve. The brain interprets these impulses as body movements.

The vestibule senses the position of the head with respect to gravity. In addition, the vestibule detects changes in velocity when traveling in a straight line. When riding in a car, for example, we can sense acceleration and deceleration. The vestibule functions in a similar way to the semicircular canals. It contains two pouches, the **utricle** and the **saccule,** both of which are filled with jellylike fluid and lined with ciliated hair cells. Granules of calcium carbonate, called **otoliths,** float on the fluid. The granules in the utricle move in response to acceleration or to tilting of the head or body, bending the cilia on the hair cells. The rate of sensory impulses to the brain increases as the cilia bend in one direction. A shift in the opposite direction inhibits the sensory neuron. The brain interprets this information as a change in velocity or a change in body position.

Motion sickness results from contradictory signals. The inner ear signals the brain that the person is not accelerating. At the same time, the eyes detect passing scenery and signal the brain that the person is moving. The result of these mixed signals is nausea.

## The Sense of Touch Includes Contact, Temperature, and Pain

Different species have different structures that sense touch—a fish uses its pressure-sensitive lateral lines, cats and rats use their whiskers, and nearly all vertebrates have diffuse touch receptors covering their bodies. Sensitivity to touch in humans comes from several types of receptors in the skin—**Pacinian corpuscles, Meissner's corpuscles,** and free nerve endings (**figure 33.16**). Both types of corpuscles are encapsulated, resembling an onion—they consist of tissue surrounding a single nerve fiber. A touch pushes the flexible sides of the corpuscle inwards, generating an action potential in the nerve fiber. High-frequency vibrations, such as those a pager or cellular telephone might produce, stimulate Pacinian corpuscles. On the other hand, a light touch stimulates Meissner's corpuscles, particularly a moving touch, as in a gentle caress.

Besides touch, other specialized receptors in the skin detect environmental stimuli. For example, free nerve endings that act as **thermoreceptors** respond to temperature changes. Heat receptors are most sensitive to temperatures above 77°F (25°C) and become unresponsive at temperatures above 113°F (45°C). Approaching this upper limit also triggers pain receptors, causing the intense pain of a burn. Cold receptors are most sensitive when the temperature ranges from 50°F (10°C) to 68°F (20°C). Temperatures below this lower limit stimulate pain receptors, causing the sensation of freezing. At the more common intermediate temperatures, the brain integrates input from many cold and heat receptors. Thermoreceptors adapt fast, which is why we quickly become comfortable after jumping into a cold swimming pool or submerging into a steaming hot tub.

**Pain receptors** are free nerve endings located nearly everywhere in the body, except the brain. They are specialized to detect mechanical damage, temperature extremes, chemicals, and blood deficiency. A particular pain sensation may be the consequence of

This camouflaged octopus, wearing algae and carrying a piece of coconut husk, walks upright on the sea bottom to confuse predators.

## Octopus Arms Find Food and Provide Protection

For some activities, eight arms can be better than two. In octopuses, constrained fluid provides the solid support against which muscles contract, generating movement.

Octopuses usually walk across sea bottoms using all eight appendages. But investigators from Hebrew University in Israel observed octopuses covered in algae walking on two arms in Australia and walking beneath coconuts on ocean bottoms in Indonesia. Their conclusion: walking on two legs is an adaptation to escape predation. The sharks, reef fishes, and stingrays that normally dine on octopuses strolling along using all eight arms do not detect the naturally camouflaged two-armed walkers.

Investigators at the University of California, Berkeley, identified "quasi-joints" in the animals' limbs that are apparently adaptations to feeding. When food is placed at a sucker on an arm, the limb bends at a point that divides the length between that sucker and the mouth approximately in two. The arm then rotates around that newly established point, and the animal brings the food to its mouth. Unlike our own bone-based skeletons or the joints of an insect's armor, the point around which the limb bends in an octopus changes with the position of the food along the arm.

The quasi-joints of the octopus arm and the joints of vertebrates and arthropods provide a compelling example of convergent evolution—similar solutions to the same challenge using different materials.

# The Musculoskeletal System

## Chapter Preview

1. The musculoskeletal system helps to maintain homeostasis by enabling an animal to move in response to environmental stimuli with supporting and protecting vital organs.

2. A braced framework, which has solid components, can be on the organism's exterior as an exoskeleton or within the body as an endoskeleton. Or these functions can be accomplished via a hydrostatic skeleton.

3. Cartilage and bone make up skeletons. Cartilage entraps a great deal of water, which makes it an excellent shock absorber. Bone has a rigid matrix and derives its strength from collagen and its hardness from minerals.

4. Muscles can be smooth, cardiac, or striated. Smooth muscle cells are spindle-shaped and involuntary. Cardiac muscle cells are striated and involuntary. Skeletal muscle cells are multinucleate, striated, and voluntary.

5. Many cells and organisms have movement mechanisms based on sliding filaments of actin and myosin. A muscle fiber is a chain of contractile units called sarcomeres.

6. Motor neurons stimulate muscle fibers by initiating an action potential at the neuromuscular junction. The complex interplay of proteins uses ATP energy to pull myosin filaments along actin fibers.

7. Muscles form antagonistic pairs, which enable bones to move in two directions. Motor units determine the delicacy of the movement.

8. Joints join one bone to another and may be freely moving or immobile. Muscles attach to bones to produce movement through lever action.

A

B

**FIGURE 35.1 Muscles and Bones Assist in Movement.** (A) The graceful movements of these dancers are possible because muscles contract against bones in a lever system. (B) The leap of this gray tree frog also depends on the musculoskeletal system.

## 35.1 Overview of Musculoskeletons and Their Functions

The dancers line up nervously, each perfectly still, waiting for the choreographer's signal. At the cue, they suddenly come alive, jumping and turning in unison, their graceful movements precisely timed to the music (**figure 35.1A**). Their ability to move depends upon connected bones and muscles, just as the leap of a gray tree frog does (figure 35.1B).

Muscle attached to bone is only one type of **musculoskeletal system.** An earthworm's skeleton, for example, isn't bone, but fluid-filled compartments. Together with perpendicular muscles in its body wall, this skeleton helps the worm cling tenaciously to its burrow.

The first organisms were probably able to move by a mechanism similar to the one that powers our muscles today—protein filaments that slide past each other, shortening and thereby contracting the cells containing them. In multicellular animals, contractile cells are organized into muscle tissues, which form organs called muscles.

Some muscles operate independently of skeletal structures. A vertebrate's digestive tract muscles, for example, mix and propel food without skeletal support. When many large animals move, however, muscles pull against rigid bones to form lever systems, which increase the strength and efficiency of movement. The skeletal and muscular systems have evolved as closely allied, interacting organ systems.

Musculoskeletal systems are under perpetual biological renovation. A person beginning an intense weight-lifting program, for example, will gradually develop larger muscles and stronger bones in response to increased use. In the opposite situation, astronauts lose bone density if they are in a prolonged weightless environment, because their musculoskeletal systems don't have to work as hard as on Earth.

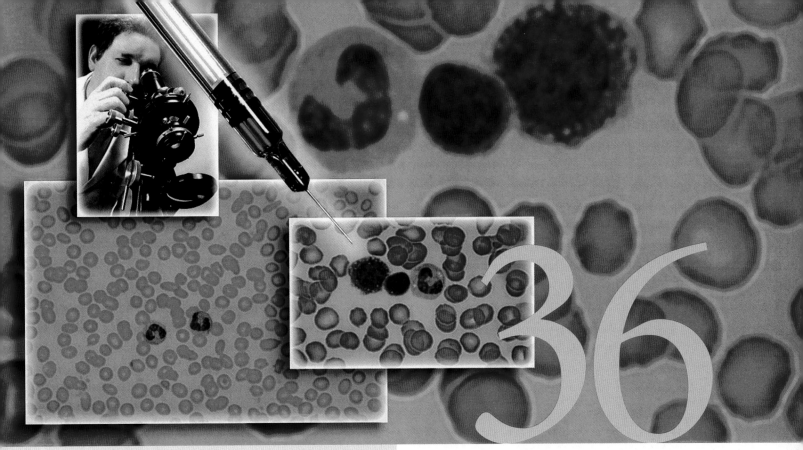

Blood is a complex mixture of cells, cell fragments, and dissolved molecules. It delivers oxygen and nutrients to cells and removes wastes.

## Using Stem Cells to Replace Hearts and Blood

Treatment for heart failure may one day entail coaxing stem cells already present in the heart to divide and produce daughter cells that can differentiate into exactly what is needed to heal the damaged tissue.

The presence of stem cells capable of this action was discovered in cases where men had received heart transplants from women donors. Later studies detected cells in the donor hearts that had the telltale Y chromosome of males, indicating that cells from the recipient had migrated to and mingled with the donor heart cells. These cells had specialized into connective tissue, cardiac muscle tissue, and epithelium—tissues needed for proper heart function.

Future red blood cell replacements may also come from stem cells. In one experiment, researchers coaxed stem cells from human umbilical-cord blood into dividing along the pathway toward forming red blood cells by using a medium with certain growth factors. Just before the final stage, in which red blood cells jettison their nuclei, the cells were infused into immune-suppressed mice. Development continued in the mice, producing mature red blood cells that made adult hemoglobin.

Stem cells could therefore respond to signals in a host to differentiate appropriately. In the case of red blood cell replacement, investigators estimate that 10 days of culture beginning with 1 million cord blood cells could yield 6 to 10 billion cells, and that once infused into a person, these cells could yield 600 billion to a trillion cells—three to five times the body's daily red blood cell output.

# The Circulatory System

# Chapter Preview

1. A circulatory system consists of fluid, a network of vessels, and a pump. The fluid delivers nutrients and oxygen to tissues and removes metabolic wastes.

2. In simple animals, open body cavities give interior structures direct contact with fluids from the environment so that nutrients and oxygen can diffuse into cells and wastes can diffuse out. In more complex animals, a heart pumps the fluid to the body cells.

3. In an open circulatory system, the fluid bathes tissues directly in open spaces before returning to the heart.

4. In a closed circulatory system, the heart pumps the fluid through a system of vessels to cells and back.

5. Vertebrates have closed circulatory systems that increase in complexity in fishes, amphibians, reptiles, birds, and mammals, reflecting the challenges of life on land. A multichambered heart and double circulation that separates oxygenated from deoxygenated blood enables land vertebrates to be active.

6. The heart is the muscular pump that drives the human circulatory system. It has two separate pumping pathways to maximize the amount of oxygen available to cells.

7. The heart beats independent of neural stimulation by responding to chemicals in the bloodstream and coordinates the beat through pacemaker cells.

8. The circulatory system leads to and from the lungs and to and from the rest of the body. Blood flows from larger vessels to increasingly smaller ones to contact each cell, then returns via increasingly larger vessels. Arteries have thicker, more elastic walls than veins.

9. The pumping of the heart and constriction of blood vessels produces blood pressure. Systole is the pressure exerted on blood vessel walls when the ventricles contract. The low point, diastole, occurs when the ventricles relax.

10. Human blood is a mixture of cells and cell fragments (collectively called formed elements), proteins, and molecules that are dissolved or suspended in plasma.

11. Control of the circulatory system helps maintain homeostasis by adjusting blood pressure.

12. The lymphatic system is a network of vessels that collect fluid from the body's tissues, purify it, and return it to the blood.

## 36.1 Overview of Circulatory Systems

The bodies of animals are like towns and cities—they require transportation systems to bring in supplies (nutrients and oxygen for energy) and to remove garbage (metabolic wastes) without disturbing their internal environments. Just as small towns can accomplish these tasks with simpler systems than can larger cities, animals with simpler bodies can sustain life with less complex systems. Diffusion across the body surface suffices in some animals; in more complex animals, circulatory systems transport nutrients.

Animals that have flattened bodies can distribute materials to and from cells without the aid of a special circulatory system. In flatworms, for example, a central cavity branches, and all cells lie close to a branch or to the body surface (see figure 25.14). This simple system enables the animal to use diffusion to meet metabolic demands. Fluid moves through the highly branched digestive tract as muscles within the body wall contract.

A sea star has a fluid-filled body cavity, a coelom, that contacts internal structures. The coelomic fluid washes through the body as cilia lining the coelom beat. Sea stars have an additional system of small channels, called the hemal system, that parallels the water vascular system and appears to assist food distribution.

In an animal thicker than a flatworm or sea star, the increase in size decreases the animal's surface-to-volume ratio. This makes diffusion inefficient for distributing raw materials throughout the organism. Circulatory systems in large, active animals deliver vital materials to cells and remove wastes efficiently enough to permit the animals to take in sufficient energy to move, yet maintain the internal environment.

### Circulatory Systems Are Open or Closed

A circulatory system transports fluid in one direction, powered by a pump that forces the fluid throughout the body. Circulatory systems evolved in conjunction with respiratory systems. That is, animals that have organs (lungs or gills) that compartmentalize gas exchange also have systems that transport oxygen throughout the body.

Circulatory systems are of two basic types: open and closed. In an **open circulatory system,** the fluid leaves the vessels. Most mollusks and arthropods have open circulatory systems. These systems consist of a heart and short vessels that lead to spaces where the fluid, called **hemolymph,** directly bathes cells before returning to the heart (**figure 36.1**). Invertebrates with open circulatory systems can be fairly active because their respiratory systems branch in a way that allows the outside environment to come close to internal tissues. Insects, for example, are among the most active animals, and their respiratory systems operate independently of their open circulatory systems.

In a **closed circulatory system** blood remains within vessels. Large vessels called **arteries** conduct blood away from the heart and branch into smaller vessels, called **arterioles,** which then

This unidentified caterpillar may have a more elaborate respiratory system than previously suspected.

## A New View of Insect Respiratory Systems

Insects exchange gases through a system of narrowing tubes. The tubules that lead into the system, termed tracheae, open to the outside through holes called spiracles that dot the integument of the insect's body from the middle section (thorax) to the end of the abdomen. The smallest internal tubules, called tracheoles, supply $O_2$ to the cells and receive $CO_2$, presumably by diffusion.

The insect's pulsating abdominal muscles are enough to power the tracheal system. Lungs and respiratory pigments such as hemoglobin aren't necessary—or so it was thought.

In the late 1990s, entomologist Michael Locke investigated the tracheae in the hindmost region of the caterpillar of a Brazilian skipper butterfly, *Calpodes ethlius*. These tubules are wider than the others and had long been termed "aerating tracheae," compared with other "conducting tracheae."

Instead of branching and infiltrating many cells, these aerating tracheae form tufts of narrow, thin-walled tracheoles. Locke discovered that circulating hemolymph, the insect's version of blood, picks up oxygen from the waving tufts, which move the hemolymph toward the heart. Certain cells found in hemolymph, called hemocytes, were found to nearly double in number when oxygen is scarce.

Locke's work suggests that insect respiration may be more complex than previously thought. In addition to the highly branching tracheal system, transfer of gases may involve hemolymph circulation and specialized cells that assist in carrying oxygen.

As we learn more, scientific theories change—one of the strengths of scientific inquiry.

# The Respiratory System

## Chapter Preview

1. Respiratory systems are designed for gaseous exchange, bringing in oxygen and removing waste $CO_2$ formed via cellular energy reactions.

2. In external respiration, oxygen and $CO_2$ are exchanged by diffusion across a moist membrane. Body size, metabolic requirements, and habitat have affected the evolution of respiratory systems.

3. Terrestrial arthropods bring the environment into contact with almost every cell through a highly branched system of tracheae.

4. Complex aquatic animals exchange gases across gill membranes, which are body surface extensions. In bony fishes, water flows over the gills in the direction opposite blood flow.

5. Vertebrate lungs create an extensive, moist internal surface for two-way gas exchange.

6. The volume of air moved in and out of the lungs during a respiratory cycle is the tidal volume. The amount of air that can be exhaled after a maximal inspiration is the vital capacity. Residual air remains in the lungs after expiration.

7. Almost all oxygen transported to cells is bound to hemoglobin in red blood cells. Increase in blood acidity (usually due to a rise in $CO_2$ level) or elevated temperature due to metabolism increases the amount of oxygen reaching cells.

8. Some $CO_2$ in the blood is bound to hemoglobin or dissolved in plasma. Most $CO_2$, however, is transported as bicarbonate ion, generated from the reaction of $CO_2$ with water.

9. Breathing rate is monitored and adjusted by sensing the chemical characteristics of blood.

## 37.1 Respiration and Gas Exchange

The first breath of life is the toughest. A baby often emerges from the birth canal with a bluish color that may be frightening to new parents. The first breath of a newborn baby requires 15 to 20 times the strength needed for subsequent breaths. The newborn must force air into millions of partially inflated air sacs for the first time. As oxygen rapidly diffuses into the bloodstream and reaches the tissues, the infant turns a robust pink and lets out a yowl (**figure 37.1**).

The lungs of an adult who has spent years breathing polluted air and smoking cigarettes are quite different from the healthy lungs of a newborn. The passageways to these damaged lungs are dotted with bare patches where dense cilia, which move particles up and out of the respiratory tract, once waved. Deep within the lungs, the pattern of bare patches continues, and sections of air sacs are deflated or altogether gone. While the newborn's lung linings are pure pink, the smoker's are a sooty black; the tissues that capture life-giving oxygen have been ravaged. The adult notices the effects with each hacking cough.

### Respiration Occurs at Three Levels

Breathing, technically termed **external respiration,** is actually one of three forms of respiration. External respiration is the process by which animals exchange oxygen and carbon dioxide ($CO_2$) between moist **respiratory surfaces** and the blood. **Internal respiration** exchanges these gases between the blood and cells. Aerobic **cellular respiration** refers to oxygen-utilizing biochemical pathways that store chemical energy as ATP. The gas exchange that breathing makes possible enables cells to harness the energy held in the chemical bonds of nutrient molecules. Without gas exchange, cells die. • **ATP, p. 96**

Recall that energy is slowly liberated from food molecules when electrons are stripped off and channeled through a series of electron carriers, each at a lower energy level than the previous one (see figure 8.12). The energy released at each step is used to form ATP. In organisms that have aerobic respiration, oxygen is the final acceptor of the low-energy electrons at the end of the chain of acceptors, where it combines with hydrogen to form water. Without oxygen, electrons cannot pass along this series of acceptors; this halts breakdown of organic molecules much earlier and leaves a great deal of energy in the nutrient molecule.

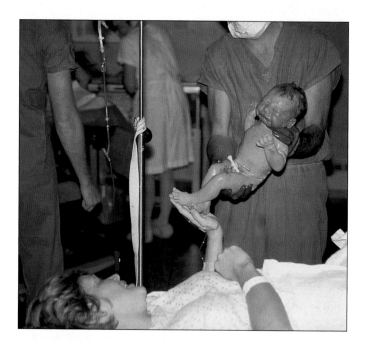

**FIGURE 37.1    The First Breath of Life Is the Toughest.**    A baby must gather great strength to fill the lungs for the first time.

Parrots and macaws exhibit many adaptive behaviors, such as tool use and geophagy. Shown is the hyacinth macaw, *Anodorhynchus hyacinthinus*.

# Digestion and Nutrition

## Why Do Animals Eat Soil?

Why do animals as diverse as rabbits, butterflies, reptiles, birds—and even some pregnant women—eat soil? Biologists have devised several hypotheses to explain geophagy—eating soil. Among them are deficiencies in certain minerals that can be ingested with soil.

In humans, compulsive eating of nonnutritive substances is called pica, and it has included not just eating soil but also ice, gravel, laundry starch, and even matches. Some people who exhibit pica have zinc or iron deficiencies. Pregnant women are more likely to have these deficiencies, which may explain why pica is more common among this group.

A study of blue-headed parrots (*Pionus menstruus*) indicates another reason for geophagy that no one had thought of: Minerals in soil deactivate toxins in their other foods.

The parrots live in the tropical rainforest of Peru, and every morning at the same time they eat claylike soil from one particular layer of sediment along the banks of the Manu River. They do not consume soil from other layers.

Researchers found that the seeds and unripened fruits the parrots eat contain high levels of alkaloids, chemicals with a bitter taste that are frequently poisonous. Producing alkaloids in seeds is an evolutionary adaptation that helps protect the plant's offspring.

The investigators hypothesized that the clay particles, which bear negative charges, bind to the positively charged alkaloids in the parrots' stomachs. The clay then passes out of the digestive tract, taking its toxic cargo along with it.

# Chapter Preview

1. Food is ingested, mechanically broken down, digested, and absorbed into the bloodstream; waste is eliminated.

2. Animals have varied mechanisms for obtaining food. The advent of heterotrophy allowed organisms to obtain nutrients from the environment.

3. The focus of digestive systems is digestion—breaking macromolecules to usable nutrients. Animals eat diverse foods but break them down into the same types of nutrients. Plant matter is the hardest to digest.

4. Protista and sponges have intracellular digestion, in which cells engulf food and digest it in food vacuoles. More complex animals have extracellular digestion in a cavity outside cells. The cavity plus accessory organs form the digestive system.

5. Digestive cavities can have one opening or two. In a system with two openings, food enters through the mouth and is digested and absorbed; undigested material leaves through the second opening, the anus.

6. A series of cavities progressively digest food and absorb nutrients. The stomach provides acid and enzymes to digest proteins, which are inactive in the absence of acid to prevent tissue damage.

7. The small intestine is the main site of nutrient absorption, and the large intestine reabsorbs water, salts, and minerals.

8. The pancreas, liver, and gallbladder aid digestion. The pancreas supplies pancreatic amylase, trypsin, chymotrypsin, lipases, and nucleases. The liver supplies bile, which emulsifies fat, and the gallbladder stores bile. The nervous and endocrine systems regulate digestive secretion.

9. Kilocalories measure energy—1 calorie is the energy needed to heat 1 gram of water by 1 degree.

## 38.1 The Art and Evolution of Eating

Animals eat to obtain energy and raw materials to build their bodies. Although the diets of humans, squids, dragonflies, and jellyfishes are certainly different, all animals break their varied foods down into the same types of nutrients. Then, the pathways of cellular metabolism extract the energy and use it to produce ATP. Digestive systems accomplish the task of breaking down foods to their constituent nutrients.

A robin eating a meal illustrates the steps of the digestive process. The robin's diet varies with the season, depending on what is available (**figure 38.1A**). It eats mostly insects, spiders, and worms in the fall, winter, and spring but relies more on fruits and berries in the summer, which makes it a "seasonal frugivore" (**table 38.1**). It is more broadly an omnivore, consuming both plants and animals. The robin's digestive system, like that of many birds, is highly adapted to its eclectic diet. Its digestive biochemicals can break down the protein and fat that make up most of its animal-based meals, yet its stomach is specialized to rapidly digest fruits. In the stomach, large seeds are separated from a fruit meal and regurgitated; the pulp is then digested quickly. The intestine absorbs the nutrients, and the fruit's peel and remaining smaller seeds move on to become a bird dropping.

For the robin to use the proteins and fats in its worm meal or the sugars in fruits, its body must dismantle these nutrient molecules into smaller molecules—proteins to amino acids, complex carbohydrates (starches) to simple carbohydrates (sugars), and fats to fatty acids and monoglycerides. Only such small molecules can enter the cells lining the digestive tract and then leave them to enter the circulation. The bloodstream delivers nutrient molecules to cells, where enzymes break them down, releasing the energy in their chemical bonds. Nonnutritive parts of food are eliminated from the body as feces.

Animals eat not only to replace energy they expend in the activities of life, but also to provide raw material for the structural components required for growth, repair, and maintenance of the body. The overall process of obtaining and using food has several major steps. Ingestion is the taking in of food. The next stage, mechanical breakdown, cuts food into smaller pieces, usually by chewing (masticating). Mechanical breakdown exposes sufficient surface area for various digestive enzymes to contact. **Digestion,** the next step, is the chemical breakdown of the food particles. Hydrolytic enzymes add water molecules between the building blocks of large nutrient molecules, splitting them apart. Released nutrients are then absorbed into the bloodstream, typically through the lining of a hollow organ specialized for this function, such as an intestine. Finally, undigested material is eliminated (egested) from the animal's body as feces.

Animals obtain food in diverse ways. Recall from chapter 26 that a hagfish spikes its worm food using tentacles on its mouth,

| TABLE 38.1 | |
|---|---|
| **Types of Feeders** | |
| **Term** | **Type of Food** |
| Carnivore | Animals |
| Frugivore | Fruits and berries |
| Herbivore | Plants |
| Insectivore | Insects |
| Omnivore | Plants and animals |
| Detritivore | Nonliving organic matter |

**FIGURE 38.1    Feeding Diversity.    (A)** A robin's digestive system, including its beak, is adapted to eating fruits and berries, as well as insects, spiders, and worms. It is an omnivore. **(B)** A baleen whale is a filter feeder. It can swim through a school of krill and filter them from 60 tons of water in a few seconds, thanks to a curtain of baleen plates. **(C)** The giant panda has a very specific diet, consuming enormous quantities of bamboo. It is an herbivore.

and some lampreys use their oral discs to attach to fishes. Many animals use teeth or other chewing mouthparts to grab food, and some, such as snakes and lizards, swallow mammal meals whole. Some animals drink their food—a mosquito alights on human skin to take a "blood meal." Other organisms digest their food alive, as a spider consumes an insect or a sea star eats a clam.

Several types of aquatic animals are **filter feeders.** They use ciliated tissue surfaces to create waves that usher food particles into their mouths. Filter feeding is particularly adaptive in animals that do not move about much, such as corals, because it brings the food to them, although the diet is limited to organisms that live in the vicinity. Bivalve mollusks such as clams and mussels sweep food toward their mouths using cilia on the feathery surfaces of their gills. Tunicates and lancelets capture particles of food suspended in water in the sticky mucus that lines the pharynx. Cilia move the food to the digestive organs.

Motile filter feeders range from tiny crustaceans to the baleen whale. This giant mammal opens its huge mouth and gulps in a few thousand fishes and crustaceans as it moves forward. Smaller prey, such as the plentiful, tiny shrimplike animals called krill, adhere to some 300 horny baleen plates that dangle from the roof of the whale's mouth like a shredded shower curtain (figure 38.1B). The whale licks the food off the baleen and swallows it.

A terrestrial counterpart to the filter feeder is the **deposit feeder.** This type of feeder strains food from soil or other sediments. The earthworm is a familiar deposit feeder.

Digestion breaks food down into nutrients. A nutrient is a food substance used in an organism to promote growth, maintenance, and repair of its tissues. Recall that obtaining nutrients from other organisms, as opposed to synthesizing them, is termed heterotrophy. • **energy use, p. 4**

The earliest heterotrophs probably engulfed organic matter and used it to supply energy and building blocks for proteins and nucleic acids. The evolution of photosynthesis and chemosynthesis, which enabled organisms to manufacture useful biochemicals from solar energy and inorganic chemicals, had two important implications for early life-forms. First, some organisms could now synthesize the complex organic molecules they required, freeing them from depending on an uncertain and possibly dwindling food supply. Second, the nutrient molecules obtained from photosynthesis and chemosynthesis, called primary production, eventually built up a reservoir of food that would make heterotrophy a sustainable lifestyle over the long term for many types of organisms.

Heterotrophy may have evolved because many organisms never acquired the ability to synthesize their own biochemicals. In terms of natural selection, heterotrophy is advantageous if it takes less energy to obtain a nutrient from the environment than it does to synthesize it. This may be especially true for a heterotroph that has a nearly constant supply of nutrients, such as an animal that has an abundance of a particular amino acid in its diet. Over time, the species could lose the ability to synthesize this amino acid with no harmful consequence and divert the energy required for its synthesis to other functions.

Over hundreds of millions of years, considering the great diversity of life, it is not surprising that many animals have become very specialized feeders, relying on one or a few kinds of food to supply nutrients. Consider the giant panda (see figure 38.1C). It has a short digestive tract characteristic of its meat-eating ancestors, with less capacity to absorb nutrients. The panda lacks cellulase (a cellulose-degrading enzyme) to break down bulky plant matter and also lacks an organ to store food. Thus, the panda must eat almost constantly. Even though it consumes 6% of its body weight in bamboo each day (compared with 2% for most plant-eating mammals), it just barely meets its basic nutritional requirements. Animals with more varied diets, such as the robin and human, can live in more varied habitats.

---

### Reviewing Concepts

- Digestion mechanically and chemically breaks down nutrient molecules, releasing the energy in the chemical bonds.
- Different species obtain nutrients in different ways, depending on their environment.

---

# 38.2   Digestive Diversity

Animals' varied ways of obtaining and digesting food reflect specializations for their nutrient requirements, their habitats, the kinds of food available, and each species' evolutionary background. Even the animals in a particular taxonomic group may have diverse feeding habits and digestive adaptations.

## Digestion Takes Place in Specialized Compartments

Because animals and nutrient molecules are composed of the same types of chemicals, digestive enzymes could just as easily attack an animal's body as its food. To prevent this, digestion occurs within specialized compartments.

**Intracellular digestion**   Even in protista such as *Paramecium*, digestion is separated from other functions. These organisms take in dissolved nutrients by endocytosis and envelop the food in a food vacuole. Some protista have specialized mouthlike openings that food passes through en route to food vacuoles. Once loaded, the food vacuole fuses with another sac containing digestive enzymes that break down large nutrient molecules. Digested nutrients exit the food vacuole and are used in other parts of the cell. The cell extrudes waste. Digestion within a cell's food vacuoles is **intracellular digestion** (**figure 38.2**).

Sponges are the only types of animals that rely solely on intracellular digestion. The cells that form the body wall, the choanocytes, capture food particles with their flagella and pass them to amoebocytes, which digest them in food vacuoles, then store the nutrients or pass them to other cells (see figure 25.8). It isn't surprising that these simplest of animals have protistlike food vacuoles for intracellular digestion, because sponges likely descended from colonial choanoflagellates (see figures 22.1 and 25.1).

**Extracellular digestion**   Intracellular digestion can handle only very small food particles. In **extracellular digestion,** hydrolytic enzymes in a cavity or system of tubes connected with the outside world (the vertebrate stomach, for example) dismantle larger food particles outside the cells that will use them. When nutrient molecules are small enough, they enter cells lining the cavity, where chemical breakdown by intracellular digestion may continue. Extracellular digestion eases waste removal and is more efficient and specialized than intracellular digestion. The cavity in which extracellular digestion occurs, along with accessory organs, together constitute a **digestive system.** (Recall from chapter 24 that fungi have truly extracellular digestion—it occurs outside their bodies. Fungi secrete enzymes that digest organic molecules in the environment, then take up the liberated nutrients.)

**Two-way systems and flow-through systems**   Digestive systems may have one or two openings (**figure 38.3**). Some animals, such as *Hydra* and flatworms, have digestive systems with a single opening. Indigestible food exits the digestive system through the same opening it entered. Consequently, food must be digested and the residue expelled before the next meal can begin. This two-way traffic makes any specialized compartments for storing, digesting, or absorbing nutrients impossible. In these organisms, the digestive cavity doubles as a circulatory system, which distributes the products of digestion to the body cells for use. For this reason, the digestive cavity is also called a **gastrovascular cavity.**

The digestive systems of many familiar animals, such as segmented worms, mollusks, insects, sea stars, and vertebrates, have

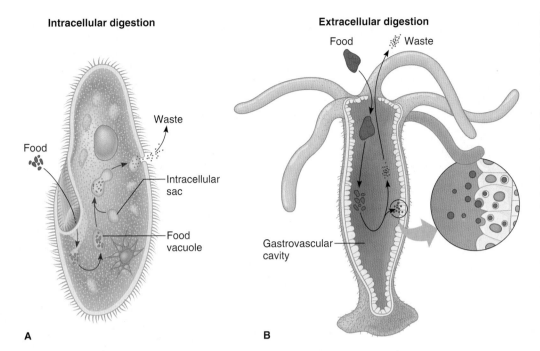

**Intracellular digestion**

Waste

Food

Intracellular sac

Food vacuole

**A**

**Extracellular digestion**

Food     Waste

Gastrovascular cavity

**B**

**FIGURE 38.2** Intracellular and Extracellular Digestion. (**A**) Intracellular digestion is an ancient form of obtaining nutrients that sponges and single-celled organisms such as *Paramecium* use as their sole means of digestion. Food is digested when the food vacuole that contains it fuses with an intracellular sac containing digestive enzymes. (**B**) In most multicellular animals, food is broken down in a cavity by extracellular digestion. The cells lining the cavity absorb the nutrients and continue the breakdown process by intracellular digestion. *Hydra* has a simple system for extracellular digestion.

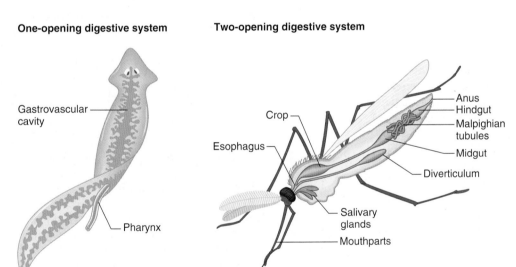

**One-opening digestive system**

Gastrovascular cavity

Pharynx

**A**

**Two-opening digestive system**

Crop

Esophagus

Anus
Hindgut
Malpighian tubules
Midgut
Diverticulum

Salivary glands

Mouthparts

**B**

**FIGURE 38.3** One-Opening and Two-Opening Digestive Systems. In a digestive system with a single opening, such as in the gastrovascular cavity of a flatworm (**A**), food and wastes mix. Cells lining the digestive cavity absorb partially digested nutrients. Further digestion occurs in these cells. (**B**) A mosquito's digestive system is open at the mouth and anus. Digestive systems with two openings are often specialized into compartments.

two openings, a separate entrance and exit in a tubelike structure that allows one-way traffic. The roundworms are the simplest animals to have digestive systems with two openings. Hydrolytic enzymes are released into the tube, and the products of digestion are absorbed and delivered to cells via a circulatory system. Indigestible food remains in the tube and leaves the body as part of the feces, from an **anus**. A major advantage of a two-opening digestive system is that regions of the tube can become specialized for different functions: breaking food into smaller particles, storage, chemical digestion, and absorption.

# From Robins to Rabbits, Digestive Systems Exhibit Adaptations to Diet

The vertebrates exhibit great digestive diversity. Birds metabolize nutrients very rapidly, and they must eat continually and digest food quickly to fuel their lightweight bodies. Their mouths have adapted to their diets (see figures 15.5 and 26.17). The stork's bill easily scoops fish; the strong short beak of the finch cracks and removes hulls from seeds; the vulture's hooked beak rips carrion into manageable chunks. A bird's esophagus has an enlargement

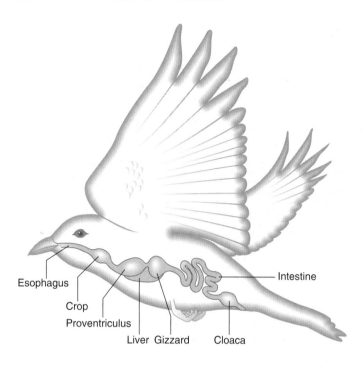

**FIGURE 38.4    Digestion in a Bird.**    Food is stored temporarily in the crop, then passed through the proventriculus and the gizzard. Absorption occurs at the intestine, and waste exits through the cloaca.

called a **crop,** which temporarily stores food. After the crop, food moves to the proventriculus, a glandular stomach that secretes gastric juice that chemically digests food. Next, food moves to the **gizzard,** a muscular organ lined with ridges. It mechanically digests food with the aid of sand and small pebbles the bird swallows. Nutrients are absorbed in the intestines. The digestive system ends at the **cloaca,** an opening common to the digestive, reproductive, and excretory systems (**figure 38.4**).

The intestines of some types of birds show a remarkable adaptation—they enlarge when the animal eats huge amounts of food, such as just before a long migration. In the wren, rufous-sided towhee, rock ptarmigan, and spruce grouse, the intestine lengthens up to 22%. This adaptation enables birds to store more food than usual. Because it takes longer for food to travel through the lengthened intestine, more nutrients are absorbed.

The digestive systems of mammals are adapted to their diets (**figure 38.5**). Many mammals have digestive system specializations that enable them to break down the cellulose in the cell walls of plant foods. The cow has a four-sectioned stomach. Grass enters the first section, the **rumen,** where bacteria break it down into balls of cud. The cow regurgitates the cud back up to its mouth, where it chews to mechanically break the food down further. The cow's teeth mash the cud forward and backward, up and down, and sideways. Then, the cow swallows again, but this time the food bypasses the rumen, continuing digestion in the other three sections of the stomach. Sheep, deer, goats, antelope, buffalo, and giraffes also have quadruple stomachs. These animals with large rumens are called **ruminants.**

Elephants have difficulty digesting the leaves and twigs they eat. Because they have only tusks and molars, a large amount of wood ends up in their massive stomachs. It remains there for over 2 days, churning about amidst digestive juices and cellulose-degrading bacteria. Still, elephant dung contains many undigested twigs. An elephant eats 300 to 400 pounds (135 to 150 kilograms) of food each day.

Rabbits eat mostly leaves, which they partially digest and egest as moist pellets. Then they eat the moist pellets, which continue digestion in the stomach and intestines. This time, the rabbits egest dry pellets, which they do not eat. The gastrointestinal tracts of rabbits and several other types of mammals have a side compartment, called a **cecum,** where plant matter is fermented and absorbed. Insectivores, such as the shrew, and carnivores, such as the fox, eat mostly protein. The digestive tracts of these animals are adapted to this diet by having short intestines and a small or absent cecum (see figure 38.5).

### Reviewing Concepts

- Some protista engulf food and break it down intracellularly.
- Animals digest nutrients in extracellular compartments.
- Simpler digestive systems have one opening, whereas more complex systems have two openings.
- Digestive systems are adapted to the frequency and composition of a species' meals.

## 38.3    Overview of the Human Digestive System

Digestion begins with eating. Hormones control appetite by affecting a part of the hypothalamus called the arcuate nucleus. These hormones can be classed by how quickly they exert their effects. Insulin (from the pancreas) and leptin (from adipocytes) regulate fat stores in the long term, whereas ghrelin (from the stomach) and cholecystokinin (from the small intestine) work in the short term. Ghrelin stimulates appetite and cholecystokinin signals satiety. Endocrine cells in the small and large intestines secrete neuropeptide Y proteins, which act over an intermediate time period, peaking in the bloodstream between meals. Neuropeptide Y proteins integrate incoming information from the other appetite-control molecules and may delay eating for up to 12 hours. Altogether, these hormones maintain homeostasis of lipid levels in the blood. Drug developers are focusing on these weight-control proteins in the never-ending search for obesity treatments.  ● **hypothalamus, p. 664**

The human digestive system consists of a continuous tube called the **gastrointestinal tract,** or alimentary canal. It includes the mouth; pharynx; esophagus; stomach; small intes-

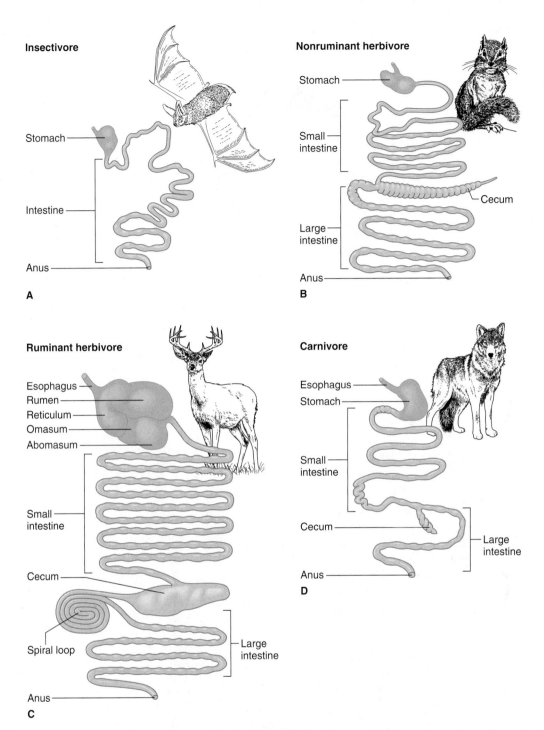

**FIGURE 38.5** **Digestive Systems Are Adapted to an Animal's Diet.** **(A)** Small mammals that are insectivores have short simple digestive systems, because they consume little plant material that would require fermentation. **(B)** The intestines of nonruminant herbivores such as rabbits and rodents harbor anaerobic microorganisms that break down cellulose in plant foods. **(C)** Ruminant herbivores have a characteristic four-part stomach and an extensive gastrointestinal tract. **(D)** The mostly protein diet of carnivores is easier to digest than plants, so the digestive system is much shorter, with a reduced cecum.

tine; and large intestine, or colon, as well as accessory structures (**figure 38.6**). The accessory structures aid food breakdown either mechanically, such as being torn apart, or chemically, with digestive enzymes. Other accessory structures, the liver and gallbladder, produce and store a substance (bile) that assists fat digestion. The mesentery, an epithelial sheet reinforced by connective tissue, supports the digestive organs and glands.

The digestive lining begins at the mouth and ends at the anus. The moist innermost layer secretes mucus that lubricates the tube so that food slips through easily, while also protecting cells from rough materials in food and from digestive enzymes. In some regions of the digestive system, cells in the innermost lining also secrete digestive enzymes. Beneath the lining, a layer of connective tissue contains blood vessels and nerves. The blood

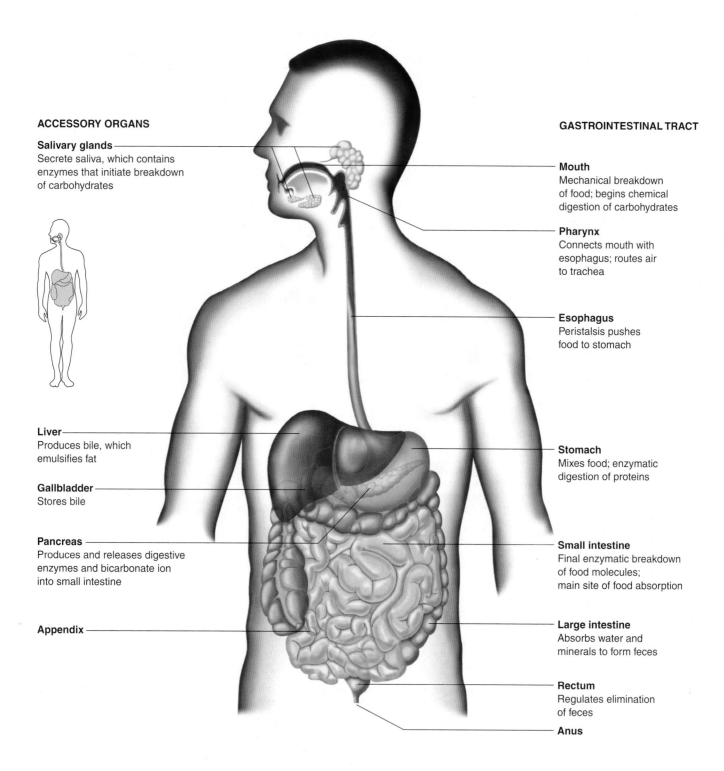

**ACCESSORY ORGANS**

**Salivary glands**
Secrete saliva, which contains
enzymes that initiate breakdown
of carbohydrates

**Liver**
Produces bile, which
emulsifies fat

**Gallbladder**
Stores bile

**Pancreas**
Produces and releases digestive
enzymes and bicarbonate ion
into small intestine

**Appendix**

**GASTROINTESTINAL TRACT**

**Mouth**
Mechanical breakdown
of food; begins chemical
digestion of carbohydrates

**Pharynx**
Connects mouth with
esophagus; routes air
to trachea

**Esophagus**
Peristalsis pushes
food to stomach

**Stomach**
Mixes food; enzymatic
digestion of proteins

**Small intestine**
Final enzymatic breakdown
of food molecules;
main site of food absorption

**Large intestine**
Absorbs water and
minerals to form feces

**Rectum**
Regulates elimination
of feces

**Anus**

**FIGURE 38.6     The Human Digestive System.**     Food is broken down as it moves through the various chambers of the gastrointestinal tract.
Accessory organs deliver digestive enzymes and other chemicals that assist the chemical digestion of food.

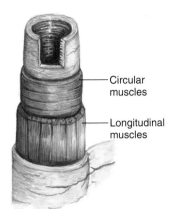

**FIGURE 38.7**   **Muscles Move Food.**   The two smooth muscle layers of the digestive tract coordinate their contractions to move food in one direction. Note that the muscle layers are perpendicular to each other. (In the stomach, a third layer lies obliquely.)

nourishes the cells of the digestive system and in some regions picks up and transports digested nutrients.

The layers against the innermost layers are muscular. Except for the stomach, which has three layers, two layers produce the movements of the digestive tract (**figure 38.7**). The muscles of the inner layer circle the tube, constricting it when they contract, and the muscles in the outer layer run lengthwise, shortening the tube when they contract. These contractions, which occur about three times per minute, churn food, mixing it with enzymes into a liquid. Waves of contraction called **peristalsis** propel food through the digestive tract. When food distends the walls of the tube, the circular muscles immediately behind it contract, squeezing the mass forward.

## Reviewing Concepts

- Hormones control appetite and digestion.
- Complex digestive tracts are long tubes that use waves of contraction called peristalsis to move food from one region to the next.

# 38.4   The Mouth and the Esophagus: Beginning of the Process

The thought, smell, or taste of food triggers three pairs of salivary glands near the mouth to secrete saliva. Chemical digestion begins when **salivary amylase,** an enzyme in saliva, starts to break down starch into maltose. Salivary amylase causes a piece of bread in the mouth to taste sweet as the starch is broken down to sugar.

Mechanical breakdown begins as the teeth chew the food, a process called mastication. Water and mucus in saliva aid the

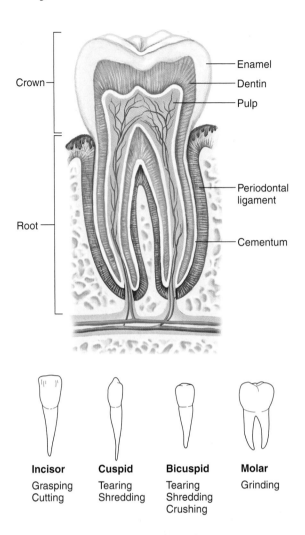

**FIGURE 38.8**   **Anatomy of a Tooth.**   For many animals, processing food begins with the teeth, which grasp, cut, tear, shred, crush, and grind food into pieces small enough to be swallowed.

teeth as they tear food into small pieces, thereby increasing the surface area available for chemical digestion. Researchers recently used a computer simulation of chewing to test the hypothesis that each food requires an optimum number of chews. Ideally, food should be torn enough to form a smooth lump, called a **bolus,** yet not become so softened that it sticks in the throat. Eating raw carrots, for example, requires 20 to 25 chews. Experiments with people confirmed the researchers' predictions.

Tooth structure and shape are adapted to particular functions. The thick **enamel** that covers a tooth is the hardest substance in the human body. Beneath the enamel is the bonelike **dentin,** and beneath that, the soft inner **pulp,** which contains connective tissue, blood vessels, and nerves (**figure 38.8**). Two layers on the outside of the tooth, the periodontal ligament and the cementum, anchor the tooth to the gum and jawbone. The visible part of a tooth is the crown, and the part below the surface is the root. Calcium compounds harden teeth, as they do bones.

The tongue rolls chewed food into a bolus and pushes it to the back of the mouth for swallowing. The bolus passes first

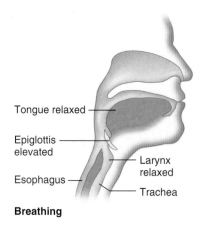

Tongue relaxed

Epiglottis
elevated

Esophagus

Larynx
relaxed

Trachea

**Breathing**

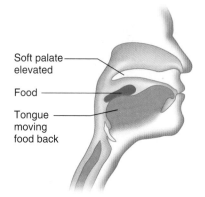

Soft palate
elevated

Food

Tongue
moving
food back

**Beginning of swallow**

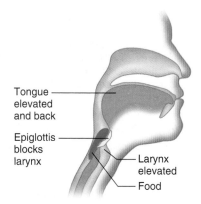

Tongue
elevated
and back

Epiglottis
blocks
larynx

Larynx
elevated

Food

**Swallowing reflex**

**FIGURE 38.9    Separating Air from Food.** During breathing, the tongue and larynx are relaxed and the epiglottis is elevated. When a swallow begins, the soft palate is elevated, and the tongue moves food to the rear of the mouth. During the swallowing reflex, the tongue and the larynx are elevated, and food is diverted down the esophagus by the epiglottis, which now blocks the larynx.

through the **pharynx** and then through the **esophagus,** a muscular tube leading to the stomach. During swallowing, the **epiglottis** covers the passageway to the lungs, routing food to the digestive tract (**figure 38.9**).

Food does not merely slide down the esophagus due to gravity—contracting esophageal muscles push it along in a wave of peristalsis. This is why it is possible to swallow while standing on your head. Muscle contractions continue, propelling the food down the esophagus toward the stomach, where the next stage in digestion takes place.

> ## Reviewing Concepts
> - Salivary amylase begins chemical digestion as the teeth mechanically break down food.
> - Structures in the throat move food to the stomach and prevent it from routing to the lungs.

## 38.5    The Stomach: Further Processing

From the esophagus, food progresses to the **stomach,** a J-shaped bag about 12 inches (30 centimeters) long and 6 inches (15 centimeters) wide. The stomach has three important functions: storage, some digestion, and pushing food into the small intestine.

The stomach is the size of a large sausage when empty, but it can expand to hold as much as 3 or 4 quarts (approximately 3 or 4 liters) of food. Folds in the stomach's mucosa, called **rugae,** can unfold like the pleats of an accordion to accommodate a large meal. Muscular rings called **sphincters** control entry to and exit from the stomach, pinching shut to contain the stomach's contents (**figure 38.10**).

Mechanical breakdown and chemical digestion occur in the stomach. Waves of peristalsis push food against the stomach bottom, churning it backward, breaking it into pieces, and mixing it with gastric juice to produce a semifluid mass called **chyme.** Fats usually remain in the stomach from 3 to 6 hours, proteins for up to 3 hours, and carbohydrates from 1 to 2 hours.

Gastric juice is responsible for chemical digestion in the stomach. About 40 million cells lining the stomach's interior secrete 2 to 3 quarts (approximately 2 to 3 liters) of gastric juice per day. Gastric juice consists of water, mucus, salts, hydrochloric acid, and enzymes. The hydrochloric acid creates a highly acidic environment, which denatures proteins in food and also activates the enzyme **pepsin** from its precursor, pepsinogen. Pepsin breaks down proteins to yield polypeptides, a first step in protein digestion. Cells called chief cells secrete pepsinogen, and parietal cells secrete hydrochloric acid from indentations in the stomach lining called gastric pits.

Nerves and hormones regulate gastric juice secretion. The thought or taste of food initiates nerve impulses that stimulate the stomach to secrete between 6 and 7 ounces (about 200 milliliters) of gastric juice. This prepares the stomach to receive food. When food arrives in the stomach, endocrine cells in the stomach's lining release a hormone, **gastrin,** that stimulates secretion of another 20 ounces (about 600 milliliters) of gastric juice.

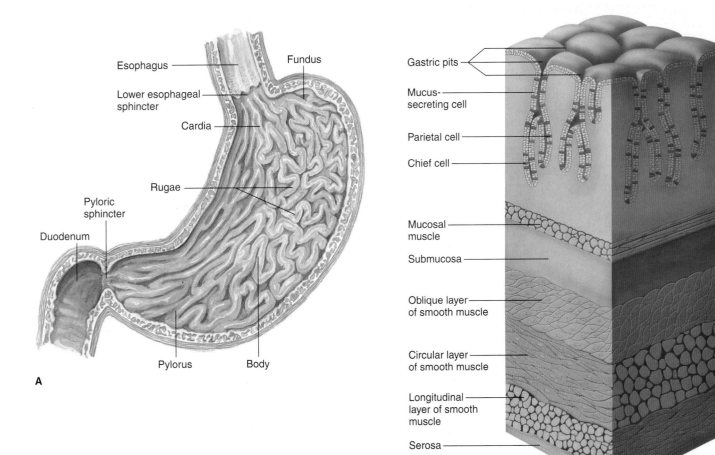

**FIGURE 38.10    The Stomach.**    The stomach is a J-shaped bag that stores, mixes, and digests food until it is fluid enough to move on to the small intestine. **[A]** The stomach has four regions (fundus, cardia, body, and pylorus), two sphincters, and folds called rugae that increase its capacity. **[B]** The lining of the stomach contains gastric pits, where chief cells secrete pepsinogen and parietal cells secrete hydrochloric acid. Other lining cells secrete the abundant mucus that coats the stomach lining, preventing it from digesting itself. The three layers of muscle that make up the stomach wall, running at different angles, enable the organ to move food around, mechanically breaking it into smaller pieces.

The stomach usually doesn't digest the protein of its own cells along with the protein in food. The organ has built-in triple protection. First, the stomach secretes little gastric juice until food is present for it to work on. Secondly, some stomach cells secrete mucus, which coats and protects the stomach lining from the corrosive gastric juice. Finally, the stomach produces pepsin in an inactive form, pepsinogen, and cannot digest protein until hydrochloric acid is present. Unusual research led to a better understanding of gastric ulcers, a common condition in which the stomach's protection fails. They are now believed to be caused by a specific bacteria and may be cured by antibiotics.

The stomach absorbs very few nutrients; most food has not yet been digested sufficiently. However, the stomach can absorb some water and salts (electrolytes), a few drugs (such as aspirin), and, like the rest of the digestive tract, alcohol. This is why aspirin can irritate the stomach lining and why we feel alcohol's intoxicating effects quickly.

After the appropriate amount of time, depending upon the composition of a meal, stretching of the stomach wall or release of gastrin triggers neural messages that relax the **pyloric sphincter** at the stomach's exit. The stomach then squirts small amounts of chyme through the sphincter into the small intestine. Nerve impulses from the first part of the small intestine then control the stomach's actions. When the upper part of the small intestine is full, receptor cells on the outside of the intestine are stimulated, and the pyloric sphincter tightens. Once pressure on the small intestine lessens, the sphincter opens, the stomach contracts, and more chyme enters.

## Reviewing Concepts

- The esophagus conducts chewed food to the stomach, propelled by peristalsis, where it liquefies under the action of churning and pepsin.
- Muscles called sphincters close the exits at various points.

## Preventing Stomach Ulcers

Medical science is often hesitant to abandon long-held ideas. So it was with the discovery that painful stomach ulcers result from bacterial infection. Ulcers were thought to stem from excess stomach acid secretion, which, in turn, reflected stress and eating spicy foods. The fact that drugs that block acid production relieve ulcer pain, albeit temporarily, supported the acid hypothesis. Over the many years that ulcers typically persisted, treatment with a bland diet, stress reduction, acid-reducing drugs, or surgery was costly. The idea that a stomach ulcer could instead be cured with a 2-week course of antibiotic drugs and antacids seemed preposterous.

*Helicobacter pylori*—the ulcer bacterium—was unveiled in the laboratory of J. Robin Warren at Royal Perth Hospital in western Australia, isolated from stomach tissue of people suffering from gastritis, an inflammation of the stomach. But were the bacteria attracted to inflamed tissue, or did they cause the inflammation? Warren's assistant, medical resident Barry Marshall, helped choose between these hypothesis. He had a healthy stomach, so on a hot July day in 1984, he drank "swamp water"—a brew of a billion or so bacteria. He suffered only a few days of gastritis, which cleared up on its own, but a second volunteer was sick for months, until antibiotics cured him.

In many developing nations, however, the antibiotic-based treatment for stomach ulcers is too costly. But learning more about the interactions between the bacterium and the human stomach wall is providing clues to prevention, rather than treatment.

About half of the world's human population is infected with *H. pylori,* but only 2% develop stomach ulcers, and 1% develop stomach cancer associated with the bacteria. Therefore, researchers reason, most of us must have some way of protecting our digestive systems from invasion by this pathogen. Part of that protection stems from glycoproteins found in the inner mucosal layer of the stomach lining, where the bacteria do not penetrate—instead, they amass in the outermost layer. Experiments mimicked the conditions in the two stomach layers in the laboratory, and introduced bacteria. In the cultures that included the glycoproteins, the normally squiggly micro-organisms stopped moving, lost their characteristic shapes, and died, as if they had been hit with cell-wall-dismantling antibiotic drugs. That is exactly what happens. The glycoproteins prevent the bacteria from producing a form of cholesterol that they require to build their cell walls. Without it, they burst.

Using these two clues—the functions of the human glycoproteins and the bacterial cholesterol—researchers hope to genetically modify soybeans and cows to produce milk that contains the human defensive glycoproteins. Drinking the milk could then prevent *H. pylori* disease.

---

## 38.6 The Small Intestine and Nutrient Absorption

The **small intestine,** a 23-foot (approximately 7-meter) tubular organ, completes digestion and absorbs the resulting nutrients. The first 10 inches (25 centimeters) of the organ forms the **duodenum;** the next two-fifths, the **jejunum;** and the remainder (almost three-fifths), the **ileum.** Localized muscle contractions carry out mechanical digestion in the small intestine in a process called **segmentation.** During this process, the chyme sloshes back and forth between segments of the small intestine that form when bands of circular muscle temporarily contract. Meanwhile, peristalsis moves the food lengthwise along the intestine.

Digestive secretions from the small intestine, the liver, and the pancreas contribute mucus, water, bile, and enzymes. Chemical digestion in the small intestine acts on all three major types of nutrient molecules. The small intestine continues the protein breakdown the stomach began. Trypsin and chymotrypsin from the pancreas break down polypeptides into peptides, and peptidases secreted by small intestinal cells break peptides into tripeptides, dipeptides, and single amino acids small enough to enter the lining cells. **Figure 38.11** summarizes the sites of digestion of major nutrients and the enzymes that dismantle them.

Fats present an interesting challenge to the digestive system. Lipases, the enzymes that chemically digest fats, are water-soluble, but fats are not. Therefore, a lipase can only act at the surface of a fat droplet, where it contacts water. **Bile** from the liver emulsifies fats, breaking them into many droplets, which exposes greater surface area to lipase. Most fats in our diet are triglycerides, which are digested into fatty acids and monoglycerides.

Carbohydrate digestion also continues in the small intestine. The pancreas sends pancreatic amylase to the small intestine, where it acts on starches that salivary amylase missed in the mouth. In addition, the small intestine produces carbohydrases, which chemically break down certain disaccharides into monosaccharides. Sucrase, for example, is a carbohydrase that breaks down the disaccharide sucrose into the monosaccharides glucose and fructose. • **enzymes, p. 99**

Deficiency of a particular carbohydrase can cause digestive distress when a person eats certain foods. Many adults have difficulty digesting milk products because of lactose intolerance, which results from the absence of the enzyme lactase in the small

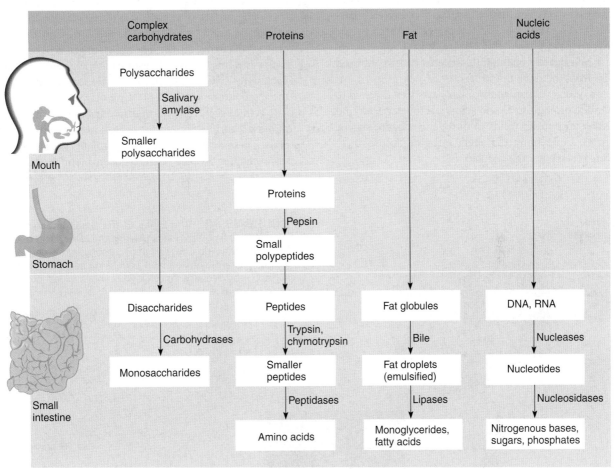

**FIGURE 38.11    Overview of Chemical Digestion.**    Digestion gradually breaks down large molecules.

intestine. This enzyme breaks down the disaccharide lactose (milk sugar) into the monosaccharides glucose and galactose. Bacteria ferment undigested lactose in the large intestine, producing abdominal pain, gas, diarrhea, bloating, and cramps. A person with lactose intolerance can avoid these symptoms by consuming fermented dairy products such as yogurt, buttermilk, and cheese instead of fresh dairy products, because lactose has already been broken down. Taking lactase tablets can also prevent symptoms of lactose intolerance.

Nucleic acids are not abundant enough to be considered a major nutrient, but they too are digested. Nucleases break RNA and DNA down into nucleotides, and then nucleosidases take apart nucleotides into their constituent bases, sugars, and phosphates.

## Hormones Coordinate Small Intestine Processes

When the stomach squirts acidic chyme into the small intestine, intestinal cells release the hormone secretin. This triggers the release of bicarbonate from the pancreas, which quickly neutralizes the acidity of the chyme. Secretin also stimulates the liver to secrete bile. Intestinal cells secrete cholecystokinin (CCK) in response to fatty chyme. In addition to affecting the hypothalamus, CCK also stimulates the liver to continue bile secretion and triggers the gallbladder to contract, sending stored bile to the small intestine. CCK also releases pancreatic enzymes, including lipase, into the small intestine. Secretin and CCK inhibit endocrine cells of the stomach wall from releasing gastrin, temporarily slowing digestion in the stomach.

Hormonal regulation of digestion protects intestinal cells because digestive biochemicals are produced only when food is present. As an additional safeguard, other glands in the small intestine secrete mucus, which protects the intestinal wall from digestive juices and neutralizes stomach acid. Mucus offers limited protection, however, and many intestinal lining cells succumb to the caustic contents. The lining persists because the epithelial cells divide so often that they replace the lining every 36 hours. Nearly one-quarter of the bulk of feces consists of dead epithelial cells from the small intestine.

At the end of digestion, carbohydrates have been digested to monosaccharides, proteins to amino acids, fats to fatty acids and monoglycerides, and nucleic acids to nitrogenous bases, sugars, and phosphates. Cells lining the small intestine absorb these products, which then enter the circulation.

## Nutrient Absorption Occurs at Villi and Microvilli

Nutrient absorption must take place at a surface. The small intestine maximizes surface area. The organ is very long and highly folded, enabling it to fit into the abdomen. In addition, the innermost layer is corrugated with circular ridges almost half an inch high. The surface of every hill and valley of the lining looks velvety due to additional folds—about 6 million tiny projections called **villi** (**figure 38.12**). The tall epithelial cells on the surface of each villus bristle with projections of their own called **microvilli**. Each villus cell and its 500 microvilli increase the surface area of the small intestine at least 600 times.

Within each villus, a capillary network absorbs amino acids, monosaccharides, and water, as well as some vitamins and minerals. These digested nutrients then enter the circulation. Recall from chapter 36 that nutrient-laden blood from the intestines

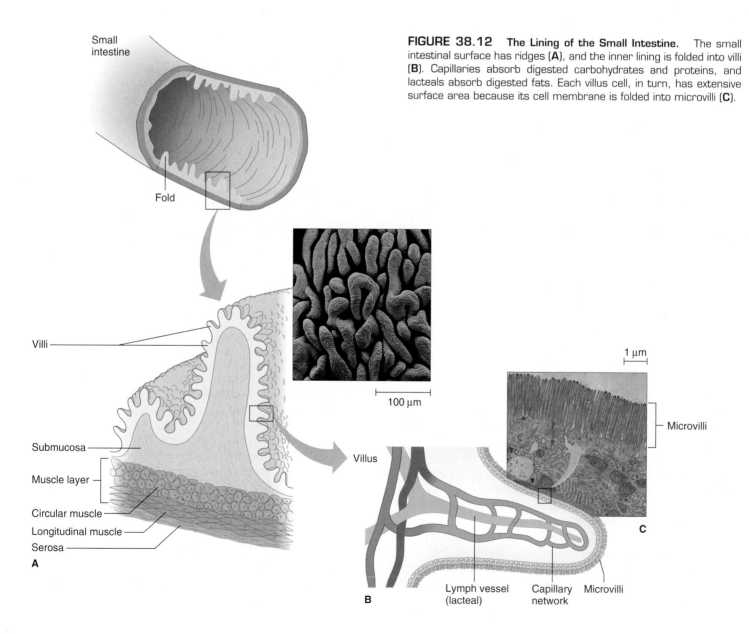

**FIGURE 38.12    The Lining of the Small Intestine.** The small intestinal surface has ridges (**A**), and the inner lining is folded into villi (**B**). Capillaries absorb digested carbohydrates and proteins, and lacteals absorb digested fats. Each villus cell, in turn, has extensive surface area because its cell membrane is folded into microvilli (**C**).

Small intestine

Fold

Villi

Submucosa

Muscle layer

Circular muscle

Longitudinal muscle

Serosa

A

100 μm

Villus

Lymph vessel (lacteal)

Capillary network

Microvilli

B

1 μm

Microvilli

C

first passes through the liver before the heart pumps it throughout the body.

Fatty acids and monoglycerides follow another route. These molecules are reassembled into triglycerides within the small intestinal lining cells, and they are coated with proteins to make them soluble before they enter the lymphatic system. The triglycerides then enter lymph vessels called **lacteals** that are surrounded by capillaries in the intestinal villi. Triglycerides enter the blood where lymphatic vessels join the circulatory system in the upper chest.

Located among the epithelial cells are scattered cells called M cells. These very rare cells bind toxins or pathogens that are eaten at one end of the cell, then "present" the toxins to immune system cells at their other end within the small intestinal lining. The alerted immune system then mounts a defense against the toxin or pathogen. In this way, the body removes poisons from food.

## Reviewing Concepts

- Nutrients are absorbed through small intestinal villi and microvilli, which greatly increase surface area.
- Different nutrients are absorbed at different points within the intestines.

# 38.7 The Large Intestine: End of the Process

The material remaining in the small intestine after nutrients are absorbed enters the **large intestine,** or **colon.** The large intestine is shorter than the small intestine, but its 2.5-inch (6.5-centimeter) diameter is greater. The 5-foot (1.5-meter) tube surrounds the convoluted mass of the small intestine, roughly in the shape of a question mark. At the start of the large intestine is a pouch called the cecum (**figure 38.13**).

Dangling from the cecum is the **appendix,** a thin wormlike tube. The appendix has a still not completely understood role in immunity. In our primate ancestors, the appendix may have helped digest fibrous plant matter. If bacteria or undigested food become trapped in the appendix, the area can become irritated and inflamed, and infection can set in, producing severe pain. If this happens, the appendix must be promptly removed before it bursts and spills its contents into the abdominal cavity and spreads the infection.

The large intestine absorbs most of the water, electrolytes, and minerals from chyme, leaving solid or semisolid feces. Billions of bacteria of about 500 different species are normal inhabitants of the healthy human large intestine. These "intestinal flora" decompose any nutrients that escaped absorption in the small intestine; they produce vitamins $B_1$, $B_2$, $B_6$, $B_{12}$, K, folic acid, and biotin; and they break down bile and foreign chemicals such as certain drugs.

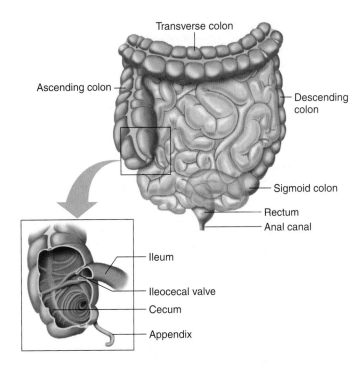

**FIGURE 38.13   The Large Intestine.**   The large intestine absorbs water, salts, and minerals and temporarily stores feces. Inset: The appendix attaches to the pouchlike cecum, which receives food from the ileum of the small intestine.

Intestinal flora produce foul-smelling compounds that cause the characteristic odors of intestinal gas and feces. These bacteria also help prevent infection by other microorganisms. Antibiotic drugs often kill the normal bacteria and allow other microorganisms, especially yeasts, to grow. This alteration in the intestinal flora causes the diarrhea that is sometimes a side effect of taking antibiotic drugs.

The remnants of digestion—cellulose, bacteria, bile, and intestinal cells—collect as feces in the **rectum,** a 6- to 8-inch- (15- to 20-centimeter-) long region. Within the rectum is the 1-inch- (2.5-centimeter-) long anal canal. The opening to the anal canal, the anus, is usually closed by two sets of sphincters, an inner smooth-muscle sphincter under involuntary control and an outer skeletal-muscle sphincter, which is voluntary. When the rectum is full, receptor cells trigger a reflex that eliminates the feces.

**Figure 38.14** summarizes the lengths of various parts of the human digestive system and the amount of time food spends in each region.

## Reviewing Concepts

- Digestion is completed in the large intestine, and waste leaves through the rectum and anus.
- Microorganisms in the large intestine aid in the digestive process and provide vitamins.

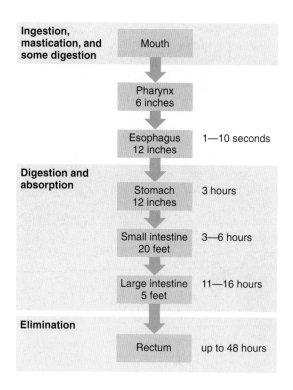

| Ingestion, mastication, and some digestion | Mouth | |
| --- | --- | --- |
| | Pharynx 6 inches | |
| | Esophagus 12 inches | 1—10 seconds |
| **Digestion and absorption** | Stomach 12 inches | 3 hours |
| | Small intestine 20 feet | 3—6 hours |
| | Large intestine 5 feet | 11—16 hours |
| **Elimination** | Rectum | up to 48 hours |

**FIGURE 38.14** Food's Journey Through the Human Digestive Tract. This flowchart indicates the length of each part of the human gastrointestinal tract and the approximate duration that food spends in each area.

## 38.8 The Roles of the Pancreas, Liver, and Gallbladder

Several other glands and organs assist digestion. The **pancreas** is a multifunctional structure associated with the digestive tract (**figure 38.15**). It sends about a liter of fluid to the duodenum each day, including trypsin and chymotrypsin to digest polypeptides, pancreatic amylase to digest carbohydrates, pancreatic

lipase to further break down emulsified fats, and nucleases to degrade DNA and RNA. Pancreatic "juice" also contains sodium bicarbonate to neutralize the acidity that hydrochloric acid produces in the stomach. The pancreas also functions as an endocrine gland, regulating blood sugar level (see figure 34.13).

- **diabetes mellitus, p. 657**

At a weight of about 3 pounds (1.4 kilograms), the **liver** is the largest solid organ in the body. It has more than 200 functions, including detoxifying harmful substances in the blood, storing glycogen and fat-soluble vitamins, and synthesizing blood proteins. The liver's contribution to digestion is the production of greenish yellow bile. Bile is stored in the **gallbladder** until chyme in the small intestine triggers its release. The cholesterol in bile can crystallize, forming gallstones that partially or completely block the duct to the small intestine. Gallstones are very painful and may require removal of the gallbladder.

Bile consists of pigments derived from the breakdown products of hemoglobin and bile salts from the breakdown of cholesterol. This colorful substance is responsible for the brown color of feces and the pale yellow of blood plasma and urine. It also creates the abnormal yellow complexion of jaundice, a condition that deposits excess bile pigments in the skin.

### Reviewing Concepts

- The pancreas secretes digestive enzymes, and the liver produces fat-emulsifying bile, which the gallbladder stores and secretes.

## 38.9 Human Nutrition

The products of digestion provide energy, which cells use to function and to synthesize compounds. Figure 8.12 depicts the fates of these products—glucose is the immediate energy source, but amino acids and fatty acids enter the energy pathways, too.

- **organic molecules, p. 30**

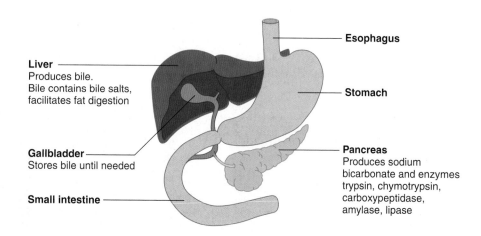

**Liver**
Produces bile.
Bile contains bile salts, facilitates fat digestion

**Gallbladder**
Stores bile until needed

**Small intestine**

**Esophagus**

**Stomach**

**Pancreas**
Produces sodium bicarbonate and enzymes trypsin, chymotrypsin, carboxypeptidase, amylase, lipase

**FIGURE 38.15** Accessory Structures. The liver produces bile, and the gallbladder stores it. Bile emulsifies fats. The pancreas secretes sodium bicarbonate and several types of digestive enzymes.

## A Variety of Nutrients Are Necessary for Good Nutrition

Nutrients are diet-derived chemical compounds whose breakdown products supply metabolic energy. Carbohydrates, proteins, and lipids are called **macronutrients** because humans require them in large amounts, or energy nutrients because they supply energy. Humans require vitamins and minerals in very small amounts, and hence, these nutrients are called **micronutrients.** Water is also a vital part of the diet.

**Tables 38.2** and **38.3** provide details on specific vitamins and minerals. Vitamins function as coenzymes, and minerals are essential to many biochemical pathways. Vitamins are classified by whether they are soluble in water or fat.

**Essential nutrients** must come from food because the body cannot synthesize them. Essential nutrients vary among species. Vitamin C is an essential nutrient in humans, guinea pigs, Indian fruit bats, and certain monkeys but not in most other animals. If a guinea pig eats only rabbit chow, which lacks vitamin C, the guinea pig will develop vitamin C deficiency. Nonessential nutrients come from foods but are also made in the body. For example, the adult human body synthesizes 11 of the 20 types of amino acids, which are therefore nonessential.

The amount of energy potentially released from a nutrient is measured in **kilocalories (kcal),** often called simply calories. A food's caloric content is determined by burning it in a bomb calorimeter, a chamber immersed in water. When burning food is placed in the chamber, the energy released raises the water temperature, and this energy is measured in kilocalories. One kilocalorie is the energy needed to raise 1 kilogram of water 1°C under controlled conditions. Bomb calorimetry studies have shown that 1 gram of carbohydrate yields 4 kilocalories, 1 gram of protein yields 4 kilocalories, and 1 gram of fat yields 9 kilocalories. (A teaspoon of dried food weighs approximately 5 grams.) Although the body cannot extract all of the potential energy in food, these values do help to explain why a fatty diet may cause weight gain; fats supply more energy than most people can use.

### TABLE 38.2
**Vitamins and Health**

| Vitamin | Function(s) | Food Sources | Deficiency Symptoms |
|---|---|---|---|
| **Water-Soluble Vitamins** | | | |
| Thiamine (vitamin $B_1$) | Growth, fertility, digestion, nerve cell function, milk production | Pork, beans, peas, nuts, whole grains | Beriberi (neurological disorder), loss of appetite, swelling, poor growth, heart problems |
| Riboflavin (vitamin $B_2$) | Energy use | Liver, leafy vegetables, dairy products, whole grains | Hypersensitivity of eyes to light, lip sores, oily dermatitis |
| Pantothenic acid* | Growth, cell maintenance, energy use | Liver, eggs, peas, potatoes, peanuts | Headache, fatigue, poor muscle control, nausea, cramps |
| Niacin | Growth, energy use | Liver, meat, peas, beans, whole grains, fish | Dark rough skin, diarrhea, mouth sores, mental confusion (pellagra) |
| Pyridoxine (vitamin $B_6$)* | Protein use | Red meat, liver, corn, potatoes, whole grains, green vegetables | Mouth sores, dizziness, nausea, weight loss, neurological disorders |
| Folic acid (folate) | Manufacture of red blood cells, metabolism | Liver, navy beans, dark green vegetables | Anemia, neural tube defects |
| Biotin* | Metabolism | Meat, milk, eggs | Skin disorders, muscle pain, insomnia, depression |
| Cyanocobalamin (vitamin $B_{12}$) | Manufacture of red blood cells, growth, cell maintenance | Meat, organ meats, fish, shellfish, milk | Pernicious anemia |
| Ascorbic acid (vitamin C) | Growth, tissue repair, bone and cartilage formation | Citrus fruits, tomatoes, peppers, strawberries, cabbage | Weakness, gum bleeding, weight loss (scurvy) |
| **Fat-Soluble Vitamins** | | | |
| Retinol (vitamin A) | Night vision, new cell growth | Liver, dairy products, egg yolk, vegetables, fruit | Night blindness, rough dry skin |
| Cholecalciferol (vitamin D) | Bone formation | Fish liver oil, milk, egg yolk | Skeletal deformation (rickets) |
| Tocopherol (vitamin E)* | Prevents oxidation of certain compounds | Vegetable oil, nuts, beans | Anemia in premature infants |
| Vitamin K* | Blood clotting | Liver, egg yolk, green vegetables | Bleeding, liver problems |

*These vitamin deficiencies are rare in humans, but they have been observed in experimental animals.

## Minerals in the Human Diet

| Mineral | Food Sources | Functions in the Human Body |
|---|---|---|
| **Bulk Minerals** | | |
| Calcium | Milk products, green leafy vegetables | Bone and tooth structure, blood clotting, hormone release, nerve transmission, muscle contraction |
| Chloride | Table salt, meat, fish, eggs, poultry, milk | Digestion in stomach |
| Magnesium | Green leafy vegetables, beans, fruits, peanuts, whole grains | Muscle contraction, nucleic acid synthesis, enzyme activity |
| Phosphorus | Meat, fish, eggs, poultry, whole grains | Bone and tooth structure |
| Potassium | Fruits, potatoes, meat, fish, eggs, poultry, milk | Body fluid balance, nerve transmission, muscle contraction, nucleic acid synthesis |
| Sodium | Table salt, meat, fish, eggs, poultry, milk | Body fluid balance, nerve transmission, muscle contraction |
| Sulfur | Meat, fish, eggs, poultry | Hair, skin, and nail structure, blood clotting, energy transfer, detoxification |
| **Trace Minerals** | | |
| Chromium | Yeast, pork kidneys | Regulates glucose use |
| Cobalt | Meat, eggs, dairy products | Part of vitamin $B_{12}$ |
| Copper | Organ meats, nuts, shellfish, beans | Part of many enzymes, storage and release of iron in red blood cells |
| Fluorine | Water (in some areas) | Maintains dental health |
| Iodine | Seafood, iodized salt | Part of thyroid hormone |
| Iron | Meat, liver, fish, shellfish, egg yolk, peas, beans, dried fruit, whole grains | Transport and use of oxygen (as part of hemoglobin and myoglobin), part of certain enzymes |
| Manganese | Bran, coffee, tea, nuts, peas, beans | Part of certain enzymes, bone and tendon structure |
| Selenium | Meat, milk, grains, onions | Part of certain enzymes, heart function |
| Zinc | Meat, fish, egg yolk, milk, nuts, some whole grains | Part of certain enzymes, nucleic acid synthesis |

Good nutrition is a matter of balance. When we take in more kilocalories than we expend, weight increases; those who consume fewer kilocalories than they expend lose weight—and, if taken to an extreme, may even starve. Balancing vitamins and minerals is important to health, too. Eating a variety of foods helps meet nutritional requirements. The dietary suggestions pictured in the food pyramid the U.S. government recommends (**figure 38.16**) can help a person make healthful food choices. The revised proportional food pyramid, available on the www.mypyramid.gov website, shows groups in a different way. The website gives information for vegetarians as well as allowing personalized plans.

The four-food-group plan devised in the 1950s suggested that humans need nearly as many daily servings of meat as of grains, dairy products, and fruits and vegetables. In the 1940s, an eight-food-group plan included groups for butter and eggs—foods now associated with the development of heart disease. In the 1920s, an entire food group was devoted to sweets!

## Nutrient Deficiencies Can Be Hidden or Obvious

Obtaining sufficient kilocalories each day from a variety of foods is sometimes difficult. A student may eat nothing but pizza during exam week; a busy working person may skip meals or eat on the run; many individuals may not be able to afford enough milk or meat. All these situations may lead to primary nutrient deficiencies, which are caused by diet. Secondary nutrient deficiencies, in contrast, result from inborn metabolic conditions that cause the body to malabsorb, overexcrete, or destroy a particular nutrient.

Symptoms of micronutrient deficiencies develop more slowly than the obvious weight loss that accompanies macronutrient deficiency. Vitamin or mineral deficiencies cause subtle changes before health is noticeably affected. For example, the blood of a vegetarian who does not eat many iron-containing foods may have an abnormally low number of red blood cells or small red blood cells with little hemoglobin. The person feels fine, temporarily, as the body uses its iron stores. After several weeks, however, the signs of anemia appear—weakness, fatigue, frequent headaches, and a pale complexion.

### Reviewing Concepts

- Carbohydrates, proteins, and lipids are required in large amounts and provide energy, which is measured in kilocalories, and raw materials for building bodies.
- Vitamins and minerals are required in small amounts and have specific functions.

ever. The endotherm must eat much more food and must use 80% of the energy from that food just to maintain its temperature.

Most endotherms maintain a relatively constant body temperature by balancing heat generated in metabolism with heat lost to the environment, and the body temperature of most ectotherms varies as external conditions change. However, there are exceptions to these trends. Reptiles are ectotherms, but many can maintain their body temperatures within fairly narrow limits by moving around to gain or lose heat. On the other hand, a hummingbird is an endotherm whose body temperature is subject to occasional fluctuations.

Adaptations that enable animals to regulate their temperatures fall into three general categories: anatomical and physiological; metabolic and hormonal; and behavioral. Clusters of neurons in the hypothalamus control many of an animal's thermoregulatory responses by reacting to feedback from sensory cells elsewhere in the body. Thermoregulation depends on the interactions of the musculoskeletal, nervous, endocrine, respiratory, integumentary, and circulatory systems. Animals have adapted to a variety of environmental extremes.

## Adaptations to Cold Seek to Maintain Core Temperature

For many fishes, water temperature determines body temperature. Yet these animals must maintain body temperatures warm enough to permit survival. To do this, many fishes have biochemicals in their blood that function as antifreeze, lowering the point at which the blood freezes so that it stays liquid even when the water is very cold.

The great white shark and bluefin tuna (see figure 26.9) have another type of adaptation to cold water, called a **countercurrent exchange system.** This is an anatomical arrangement that transfers heat between blood vessels. In these animals, venous blood coming from warm interior muscles warms colder blood in arteries from near the body surface. In another type of circulatory adaptation, marine reptiles such as sea turtles and sea snakes shunt warmer blood to the centers of their bodies, preferentially protecting interior vital organs. Amphibians, which tend to lose what little metabolic heat they generate through their moist skins, conserve heat by restricting their habitats to warm, moist places.

Birds and mammals can function for long periods at low temperatures, thanks to their adaptations to the cold. Feathers retain warmth by creating pockets that trap air near the skin. In many mammals countercurrent exchange systems transfer the heat in some blood vessels to cooler blood in nearby vessels, retaining heat in the body rather than allowing it to escape to the environment through the extremities. Countercurrent exchange systems allow penguins to spend hours on ice or snow or in frigid water, and they enable arctic mammals such as wolves to hunt in extreme cold (**figure 39.2**).

Blubber helps whales and seals maintain a constant body temperature. In their flippers and tails, which lack blubber, countercurrent-heat-exchange blood vessels keep the heat in.

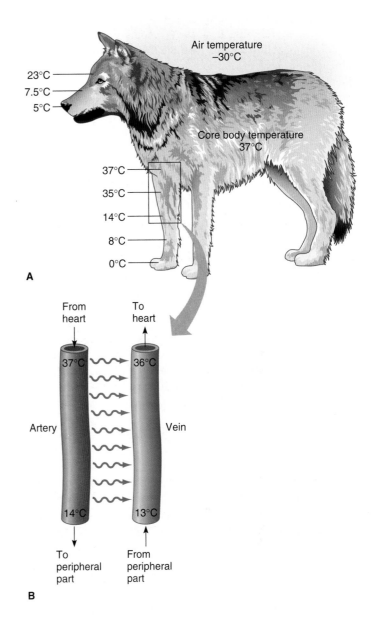

**FIGURE 39.2 Countercurrent Heat Exchange.** **(A)** A mammal's extremities chill more easily than interior body parts. **(B)** In a countercurrent exchange system, arteries carrying warm blood lie in close proximity to veins carrying cooler blood, thereby conserving heat, rather than losing it to the environment.

Mammals also have several behavioral strategies for retaining heat, including migrating, hibernating, and simply huddling together.

Mammals generate much of their metabolic heat in a process called **thermogenesis.** In one type of thermogenesis, fatty acids in brown fat are oxidized to release heat energy. Brown fat is a type of adipose tissue with a rich blood supply and it is specialized for fast heat production. It is abundant in animals that hibernate, and it also helps many newborn mammals, including humans, to stay warm.

Shivering is another form of thermogenesis. Shivering contracts muscles, releasing heat as ATP splits during actin and myosin filament interaction (see chapter 35). Hormones also influence body heat. Falling body temperatures stimulate the thyroid gland to produce more thyroxine, which increases cell membrane permeability to sodium ions ($Na^+$). When cell membranes pump $Na^+$ out of the cells, ATP is split, releasing heat energy. Hormone-directed internal heating along with brown fat utilization is called nonshivering thermogenesis.

## Adaptations to Heat Include Evaporative Cooling and Other Strategies

Invertebrates and vertebrates share many adaptations for surviving in extreme heat and drought. One common strategy for lowering body temperature is **evaporative cooling,** in which air moves over a moist surface, evaporating the moisture and cooling that part of the body. A coyote pants and an owl flutters loose skin under its throat to move air over moist mouth surfaces. When a honeybee's temperature reaches 113°F (45°C), a heat sensor in its brain stimulates it to regurgitate nectar stored in an organ called a honey-crop. The nectar spills onto a tonguelike structure, and the bee licks itself, cooling as its nectar bath evaporates. Australian sawfly larvae smear their bodies with rectal fluid, which evaporates and enables them to survive another hour in 115°F (46°C) heat.

Humans sweat. Some invertebrates sweat in a sense, too. Consider a type of cicada (an insect) that lives in the Sonoran Desert in Arizona, where the temperature can reach 105°F (41°C) in the shade. Water from the cicada's hemolymph moves through tubules to the body's surface, where it exits through ducts and evaporates (**figure 39.3**). By "sweating," the cicada can lower its body temperature as much as 18°F to 27°F (10°C to 15°C) below that of surrounding air.

A cicada can lose 20 to 35% of its body weight in fluid when the outside temperature hits 115°F (46°C). Most insects perish at a 20% loss, and humans cannot survive a 10% loss of body fluid. The Sonoran cicada can survive such extreme dehydration because fluid loss is balanced by intake—the insect imbibes tremendous volumes of fluid in its succulent plant food.

In many mammals, evaporative cooling is linked to circulatory system specializations that route cooled blood past warmer blood. An anatomical organization called a **carotid rete** in many mammals consists of warm-blood-bearing arteries that run through a pool of cool venous blood at the base of the brain (**figure 39.4**).

The ancient Greeks, who discovered the carotid rete, named it the *rete mirabile,* which means "wonderful net." Cats, dogs, seals, sea lions, antelope, sheep, goats, and cows all have them. These animals also have long snouts and tend to pant. The snout interiors have a rich blood supply and many glands that secrete fluids to keep the area moist. Each time the animal pants, water from the secretions evaporates, cooling blood in nearby vessels. Veins take the cooled blood to the pool beneath the brain where the carotid arteries bring warm blood. Cooling just the brain—the area that needs it most—requires less energy than cooling the entire body.

Humans have a different circulatory adaptation for keeping a cool head, which enables us to survive the heat of high fevers. Tiny "emissary" veins in the face and scalp reroute blood cooled near the body's surface through the braincase, which cools brain tissue. This adaptation, called a "cranial radiator," causes the facial flushing of a person who is vigorously exercising.

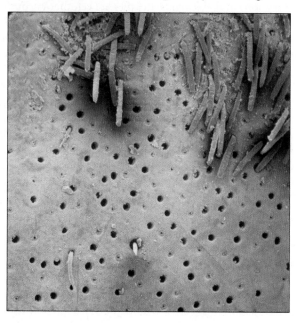

**FIGURE 39.3**  **Cicadas Living in the Desert "Sweat."**  These sweat pores in the cicada's cuticle are magnified 330 times.

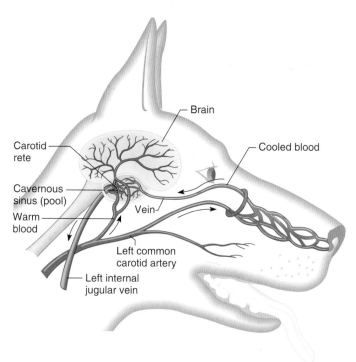

**FIGURE 39.4**  **The Carotid Rete.**  The brain of a hoofed mammal or carnivore stays cool thanks to the carotid rete, a network of arteries that cools the blood they deliver to the brain by passing first through a pool of cool venous blood beneath the brain. The venous blood comes from veins in the animal's muzzle.

The role of genes and the protein gradients they produce is well studied in the fruit fly *Drosophila melanogaster*. For example, a gene in a female fly produces a protein in the egg called bicoid ("two tails"), which instructs the embryo to develop a head end and a tail end. When the gene is mutated, the embryo develops two rear ends! Normally, a greater concentration of bicoid protein in the anterior end of the embryo causes the tissues characteristic of the head to differentiate. At the posterior of the animal, higher concentrations of a protein called nanos signal abdominal tissues to differentiate. Therefore, high levels of bicoid and low levels of nanos indicate "head" formation, and the reverse specifies "tail." Proteins such as bicoid and nanos, present in gradients that influence development, are called morphogens (morphology means "form").

Gradients of several different morphogens distinguish parts of the developing embryo, like using increasing intensities of several colors in a painting (**figure 41.B**). The morphogen gradients are signals that stimulate cells to produce yet other proteins, which ultimately regulate the formation of a specific structure. A cell destined to be part of the adult fly's antenna, for example, has different morphogen concentrations than a cell whose future lies in the eye. Sometimes, when morphogen genes mutate, developmental signals go off in the wrong part of the animal—with striking results. The fly in **figure 41.C** is an example. Called, appropriately, *Antennapedia*, it has legs in place of its antennae. Genes that, when mutated, lead to organisms with structures in the wrong places are called homeotic.

Homeotic genes are widespread. Cows, earthworms, humans, frogs, lampreys, corn, mice, beetles, locusts, bacteria, yeast, chickens, roundworms, and mosquitoes have them. Homeotic mutations cause diseases in humans, including a blood cancer, cleft palate, extra fingers and toes, lack of irises, and thyroid cancer. The fact that homeotic genes are found in diverse modern species indicates that these controls of early development are both ancient and very important.

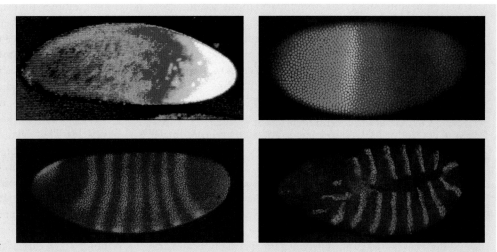

**FIGURE 41.B** **Biochemical Gradients Control Early Development in Fruit Flies.** Very early in the development of a fruit fly, a gradient of bicoid protein distinguishes the embryo's head from its rear. In the first panel of this computer-enhanced image (*upper left*), different colors represent different concentrations of bicoid. Bicoid protein directs the synthesis of two other proteins, one shown in red and one in green (*upper right*). A half hour later, another gene directs production of a protein that divides the embryo into seven stripes (*lower left*). Yet another gene oversees dividing each existing section of the embryo in two (*lower right*). Overall, genes produce protein gradients that set up distinct biochemical environments in different parts of the embryo, which profoundly influence differentiation. (The egg is about 0.5 millimeter long.)

**FIGURE 41.C** **Homeotic Mutations.** A homeotic mutation sends the wrong morphogen signals to the antenna of a fruit fly, directing the tissue to differentiate as leg. A normal fly is shown at left; a homeotic fly is on the right. Originally studied in fruit flies, mutant homeotic genes are now known to cause various disorders in humans. (A fly's head is about 1 millimeter wide.)

of names, but we will call this period the **preembryonic stage.** It includes fertilization; rapid mitotic cell division of the zygote as it moves through the uterine tube toward the uterus; implantation into the uterine wall; and initial folding into layers.

The **embryonic stage** lasts from the end of the second week until the end of the eighth week. Cells of the three layers continue to divide, differentiate, and interact, forming tissues and organs. Structures that support the embryo—the placenta, umbilical cord, and extraembryonic membranes—also develop during this period.

The third stage of prenatal development is the **fetal period,** lasting from the beginning of the ninth week through the full 38 weeks of development. Organs begin to function and coordinate to form organ systems. Growth is very rapid. Prenatal development ends with labor and parturition (birth).

In the reproductive systems of both sexes, gametes are set aside and nurtured, then transported to where they can meet. This union occurs following sexual intercourse. The reproductive systems of the human male and female are similarly organized. Each system has paired **gonads** in which the sperm or oocytes are manufactured; a network of tubes to transport these cells; and hormones and glandular secretions that control the entire process. The details of sperm and oocyte production are covered in chapter 10.

### Reviewing Concepts

- Before biologists could observe early development, they thought that a fertilized ovum contained a small but complete organism.
- The idea that cells specialize gradually grew as researchers compared embryos in different species and throughout development.
- Experiments in the twentieth century revealed that cell specialization is a consequence of differential gene action.
- Research in the twenty-first century is identifying the sets of genes that turn on and off as development proceeds.

## 41.2  The Male Reproductive System: Geared for Billions of Sperm

Sperm cells are manufactured within a 410-foot-long (125-meter-long) network of tubes called **seminiferous tubules,** which are packed into paired, oval organs called **testes** (sometimes called testicles) (**figure 41.6**). The testes are the male gonads. They lie outside the abdomen within a sac called the **scrotum.** Their location outside of the abdominal cavity allows

the testes to maintain a lower temperature than the rest of the body, which is necessary for sperm to develop properly. Leading from each testis is a tightly coiled tube, the **epididymis,** in which sperm cells mature and are stored. Each epididymis continues into another tubule, the **vas deferens** (plural: vasa deferentia). Each vas deferens bends behind the bladder and joins the **urethra,** the tube that also carries urine out through the **penis.**

- **sperm development, p. 173**

Along the sperm's path, three glands contribute secretions. The **prostate gland,** which produces a thin, milky, alkaline fluid that activates the sperm to swim, wraps around the vasa deferentia. Opening into each vas deferens is a duct from the **seminal vesicles.** These glands secrete the sugar fructose, which supplies energy, plus prostaglandins, which may stimulate contractions in the female reproductive tract that help propel sperm. The **bulbourethral glands,** each about the size of a pea, attach to the urethra where it passes through the body wall. These glands secrete alkaline mucus, which coats the urethra before sperm are released. All of these secretions combine to form the seminal fluid, which carries the sperm cells.

During sexual arousal, the penis becomes erect so that it can penetrate the vagina and deposit sperm in the female reproductive tract. At the peak of sexual stimulation, a pleasurable sensation called **orgasm** occurs, accompanied by rhythmic muscular contractions that eject the sperm from the vasa deferentia through the urethra and out the penis. The discharge of sperm from the penis is called **ejaculation.** One human ejaculation typically delivers about 100 million sperm cells (see figure 10.12).

### Reviewing Concepts

- The human male reproductive system provides for a relatively constant production of sperm, with the ability to deliver those gametes at any time.
- Nutrient fluids are also produced to protect and provide energy to the sperm.

## 41.3  The Female Reproductive System: One Egg at a Time

The female sex cells develop within paired gonads, called the **ovaries,** in the abdomen (**figure 41.7**). Within each ovary of a newborn female are about a million oocytes. Nourishing follicle cells surround each oocyte. Like a testis containing sperm cells in various stages of development, an ovary houses oocytes in different stages of development. Approximately once a month, beginning at puberty, one ovary releases the most mature oocyte. Beating cilia sweep the mature oocyte into the fingerlike projections of one of the two **uterine tubes** (also called fallopian tubes). The tube carries the oocyte into a muscular saclike organ, the **uterus,** or womb.

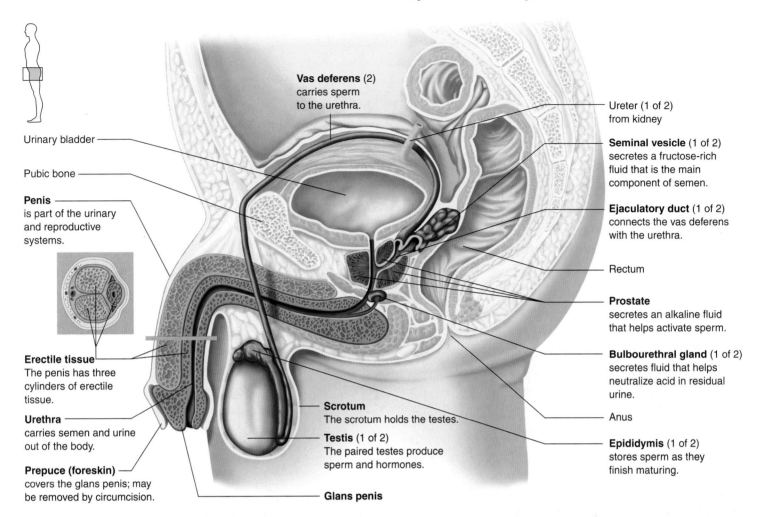

**FIGURE 41.6    The Human Male Reproductive System.**    Sperm cells are manufactured in seminiferous tubules (not shown) within the paired testes. Sperm mature and are stored in the epididymis and exit through the vas deferens. The paired vasa deferentia join in the urethra, through which seminal fluid exits the body. Secretions are added to the sperm cells from the prostate gland, the seminal vesicles, and the bulbourethral glands. During sexual arousal, three cylinders of erectile tissue (inset) fill with blood and cause the penis to become erect.

The lower end of the uterus narrows to form the **cervix,** which opens into the tubelike **vagina,** which opens to the outside of the body. Two pairs of fleshy folds protect the vaginal opening on the outside: the labia majora (major lips) and the thinner, underlying flaps of tissue they protect, called the labia minora (minor lips). At the upper junction of both pairs of labia is a 1-inch-long (2-centimeter-long) structure called the **clitoris,** which is anatomically analogous to the penis. Rubbing the clitoris stimulates females to experience orgasm.

Hormonal fluctuations regulate the timing of oocyte release and the body's preparation for pregnancy. Once released from the ovary, an oocyte can live for about 72 hours, but a sperm can penetrate it only during the first 24 hours of this period—possibly even less. If the oocyte encounters a sperm cell in a uterine tube and the cells combine and their nuclei fuse, the oocyte completes development and becomes a fertilized ovum, or zygote. It then reaches the uterus and implants in the blood-rich lining that has built up. The uterine lining is called the **endometrium.** If the

oocyte is not fertilized, both endometrium and oocyte are expelled as the menstrual flow.

Hormones control the cycle of oocyte maturation in the ovaries and prepare the uterus to nurture a zygote. On the first day of the cycle, when menstruation begins, low blood levels of estrogen and progesterone signal the hypothalamus to secrete gonadotropin-releasing hormone (GnRH). This prompts the anterior pituitary to release abundant follicle-stimulating hormone (FSH) and small amounts of luteinizing hormone (LH), setting into motion hormonal interactions that thicken the uterine lining. A midcycle surge in the level of LH in the bloodstream triggers the ovary to release an oocyte, an event called **ovulation.** The follicle cells left behind in the ovary change into a structure called a corpus luteum. If pregnancy occurs, the implanted fertilized ovum produces human chorionic gonadotropin (hCG), the hormone that is the basis of pregnancy tests. For a while, hCG keeps the cells of the corpus luteum producing progesterone, which maintains the uterine lining.

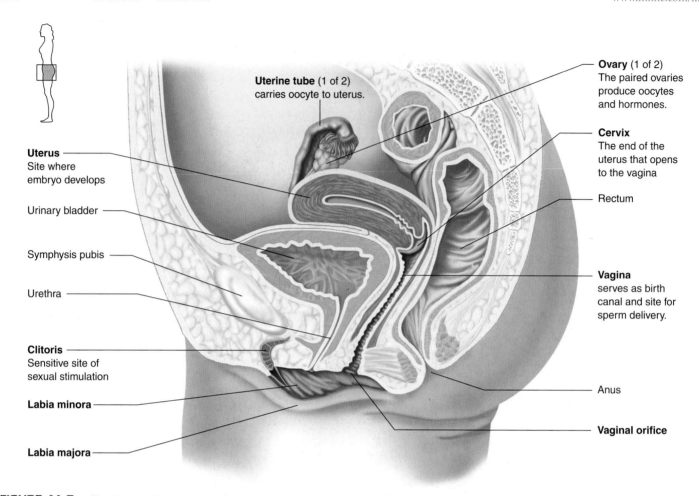

**Uterine tube** (1 of 2)
carries oocyte to uterus.

**Ovary** (1 of 2)
The paired ovaries
produce oocytes
and hormones.

**Cervix**
The end of the
uterus that opens
to the vagina

Rectum

**Uterus**
Site where
embryo develops

Urinary bladder

Symphysis pubis

Urethra

**Vagina**
serves as birth
canal and site for
sperm delivery.

**Clitoris**
Sensitive site of
sexual stimulation

Anus

**Labia minora**

**Vaginal orifice**

**Labia majora**

**FIGURE 41.7    The Human Female Reproductive System.**    Immature egg cells (oocytes; not shown) are packed into the paired ovaries. Once a month, one oocyte is released from an ovary and is drawn into the fingerlike projections of a nearby uterine tube by ciliary movement. If the oocyte is fertilized by a sperm cell in the uterine tube, it continues into the uterus, where it is nurtured for 9 months as it develops into a new individual. If the ovum is not fertilized, it is expelled, along with the built-up uterine lining, through the cervix and then through the vagina. The external genitalia consist of the labia minora and majora and the clitoris.

If pregnancy does not occur, the corpus luteum degenerates, and levels of progesterone and estrogen decline. The reduced levels of these hormones are no longer able to maintain the endometrium, which is then shed through the cervix. Lowered progesterone and estrogen levels also release their inhibition of LH and FSH in the brain, and the cycle begins anew. **Figure 41.8** tracks changes in the ovarian follicle, the uterine lining, and the levels of four hormones during the menstrual cycle.

## Reviewing Concepts

- The female reproductive system provides an ovum to be fertilized and manufactures hormones that synchronize the system to support a fertilized ovum.
- These hormones cycle monthly, preparing one ovum each month and replacing the endometrium for a newly fertilized ovum.

## 41.4    Fertilization and Cleavage

The first step in prenatal development is the initial contact between sperm and secondary oocyte. Recall from chapter 10 that the oocyte arrests in metaphase II. A thin, clear layer of proteins and carbohydrates, the **zona pellucida,** encases the oocyte, and a layer of cells called the **corona radiata** surrounds the zona pellucida. The sperm must penetrate these layers to fertilize the ovum (**figure 41.9**). • **ovum development, p. 176**

### Sperm Seek an Oocyte to Join Genetic Material

Ejaculation deposits about 100 million sperm cells in the woman's vagina. A sperm cell can survive here for up to 6 days, but it can fertilize the oocyte only in the 12 to 24 hours after ovulation. A process called **capacitation** activates the sperm inside the woman, altering sperm cell surfaces in a way that enables

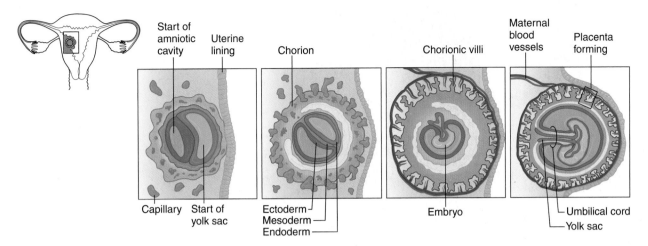

**FIGURE 41.11    Implantation.**    As the inner cell mass settles against the uterine lining, the trophoblast sends out fingerlike extensions that begin to form the chorion, a membrane that develops into the placenta. Meanwhile, the enlarging embryo folds into the three primary germ layers. The yolk sac forms blood cells, immune system stem cells, and part of the embryo's digestive system; it also becomes incorporated into the umbilical cord.

digestive and respiratory systems. However, small packets of stem cells that persist in organs in the adult may retain the ability to specialize as many cell types.

The preembryonic stage ends after the second week of prenatal development. Although the woman has not yet missed her menstrual period, she might notice effects of her shifting hormones, such as swollen and tender breasts and fatigue. By now, her urine contains enough hCG for an at-home pregnancy test to detect. Highly sensitive blood tests can detect hCG as early as 3 days after conception.

### Reviewing Concepts

- Within a few days after conception, the preembryo implants into the endometrium, and the three major tissue layers form: the endoderm, ectoderm, and mesoderm.
- These give rise to the rest of the body's tissues and organ systems.

## 41.6    Organogenesis: The Human Body Takes Shape

During the embryonic stage, organs begin to develop and structures form that will nurture and protect the developing organism. As the days and weeks proceed, different rates of cell division in different parts of the embryo fold tissues into intricate patterns. In a process called embryonic induction, the specialization of one group of cells causes adjacent groups of cells to specialize. Gradually, these changes mold the three primary germ layers into organs and organ systems. **Organogenesis** is the term that describes the transformation of the structurally simple, three-layered embryo into a body with distinct organs. Developing organs are particularly sensitive to damage by environmental factors such as chemicals and viruses.

During the third week of prenatal development, a band called the **primitive streak** appears along the back of the embryonic disk. It gradually elongates to form an axis, which is an anatomical reference point that other structures organize around as they develop. The primitive streak eventually gives rise to connective tissue precursor cells and the **notochord,** a structure that forms the basic framework of the skeleton. The notochord induces overlying ectoderm to differentiate into a hollow **neural tube (figure 41.12),** which develops into the brain and spinal cord (central nervous system). Formation of the neural tube, or **neurulation,** is a key event in early development because it marks the beginning of organ formation. Soon after neurulation ensues, a reddish bulge containing the heart appears. It begins to beat around day 18. Then the central nervous system begins to elaborate. **Figure 41.13** shows some early embryos undergoing these changes.

The fourth week of the embryonic period is a time of rapid growth and differentiation. Blood cells begin to form and to fill primitive blood vessels. Immature lungs and kidneys appear. If the neural tube does not close normally at about day 28, a neural tube defect results, which leaves open an area of the spine from which nervous tissue protrudes, causing paralysis from the site downwards. Small buds appear that will develop into arms and legs. The 4-week embryo has a distinct head and jaw and early evidence of eyes, ears, and nose. The rudiments of a digestive system appear as a long hollow tube that will develop into the intestines. A woman carrying this embryo, which is now only 1/4 inch (0.6 centimeter) long, may suspect that she is pregnant because her menstrual period is about 2 weeks late.

By the fifth week, the embryo's head appears disproportionately large. Limbs extending from the body end in platelike structures. Tiny ridges run down the plates, and by week 6, the ridges deepen as certain cells die, molding fingers and toes. The eyes

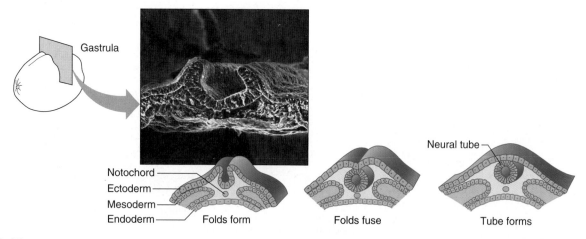

**FIGURE 41.12    Neurulation.**    At a signal from the notochord, ectoderm folds into the neural tube, which will gradually form the brain and spinal cord. The micrograph shows a chick embryo at the neural fold stage. Some nations allow research on human embryos up until this point, because before neural tissue forms, a developing organism cannot sense or be aware.

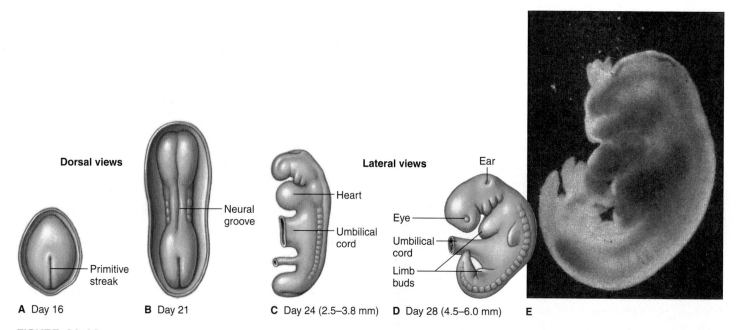

**FIGURE 41.13    Early Embryos.**    It takes about a month for the embryo to look like a "typical" embryo. At first, all that can be distinguished is the primitive streak (**A**), but soon the central nervous system begins to form (**B**). By the 24th day, the heart becomes prominent as a bulge (**C**), and by the 28th day, the embryo is beginning to look human (**D**) and (**E**).

open, but they do not yet have eyelids or irises. Cells in the brain are rapidly differentiating. The embryo is now about 1/2 inch (1.3 centimeters) from head to buttocks.

During the seventh and eighth weeks, a cartilage skeleton appears. The placenta is now almost fully formed and functional, secreting hormones that maintain the blood-rich uterine lining. The embryo is about the size and weight of a paper clip. The eyes now seal shut and will stay that way until the seventh month. The nostrils are closed. A neck appears as the head begins to make up proportionally less of the embryo, and the abdomen flattens somewhat.

## Reviewing Concepts

- The embryonic stage experiences the first organogenesis.
- The nervous system develops first, starting with the neural tube.
- Within a few weeks, distinctly human features begin to appear and develop.

tend to be healthy, the female's choice of the strongest partners reflects the process of sexual selection.

Unlike the Sierra dome spider, the hedge sparrow displays all types of mating systems. Females maintain territories in thick hedges, but one territory may house a monogamous pair, another a female with two males, yet another a single male who flits between several females' lairs. Two or three males may mate in groups with two to four females. Parental investment in these birds varies with type of mating behavior. In monogamous pairs of hedge sparrows, the male cares for all of the chicks—and they are all his. In groups of two males and one female, the proportion of time each male spends feeding young is about the same as the proportion of time that he mates with the female.

The diversity of mating styles among hedge sparrows arose, ethologists hypothesize, because the sexes seek different strategies. For a female hedge sparrow, mating is most adaptive when male participation in chick-rearing is maximized, because more chicks will survive. Monogamy or polyandry, where she is the sole female, each accomplishes this. For a male, however, mating with several females (polygyny) maximizes the number of offspring he produces. Polyandry is the least efficient mating style for a male, because he may care for chicks that aren't his.

Male and female hedge sparrows are pitted against each other in a dominance struggle. If the male wins, polygyny proceeds. If the female wins, polyandry rules—and she takes a second mate. If neither male nor female can get a second mate, monogamy is a default option. Finally, polygynandry may represent a stalemate situation, when one female cannot chase away another, and one male can't get rid of another. (Of course, this is just a human view of the birds' mating styles.) **Figure 42.14** shows another interesting sexual behavior of hedge sparrows— males can ensure that it is their sperm that fertilizes eggs.

Primate species have diverse sexual behaviors too. They form male-female pairs, one-sex bands, mixed-sex troops, harems, homosexual pairs, loners, and social groups with frequently changing members. Devoted monogamous gibbon pairs sing together in the early morning to defend their territory. In bonobo societies, females are dominant and often form intense relationships with each other. Orangutans are polygynous, with a female mating only once every few years.

Primate sexual behavior varies even within species. An aggressive male mountain gorilla mates with and kills an older female who cannot keep up with the troop; yet other mountain gorillas form long-lasting bonds. Some chimpanzee males prefer to mate with an unfamiliar female. Other males take turns mating with a single female. Still other chimpanzees spend their lives in monogamy. Humans have also practiced every mating system observed in nonhuman primates.

### Reviewing Concepts

- Mating behaviors provide one of the mechanisms for sexual selection and for the separation of distinct species.
- They also can have some adaptive advantage when they improve offspring survival.

Female displays cloacal opening

Male pecks opening

Female releases droplet containing sperm from past matings

Male continues mating

**FIGURE 42.14 Fatherhood Assured.** Before a male hedge sparrow mates, he pecks the female on her cloacal opening. In response, she releases a droplet containing sperm stored from previous matings. Only then, with his competition for fatherhood reduced, does the male complete mating.

## 42.7 Altruistic Behavior: Helping One's Kin

Dictionaries define altruism as concern for the welfare of others. In the study of animal behavior, **altruism** has a more specific meaning—increasing another's fitness (ability to pass genes to the next generation) at a cost of one's own fitness. Charles Darwin noted various examples of animals helping their kin to raise young, while not having young of their own. For example, the white-fronted bee-eater lives in extended family groups of five to nine birds. In each group, "cooperative breeders" do not raise their own young, but rather assist parents, who are usually their siblings or half-siblings, in rearing offspring. Ants, bees, wasps, and naked mole rats also display such altruistic behavior, some individuals even losing their ability to reproduce.

Darwin wondered why natural selection did not remove altruistic individuals from populations, since they leave no progeny. An explanation came with a hypothesis called **kin selection,** proposed by geneticist W. D. Hamilton in 1964. In kin selection, an individual helps a relative, either by assisting it to survive or to reproduce. Hamilton viewed selection at the level of genes, instead of individuals. An animal helping a relative reproduce may not become a parent itself, but nonetheless assists its own genes in staying in the population through nieces, nephews, and cousins. Even earlier, evolutionary biologist J. B. S. Haldane said, in 1932, "I would lay down my life for two brothers or eight cousins," referring to the fact that an individual shares half its genes with siblings, and one-eighth with first cousins. Another route to kin selection is cannibalism of nonrelatives, while sparing relatives.

Hamilton developed the concept of **inclusive fitness,** which maintains that two selective mechanisms operate. Direct fitness is the usual route of passing on one's genes through offspring. Indirect fitness, or kin selection, is the transmission of genes by helping relatives to reproduce. A mathematical expression describes the conditions under which kin selection is likely to occur. In words, kin selection occurs if certain genes are more likely to be passed on if the relatives reproduce than if the altruistic individual does. For example, if a naked mole rat does not reproduce, two alleles of a particular gene are lost from the population. If the rat helps a sister to reproduce and she has eight pups, then 16 copies of the gene are passed to the next generation. The gain of 16 alleles exceeds the loss of two alleles.

Inclusive fitness predicts that animals can recognize close relatives. Many observations and experiments have shown that animals recognize kin through pheromones, taste, hormone secretions, appearance, and by somehow knowing that relatives are in a particular place at a particular time. Flour beetles, for example, do not cannibalize other flour beetles if they are in a place where they have just laid eggs; this way they do not kill their relatives.

The theory of kin selection also predicts that the degree of altruism should be directly proportional to the degree of relatedness. The tiger salamander *Ambystoma tigrinum* illustrates the ability of cannibals to recognize kin. Like certain nematodes, rotifers, and wasp larvae, tiger salamander tadpoles can be cannibals or noncannibals. Crowding stimulates certain individuals to develop a wider mouth and prominent teeth and jaws, becoming cannibals (**figure 42.15**). (In some other species, the chance ingestion of meat triggers a similar transformation.) The killer tadpoles consume the noncannibals, but preferentially those that are not relatives. Researchers replicated this behavior in tanks in the laboratory, controlling the numbers of relatives and nonrelatives and tracking who was related to whom. They crowded the tanks and observed that the more closely related a cannibal is to a particular noncannibal tadpole, the less likely it is to eat it.

**FIGURE 42.15    Cannibalism and Kin.**    Noncannibal tiger salamanders develop into killers under crowded conditions, but they tend not to eat their relatives.

## Reviewing Concepts

- Altruism helps some populations through providing support systems for offspring, even though they do not belong to the parents involved in the supportive behavior.
- The theory of kin selection provides for a mechanism whereby genes for such behavior are selected.

# 42.8    Advantages and Disadvantages of Group Living

Many animals live in organized groups. Such a group is termed **eusocial,** meaning that it constitutes a biological society, if it exhibits three characteristics:

1. cooperative care of the young;
2. overlapping generations;
3. division of labor.

Communication among members is also a requirement of eusocial groups, but species that do not have such groups also communicate. The best-studied eusocieties are those of ants, termites, and some species of bees and wasps. Fossil evidence of insects in groups suggests that insect societies existed 200 million years ago.

Division of labor may be the key to the evolutionary success of social insect colonies, such as those seen with fire ants. Coordinated groups of individuals make up the colony, each group with a specific function. One group might locate new food sources, while another assesses the sugar content of prospective foods. Other groups care for young or dispose of dead members. Many individuals each performing only a few tasks minimizes errors. It is similar to the way a large company operates.

# 42.11   Three Animal Societies

Animal societies are common among insects and primates. Ethologists are learning more about well-understood social animals and are discovering new ones. Here is a look at a newly described animal society, one known for a few years, and a long-known society.

## Snapping Shrimp Live Colonially in Sponges

A newly recognized social species is the snapping shrimp, *Synalpheus regalis,* named for their powerful fighting claws. These crustaceans occupy cavities within two types of sponges that live among coral reefs in the Caribbean. Shrimp colonies within sponges average 149 individuals but may include up to 300.

Each shrimp colony has only one female that reproduces, which she does very often (**figure 42.19A**). The members of the colony are therefore half or full siblings. The spongy home protects the shrimp and provides food, which floats into the colony from the sponge's internal canal system.

Collecting and dissecting sponges reveal two of the three defining characteristics of an animal society. First, generations overlap, as shown in the groups of young of different size and presumably different age. Second, shrimp colonies have a division of labor, as evidenced in the lone reproductive female. Demonstrating the third characteristic of an animal society—cooperative care

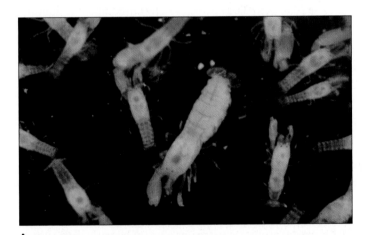

A

B

C

**FIGURE 42.19**   **Three Animal Societies.**    **(A)** Snapping shrimp (*Synalpheus regalis*) live in social groups in sponges in Caribbean coral reefs. Each colony has a single reproductive queen, shown here. Some of the shrimp around her display their characteristic claws, which they use to protect the colony, because sponge homes are in high demand. **(B)** Naked mole rats from different colonies attack each other viciously, each trying to drag the enemy into its own territory to sink its teeth into the other animal's flesh. **(C)** Honeybee workers build a chamber called a queen cell in which the potential queen develops.

of young—required experiments that reveal a way that adults might help juveniles.

Young shrimp obtain food without adult assistance—it just floats in. Nor must adults build a nest—the host sponge provides a secluded habitat. Biologist J. Emmett Duffy, of the Virginia Institute of Marine Science, hypothesized that adults care for young by protecting their hold on the sponge. He based this hypothesis on the observation that he could not find any sponges that did not house shrimp.

If competition exists for sponge homes, then introducing a threat should provoke protective behavior among resident shrimp. Duffy set up two experimental situations in sponges reared in his laboratory. In the first experiment, a shrimp colony encountered shrimp of another species. In the second experiment, a shrimp colony encountered shrimp of the same species that had been removed from the colony beforehand. Results were striking. When confronted with unrelated shrimp that competed for sponge space, the resident snapping shrimp fought valiantly, chopping up the competition. In contrast, the resident shrimp did not attack their former colony mates. More telling, only the larger individuals defended the colony—which Duffy concluded is evidence of cooperative care of young. The shrimp are the first eusocial invertebrate discovered in the oceans.

## Naked Mole Rats Fulfill Roles in Their Community

They live in vast, subterranean cities, digging tunnels and chambers in the soils of Kenya, Ethiopia, and Somalia. Roads above them collapse; crops that get in their way, such as yams, they eat. The naked mole rats of Africa live in societies remarkably like those of eusocial paper wasps and sweat bees, with a powerful queen ruling a colony of up to 300 individuals. Unlike most mammalian social groups, in which many females reproduce, the naked mole rat queen is the sole sexually active female.

Other roles in the naked mole rat community are well defined. The "janitors," the smallest and youngest males and females, patrol the colony's tunnels. They keep the walls smooth by rubbing them with their soft, hairless bodies. Janitors eat roots hanging from ceilings and confine excrement to widened dead ends.

The oldest and largest mole rats serve sentinel duty, their senses sharpened to detect intruders (figure 42.19B). A sentinel will stop and feel the air currents on its nearly sightless eyeballs, a change in current indicating an invader. The sentinel also senses the low-frequency sound of footsteps overhead and detects many odors. If these signs, for example, indicate that a snake has poked its head into a tunnel, the sentinels signal to the others to descend to wider tunnels.

Mole rats dig new tunnels in early morning and late afternoon. "Dirt carriers" form groups; the head animal is a "digger," and a "kicker" works at the back to get rid of the dirt. Food carriers use their long incisors to break off bits of roots and tubers, consuming some of the food before returning to the group.

The queen patrols the tunnels several times an hour. Every 3 months, she gives birth to a litter of up to 27 pups, nursing them for a month while three male workers bring her food. For the next 2 months, these workers care for the young, who then become workers. Apparently the queen's presence is crucial to the integrity of this social colony. In the laboratory, when the queen is removed, chaos ensues. Jobs vanish, and the cooperation and interaction that define the colony break down.

## Honeybees Exhibit Complex Communication Behaviors

The inside of a honeybee (*Apis mellifera*) hive is an efficient living machine, with each individual taking a specific role. At the summit of the organizational ladder is the queen, a large, specialized female who lays about 1,000 eggs each day. Eggs fertilized by a male drone develop through larval and pupal stages into female workers. Unfertilized eggs develop into males. Workers secrete a substance called royal jelly onto a very few eggs. These eggs hatch to yield larvae that eat much more royal jelly than do other larvae, which places them on a developmental pathway toward eventual queendom.

The 20,000 to 80,000 workers vastly outnumber the hundred or so drones, and only one queen reigns per hive. If the queen leaves, workers chemically detect her absence and hasten development of the young potential queens (figure 42.19C). The first new queen to emerge from her pupa case kills the others and then takes a "nuptial flight" to attract drones. She collects and stores their sperm and then later uses it to fertilize eggs. Males die when they inseminate the queen, or if they survive, they are forced out of the hive when food becomes scarce or autumn arrives.

A worker's existence is more complex, including several stages and specializations. Newly hatched workers feed the others. One-week-old females make and maintain the wax cells of the hive, where larvae and pupae develop. Some females are undertakers, ridding the hive of dead bees. Older workers forage for food.

Honeybees must store large amounts of honey in their hives to provide energy stores to survive the winter, when nectar, the source of the honey, is scarce. Because a hive may house up to 15,000 insects, a division of labor, plus a highly effective communication system, are necessary to collect and store sufficient nectar. The bees' activities must also be highly coordinated with the life cycles of flowering plants. On a spring day, a bee colony can gather up to 10 pounds of nectar, which ultimately yields half that amount of honey after drying out. About a third of the hive's members are foragers, whose task it is to locate and obtain nectar and communicate its source. Foragers are infertile daughters of the queen, who spend the last 10 days of their 30-day life spans searching for nectar.

Much of what we know about honeybee foraging behavior comes from the extensive work of Austrian zoologist Karl von Frisch. In 1910, he read a paper stating that bees are color-blind. Why, he wondered, are flowers so brightly colored if not to attract the bees that pollinate them?

To demonstrate color vision in bees, von Frisch placed sugar water on a blue cardboard disc near a hive. The bees drank the water. Next, he placed blue and red discs near the hive; the bees went only to the blue disc. Similar trials with different colors and shades showed that bees not only see color (except red) but they can also detect ultraviolet and polarized light, which humans cannot see.

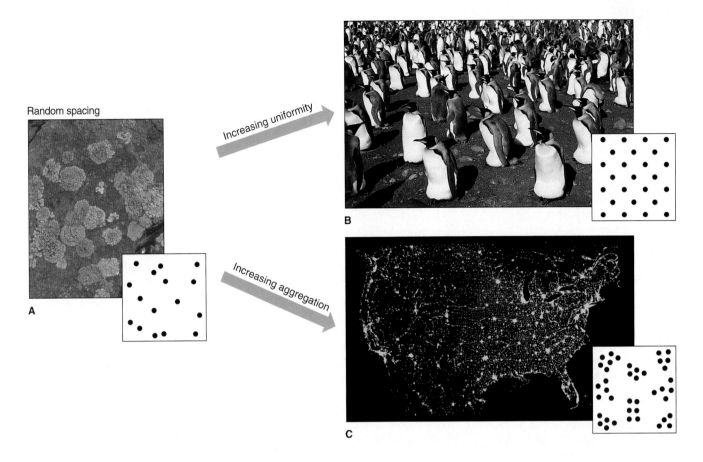

**FIGURE 43.1   Population Dispersion Is a Basic Feature of Populations.**   Random spacing, as illustrated by these lichens (**A**), is rare because it requires an environment in which resources are very evenly distributed. Social interactions such as territoriality (**B**) can lead to increasing uniformity, whereas an uneven distribution of resources or other social interactions may yield increasing aggregation, as in the human population of the United States (**C**).

The pattern in which individuals are scattered through habitat space is called **population dispersion** (**figure 43.1**). Organisms may be dispersed in a random pattern if the environment is relatively uniform (such as the deep shade under a dense forest canopy) and individuals neither strongly attract nor repel one another. A tendency toward uniform spacing may occur if individuals respond negatively to each other's presence. Tree species whose seedlings require abundant sunlight, for example, are more or less uniformly distributed because offspring cannot survive in the shade of their parents. Strongly territorial animals may also spread themselves evenly throughout a habitat. Most often, however, organisms show some degree of aggregation, coexisting where habitat conditions are most favorable, seeking out each other due to social attraction, or clumping in the vicinity of their parents. Recall the three zebra species described in chapter 42. Mountain and plains zebras are fairly uniformly distributed because their habitats usually have abundant food. Grevy's zebras, in contrast, dot the landscape in small groups near sparse grassy areas (see figure 42.17).

A species is rarely distributed continuously and in the same density over broad geographic areas within its range. For example, **figure 43.2** shows a range map for the red imported fire ant,

*Solenopsis wagneri* (also called *S. invicta*), in the southeastern United States. These aggressive, stinging ants nest in open, sunny areas such as pastures and cultivated fields. Their range in the United States corresponds closely to the relatively mild winter climate to which the ants are adapted. Since the ants' preferred habitat type is not distributed continuously throughout the entire region, however, a map of the population viewed at a closer scale would appear much patchier than the range map suggests. These scattered local populations may exchange individuals by migration—immigration is migration into a population and emigration is migration out. Such semi-isolated populations form regional "metapopulations."

**Reviewing Concepts**

- Ecology considers interactions between organisms and the living and nonliving parts of the environment.
- Populations, which are composed of members of a single species, exist in communities of interacting organisms.
- Several parameters are used to describe populations.

## Conducting a Wildlife Census

In Washington, D.C.'s Lafayette Park, biologist Vagn Flyger and his helpers place nest boxes in trees and then collect them when gray squirrels wander in. They anesthetize the animals and record their vital statistics—weight, sex, age, and health status. The researchers also tattoo the squirrels so their whereabouts can be monitored and then release them. Once a certain number of squirrels have been tattooed ($M$), a "recapture" sample is taken and the fraction of tattooed animals ($f$) noted. The number of marked squirrels ($M$) divided by this fraction ($f$) estimates the total squirrel population size.

For example, suppose researchers mark and release a total of 50 squirrels ($M = 50$). Subsequently, they capture 25 squirrels, of which 5 bear the tattoo ($f = 0.20$). For the area sampled, then, the estimated total squirrel population is $M/f = 50/0.20 = 250$. Or, set up a ratio: $5/25 = 50/x$.

Wildlife biologists also use a variation of this "capture-mark-recapture" approach on deer. Researchers place bright orange collars on a small number of deer. Then, an observer spots a group of deer from a plane and takes an aerial photograph. From the easily seen percentage of collared animals in the photo, biologists calculate the total number of animals in the group.

Counting droppings is another way to take a wildlife census, as is done for deer and moose in the Adirondack mountains of upstate New York. The numbers of excrement piles are correlated with the number of animals. A wildlife refuge on the Gulf Coast combines the capture-mark-count approach with dropping detection. Scientists capture otters and inject them with a harmless radioactive chemical, which the animals eliminate in feces. The scientists then collect droppings and extrapolate population size from the proportion of radioactively labeled droppings.

Many wildlife surveys count migrating herds of mammals and flocks of waterfowl from the air (**figure 43.A**). Counts are also done from the ground. Every year, the North American Breeding Bird Survey obtains counts along more than 2,000 routes. Perhaps most inventive is a device used in Montana. Rotting meat or fresh sardines are draped in a tree, along with an infrared heat detector attached to a camera. When a bear saunters over to collect the treat, the detector picks up the bear's body heat and snaps its picture (**figure 43.B**). Tags are used to trace bear identities.

Describing wildlife populations is useful in evaluating the effects of environmental catastrophes, such as fires, volcanic eruptions, or oil spills, by providing information on population dynamics and habitats before the disaster. Hunting, trapping, and fishing regulations and schedules are based on these data, as well as decisions on where to build—or not to build—houses, dams, bridges, and pipelines. The status of wildlife populations also reflects the overall well-being of an ecosystem.

**FIGURE 43.A** **Bird Count.** Each spring, U.S. and Canadian waterfowl spotters traverse 1.3 million square miles (3.4 million square kilometers) trying to count the number of animals in flocks, such as these snow geese.

**FIGURE 43.B** **Bear Count.** A male grizzly bear takes his own picture while stealing food from a tree with a camera triggered by the bear's body heat. From tags or physical features visible in such photos, biologists can obtain information on the overall bear population in a region.

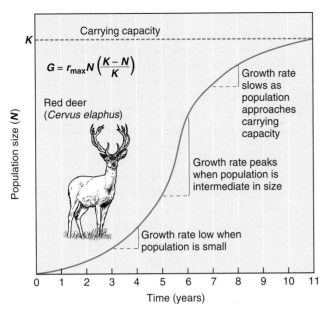

$$G = r_{max}N\left(\frac{K-N}{K}\right)$$

Red deer
(*Cervus elaphus*)

Carrying capacity

$K$

Population size ($N$)

Growth rate slows as population approaches carrying capacity

Growth rate peaks when population is intermediate in size

Growth rate low when population is small

Time (years)

$G$ = Growth rate (number of individuals added per unit time)

$r_{max}$ = Maximum per capita rate of increase

$N$ = Number of individuals at start of time interval

$K$ = Carrying capacity

**FIGURE 43.8** **Logistic Growth.** Populations of red deer on the Scottish island of Rum can grow quite large when hunting is not allowed. When food becomes scarce, pregnant does become malnourished, and fetuses die. Population growth slows as it approaches the carrying capacity, $K$. *Source:* Data from Andrew Cockburn, June 3, 1999, "Deer Destiny Determined by Density," in *Nature*, vol 399, p. 407.

## Crowding, Competition, Predation, and Pathogens Limit Population Size

Environmental resistance sometimes depends on the size of a population. Density-dependent factors are conditions whose growth-limiting effects increase as a local population grows. **Density-dependent factors** include crowding, competition, predation, and infectious disease. In response to crowding, for example, some animals cease mating, neglect their young, and become aggressive. Physiological responses to crowding include increased rate of spontaneous abortion, delayed maturation, and hormonal changes. Such responses ultimately slow population growth. Infectious disease is also considered to be a density-dependent factor because crowding facilitates the spread of disease.

Competition for space and food is a common density-dependent factor that affects population growth. When many individuals share limited food, none of them may eat enough to be able to reproduce, and population growth slows. For example, if several hundred fruit fly eggs hatch on a very small apple already riddled with larvae, they soon deplete the food supply. Larvae that do not eat enough starve to death or develop into small adults too weak to reproduce. Members of a population may also compete for a resource indirectly. Animals may compete for social dominance or possession of a territory, factors that guarantee the winners an adequate supply of a limited resource. The losers get less of the resource or none at all. Population growth slows because losers of the competition breed less successfully than the winners. Since multiple species often share a habitat, competition between species may also limit population growth (see figure 44.2). • **interspecific competition, p. 866**

Predation is another density-dependent factor that limits the population growth of both predator and prey. Effects on the prey population are obvious—their numbers diminish as they are eaten. Predators also influence prey populations by eliminating the weakest individuals, who are more easily captured. Prey populations are often maintained by high reproduction rates, which help compensate for the loss to predation. However, the introduction of alien species to an ecosystem may disrupt this delicate balance between predator and prey. As described at the beginning of the chapter, the resident species on Guam were decimated after the accidental introduction of the predatory brown tree snake, which has itself flourished in the absence of natural predators.

Removing predators may enable prey populations to swell and expand into new ranges. For example, people in Canada and the northern United States are reporting more frequent encounters with moose (**figure 43.9**). In the nineteenth century, many moose died from hunting and conversion of their forest habitat to farms. By the end of World War II, about 350,000 moose lived in Canada and the United States. Moose numbers have recently begun to recover due to a combination of factors—stricter hunting regulations, a return of forest, and fewer wolves and grizzly bears, which are their primary predators. Today, the animals top the 30,000 mark in Maine alone, and the metapopulation on the continent exceeds 1 million animals.

For the predator, inability to find food may mean death, and reproduction among predators is intimately tied to the density of their prey population. Many animals that cannot find sufficient food become temporarily infertile or have small litters. A hungry female bobcat has only one or two kittens, instead of the usual three, and those few may not survive their first 10 months if the mother cannot find enough food for them.

How can the brown tree snake continue to thrive on Guam as it depletes its food supply? Perhaps enough small reptiles such as geckos and skinks remain to support the snake population in

**FIGURE 43.9** **Expanding Moose Populations.** A decline in predator populations and changing human activities account for the present resurgence among moose populations. Their range creeps ever southward.

the short term. Ecologists do predict, however, that snake reproduction should slow as its food becomes less abundant. Chapter 44 further explores predator-prey interactions in communities.

## Environmental Disasters Kill Regardless of Population Size

**Density-independent factors** exert effects that are unrelated to population density. A severe cold snap, for example, might kill a certain percentage of fishes in two populations, even if the populations differ in size. Natural disasters such as earthquakes and severe weather conditions are typical density-independent factors. Consequences of human-caused environmental disasters are also independent of population density. Oil spills, for example, kill whatever life becomes entrapped.

A spectacular density-independent event occurred on May 18, 1980, when the north face of Mount St. Helens in Washington State began to bulge and rumble. Underground, water superheated by molten rocks was turning to steam. At 8:32 a.m., Mount St. Helens erupted. The 900°F (477°C) blast pulverized rocks and trees, killing nearly everything within a 6-mile (9.7-kilometer) radius and triggering earthquakes and avalanches. A 25-foot- (nearly 8-meter-) high wave of mud and molten rock, joined by melting snow and ice in the higher elevations, buried everything in its path. A forest of five-century-old fir trees disappeared instantly. The animal death toll included

some 5,200 elk, 6,000 black-tailed deer, 200 black bears, 11,000 hares, 15 mountain lions, 300 bobcats, 1,400 coyotes, 27,000 grouse, 11 million fishes, and an uncountable number of insects. The recovery of biotic communities following such disasters occurs by a sequence of events called succession.

● **communities change over time, p. 870**

A density-independent event that is disastrous for some species might benefit others. In 1993, a flooding Mississippi River sent deer, raccoons, rodents, skunks, and opossums scurrying to higher ground, but provided rich, new shallow-water habitats for microscopic algae and plankton, and the fishes and insects that feed on them. Wading birds ate fishes trapped in the shallower floodwaters and mud. Floodwaters carried wild gourds and zebra mussels to new habitats. Duck populations soared as the ravaged riverbanks provided more nesting sites and places to hide from predators. Other birds weren't as lucky. The least tern, for example, already endangered, became even more so as the waters washed away its nests, eggs, and fledglings.

Sometimes a population decline is due to both density-dependent and density-independent factors. This is the case for the spectacular die-off of monarch butterflies in parts of Mexico where a snowstorm struck in January, 2002. Millions of migrating butterflies froze to death mid-air, carpeting the forest floor in a foot-deep pile of black and orange. Some of the butterflies, however, did not freeze but were trapped by the bodies of others that had frozen—a density-dependent event. (Enough butterflies survived to restore the population.)

## Populations May Enter a Boom-and-Bust Cycle

In populations with high reproductive rates, exponential growth may overshoot the carrying capacity. Then the population drastically drops or crashes. A population that repeatedly and regularly increases (booms) and decreases (busts) in size over a given period of time exhibits a **boom-and-bust cycle.** Food supply, predation, and changes in a population's age structure influence boom-and-bust cycles. These cycles can occur when density-dependent factors are delayed relative to the reproductive rate, which allows population size to reach a level that cannot continue.

Snowshoe hare populations in Canada and Alaska follow a boom-and-bust cycle, with their numbers peaking every 8 to 11 years. When a snowshoe hare population is at its most dense, up to 10 hares may live on a hectare of land. In the leanest years, one hare may be the only one of its species on 80 hectares. The population densities of animals that eat hares fluctuate along with the changing hare population.

Ecologists have tracked boom-and-bust cycles among Soay sheep on the island of Hirta, near Scotland (**figure 43.10**). In 1932, the Marquis of Bute brought 107 feral (wild) sheep to the island. By 1942, their numbers had grown to 500, and by 1952, to 1,114 animals. A few summers later, when the population exceeded 1,400, the tiny island became littered with sheep corpses. More than a third of the animals had starved to death. Curiously, populations of sheep on other islands, and of other large animals on Hirta, did not decline.

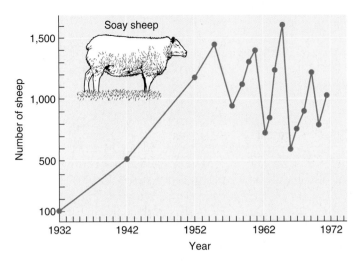

**FIGURE 43.10    Boom-and-Bust Cycles.** Soay sheep on the island of Hirta graze in rocky sheltered areas. As their population explodes, due to early and frequent pregnancies, the sheep eat all the vegetation. Every 4 years, many sheep starve, but the population recovers rapidly.

The nature of the Soay sheep's reproductive cycle explained the boom-and-bust cycle. A combination of factors enables the population to grow much too quickly for its own good. Females can become pregnant before they are a year old, and 20% of them deliver twins. Infant mortality is low. Babies stop nursing in June, when food is abundant, which gives their mothers sufficient time to gain enough weight to conceive in the fall. The Soay sheep also lack predators and cannot leave the island. Although many sheep starve, the population recovers because reproduction rates are high.

## Natural Selection Influences Population Growth

Species vary in their life history characteristics. Species with Type III survivorship curves, for example, tend to produce abundant offspring, each of which has a very low probability of surviving to reproduce (see figure 43.4). Other species, such as those with Type I survivorship curves, produce far fewer young, but parental investment is high so most offspring live and reproduce. Natural selection has adapted the life histories of these species to reflect the trade-off between reproductive output and parental investment. The differences between these types of life histories reflect different population growth rates.

At one extreme are **r-selected species,** in which individuals tend to be short-lived, reproduce at an early age, and have many offspring that receive little care. Weeds, crop pests such as insects, and many other invertebrates live in strongly r-selected populations. These species are usually excellent colonizers because of their high reproductive rates and effective dispersal strategies.

**K-selected species** are at the other extreme. Density-dependent factors regulate K-selected populations close to the carrying capacity (recall that K stands for carrying capacity). Individuals tend to be long-lived, to be late-maturing, and to produce a small number of offspring that receive extended parental care. Many

birds and large mammals live in K-selected populations. These species tend to be good competitors that exploit relatively stable environments. Whereas many insects that consume crops are r-strategists, many parasites and predators are K-selected. Perhaps this difference in life history is one reason that the introduction of exotic natural enemies rarely works for control of introduced pests. **Table 43.2** compares the characteristics of K-selected and r-selected species.

For decades, ecologists thought that logistic growth and close regulation of populations to a stable carrying capacity was the general rule in nature. Long-term studies of populations of many organisms, however, reveal that such patterns are the exception rather than the rule. Even populations of large animals that show K-selected life histories fluctuate greatly in response to changes in their environments that range from gradual and progressive to sudden and catastrophic.

### Reviewing Concepts

- Environmental resistance prevents unlimited population growth and may or may not depend upon population density.
- The carrying capacity, which may fluctuate over time, is the maximum number of individuals that can be supported indefinitely in a particular environment.
- Populations may exceed the carrying capacity and crash or remain near the carrying capacity.

## 43.4    Human Population Growth

The principles of population ecology apply to all species, including humans. From the appearance of modern humankind some 40,000 years ago until 1850, the beginning of the Industrial Revolution, our population grew slowly, reaching 1 billion. Yet only 80 years later, by 1930, we doubled our number to 2 billion. In another 30 years, by 1960, 3 billion humans populated Earth. The human population hit 4 billion in 1974,

### TABLE 43.2
### Characteristics of r-Selected and K-Selected Species

| r-Selection | K-Selection |
| --- | --- |
| Small individuals | Large individuals |
| Short life span | Long life span |
| Fast to mature | Slow to mature |
| Many offspring | Few offspring |
| Little or no care of offspring | Extensive care of offspring |

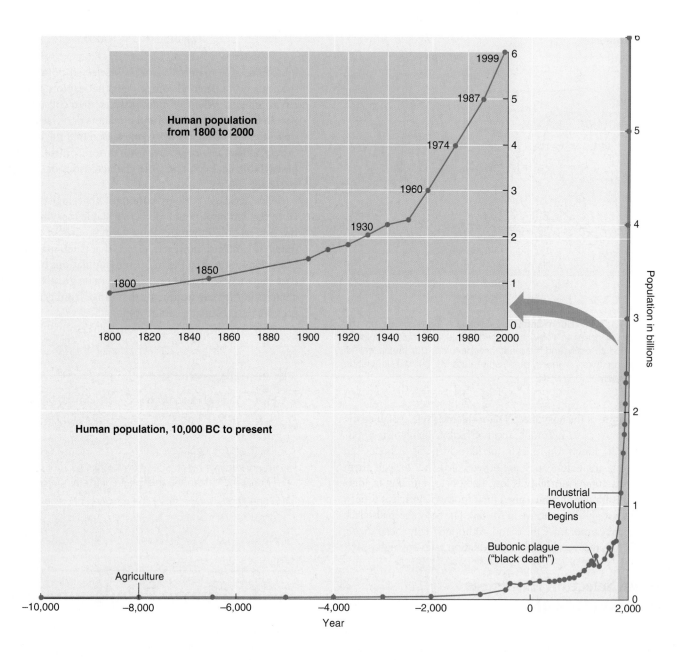

**FIGURE 43.11** **The Human Population Is Growing Exponentially.** Human population growth is on the rise, as indicated by the J–shaped curve. The most rapid population growth has occurred in the past 200 years.
*Source:* Data from U.S. Census Bureau.

5 billion in 1987, 6 billion in October 1999, and 6.2 billion by 2002 (**figure 43.11**). It is expected to reach 8 billion by 2030 and to level off at about 10 billion by 2200. The rate of increase may be slowing. In the late 1980s, nearly 90 million people were added each year. By the end of the 1990s, the rate had slowed to 78 million people per year, and by 2002, to 77 million per year.

## Average Fertility Has Been Declining

Part of the reason for the global slowdown in the overall rate of human population growth is a decline in average fertility. Worldwide, the average number of children has declined from 5 per female in 1950 to 2.8 today, partly because of increasing

access to contraceptives in developing countries. Family size has declined even in some countries where it was not expected to, such as Iran and Italy. Still, some populations of Latin America, Asia, and Africa will continue to grow for many years, as each generation exceeds the replacement rate of 2.1 children per female. Today, 85% of human population growth is in developing nations. In some countries, declining fertility is not noticeable because the population is already so large. This is the case for India, which adds more individuals per year than China, Nigeria, and Pakistan combined.

Fertility, mortality, and age structure differ worldwide to produce a mosaic of regional human subpopulations—some growing, some stable, and some declining (**figure 43.12**). The overall growth rate of the human population is 1.2% per year, but it varies from

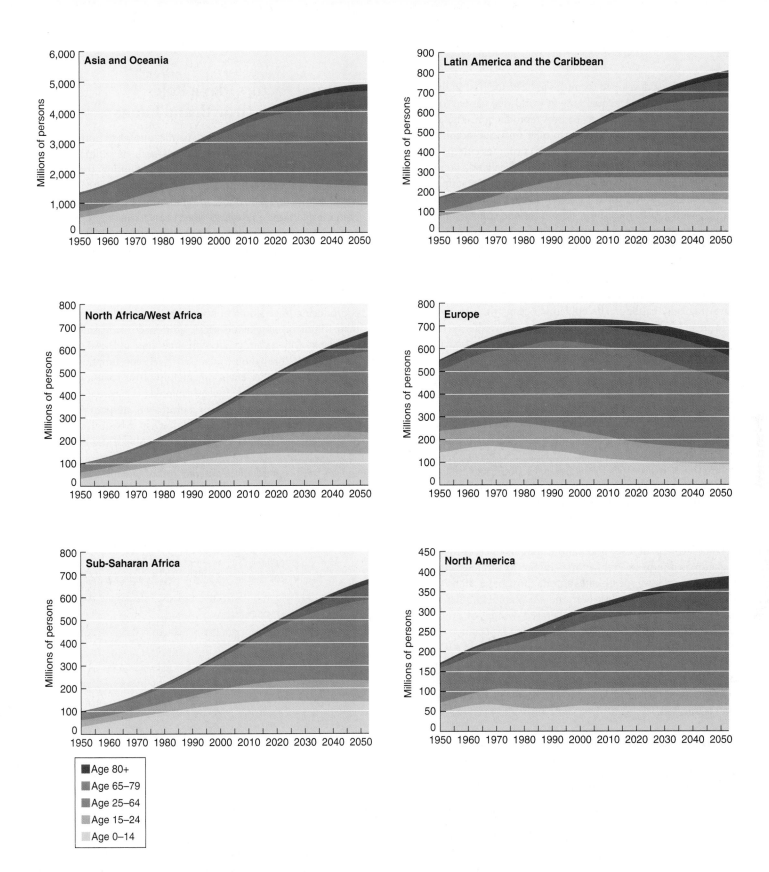

**FIGURE 43.12** **Regional Trends in Future Population Growth.** As human population growth continues in the first half of the twenty-first century, differences in age structure will produce a mix of growing, stable, and declining regional populations. Note that the graphs are plotted on different scales to account for regional differences in population size.

*Source:* Population Division of the Department of Economic and Social Affairs of the United Nations Secretariat (1999). *World Population Prospects: The 1998 Revision.* Vol. II: The Sex and Age Distribution World Population. (United Nations publication, Sales No. E.99.XIII.8). The United Nations is the author of the original material.

2.6% in the least developed countries to 0.3% (or less) in the most developed countries. Unequal growth worldwide has led to opposite governmental attempts to control reproduction. The governments of France, Canada, and Japan offer financial incentives and extended new parent leaves from work to encourage citizens to have children to bolster aging populations. In Thailand, population growth has fallen in the past 20 years because of increasing availability of contraceptives. China controlled runaway population growth with drastic measures, rewarding one-child families and revoking the first child's benefits if a second child is born.

Although China's "one-child policy" has prevented hundreds of millions of births, it also has had negative effects. Children have few siblings, cousins, aunts and uncles as families shrink. In Henan province, the one-child policy and an AIDS epidemic have tragically clashed. The epidemic began when many adults sold their blood in the early 1990s; unclean needles were used. Today, many children are alone, lacking siblings and having lost older relatives to AIDS.

## Life Span Is Increasing

Increasing life spans will contribute to the growing population. Average life expectancy is increasing in most parts of the world, thanks in part to medical advances. Life expectancy for the cohort born from 1995 to 2000 ranges from 50.3 in the least developed countries to 74.9 in most developed countries. In some countries, the fastest growing age group consists of those over 80. In Sweden, for example, the maximum age at death increased from 101 in the 1860s to 108 in the past decade. The current growth of this population, and its increasingly older makeup, is due to several factors—higher birthrates, better survivorship, and lower death rates among those older than 70 years. In sub-Saharan Africa, however, the AIDS epidemic has significantly affected mortality rates and erased decades of progress in increasing life expectancy. For example, life expectancy in Zimbabwe would be 70, if not for AIDS; instead it is 36! AIDS is so prevalent in some areas of Africa that entire countries will be decimated in a few years. Health officials are concerned particularly about sub-Saharan Africa, where nearly 80% of the population is infected.

## Cultural Factors Play a Role in Population Growth and Limits

When the study of population dynamics applies to humans, fields other than biology come into play—such as politics, history, economics, anthropology, and sociology. Shifting the focus of studying populations from a global view to a family view helped reveal factors that contribute to rapid human population growth. For example, families are large when people want many children to help obtain increasingly scarce resources; also, in some areas, women's roles in society are limited to having children.

Birth and death rates affect the human population, but migration also affects the quality of life regionally. For example, migration from rural to urban areas lacking adequate economic opportunities is producing huge slums in several "megacities" (cities with more than 10 million inhabitants each). Worldwide, increasing numbers of people will mean greater pressure on land, water, and air resources. Deforestation, increased fuel consumption, species extinctions, and other environmental problems resulting from the expanding human population are explored in chapter 46.

Despite the influences of culture on human population dynamics, our population ups and downs are basically like those of other species. Changing numbers reflect birth and death rates, as well as diverse environmental influences. Chapters 44 and 45 explore in greater detail the interactions of organisms and the living and nonliving environment—ecology.

### Reviewing Concepts

- Human population growth increased following acquisition of tool use and the rise of agriculture.
- The population growth rate has exploded since the Industrial Revolution.
- Today, the human population growth rate varies greatly among nations.
- Human population biology is complex because it reflects not only ecological principles but also diverse cultural, social, and economic influences.

# Connections

We learn much about a species by examining how members interact and grow as a population. Populations evolve but individuals do not. The genes in each member of a population represent the genetic possibilities in each successive generation. A gene is information for building a protein, and the presence of a single, small protein may translate to the actions of an entire population of organisms. In a way, a population acts as a type of "buffer" to changes in the environment. Because genetic variation exists within a population, it can respond to most changes in the world around it. But it is nearly impossible to consider a population without including the other species, along with the nonliving components, in the area occupied by that population. No population lives in absolute isolation. As humans we need to recognize this fact—we interact with many other species that can affect us, or can be affected by us. Chapters 44 through 46 explore this dependence. As you read them, remember that we can, and should, apply the principles of populations even to humans.

# Student Study Guide

## Key Terms

age structure  *853*
biotic potential  *855*
boom-and-bust cycle  *858*
carrying capacity (*K*)  *856*
cohort  *853*
community  *850*

demography  *850*
density-dependent factor  *857*
density-independent
   factor  *858*
ecology  *850*
environmental resistance  *856*

exponential growth  *853*
fecundity  *853*
*K*-selected species  *859*
life table  *853*
logistic growth  *856*
per capita rate of
   increase  *853*

population  *850*
population density  *850*
population dispersion  *851*
*r*-selected species  *859*

## Chapter Summary

### 43.1 Overview of Populations

1. Ecology considers interactions between organisms and the living and nonliving parts of the environment.
2. Populations, which are composed of members of a single species, exist in communities of interacting organisms and are described using different parameters.

### 43.2 How Populations Grow

3. The size of a population reflects birth and death rates and immigration and emigration.
4. With unlimited resources, population growth is exponential and can be described with a J-shaped curve.
5. Since an organism's chances of reproducing and dying vary with its age, the growth rate of a population reflects its age structure.

### 43.3 Factors That Limit Population Size

6. Environmental resistance prevents unlimited population growth and may or may not depend upon population density.
7. The carrying capacity, which may fluctuate over time, is the maximum number of individuals that can be supported indefinitely in a particular environment.
8. Populations may exceed the carrying capacity and crash or remain near the carrying capacity.

### 43.4 Human Population Growth

9. Human population growth surged following the Industrial Revolution.
10. Today, the human population growth rate varies greatly among nations.
11. Human population biology is complex because it reflects not only ecological principles but also diverse cultural, social, and economic influences.

# What Do I Remember?

1. Distinguish between a population and a community.
2. Define fecundity, survivorship, and age structure, and indicate how they influence population growth.
3. What is the difference between *K*-selected and *r*-selected species?
4. What conditions result in exponential versus logistic population growth?
5. In what ways have human populations exceeded the carrying capacities of their environments?

**Fill-in-the-Blank**

1. _____ is the number of offspring an individual produces in a lifetime.
2. Factors that check population growth are termed _____.

3. Exponential growth is exhibited by a(n)_____curve.
4. The human population is currently exhibiting _____ growth.
5. When density-dependent factors are delayed relative to reproductive rate, the population experiences _____ cycles.

**Multiple Choice**

1. Which of the following causes a population to be uniformly dispersed?
   a. the presence of a predator
   b. a negative response between individuals
   c. the location of scarce resources, such as a water hole
   d. exceeding a carrying capacity

2. A species that shows a constant survivorship rate throughout life exhibits what type of survivorship curve?
   a. type I
   b. type II
   c. type III
   d. type I and II
   e. type II and III

3. If there are no restrictions on the reproduction and survival of a population, what is the result?
   a. It will reach and maintain its carrying capacity.
   b. Its fecundity will dramatically decrease.
   c. It will reach its biotic potential.
   d. It will exhibit an inverted age structure and collapse.
   e. More than one of these will occur.

4. Which of the following is an example of an *r*-selected species?
   a. animals that produce many offspring that live brief lives, reproducing quickly
   b. animals that produce few offspring and nurture them for many years
   c. a species that produces two offspring and mates for life
   d. animals that produce several offspring that are mature within a year
   e. animals that carefully maintain their carrying capacity year after year

5. What would be the effect of taking a *K*-selected species and placing them on an island with abundant resources, no disease, and no predation?
   a. The species would quickly overshoot its carrying capacity, and the entire population would starve.
   b. The species would establish a balance with its resources and maintain a steady number of individuals.
   c. The population size would remain the same as the number introduced.
   d. The population size would fluctuate wildly forever in boom-and-bust cycles.

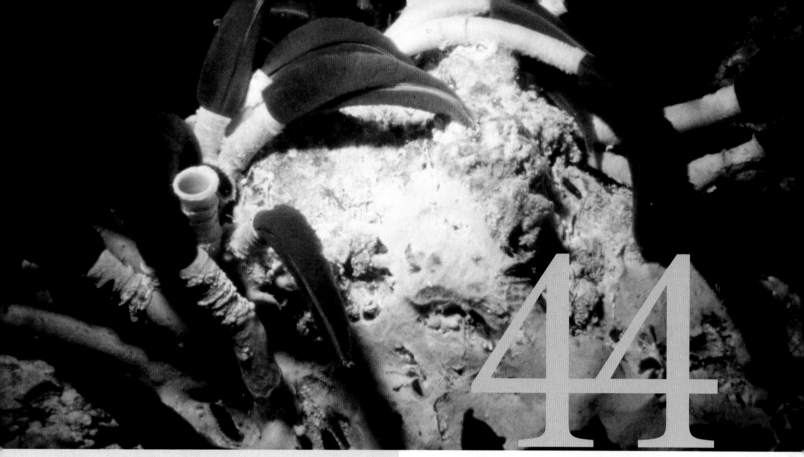

Energy in the bonds of inorganic chemicals spewing from Earth's interior supports this ecosystem, in which tube worms are abundant consumers.

# Communities and Ecosystems

## Life in Deep-Sea Hydrothermal Vents

In 1977, the submersible *Alvin* took a group of geologists to the sea bottom near the Galápagos Islands, directly above cracks in the Earth's crust called deep-sea hydrothermal vents. The researchers were astounded to see abundant life around these vents—tubelike worms, anemones, crabs, shrimp, and others, some of which were unknown to marine biologists.

Too far below the surface to use sunlight, the producers in these communities tap energy from the chemicals released from the vents, primarily hydrogen sulfide. In 1998, researchers were able to analyze life on chimney-like structures termed "black smokers," which exude intensely hot, mineral-laden water from the sea floor. The primary producers of the black smoker ecosystem are chemoautotrophic, hyperthermophilic bacteria and archaeans. From 10,000 to 100,000 of these microorganisms inhabit each gram of rock in the chimney.

Consumers include a variety of worms and clams that harbor symbiotic, chemosynthetic prokaryotes. Other organisms prey on the tube worms, and crabs are scavengers and decomposers.

So far, biologists have identified several hundred of the thousands of types of microorganisms present, but only about 5% of the 300 or so species are visible to the unaided eye.

Deep-sea hydrothermal vents are by no means rare—scientists simply were not able to look on the bottom of the ocean for living communities until the last 30 years.

# Chapter Preview

1. Biotic communities are made up of coexisting species that are specialized in where, when, and how they live. An ecosystem consists of biotic communities and the abiotic environment.

2. Each species has characteristic conditions where it lives (habitat) and resources necessary for its life activities (niche). Slight differences in niche allow different species to share surroundings.

3. Symbiotic relationships and predation influence community structure. In coevolution, the interaction between species is so strong that genetic changes in one population select for genetic changes in the other.

4. As species interact with each other and their physical habitats, they change the community, a process called ecological succession.

5. An ecosystem can range from a very small area to the entire biosphere. Ecosystems interact and change through time.

6. A food chain begins when primary producers harness energy from the sun or inorganic chemicals, forming the first trophic level. The total amount of energy converted to chemical energy is gross primary production.

7. Consumers comprise the next trophic levels, getting their energy ultimately from eating producers. Decomposers break down nonliving organic material (detritus) into inorganic nutrients.

8. Ecological pyramids measure energy, numbers of organisms, or biomass in a food chain. Food chains interact, forming food webs.

9. Biogeochemical cycles are geological and chemical processes that recycle chemicals essential to life.

10. Water cycles from the atmosphere as precipitation over land or water, then into organisms that release water in transpiration, evaporation, or excretion.

11. Autotrophs use atmospheric carbon in $CO_2$ to manufacture carbohydrates. Cellular respiration and burning fossil fuels release $CO_2$. Decomposers release carbon from once-living material.

12. Nitrogen-fixing bacteria convert atmospheric nitrogen to ammonia, which plants can incorporate into their tissues. Decomposers convert the nitrogen in dead organisms back to ammonia. Nitrifying bacteria convert the ammonia to nitrites and nitrates. Denitrifying bacteria convert nitrates and nitrites to nitrogen gas.

13. As rain falls over land, rocks release phosphorus as usable phosphates. Decomposers return phosphorus to the soil.

14. Bioaccumulation concentrates chemicals in cells relative to their surroundings. Biomagnification concentrates chemicals to a greater degree at successive trophic levels as the chemical passes to the next consumer rather than being metabolized or excreted.

# 44.1 Overview of Biotic Communities

All of the populations in a given area constitute a **biotic community.** One or more communities plus the nonliving, or abiotic, environment constitute an **ecosystem.** A downed Douglas fir, for example, is a bustling biotic community (**figure 44.1**). Some resident species inhabit dry, shaded branches; others, the soggy, rotting roots; and still others, the deepest heartwood. Soon after the tree falls, certain insects invade the inner bark, and then other species enter the sapwood. Still later, different species attack the heartwood. Deep within this decomposing tree, bark beetles whittle a labyrinth-like "egg gallery" where they rear the next generation; the mother beetle guards the entrance. This chamber also houses more than 100 other types of beetles. Tiny wasps drill in from the outside and lay eggs on the bark beetle larvae. When wasp larvae hatch, they eat the beetle larvae. Scavenging beetles eat any remaining dead bark beetle larvae. Other insects eat the fungus that grows in indentations in the bark beetles' exoskeletons.

Communities can range greatly in size. Right Whale Bay in the South Georgia region of Antarctica is a community of fur seals, king penguins, macaroni penguins, Dominican gulls, elephant seals, and all the marine organisms these animals eat. Yet a dead and decaying squirrel riddled with insects, worms, and microorganisms is also a biotic community. The boundaries that delimit a community or ecosystem are arbitrary, just as they are for populations.

The planet's land and waters house a patchwork of different communities. Biodiversity is highest near the equator, possibly due to a favorable climate for year-round plant growth, and declines as the climate fluctuates more toward the poles. In general, similar types of biological communities appear at corresponding latitudes because they have similar climates. The steamy heat of South American and African rain forests is likely to be home to species with similar adaptations. Differences among communities at the same latitude in different parts of the world reflect regional climatic characteristics or geographical influences. • **climate influences biomes, p. 890**

## Interspecific Competition Prevents Species from Occupying the Same Niche

Each species in a community has a characteristic home and way of life. Recall from chapter 43 that a **habitat** is the physical place

**FIGURE 44.1** **A Fallen Tree Holds a Thriving Community of Organisms.** Rotting wood houses a changing community of insects, worms, fungi, and microorganisms.

where members of a population typically live, such as the bottom of a river or a rain forest canopy. Many different plants, fungi, animals, and microorganisms live together in major habitats, yet species that are very similar often do not live in the same places. To understand the reasons for this, it helps to distinguish between a species' habitat and the way it makes its living, called its ecological **niche.** The niche is the total of all the resources a species exploits for its survival, growth, and reproduction. It also includes the range of abiotic environmental conditions (such as salinity, temperature, and water availability) where the species lives. As part of the resources that comprise a species' niche, the habitat is actually a subset of the niche.

Two terms describe a species' resource use in a niche. The **fundamental niche** includes all the resources it could possibly use. If two or more species with similar but not identical fundamental niches live in the same area, the **interspecific competition** between them may restrict each species to only some of the resources available. Or, the species may be forced to use resources in different habitats. Therefore, due to competition, a species'

**realized niche,** the one it actually fills, may be smaller than its fundamental niche. Two species of barnacles, for example, live in the intertidal zone along Scotland's shoreline. Both species share the same fundamental niche, the entire intertidal zone. When present alone, either type of barnacle can survive throughout the entire zone. When both are present, however, one species grows faster than the other and crowds out its competitor in the lower, wetter region of the intertidal zone. The slower-growing species, however, better tolerates dehydration when exposed to the air while the tide is out and can more efficiently use the resources of the upper region of the intertidal zone. Consequently, the two populations of barnacles have different realized niches (**figure 44.2**).

The barnacle example illustrates the **principle of competitive exclusion,** which says that two species cannot coexist indefinitely in the same niche. The two species will continually compete for the limited resources required by both, such as food, nesting sites, or soil nutrients. If one species acquires more of the resources that are in short supply, it will eventually replace the other, less successful species. For example, on Africa's Serengeti Plain, wild dogs and hyenas hunt zebras, wildebeests, and antelopes. Hyenas are larger and more aggressive than wild dogs, and they sometimes displace wild dogs from their kills. The hyena's superior hunting skills have contributed to the decline of the wild dog population (**figure 44.3**).

When a new species enters an area, competition may drive a native population, or an entire species, to extinction. This happened when starlings were released in New York City's Central Park in 1891. The starlings robbed the native bluebirds of their nesting sites. The displacement of native species by introduced species occasionally disrupts whole communities. Zebra mussels (*Dreissena polymorpha*), for example, are native to the Caspian Sea in Asia. They were accidentally introduced to the Great Lakes in the 1980s and since then have spread to many waterways in the United States and Canada. Not only have they crowded out native mussel species along the way, but the tiny filter-feeders have also changed plant communities by greatly increasing the clarity of the waters they inhabit. The new plants provide a different type of habitat than the previous species, so the community of fishes and other organisms has also changed in waterways that zebra mussels inhabit.

The competitive exclusion principle says that species can coexist in a community if they use different resources within the habitat. Alternatively, competing species may use the same resource in a slightly different way or at a different time. This specialization is called **resource partitioning.** For example, in New England forests, five species of warblers, all small insect eaters, coexist in the same trees, each type feeding in different ways and on different parts of the tree (**figure 44.4**).

## Symbiosis and Predation Involve Two-Way and One-Way Exchanges

Competition is just one of many ways populations can interact. Some species live in or on another, a phenomenon called **symbiosis** (literally, "living together"). As table 44.1 shows, the

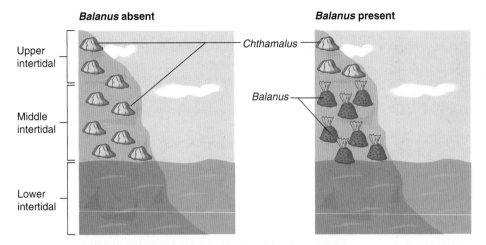

**FIGURE 44.2** Fundamental and **Realized Niches.** When the barnacle *Balanus* is absent, *Chthamalus* adults occupy the entire intertidal zone. *Balanus* grows faster than *Chthamalus*, however, and when *Balanus* is present, interspecific competition for space limits *Chthamalus* to the upper intertidal zone.

**FIGURE 44.3** Interspecific Competition for Food. This spotted hyena is carrying a scavenged rib cage away from the site of another animal's kill.

relationship between symbiotic species may be a **mutualism,** or mutually beneficial, such as when a cow derives simple sugars from the cellulose-digesting microorganisms living in its rumen. In a rare type of symbiosis, called **commensalism,** one species benefits, and the other is unaffected. Most humans, for example, never notice the tiny mites that live, eat, and breed in their hair follicles. A more common form of symbiosis is **parasitism,** in which one species derives

### TABLE 44.1

### Types of Symbiosis

| Type | Definition | Example |
|---|---|---|
| Mutualism | Both partners benefit | Mycorrhizae (see chapter 24) |
| Commensalism | One partner benefits with no effect on the other | Moss plants on tree bark |
| Parasitism | One partner benefits to the detriment of the other | Disease-causing organisms (see chapters 22, 23, and 24) |
| Parasitoidism | One partner benefits and the other dies as a result | Wasps that lay eggs in other insect larvae |

nutrients or other resources at the expense of another. Microorganisms, fungi, plants, and animals may be parasites.

Mistletoe, for example, is a parasitic plant that acquires nutrients by tapping into the vascular tissue of a host plant. Whereas parasites rarely kill their hosts outright, another type of symbiotic partner, called a **parasitoid,** does kill. Some wasps are parasitoids, laying their eggs in the larvae of other insects. When the wasp eggs hatch, the tiny larvae eat the host alive from the inside out.

Species also interact when one, a **predator,** eats another, the **prey.** Recall from chapter 43 that predator-prey relationships may strongly influence the population size of both the hunter and the hunted. Prey species often have obvious adaptations that help them avoid being eaten (**figure 44.5**). **Camouflage,** for example, is an adaptation of shape, color, or behavior that allows a species to "hide in plain sight." Adaptations in many animals enable them to blend into their surroundings. At the opposite end of the spectrum, some toxic or well-defended prey species, such as stinging bees and poison dart frogs, produce bright or distinctive **warning coloration** to advertise their special defenses. Predators quickly learn to avoid prey with warning coloration. An interesting variation on the theme of warning coloration is **mimicry,** in which different species develop similar appearances. For example, a harmless species may deter predators with coloration similar to that of a noxious or inedible species. In another type of mimicry, several distantly related noxious species may share a similar type of warning coloration, such as the yellow and black stripes on many stinging bees and wasps. Such mimicry is mutually beneficial to all species with the coloration.

Natural weapons and structural defenses are also commonly used among prey species. Some plant species, for example, produce distasteful or poisonous chemicals that deter most herbivores. Figure 25.10 illustrates the nematocysts that are a defensive trademark of the cnidarians, and figure 25.24 shows the characteristic protective "spiny skin" from which the echinoderms take their name.

Predators are adapted to circumvent the defensive strategies of prey. Monarch butterflies, for example, can tolerate the noxious chemicals in milkweed plants that repel other herbivores. Some predators, such as tigers and other big cats, have markings that hide their shape against their surroundings, which helps them sneak up on their prey. Natural selection may also favor predators with superior senses of smell or vision. Many predators hunt in groups. The Harris hawks perched in a tree in **figure 44.6,** for example, are scan-

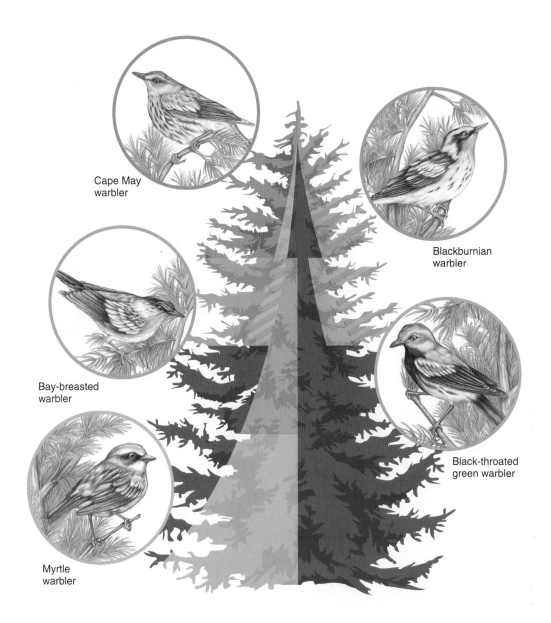

**FIGURE 44.4  Populations Coexist by Reducing Competition.** These five species of North American warblers live together in conifer forests because each forages in a different portion of the tree and uses different foraging behavior. Different-colored areas of the tree indicate primary feeding zones for each warbler species.

ning the area for prey. When they spot a rabbit, the hawks take turns diving at it. Should the rabbit escape into a burrow, the hawks again take turns landing on the ground and trying to scare it out of its hiding place. When the rabbit runs out, the other hawks, waiting in a circle, kill it. Afterward, the hawk hunting team shares the meal.

Some connections between species are so strong that the species directly influence each other's evolution. In **coevolution,** a genetic change in one species selects for subsequent change in the genome of another species. Of course, all interacting species in a given community have the potential to influence one another, and they are all "evolving together." However, in the strictest sense, such genetic changes are said to result from coevolution only if scientists can demonstrate that the traits in the coevolving species specifically result from their interactions.

One example of coevolution is the mutualistic relationship between ants (*Pseudomyrmex* species) and acacia trees. The ants defend their acacia trees from potential herbivores, seemingly in

exchange for food the acacia produces. Each species has special adaptations not found among related species that lack the mutualistic relationship. For example, the acacia produces swollen, hollow thorns in which the ants live. The acacia also produces nectar and nutritious nodules on its leaflets, which the ants consume. These adaptations have, in turn, selected for ant behaviors that defend the acacia tree against potential herbivores.

Sometimes, many species in a community depend on one type of organism, called a **keystone species.** These critical keystone organisms are not the dominant producer species that support all communities. In contrast, keystone species make up a small portion of the community by weight, yet exert a disproportionate influence on community diversity. Many keystone species are versatile predators that maintain competition, such as sea stars that prey on diverse tidepool invertebrates, or birds that eat herbivorous insects. Mycorrhizal fungi are keystone species that influence the composition of certain plant communities. • **mycorrhizal fungi, p. 452**

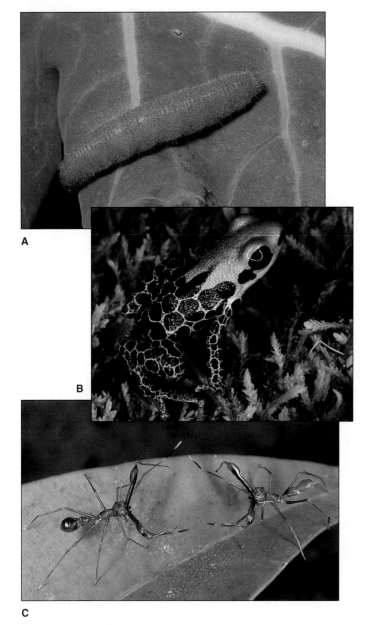

A

B

C

**FIGURE 44.5** **Prey Defenses.** Prey species may use camouflage to hide from predators (**A**) or warning coloration to advertise their defenses (**B**). Other species may, in turn, mimic those displays. Although the animals in (**C**) look like ants, each has eight—not six—legs. They are ant-mimicking jumping spiders with none of the weaponry of the ants they resemble.

### Reviewing Concepts

- A species' habitat is where it lives; its niche includes the resources it requires and all other factors that limit its distribution and abundance.
- Interspecific competition may lead to specialized adaptations that allow a species to exploit a subset of the resources in its habitat.
- Other interactions among species include symbiosis and predation.
- Coevolution is a result of highly dependent interactions between species.

**FIGURE 44.6** **Predators Sometimes Cooperate.** The hunt begins when the group of Harris hawks assembles on a lookout. Smaller teams of two or three birds make short flights to nearby perches, scanning the area for prey. After the hunt, the hawks share the meal.

## 44.2 Succession: Change Over Time

Communities may appear static, but they do change throughout time in response to many influences—climatic change, disturbance, and species invading from other areas. A community also changes as a result of its interaction with its physical environment. This gradual and directional process of change in the community, in which species replace each other and the biotic structure may change considerably, is called **ecological succession.**

Succession often leads toward a **climax community,** which is a community that remains fairly constant as long as climate does not change and major disturbances do not strike. However, few communities ever reach true climax conditions. Pockets of local disturbance, such as the area affected when a large tree blows over, create a patchy distribution of successional stages across a landscape. Factors such as fire, disease, and severe storms can influence successional patterns for centuries. Usually, the rate of change slows late in the successional process, but never ceases. In the Pacific Northwest, for example, old-growth forests are 500 to 1,000 years old, yet they are still changing in their structure and composition.

### Primary Succession Is the Development of a New Community

Ecologists define two major types of succession: primary and secondary. **Primary succession** occurs in an area where no community previously existed. Volcanoes, for example, afford won-

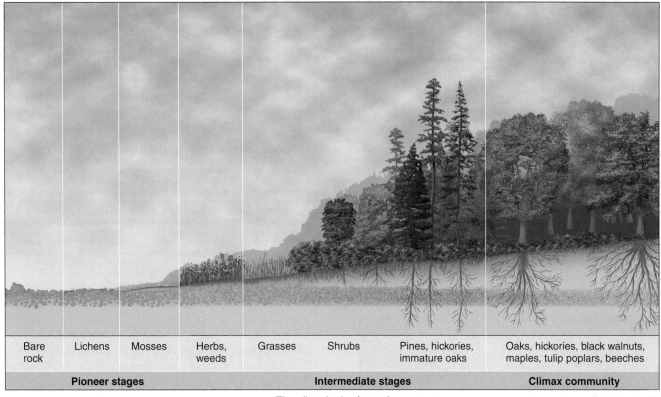

| Bare rock | Lichens | Mosses | Herbs, weeds | Grasses | Shrubs | Pines, hickories, immature oaks | Oaks, hickories, black walnuts, maples, tulip poplars, beeches |

**Pioneer stages**        **Intermediate stages**        **Climax community**

Time (hundreds of years)

**FIGURE 44.7** **Primary Succession Can Yield a Deciduous Forest After Centuries.** First, lichens colonize bare rock and produce acids that begin to break down the rock into a thin layer of soil. Next, mosses, fungi, small worms, insects, bacteria, and protista colonize the scanty soil, contributing organic matter as they live and die. Tiny herbs and weeds grow, grasses and shrubs move in, and trees come many years later after the soil is well developed. The community shown here includes species typical of New England.

derful opportunities to observe primary succession as lava flows obliterate existing life, a little like suddenly replacing an intricate painting with a blank canvas. Road cuts and glaciers that scour the landscape also expose virtually lifeless areas on which new communities eventually arise.

A patch of bare rock in New England also provides a clear view of primary succession. The first species to invade, the **pioneer species,** are hardy organisms, such as lichens and mosses that can grow on smooth rock. The fungal component of a lichen attaches to the rock and obtains water and minerals, while the algal part photosynthesizes and provides nutrients. As lichens produce organic acids that erode the rock, sand and dust accumulate in the crevices. Decomposing lichens add organic material, eventually forming a thin covering of soil. Microorganisms, worms, and insects colonize the forming soil. Then, rooted plants such as herbs and grasses invade. Soil continues to form, and larger plants, such as shrubs, appear. Larger animals move into the area. Next, aspens and conifers grow, such as jack pines or black spruces. Finally, hundreds of years after lichens first colonized bare rock, the soil becomes rich enough to support other deciduous trees, and a climax community of an oak-hickory forest may develop (**figure 44.7**). • lichens, p. 454

## Secondary Succession Occurs When a Community Has Been Disturbed

**Secondary succession** occurs where a community is disturbed but not decimated, with some soil and life remaining. Because

the area isn't completely devastated, secondary succession occurs faster than primary succession. Fires, hurricanes, and agriculture are common triggers of secondary succession.

An abandoned farm in the eastern United States illustrates secondary succession. This "old field" succession begins when the original deciduous forest is cut down for farmland. As long as crops are cultivated, natural succession stops. When the land is no longer farmed, fast-growing pioneer species, such as black mustard, wild carrot, and dandelion, move in, followed by slower-growing, taller goldenrod and perennial grasses. In a few years, trees such as pin cherries and aspens arrive and are eventually replaced by pine and oak. A century or so later, the climax community of beech and maple may again be well developed.

Ecologists have monitored secondary succession in the aftermath of hurricanes. Puerto Rico has seen 15 major hurricanes over the past three centuries. **Figure 44.8** depicts the secondary succession following Hurricane Hugo, which struck Puerto Rico on September 18, 1989. The storms there tend to keep the forest clipped to a low level. If 60 years go by without a major hurricane, a climax community of taller, denser trees grows. Typically, a large hurricane lowers the height of the forest canopy by 50%, the winds stripping leaves and branches, breaking stems, and uprooting trees. Usually, fewer than half of the trees die, most of them young and low, their limbs snapped off. Unlike primary succession, which relies on the arrival of new organisms to start growth, secondary succession here

**A**

**B**

**FIGURE 44.8    Secondary Succession.** Secondary succession occurs in the Luquillo Experimental Forest, Puerto Rico, following Hurricane Hugo. The storm destroyed about half of the forest canopy in 1989 (**A**), but 5 years later, the forest had recovered (**B**).

occurs in several ways. The Luquillo Experimental Forest in Puerto Rico recovered via resprouting and repopulation of the canopy that had been opened up by tree deaths. The wind also brought seeds that contributed to renewed growth.

## Succession Can Be a Complex Process

Succession occurs in a variety of different settings, but ecologists recognize a common set of processes in both primary and secondary succession. First, pioneer species colonize a bare or disturbed site. Recall from chapter 43 that the hardy pioneer species are usually *r*-selected, with prolific reproduction and efficient dispersal. Interestingly, these early colonists often alter the physical conditions in ways that enable other species to become established. These new arrivals continue to change the environment. Some early colonists do not survive the new challenges, which further alters the community. In contrast to the pioneers, species that appear in the climax community are usually long-lived, late-maturing, *K*-selected species that are strong competitors in a stable environment.

In addition to changes in the physical environment, interactions among species also change the composition of the biotic community. When pine trees invade a site, for example, they shade out lower-growing plants that appeared earlier in succession, and they attract species that grow or feed on pines.

Primary and secondary succession can occur simultaneously in a large area that has faced a major disturbance, such as a volcanic eruption or hurricane. Consider the return of life to Mount St. Helens, a volcano that erupted in Washington State in May 1980. Regions of primary and secondary succession dot the landscape, depending on the extent of devastation. In some areas, scorchingly hot gases and steam killed everything, so primary succession occurred, eventually, when the wind brought in pioneer species and the ground had cooled enough for them to survive and grow. An avalanche that triggered a mudslide also ushered in primary succession, for it buried everything in its path. Secondary succession occurred in the "blowdown zone," where many trees were knocked down but small plants survived. In these areas, insects, birds, and mammals returned relatively rapidly as the forest plants began to regrow.

### Reviewing Concepts

- A disturbed habitat will support a gradual reintroduction of life via succession.
- In primary succession, a new community forms; in secondary succession, the types of resident species change after a disturbance to the area.
- In both types of succession, pioneer species pave the way for changes in the community and the physical environment.

lay eggs. If a pool vanishes, the salamanders will usually not mate. To protect the animals from being crushed by traffic on roads they must traverse in their migration, some human communities post "salamander crossing" signs or even dig tunnels to assist them.

When the heat of summer dries up a vernal pool, its residents may either move on or display other adaptations to drying. Most insects lay eggs in the drying mud of summer and leave, their offspring hatching the next spring. Midges and mayflies wait until spring to lay their eggs. Snails, small invertebrates called fairy shrimp, and tiny clams can enter a dormant stage and survive until the pool forms again, even if that is years later. African bullfrogs dig channels from drying-up vernal pools to larger puddles, saving their young.

## The Niagara Escarpment Is a Vertical Ecosystem

A sheer 98-foot (30-meter) cliff face doesn't seem a likely habitat, but several hardy species occupy the Niagara Escarpment in Ontario, Canada. The climate is much more like the dry cold of the Arctic tundra than the surrounding wet and temperate Canadian forest.

Eastern white cedars of species *Thuja occidentalis* emerge from cracks in the cliff face. These trees are superbly adapted to their environment. All of the trees are stunted, and as a result, they cannot easily blow over. They grow very slowly, which may be a response to scarce nutrients and little space for root growth. A tree may widen only 0.002 inch (0.05 millimeter) in a year.

Life exists not only between the rocks of the escarpment but within them. Colored layers just beneath the rock surface consist of algae, fungi, and lichens. This sparse, cryptoendolithic ("hidden inside rocks") ecosystem survives because the translucent rock lets sufficient sunlight through to allow photosynthesis to occur.

## Fjord Ecosystems Exist Where Freshwater Meets Seawater

In the southwest corner of New Zealand's national park, narrow glacier-cut valleys called fjords (or fiords) extend from the Tasman Sea 10 miles (16.1 kilometers) distant, ending between the Southern Alps. The fjords flood, but in an unusual way—seawater pours in, but near-daily rains send rivulets of freshwater down the steep mountain slopes. The result is a body of water that is freshwater for 10 to 12 feet (3.1 to 3.7 meters) on top, but seawater below. Freshwater rises because it is less dense and because the fjords are so narrow that waves, which would mix the layers, cannot form.

The spectrum of life in the fjords ranges from a freshwater to a seawater ecosystem. In the surface layer, freshwater mussels and barnacles abound. The water here is stained yellow-brown from leaf litter. The color blocks much sunlight, so the sealike portion of the fjord lacks the light-dependent seaweeds and other algae common in the ocean. But dark-adapted sea life is plentiful and includes sea slugs, corals, and sponges. Brachiopods, crinoids ("sea lilies"), and other shelled animals cling to the cliff walls.

Predatory sea stars hover at the interface between fresh and salty water and shift position as the interface moves. When a day without rain raises the boundary, freshwater animals are trapped in salt water—giving the waiting sea stars an easily captured meal. As soon as the rain returns, the sea stars move down, because they cannot tolerate freshwater.

Chapter 45 continues our look at the environment from a broader perspective.

### Reviewing Concepts

- Earth has many different ecosystems that contain a variety of organisms and environmental conditions.
- Literally every location on Earth has been shown to support life in some fashion.
- We can learn general principles from specific ecosystems.

## Connections

In earlier chapters, we have considered the human body by examining the different systems that make life possible. In many ways, the planet is also a living entity, with systems that make life possible. These systems have common features that form their basis and variations that allow distinction and diversity. Just as the human body cannot function without even one of the organ systems, the world may not be able to survive the loss of any of its major ecosystems. The more we learn about the living world around us, the more we come to appreciate just how fragile, yet adaptable that world is. We are coming to realize that we all may suffer the loss of even a small part of the world's diversity—whether it has any commercial value or not. One of the lessons of studying ecosystems is that all species rely on many others for survival, and humans are no different. Chapters 45 and 46 touch on these subjects.

# Student Study Guide

## Key Terms

bioaccumulation  *882*
biogeochemical cycle  *878*
biomagnification  *882*
biomass  *878*
biome  *875*
biosphere  *875*
biotic community  *866*
camouflage  *868*
climax community  *870*
coevolution  *868*
commensalism  *868*
consumer (heterotroph)  *875*
decomposer  *875*

denitrifying bacteria  *880*
detritus  *875*
ecological succession  *870*
ecosystem  *866*
food chain  *875*
food web  *875*
fundamental niche  *867*
habitat  *866*
interspecific competition  *867*
keystone species  *869*
mimicry  *868*
mutualism  *868*
net primary productivity  *875*

niche  *867*
nitrifying bacterium  *880*
nitrogen-fixing
  bacterium  *880*
parasitism  *868*
parasitoid  *868*
pioneer species  *871*
predator  *868*
prey  *868*
primary producer
  (autotroph)  *875*
primary succession  *870*

principle of competitive
  exclusion  *867*
pyramid of biomass  *878*
pyramid of energy  *875*
pyramid of numbers  *875*
realized niche  *867*
resource partitioning  *867*
secondary succession  *871*
symbiosis  *867*
trophic level  *875*
vernal pool  *884*
warning coloration  *868*

## Chapter Summary

### 44.1  Overview of Biotic Communities

1. A species' habitat is where it lives; its niche includes the resources it requires and all other factors that limit its distribution and abundance.

2. Symbiosis, predation, and interspecific competition may lead to specialized adaptations that allow a species to exploit a subset of the resources in its habitat.

3. Coevolution is a result of highly dependent interactions between species.

### 44.2  Succession: Change Over Time

4. A disturbed habitat will support a gradual reintroduction of life via succession.

5. In primary succession, a new community forms; in secondary succession, the types of resident species change after a disturbance to the area.

6. In both types of succession, pioneer species pave the way for changes in the community and the physical environment.

### 44.3  Ecosystems

7. Ecosystems range greatly in size and are constantly changing and interacting.

8. Food chains and webs are built on primary producers that pass energy to successive levels of consumers.

9. Pyramid diagrams represent energy flow or numbers or biomass of organisms in an ecosystem.

### 44.4  Biogeochemical Cycles

10. Biogeochemical cycles describe how elements move between organisms and the physical environment.

11. Elements from the environment ascend food webs and return to the physical environment when an organism decomposes.

12. Some chemicals are concentrated as they ascend food webs, with toxic effects.

### 44.5  Food Webs and Chemical Magnification

13. Some substances enter living tissues that lack the ability to deal with those substances.

14. Bioaccumulation and biomagnification represent significant problems for species as toxic substances are concentrated in living tissue.

### 44.6  A Sampling of Ecosystems

15. Earth has many different ecosystems that contain a variety of organisms and environmental conditions, and we can learn general principles from these ecosystems.

16. Literally every location on Earth has been shown to support life in some fashion.

etrate. Organisms here, which rely on falling organic material from above, include mostly scavengers and decomposers such as insect larvae and bacteria. The sediment at the lake bottom comprises the **benthic zone.**

Oxygen and mineral nutrients in a lake are distributed unevenly. The concentration of oxygen is usually greater in the upper layers, where it comes from the atmosphere and from photosynthesis. As dead organic matter sinks to the bottom, decomposers consume oxygen and release phosphates and nitrates into the lower layers of the lake. In a shallow lake, wind blowing across the surface mixes the water, redistributes nutrients, and restores oxygen to bottom waters.

Deeper lakes in temperate regions often develop layers with very different water temperatures and densities. This thermal stratification prevents the free circulation of nutrients and oxygen in the lake. The degree of thermal stratification varies with the season.

In the summer, the sun heats the surface layer of the lake, but the deepest layer remains cold. Between these two layers is a third region, the thermocline, where water temperature drops quickly. In the fall, the temperature in the surface layer drops as the air cools. Gradually, water temperature becomes the same throughout the lake. Wind then mixes the upper and lower layers, creating a **fall turnover** that redistributes nutrients and oxygen throughout the lake. During winter surface water cools. When water cools to 39°F (4°C), the temperature at which it is most dense, it sinks. Water colder than this floats above the 39°F layer and may freeze, giving the lake an ice cover. In the spring, when the surface layer warms to 39°F, a **spring turnover** occurs, again redistributing nutrients and oxygen. After the spring turnover, algae thrive in the warming, nutrient-rich surface water.

Lakes age. Younger lakes are often deep, steep-sided, and low in nutrient content. The deep zone of bottom water stores a large quantity of oxygen, which is rarely depleted. These lakes are termed **oligotrophic,** which means they are low in fertility and productivity. They are clear and sparkling blue because phytoplankton aren't abundant enough to cloud the water. Lake trout and other organisms that thrive in cold, oxygen-rich deep water are numerous.

As a lake ages, organic material from decaying organisms and sediment begins to fill it in, and nutrients accumulate. These lakes are termed **eutrophic,** which means they are nutrient-rich and high in productivity. The rich algal growth turns the water green and murky. Decomposing organisms in the deeper waters deplete oxygen during the summer. Fish and plankton communities change, and fishes that can tolerate low oxygen conditions replace species such as lake trout. In time, the lake becomes a bog or marsh and, eventually, dry land. Discharge of nutrient-rich urban wastewater and runoff carrying phosphate-rich fertilizers from cultivated lands can speed conversion of oligotrophic lakes to eutrophic lakes. This transformation is termed eutrophication. In extreme cases, the nutrients promote excessive algal growth. When the algae die, they sink to the lake bottom, where decomposers deplete the water of oxygen. Fish kills and unpleasant odors often result.

## Rivers and Streams Provide a Moving Environment

The rivers and streams that flow across the terrestrial landscape carry rainwater, groundwater, snowmelt, and sediment from all portions of the land toward the ocean or an interior basin (such as the Great Salt Lake). The flow is not constant, however. Where the landscape flattens, the water may slow to a virtual standstill, forming pools. Elsewhere, the water flows in shallow runs or bends called riffles. Rapids are fast-moving, turbulent parts. Along the way, rivers provide moisture and habitat to a variety of aquatic and streamside organisms, which are adapted to both flooding and drying.

Rivers change, physically and biologically, as they move toward the ocean (**figure 45.15**). At the headwaters, the water is relatively clear and the channel is narrow. Where the current is swift, turbulence mixes air with water, so the water is rich in oxygen. In fast-moving streams, some organisms cling to any available stationary surface, such as rocks or logs. Algae, diatoms, mosses, and snails that graze on them, live here. Larval and adult insects burrow into sediments or adhere to the undersides of rocks with hooks or suckers. Many of these invertebrates eat decaying plant material that drops in from streamside vegetation, which often provides the bulk of the energy that fuels the headwater stream food chain.

As the river flows toward the ocean, it continues to pick up sediment and nutrients from the channel. Tributaries contribute to water flow, so the river widens, and as the land flattens, the current slows. The river is murky, restricting photosynthesis to the banks and water surface. As a result, the oxygen content is low

**FIGURE 45.15    Rivers Change Along Their Course.**    A narrow, swift stream in the mountains becomes a slow-moving river as it accumulates water and sediments and approaches the ocean.

relative to the river upstream. Such slower-moving rivers and streams support more diverse life, including crayfish, snails, bass, and catfish. Worms burrow in the muddy bottom, and plants line the banks.

Rivers and streams depend heavily on the land for water and nutrients. Dead leaves and other organic material that fall into a river add to the nutrients that resident organisms recycle. Rivers also return nutrients to the land. Many rivers flood each year, swelling with meltwater and spring runoff and spreading nutrient-rich silt onto their floodplains. When a river approaches the ocean, its current diminishes, which deposits fine, rich soil that forms new delta lands. Incredible disruption to ecosystems occurs when people alter a river's course.

### Reviewing Concepts

- Life in lakes, rivers, and streams must be adapted to water velocities, changing nutrient and oxygen concentrations, and drought and flooding conditions.
- Lakes are divided into different zones of oxygen availability.

## 45.7    Marine Ecosystems

The ocean, covering 70% of Earth's surface and running 7 miles (11.2 kilometers) deep in places, is the largest and most stable aquatic ecosystem. Specific regions are based on proximity to land.

Several types of aquatic ecosystems border shorelines. **Figure 45.16** illustrates these coastal areas.

### Estuaries Are Rich, Changeable Regions

At the margin of the land, where the freshwater of a river meets the salty ocean, is an **estuary.** Life in an estuary must be adapted to a range of chemical and physical conditions. The water is brackish, which means that it is a mixture of freshwater and salt water; however, the salinity fluctuates. When the tide is out, the water may not be much saltier than water in the river. The returning tide, however, may make the water nearly as salty as the sea. As the tide ebbs and flows, nearshore areas of the estuary are alternately exposed to drying air and then flood.

Organisms able to withstand these environmental extremes enjoy daily deliveries of nutrients from the slowing river as well as from the tides. Photosynthesis occurs in shallow water. An estuary houses a very productive ecosystem, its rocks slippery with algae, its shores lush with salt marsh vegetation, and its water teeming with plankton. Almost half of an estuary's photosynthetic products go out with the tide and nourish coastal communities.

Estuaries are nurseries for many sea animals. More than half the commercially important fish and shellfish species spend some part of their life cycle in an estuary. Migratory waterfowl feed and nest here as well. Human activities can threaten these important ecosystems. • **endangered estuaries, p. 923**

### Mangrove Swamps Contain Salt-Adapted Species

Another type of aquatic ecosystem where salinity varies is a **mangrove swamp,** which is distinguished by characteristic salt-tolerant plants. The general term "mangrove" refers to plants that are adapted to survive in shallow salty water, typically with aerial roots. About 40 species of trees are considered to be mangrove. Mangrove swamps mark the transitional zone between forest and ocean and are located in many areas of the tropics. Within them, salinity varies from the salty ocean, to the brackish estuary region, to the freshwater of the forest.

A mangrove swamp is home to a diverse assemblage of species because it provides a variety of microenvironments, from its treetops to deeply submerged roots in its own version of vertical stratification. Life is least abundant in the treetops, where sun exposure is greatest and water availability the lowest. Snakes, lizards, birds, and many insects live here. A hollow elevated mangrove branch may house a thriving community of scorpions, termites, spiders, mites, roaches, beetles, moths, and ants.

Aerial roots of mangroves provide the middle region of the swamp's vertical stratification. Here, roots are alternately exposed and submerged as the tide goes in and out. Barnacles, oysters, crabs, and red algae cling to the roots. Lower down lies the root region of the mangrove swamp, populated by sea anemones, sponges, crabs, oysters, algae, and bacteria. The algal slime that coats roots discourages hungry animals.

Submerged roots form the lowest region of the mangrove swamp. Here live sea grasses, polychaete worms, crustaceans, jellyfishes, the ever-present algae, and an occasional manatee. Ecologists estimate that up to 30% of the resident species here are unknown.

Unfortunately, many mangrove swamps are in prime vacation spots for humans—which means habitat destruction. When people cut down mangrove trees, small shrubs that can tolerate salt grow in the area, and trees cannot grow back. The diverse mangrove ecosystem shrinks and may vanish.

### The Intertidal Zone Favors Organisms That Can Hold Tight or Dig Deep

Along coastlines, in the littoral zone, lie the rocky or sandy areas of the **intertidal zone.** This region is alternately exposed and covered with water as the tide ebbs and flows.

The organisms in a rocky intertidal zone often attach to rocks, which prevents wave action from carrying them away. Holdfasts attach large marine algae (seaweeds) to rocks. Threads and suction fasten mussels to rocks. Sea anemones, sea urchins, snails, and sea stars live in pools of water that form between rocks as the tide ebbs. The organisms of the sandy beach, such as mole crabs, burrow to escape the pounding waves that would wash them away. Sandy beaches have very little primary production.

**FIGURE 45.16** **Coastal Ecosystems.** Coastal ecosystems include estuaries (**A**), where salt water and fresh water meet. Some coastal areas have mangrove swamps (**B**), rocky intertidal zones (**C**), or coral reefs (**D**).

## Coral Reefs Provide Habitat for Vast Numbers of Organisms

Colorful and highly productive **coral reef** ecosystems border some tropical coastlines. Coral reefs are vast, underwater structures of calcium carbonate whose nooks and crannies collectively provide habitats for a million species of plants and animals and an unknown variety of microorganisms. The Great Barrier Reef of Australia, for example, is composed of some 400 species of coral and supports more than 1,500 species of fishes, 400 of sponges, and 4,000 of mollusks. Other residents include algae, snails, sea stars, sea urchins, and octopuses. Food is abundant because the sun penetrates the shallow water, allowing photosynthesis to occur, and constant wave action brings in additional nutrients.

Coral animals have colorful popular names based on their varied forms—brain, staghorn, lace, vase, bead, button, and organpipe corals are just a few. Recall from chapter 26 that individual animals, called polyps, build the reefs and house symbiotic algae that are essential for the coral's, and the ecosystem's, survival (see figure 26.12). The living coral is but a thin layer atop the remains of ancestors. A coral reef, then, is at the same time an immense graveyard and a thriving ecosystem. It is rich in biodiversity, yet fragile. Chapter 46 considers threats to coral reefs.

## The Oceans Are a Vast Ecosystem

The oceans cover 70% of Earth's surface, but we know less about biodiversity there than we do about biodiversity in a single tree in a tropical rain forest. The reason for our scant knowledge of life in the oceans may be simply the vastness of this aquatic ecosystem. Populations are sometimes small, usually very dispersed, and nearly always difficult for us to observe. Biologists have explored only 5% of the ocean floor and 1% of the huge volume of water above. Yet by the year 2010, 65% of the world's human population will reside within 10 miles (16.1 kilometers) of an ocean.

We can, however, describe the ocean's physical characteristics, which determine the nature of its biological communities. The planet's five oceans and many seas, plus the bridging waters that interconnect them, hold about 95 trillion gallons (360 cubic kilometers) of water. The temperature ranges from 35°F (1.7°C) in the Antarctic Ocean to 81°F (27°C) near the equator. Sunlight quickly dissipates with depth. Within the first 10 meters, the water absorbs 80% of incoming sunlight, reflecting the blue wavelengths. The blue deepens with depth, and from about 600 meters and lower, everything is black, except for the occasional glow from bioluminescent organisms.

The sun heats the surface water, causing its molecules to move faster than molecules below. This warm upper layer is separated from denser, colder water below by a thin thermocline layer where the temperature changes rapidly. Tropical oceans and seas have a thermocline year round, but it appears only in the summer in temperate waters.

Very productive ocean environments arise where cooler nutrient-rich bottom layers move upward in a process called **upwelling.** The resulting sudden influx of nutrients causes phytoplankton to "bloom," and with this widening of the food web base, many ocean populations grow. Upwelling generally occurs on the western side of continents, where wind pushes surface waters offshore, such as along the coasts of southern California, South America, parts of Africa, and the Antarctic.

Like other aquatic ecosystems, the ocean is considered in zones (**figure 45.17**). These designations are horizontal and vertical (depth). The horizontal zones describe the relationship between the ocean and the land. The intertidal zone, discussed in the previous section, is the shoreline. The **neritic zone** is the area from the coast to the edge of the continental shelf, and its waters reach a depth of 200 meters. The remainder of the ocean is the **oceanic zone,** which is, in turn, subdivided according to depth.

Two general divisions of the oceanic zone that refer to habitats are the **benthic zone,** or bottom zone, and the waters above, collectively called the **pelagic zone.** The pelagic zone is, in turn, subdivided according to depth. From about 200 meters to the surface lies the epipelagic zone, the only area where photosynthesis can occur. Beneath that is the mesopelagic zone (200 to 1,000 meters), and then the bathypelagic zone (1,000 to 4,000 meters). Beneath these zones lies the abyssal zone, from 4,000 to 6,000 meters. The hadal zone describes the areas of the ocean that dip down even more, as far as 10 kilometers below sea level.

Chapter 44 considered life at the very top of the ocean—the surface microlayer—as well as life in deep-sea hydrothermal vents. The very top of the epipelagic zone is rich in microscopic species that photosynthesize and form the bases of the great oceanic food webs. Other microorganisms and small animals feed on them. These top-dwelling organisms die and provide a continual rain of nutrients to the species below. The diverse living communities of the hydrothermal vents are fueled by chemosynthetic bacteria that harness the chemical energy inside Earth. In between lies a universe of diverse species, including great numbers and varieties of fishes, mollusks, echinoderms, crustaceans, and organisms yet to be discovered and described. One estimate of biodiversity in the benthic zone is 10 million species; it may be even higher in the vast pelagic zone.

We have much to learn about the oceans and seas, which are the environmental descendants of the aquatic setting where life probably arose. But we may never do so unless the current state of the oceans changes, and soon. Already the oceans include 50 "dead zones," areas devoid of life. In some cases, the cause of the dead zone is obviously human intervention, or population shifts such as algal blooms or the effects of dinoflagellate toxins. But in some dead zones, the trigger for the unraveling of food webs and ecosystems is a mystery. Chapter 46 explores some of the ways that the oceans, and other biomes and aquatic ecosystems are struggling.

### Reviewing Concepts

- In areas where salt water meets freshwater, organisms are adapted to fluctuating salinity.
- In the intertidal zone, the ebb and flow of the tide challenges organisms.
- Life in the oceans is abundant and diverse, but we know little about it because oceans are vast and mostly inaccessible.

The Everglades is a vast ecosystem that consists of marsh grasses and a variety of other organisms adapted to flooded soils.

## The Endangered Everglades

Before 1900, the southern half of the Florida peninsula was a large, continuous waterway. The Everglades, a "sea of grasses," seeped southward from Lake Okeechobee in a vast area that included estuaries, saw grass plains, mangrove swamps, and tropical hardwood forests. This area was prone to droughts, floods, hurricanes, and fire, but native plants and animals had adapted to these frequently changing conditions.

In 1905, Governor Napoleon Bonaparte Broward began an effort to drain the Everglades. Large tracts of swamp became farmland, and Miami and Ft. Lauderdale grew up on the east coast. A 1,000-mile system of canals, levees, and pumps directed water for agriculture and urban use.

In the early 1960s, a 15-year drought began. The great efforts to prevent flooding now backfired, causing a water short- age. More than 740,000 acres of land burned in lightning- caused fires. In 1986, 20% of Lake Okeechobee experienced an algal bloom, fueled by nitrogen and phosphorous runoff from fertilizer use and animal waste from dairy operations. Today, the Everglades is half the size it had been a century ago.

In the last 20 years, massive efforts have been underway to undo the damage of the past. In 2000, the Comprehensive Everglades Restoration Plan began, a 30-year, $8-billion plan to restore the Everglades based on scientific principles. The cost is shared 50/50 between the federal government and the State of Florida. The goal is to reestablish a more natural flow of water throughout south Florida, restoring the ecosys- tem while ensuring water for human use.

# Environmental Challenges

**46.1  Sustainability**

**46.2  Air Pollution**

**46.3  Acid Deposition**
  - Acid Deposition Has Far-Ranging Effects
  - Air Quality Can Be Improved Through Implementation of Standards

**46.4  Ozone Thinning and Global Warming**
  - Stratospheric Ozone Can Be Broken Down by CFC Emissions
  - Global Warming Results from Effects of Greenhouse Gases

**46.5  Deforestation and the Loss of Food Webs**

**46.6  Desertification and the Expansion of Lifelessness**

**46.7  The Global Water Crisis**
  - Chemical Pollution Destroys Communities
  - Alteration of Waterways Can Have Unintended Effects
  - Estuaries Are Vulnerable to Pollution and Development
  - Coral Reefs Are Critically Endangered
  - Ocean Pollution Is Increasing

**46.8  Loss of Biodiversity**
  - Diversity Contributes to Ecosystem Stability
  - Biological Invaders Lessen Diversity and Use Up Resources

**46.9  The Resiliency of Life**

## Chapter Preview

1. Sustainability is the careful use of natural resources in a way that does not deplete them for future use.

2. Pollution is any change in the environment that harms living organisms. The Clean Air Act improved air quality in the United States.

3. Acid deposition forms when nitrogen and sulfur oxides react with water in the upper atmosphere to form nitric and sulfuric acids. These acids return to Earth as dry particles or in precipitation.

4. Use of chlorofluorocarbon compounds has thinned the stratospheric ozone layer, which protects life from ultraviolet radiation. These compounds are among thousands of persistent organic pollutants that harm the environment.

5. The greenhouse effect contributes to global warming. Shifting vegetation patterns, species ranges, and egg-laying schedules are responses to global warming.

6. Deforestation is the destruction of tree cover from forested areas. Subsistence-level agriculture contributes to the removal of tropical rain forests.

7. Draining lakes to irrigate crops and cattle grazing cause and hasten desertification, the expansion of desert into surrounding areas.

8. Organic and inorganic toxins, sediments, heat, heavy metals, and excessive nutrient levels pollute aquatic ecosystems.

9. Human activities and climatic fluctuations have tremendous destructive potential on fragile ecosystems.

10. The diversity-stability hypothesis proposes that an ecosystem with many diverse species is better able to survive stresses than a species-poor ecosystem.

11. After habitat destruction, biological invasion is the top cause of decreasing biodiversity as nonnative species displace native organisms.

## 46.1 Sustainability

Humans are just one of many millions of species, but we have a disproportionately large impact on the environment (**figure 46.1**). Consider some sobering facts:

- Humans have altered nearly 50% of the land, replacing grasslands with wheat fields, wetlands with suburbs, forests with grazing areas.

- By burning fossil fuels and using fertilizer on farmland, we have doubled the amount of available nitrogen in the environment.

- Overfishing, introducing nonnative species, and destroying habitats have caused or hastened many extinctions.

- Humans consume 40% of all terrestrial biomass, 25% of all marine biomass, and half of the planet's freshwater.

- In the United States, 161 million people live within 75 miles of nuclear reactor waste stored aboveground. The waste will be transported to and buried under Yucca Mountain, in Nevada, by 2010.

- Half of the world's mangrove swamps have been cleared for building or for aquaculture (farming seafood, especially shrimp). Ocean fishes are used to feed farmed shrimp, depleting marine food webs.

Humans use other types of organisms for food, shelter, energy, clothing, and drugs. When we kill enough individuals to deplete populations and reduce genetic diversity, species can become endangered—at first locally, then perhaps on a growing and global scale (**table 46.1**). Legislation has attempted to save species in danger of extinction. The Endangered Species Act of 1973 requires that the U.S. Secretary of the Interior identify threatened and endangered species and provide for their conservation. Since the act was implemented, more than 500 species of vertebrate and invertebrate animals and more than 700 species of plants and lichens have been classified as threatened or endangered. Only a few dozen species have been removed from the list because they have either recovered or become extinct or because new information revealed that their populations are larger than had been thought.

To counter loss of biodiversity, we can attempt to use living resources in a way that does not irreversibly destroy them for the future, an approach called **sustainability.** It is a guiding principle of modern agriculture. The Everglades ecosystem and other areas discussed in this chapter illustrate how misguided human activity counters sustainability.

A famous essay written in 1968 timelessly described sustainability. In "The Tragedy of the Commons," Garrett Hardin used a metaphor of a common (shared) area, such as a water supply. If each person takes from the resource without concern for the negative effects on others, eventually the resource will disappear. Hardin's scenario envisioned a town. Today, the idea of sustainability has global significance. Scientists and citizens alike are joining forces to clean and preserve the environment. Researchers at several government laboratories and universities, for example, are analyzing environmental interactions among major pollutants to better understand how industrial chemicals harm ecosystems. In another project, called All Species, nature lovers are cataloging biodiversity in their backyards and surrounding areas. We can all be part of maintaining the health of the planet. This final chapter considers the problems that plague the global environment and ways that we can help.

### Reviewing Concepts

- As we alter habitats, we may irreversibly alter ecosystems, with unpredictable consequences.

- The concept of sustainability holds that we should use natural resources in a way that does not deplete them for future generations.

A

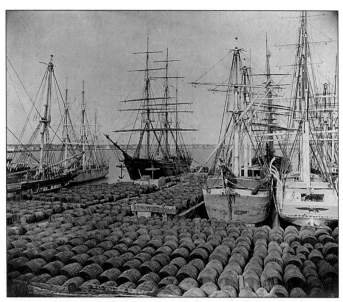

B

C

**FIGURE 46.1 Depletion of Natural Resources.** (A) A housing development in California replaces a natural ecosystem. (B) Until the mid-twentieth century, demand for whale oil to manufacture soaps, candles, margarine, and a variety of other products depleted many populations, which have yet to recover. These casks are filled with sperm whale oil. (C) This cattle pasture in western Brazil was not so long ago a lush tropical rain forest.

## 46.2 Air Pollution

Air pollution, acid deposition, depletion of the ozone layer, and global warming are some of the problems affecting the atmosphere. **Pollution** is any chemical, physical, or biological change in the environment that adversely affects living organisms. Pollution impacts air and water worldwide.

People have been polluting the air since ancient times. Analysis of lake sediments in Sweden and peat deposits in England reveal local air pollution during the Roman Empire, from 500 BC to AD 300. At that time, lead mining and metal smelting were as common as during the Industrial Revolution, a period previously credited with introducing air pollution (**figure 46.2**). Analysis of Greenland ice indicates that the smelters of the Roman Empire polluted the air with lead on a global scale.

December 5, 1952, stands out as a more recent time when we became particularly aware of air pollution. On this day, a

## TABLE 46.1

### Some Endangered Species

| Species | Habitat | Threat |
|---------|---------|--------|
| American burying beetle (*Nicrophorus americanus*) | Arkansas, Kansas, Oklahoma, Nebraska, South Dakota | Small mammals ate its food (carrion) |
| Kihansi spray toad (*Nectophrynoides asperginis*) | Mist of a waterfall in Kihansi River Gorge in Tanzania | Water diversion from hydroelectric project |
| Munz's onion (*Allium munzii*) | California clay soils | Diminishing habitat from development, water projects |
| Spix's macaw (bird) (*Cyanopsitta spixii*) | Arid savanna scrubland in northeastern Brazil | Excessive trapping |
| Sumatran rhino (*Dicerorhinus sumatrensis*) | Indonesia, Malaysia | Poachers |
| 3-striped box turtle (*Chrysemys trifasciata*) | South China, North Vietnam | Use in drugs |
| Vaquita (cetacean) (*Phocoena sinus*) | North end of Gulf of California | Caught in fishing nets |

**A**

**B**

**FIGURE 46.2    Air Pollution.    (A)** During the Industrial Revolution in England, soot-spewing smokestacks were a status symbol. **(B)** Air pollution continues to plague many cities. This is Mexico City.

temperature inversion occurred over London. A high-level layer of warm air blanketed the city, trapping cold air at ground level and preventing the particulates and gases from burning coal and other fossil fuels from dispersing upward. Sulfur dioxide ($SO_2$), which reacts in fog to form sulfuric acid, was the main pollutant. This "London fog" sent hundreds of people to hospitals with respiratory symptoms, and the death rate soared. London-type air pollution, also called **industrial smog,** occurs in urban and industrial regions where power plants, industries, and households burn sulfur-rich coal and oil.

Another type of air pollution is **photochemical smog,** which forms when vehicle emissions undergo chemical reactions in the presence of light. Photochemical smog forms in warm, sunny

areas where automobile use is heavy. The ultraviolet radiation in sunlight causes nitrogen oxides and incompletely burned hydrocarbons to react to form ozone ($O_3$) and other oxidants. These compounds poison plants and cause severe respiratory problems in some animals, including humans.

### Reviewing Concepts

- Pollution is the presence of any substance that exceeds the capacity of an ecosystem or that is unknown to the system.
- Atmospheric pollution affects breathing and vision.

## 46.3    Acid Deposition

Patterns of air circulation and geography can interact in ways that harm organisms far from the source of air pollution. In 1852, British scientist Angus Smith coined the phrase "acid rain" to describe one effect of the Industrial Revolution on the clean British countryside. In the mid-1950s, studies on clouds in the northeastern United States, farmlands in Scandinavia, and lakes in England rediscovered the effects of **acid deposition.**

Today, acid deposition in the United States is traced mostly to coal-burning power plants in the Midwest that release sulfur and nitrogen oxides ($SO_2$ and $NO_2$) into the atmosphere. Gasoline and diesel fuel burned in internal combustion engines and fumes from heavy metal smelters add to the problem. Westerly winds carry the pollution hundreds of miles to the east and northeast of their sources (**figure 46.3**). In the atmosphere, these oxides join water and fall over large areas as sulfuric and nitric acids ($H_2SO_4$ and $HNO_3$), forming acid rain, snow, fog, and dew. The acids can also return to Earth as dry particles.

Because the atmosphere contains carbon dioxide ($CO_2$) and water, all rainfall includes some carbonic acid and is therefore slightly acidic (pH about 5.6). Burned fossil fuels, however, have made the average rainfall in the eastern United States 25 to 60 times more acidic than normal. Acid deposition also affects the Pacific Northwest, the Rockies, Canada, Europe, East Asia, and the former Soviet Union.

### Acid Deposition Has Far-Ranging Effects

Acid deposition impacts lake life dramatically because it may lower the pH range by 3.0 to 5.0 units. The acid can leach toxic metals such as aluminum or mercury from soils and sediments. In this abnormal environment, fish eggs die or hatch to yield deformed offspring. Amphibian eggs do not hatch at all. Crustacean exoskeletons fail to harden. In very acidified lakes, most aquatic invertebrates and protista die, and lake-clogging mosses, fungi, and algae replace aquatic flowering plants. Clinging algae outcompete photosynthetic microorganisms and coat shallow

Physical effects of global warming were obvious before biological effects. Precipitation has increased at mid-latitudes and decreased at low latitudes over the past 30 to 40 years. Average global temperatures have increased since 1900 and especially since 1980. Since the mid-1990s, many of the 2,000 or so glaciers in Alaska and the Yukon have melted sufficiently to raise sea level by 0.2 millimeter per year, which is twice the increase from melting of the much larger Greenland ice sheet. These North American "ice rivers" are melting because average summer temperature has increased 3.5° Celsius (5°F) over the past 30 years. Growing seasons in temperate areas are lengthening. Computer models predict these trends will continue and will shift the distributions of species.

Evidence that species' ranges are shifting is beginning to mount. For example, the northern boundaries of the ranges of at least 34 species of butterflies are moving northward. At the southern ends of the ranges, where temperatures are rising, some species have become locally extinct. Another way to assess biological responses to global warming is to track events known to occur at particular times of the year. For example, in the United Kingdom, butterflies are emerging and amphibians are mating a few days earlier than usual; in North America, many plants are flowering and birds are migrating earlier.

Continued global warming will affect agriculture and public health. The U.S. south may become too dry to sustain many traditional crops. Water shortages worldwide may affect more than a billion people. Infectious disease patterns may also shift. Outbreaks of paralytic shellfish poisoning and algal blooms are already spreading and appearing where they haven't before. Tropical diseases, such as malaria, African sleeping sickness, and river blindness may affect more northern areas. The United Nations Intergovernmental Panel on Climate Change suggests that to limit the effects of global warming, nations learn to desalinate water; train people for jobs other than agriculture; encourage people to move inland; and stop destroying habitats that could protect against flooding, such as mangrove swamps.

### Reviewing Concepts

- Chlorine-containing compounds can destroy ozone, removing the Earth's protection from radiation.
- Carbon compounds contribute to trapping solar energy, warming the planet.

# 46.5 Deforestation and the Loss of Food Webs

The removal of all the tree cover from a forested area is called **deforestation.** Both tropical and temperate forests are shrinking at an average of about 1% annually. Satellite data indicate that

**FIGURE 46.7**   **Tropical Deforestation.**   Dark green areas in this satellite image of a forest in Brazil represent natural forest, whereas the light yellowish and pink areas indicate areas of deforestation.

tropical deforestation affects more than 17.4 million acres (7 million hectares) per year (**figure 46.7**), mostly in South America, Africa, and Southeast Asia. Nearly half of the world's moist tropical forests have already been cleared.

The disappearance of the native North American temperate forest has clearly paralleled settlement by Europeans. In the first half of the eighteenth century, settlers began clearing the land from east to west to create farmland and obtain fuel. Logging and draining of the swamps in the lush Tidewater region of Virginia and the Carolinas began in the late 1760s. By 1840, settlers had reached Louisiana and Arkansas. Following the Industrial Revolution, forests were cleared for timber and turpentine and to make room for railroads, towns, and fields of soybeans, cotton, and other crops. By 1880, virgin forest remained in less than 35% of the South. Today, although vast areas of managed pine forests and plantations occupy the region, less than 1% of the original temperate forest survives (**figure 46.8**).

When trees die, food webs topple. Destroying the *Casearia corymbosa* tree in the tropical rain forest of Costa Rica, for example, starves many of the 22 bird species that eat its fruits. Other trees whose seeds these birds disperse are also threatened, as are monkeys and other animals that eat and live in the trees.

Most tropical forests are cleared to make room for subsistence agriculture. Ironically, the same soils that support the lush tropical rain forest yield poor agricultural returns. The warm temperatures promote rapid decomposition of organic matter, while heavy rains deplete the soil of nutrients. Therefore, most of the nutrients in the intact forest are bound up in the plant

1620

1920

**FIGURE 46.8    Temperate Deforestation in the United States.** Since 1620, nearly all the native forests in the continental United States (dark green areas on the map) have been cleared.

growth. Once native plants are cleared to make room for crops or grazing animals, the nutrient-poor soils harden into a cementlike crust.

Deforestation contributes to diverse social and environmental problems. Tropical forest destruction threatens many native people as well as other inhabitants. Forests also harbor a tremendous diversity of plant, animal, and microbial life, which may become extinct as the forests are destroyed. Deforestation promotes soil erosion, which reduces soil fertility and contributes to water pollution. Transpiration by forest plants contributes abundant water vapor to the atmosphere, affecting global climates. Burning these forests not only removes an important component of the global water cycle, but also releases stored carbon into the atmosphere, contributing to the greenhouse effect.

## Reviewing Concepts

- Lumber harvests and practices have destroyed forests, eliminating habitats and leading to soil erosion and water pollution.

# 46.6    Desertification and the Expansion of Lifelessness

The expansion of desert into surrounding areas is called **desertification.** It can be a natural process, a human-driven one, or a combination of the two. Current desertification problems in Africa and Asia illustrate how short-sighted agricultural practices rob land of water. In Africa, the problem is cattle grazing; in Asia, it is growing cotton.

Africa's Sahara is the driest and largest desert in the world. Its dust travels as far away as Florida. Along the Sahara's southern edge and in countries farther south, natural drought and human activities are eroding productivity. The most severely affected nations are those immediately south of the Sahara in an area called the Sahel. The soil here is so dry that seeds cannot germinate. Wells are dry. Birds, reptiles, and desert shrubs are rare, and some parts of the Sahel appear to be lifeless.

As in many ecological events, desertification of the Sahel is the culmination of several steps. After 10 years of plentiful rain, a great drought began in 1968. As grasses shriveled and trees died in the Sahel, farmers let their cattle browse farther to find food. The animals ate everything growing near oases, where grasses have deep roots that tap into the low water table. When their leaves were eaten, the plants redirected their energy into replacing the leaves rather than extending their roots. The grasses died of lack of water. In place of the grasses grew plants that cattle would not eat. The short roots of the new plants, plus the movements of the cattle, eroded the soil. At first, patches of desert appeared where the plants died and grazing cattle compacted the soil. Then the patches enlarged and joined, and soon, the desert had spread. Farmers cut down more forest to create more grazing land, which is destined to become desert, too, if these practices continue.

Drought is not unusual in this part of the world, and desertification is not limited to Africa. In many nations, overgrazing, improper irrigation, unregulated vehicle activity, and other human impacts are degrading arid land. In the former Soviet Union, the catalyst for desertification was irrigating land to grow cotton. As a result, the city of Nukus, in the Republic of Karakalpakstan, unexpectedly turned into a dust bowl. When the wind picks up, residents hurry indoors to avoid the toxic dust storms that carry residues of years of uncontrolled fertilizer and pesticide use. Rates of respiratory illness and anemia are high.

The death of central Asia's Aral Sea—once the world's fourth-largest lake—began in the 1920s when its water was first used to irrigate cotton. In the 1950s, the program accelerated, with the Soviet government building a system of canals to tap and divert the water from the two rivers that fed the sea. But the water was being removed much faster than it was replenished by rainfall in this naturally dry area, and by the early 1960s, the Aral Sea was clearly shrinking. Today, it is but three small remnants of what it once was, with 9 million acres (3.6 million hectares) of seabed exposed (**figure 46.9**). Only a fifth of the original volume of water remains, and it has grown so salty from the fertilizer and pesticides used on nearby crops that all 24 native fish species have become locally extinct. Salt deposits and stranded boats can be seen, painting an eerie landscape.

The entire Aral Sea ecosystem and its surrounding lands are on their way toward lifelessness or selection for salt- and

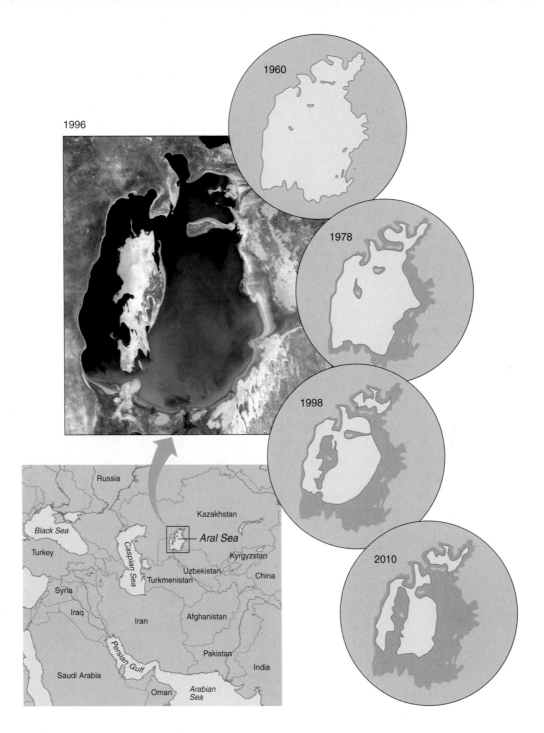

**FIGURE 46.9   A Desert Spreads.**   Due to a short-sighted irrigation plan, the once-huge Aral Sea is disappearing into desert. Reprinted with permission from Richard Stone, "Coming to Grips with the Aral Sea's Grim Legacy" in *Science,* April 2, 1999, vol. 284. *Source:* Philip Micklin, Western Michigan University, Copyright © 1999 American Association for the Advancement of Science.

drought-tolerant species. Efforts are underway to correct the damage or at least prevent it from worsening. Total restoration of the Aral Sea to a productive aquatic ecosystem would take 50 years of no irrigation. Because some farming must continue to feed human residents, smaller volumes of water are being diverted from the tributaries. The hope is that the water table will rise and the Aral Sea might begin to refill, albeit slowly, with rainwater.

## Reviewing Concepts

• Short-sighted agricultural practices combined with natural weather events and climatic conditions have led to the conversion of fertile cropland to desertlike lands that are difficult to reclaim.

• Species are lost, and food sources are no longer available.

## 46.7   The Global Water Crisis

Pollution with toxic chemicals, excessive nutrient input, and overharvest of aquatic resources threaten ocean and freshwater ecosystems.

### Chemical Pollution Destroys Communities

Water can become polluted with chemicals that adversely affect organisms. Factories, municipal wastewater treatment plants, the shipping industry, and agriculture are important sources of surface water pollution. Groundwater contamination often comes from landfills, leaking storage tanks, and agricultural operations.

The chemical pollutants that affect rivers, lakes, and groundwater are very diverse. Poisons such as cyanide and organic mercury compounds from mining operations kill key members of food webs. Over a longer term, some persistent organic pollutants such as PCBs and polycyclic aromatic hydrocarbons (PAHs) may cause cancer and disturb reproduction in some species. An essential nutrient such as phosphorus may become a pollutant if it accumulates. On land, nitrates from cattle feedlots seep into the soil and move readily into groundwater; high nitrate concentrations in drinking water can kill babies. **Table 46.2** lists some examples of chemical water pollutants, both organic and inorganic. • **lakes and ponds, p. 902**

Pollutants may also be as seemingly innocuous as sediments, which may reduce photosynthesis by blocking light penetration into water. Even heat can be a pollutant. Hot water discharged from power plants reduces the ability of a river to carry dissolved oxygen, affecting fishes and other aquatic organisms.

Organisms living in polluted areas are often exposed to several toxins. The Rhine River, which flows through Western Europe, has until recently been a chemical soup of industrial and

**FIGURE 46.10   The Polluted Rhine River.** Efforts to fight the 1986 fire in a chemical warehouse in Switzerland caused tons of poisons to flow into the already-polluted Rhine, leading to massive fish kills. This man is scooping up dead eels.

agricultural pollutants that have flowed in since Europe recovered from World War II. The Rhine was contaminated with detergents, chlorine-containing pesticides and bleaching products, petroleum products, sewage, heavy metals, nitrogen, phosphorus, and other chemicals. In 1986, following a fire in a Swiss chemical warehouse, tons of poisonous liquids flowed into the Rhine and killed hundreds of thousands of fishes (**figure 46.10**). In response, the Rhine Action Program identified 30 types of pollutants and then developed regulations and sewage treatments that had, by 1995, lowered the levels of discharge of all the pollu-

### TABLE 46.2

**Examples of Chemical Water Pollutants**

| Organic | Inorganic |
|---------|-----------|
| Sewage | Chloride ions |
| Detergents | Heavy metals (mercury, lead, chromium, zinc, nickel, copper, cadmium) |
| Pesticides | Nitrogen from fertilizer |
| Wood-bleaching agents | Phosphorus from fertilizer and sewage |
| Hydrocarbons from petroleum | Cyanide |
| Humic acids | Selenium |
| Polychlorinated biphenyls (PCBs) | |
| Polycyclic aromatic hydrocarbons (PAHs) | |

**FIGURE 46.11** **Oil Spills Can Cause Massive Destruction.** The crude oil fouling this beach in the Kenai Fjords area of Alaska was spilled hundreds of miles away when the Exxon *Valdez* ran aground in Prince William Sound.

tants by at least 50%. Salmon are back in the Rhine for the first time since the 1950s, and the river appears to be recovering.

Devastating chemical pollution blanketed Alaska's Prince William Sound on March 24, 1989. The sound once harbored millions of birds, land mammals, marine mammals, invertebrates, and fishes. An oil tanker, the Exxon *Valdez,* ran aground on Bligh Reef, spilling 11 million gallons (more than 50 million liters) of oil (**figure 46.11**). Some well-meant attempts to clean the mess backfired. For example, of 14,000 sea otters, 2,800 perished directly from the oil. Of the 1,200 others that were captured, tranquilized, and scrubbed, only 197 were alive 2 months later to be released into a still-polluted inlet. Some of the released otters carried a deadly herpesvirus to healthy wild animals.

Scrubbing oil-covered rocks with very hot water under high pressure also had unexpected repercussions. Although the treatment cleansed rocky habitats, it harmed many organisms that had survived the oil pollution. This was the fate of *Fucus,* a light brown alga also called rockweed, that had dominated parts of the sound. *Fucus* has grown back sporadically in some places, but it has yet to cover many areas in the shaggy carpet that it did before the oil spill.

Just as with air quality, the United States has seen improved water quality in recent decades. Among other provisions, the Clean Water Act of 1972 required nearly every city to build and maintain a sewage treatment plant, dramatically curtailing discharge of raw sewage into rivers and lakes. The 1987 Water Quality Act followed up on the Clean Water Act, regulating water pollution from industry, agricultural runoff, sewage overflows during storms, and runoff from city streets. Although water-quality problems still exist in the United States, many of the nation's surface waters have largely recovered from past unregulated discharge of phosphorus, other nutrients, and toxic chemicals.

## Alteration of Waterways Can Have Unintended Effects

Because a river provides diverse habitats and can cover a large area, human intervention can have profound effects on life. For example, along the banks of the Mississippi River, levees built to prevent damage from flooding alter the pattern of sediment deposition. Nutrients and sediments that once spread over the floodplain during periodic floods are now confined to the river channel, which carries them to the Gulf of Mexico. There, the nutrients feed protista, causing their populations to bloom. Red and brown tides, the result of such blooms, may kill fishes, manatees, and other sea life.

Damming alters river ecosystems by changing water temperature and nutrient levels, which affects protista, aquatic insects, and fishes, creating great gaps in food webs. Dams can flood some areas and rob others of water. The number of large dams worldwide has increased from 5,000 in 1950 to more than 45,000 today. Channelization, or straightening the course of a meandering river, has ecological consequences, too. It increases the water's flow rate, which erodes channel sediments and carries them downstream, where they can choke out stream communities.

## Estuaries Are Vulnerable to Pollution and Development

Estuaries, as links between freshwater and salt water, are pivotal ecosystems. Many fishes and shellfishes spend part of their lives in estuaries, and diverse algae and flowering plants support the food web. Yet humans have drained and filled estuaries to build houses and dumped garbage and other pollutants in their waters. Laws now protect estuaries in some areas.

Pollutants pour into estuaries from rivers and streams that drain surrounding watersheds. Animal droppings, human sewage and medical waste, motor oil, fertilizer from large farms and millions of lawns, plus industrial waste, oil spills, and garbage, all may end up in estuaries. In salt water, chemical reactions encapsulate pollutants into particles, making them heavy enough to sink. Here they remain, unless disturbed by human action or extreme weather.

Once contaminated sediments are disturbed, a deadly chain reaction ensues. Released nitrogen and phosphorus trigger algal and dinoflagellate blooms (see figure 22.10), which sometimes produce toxins that poison aquatic animals. Decomposers feeding on dead algae and dinoflagellates deplete the water's dissolved oxygen.

The 64,000-square-mile (166,000-square-kilometer) Chesapeake Bay basin, on the eastern coast of the United States, is changing in response to an influx of sewage, agricultural fertilizers, heavy metals, pesticides, and oil. Each day, millions of gallons of nutrients flow into the bay, causing phytoplankton populations to bloom and changing the composition of their communities. Dinoflagellate and green algae populations are growing, while the numbers of diatoms are falling. Reduced light penetration has killed eelgrass. Eutrophication has sharply reduced the oxygen content of the deep water. As a result of these changes in food sources, the fish and crustacean populations are declining.

Cleanup efforts are helping to restore the Chesapeake's lost biodiversity. New sewage treatment facilities are being built and old ones repaired. Floating plants introduced in some areas use the excess nitrogen. So far, efforts to reduce the phosphorus and nitrogen flowing into the estuary have succeeded in some areas.

## Coral Reefs Are Critically Endangered

Besides being home to a quarter of all marine species, coral reefs play other roles in the environment. They serve as barriers between powerful ocean currents and beaches, halting erosion. Some coral reefs filter outgoing material from mangrove swamps, controlling nutrient flow to the oceans. Coral reef inhabitants feed millions of people, and chemical compounds derived from reef organisms are the basis for many drugs.

Threats to reef ecosystems are both natural and human-induced. Several violent hurricanes in recent years have destroyed parts of coral reefs. Infectious diseases can shrink populations, affecting food webs. For example, in the Caribbean, local extinction of the sea urchin *Diadema antillarum* led to overgrowth of their algal food. The algae clogged and killed many coral reefs. Also in the Caribbean, viral infections have killed many elkhorn and staghorn corals.

Human activities have harmed coral reefs. In "cyanide fishing," divers squirt cyanide into the crevices within reefs, momentarily stunning the fishes within so that they emerge and are easily caught (**figure 46.12**). Cyanide fishing provides tropical fishes for collectors and for diners in Asia who like to select their meals from tanks displayed in restaurants. "Dynamite fishing" uses underwater explosives to blast fishes out of the water.

Overbuilding of resort facilities along shorelines pumps municipal waste into coastal waters, joining nutrient-packed sediment and runoff from agriculture. As populations of algae explode, they smother the coral polyps in the living layer of the reef. In Phuket Island, Thailand, such nutrient pollution invited a new predator to the local coral reef, the crown-of-thorns sea star.

**FIGURE 46.12    Cyanide Fishing Is Killing Corals.**    In the Philippines, divers use cyanide to stun fishes that live in coral reefs so that the fishes can be caught alive. The corals may die.

The algal bloom fed the sea star's larvae. As adults, the animals evert their stomachs through their mouths, releasing digestive enzymes that dissolve the coral's polyps; they then absorb the coral animals' remains.

Conservation efforts are saving some coral reefs, which have lost 27% of their mass since 1950. Since 1975, the Great Barrier Reef has served as a model to other coral reef areas. The Barrier Reef is divided into four sectors, and each of these is subdivided into three areas: open only to scientists, open to divers and snorkelers who can observe but not touch the coral, and general use areas where some other activities are permitted. The International Coral Reef Initiative began in 1995 to try to coordinate efforts, from local to global levels, to preserve these ecosystems. Today, researchers at the Barrier Reef and in other threatened ecosystems increasingly use marine reserve modeling, in which mapping software enables planners to predict the effects of specific interventions intended to preserve species diversity. The software uses databases assembled from field data.

## Ocean Pollution Is Increasing

The destruction of estuaries and other coastal habitats by urbanization, housing, tourism, dredging, mining, and agriculture affects life in the oceans too. Many marine animals live parts of their lives in coastal areas, and the loss of these habitats can threaten populations of commercially important species, such as bluefin tuna, grouper, and cod. These populations are further depleted as fishing crews harvest the adults faster than the fishes can reproduce. In addition, many marine mammals, seabirds, sea turtles, and nontarget fish species are killed accidentally as they are caught up in the nets set for the target species (**figure 46.13**).

Toxic chemicals and nutrients from shipping, urban and agricultural runoff, oil drilling, and industrial and municipal wastewater discharges pollute oceans. In the top layer of the ocean, petroleum-based pollutants choke out life from above. Fishing nets and plastic also entrap and kill millions of seabirds and marine mammals, including 50,000 Alaskan fur seals. Some birds build their nests with plastic and feed their nestlings plastic bits. Sea turtles mistake floating blobs of plastic for their natural food, jellyfishes, and swallow them. The plastic lodges in their intestines and kills them. Some organisms float on plastic debris to new areas. ● **sea surface microlayer, p. 882**

Human activities, El Niño, and global warming have contributed to an increase in infectious disease among some ocean species. Victims include kelp, sea mammals, fishes, corals, sea grass, clams, and sea urchins, and the infectious agents include viruses, bacteria, protista, fungi, and nematode worms. Some infections appear to be new, while others have spread because exposure to pollutants impairs host immune systems. Infections have also been transferred to new marine hosts. For example, a canine distemper virus transmitted from sled dogs has killed seals in Antarctica and Siberia. Global warming has shifted the migration routes of birds whose excrement passes influenza viruses, infecting seals and whales.

Like other biomes, oceans have natural defenses. Many toxins remain sequestered if undisturbed. The sheer volume of the

**FIGURE 46.13    Ocean Problems.**    This green turtle is trapped in a fishing net.

oceans dilutes some pollutants, and microorganisms naturally degrade certain compounds. People are attempting to aid this natural healing. Cardboard devices are replacing plastic beverage carriers, ships are no longer supposed to dump waste at sea, cattle are being kept from rivers, and many people are more aware of what they discard and pour down the drain.

### Reviewing Concepts

- Human activities are fouling the waters and destroying and displacing populations before we have a chance to fully understand the scope of biodiversity in aquatic ecosystems.
- Pollution is leading to the loss of usable water resources and affecting the world's oceans.

# 46.8    Loss of Biodiversity

Genetic diversity in a population provides insurance against disaster—should the environment change, individuals with certain adaptive phenotypes (and their underlying genotypes) are more likely to survive. On a broader scale, species diversity in ecosystems provides some assurance that at least some species will survive environmental change.

## Diversity Contributes to Ecosystem Stability

The **diversity-stability hypothesis** holds that the more diverse and the greater the number of species in an ecosystem, the more it will be able to withstand a threat. Long-term experiments conducted in a Minnesota grassland support the hypothesis, but not all ecologists agree that the experiment accurately mirrors a natural ecosystem.

Grassland threats include fire, frost, hail, drought, herbivores, and extreme temperatures. To assess an ecosystem's response to disaster, researchers measure the biomass of plants and catalog the number of species. Starting in 1982, researchers monitored 207 plots of land in the Minnesota grassland and determined plant biomass to measure ecosystem stability (**figure 46.14**). They controlled the number of species by supplying specific amounts of nitrogen in fertilizer—the more nitrogen, the more plant species a particular plot supports.

After a severe drought in 1987–88, the researchers measured the plant biomass in the plots. Plots with the most species lost half their biomass, but plots with the fewest species lost seven-eighths of their biomass! The conclusion based on this experimental ecosystem: when ecosystems are stressed, biodiversity provides stability. Under favorable environmental conditions, however, a species-poor area will also produce a large biomass.

Even though other experiments using controlled ecosystems have confirmed the finding that species diversity promotes ecosystem stability, and the idea of a "diversified portfolio" or "not putting all your eggs in one basket" seems logical, the hypothesis remains controversial. One criticism of the Minnesota grassland work is that applying nitrogen to control the number of species selects for those that grow the fastest. This might occur at the expense of slower-growing plants that have drought-resistance genes and that might have a better chance in a natural ecosystem, with its unpredictable mixes of stresses.

Another criticism is that the conclusion that species number alone ensures ecosystem survival is an oversimplification. An ecosystem remains stable not necessarily because of a large number of species, but because of their diverse roles. A forest of plants of very different heights, for example, might withstand fire better than a forest of trees of similar height, even if both areas include

**FIGURE 46.14    The Diversity-Stability Hypothesis.**    Each plot in this biodiversity experiment is planted with different numbers and combinations of native prairie plant species. The plot in the foreground has four plant species; the plot just to the right has 16. Mowed areas are walkways between the plots.

the same number of species. Another oversimplification of equating biodiversity with ecosystem stability is that species do not equally influence composition of a community. Keystone species are by definition more influential than other members of the same community. Ecologists are continuing the monitored grassland study, along with "combinatorial biodiversity experiments" that use computer simulations of species changes.

## Biological Invaders Lessen Diversity and Use Up Resources

Habitat destruction is the primary cause of diminishing biodiversity. The second most common cause is biological invasion—the introduction of nonnative species, which then displace the native species. Introducing brown tree snakes to the island of Guam decimated native species (see chapter 43's opening essay), and section 44.1 describes the invasion of zebra mussels in the Great Lakes. Similarly, 75% of Florida's current plant life is not part of the lush ecosystem it once had, but rather is composed of invaders introduced from agriculture and urbanization.

Humans are the conduits of species invasions when we take plants and animals with us when we settle new areas. Examples include crops, pets, and livestock, plus the microorganisms and parasites on and in them, and stowaways such as rodents and insects on ships. A successful invasion, though, isn't highly likely. Of every 100 species introduced, about 10 survive in the short term, competing with native species, and only 1 persists to take over a niche. Invasions are more likely to occur in areas that have been disturbed—which suggests a dangerous link to habitat destruction. They are also more common on islands, where diversity is naturally low.

A successful invasion can have severe repercussions on an ecosystem. One famous case of species invasion is that of Australian rabbits. In 1859, wealthy Thomas Austin imported a few rabbits to his estate so that he might enjoy hunting them. He apparently did not shoot them all, and soon they had taken over and spread, eating much of the vegetation and displacing native animals. The government took action in the 1950s by unleashing the myxomatosis virus, which kills only rabbits. When that virus became ineffective by 1995, they substituted another. Today, the arid lands of Australia show signs of recovery from the rabbit attack.

A more recent example of species invasion is occurring now in Europe. In the early 1990s, a plane from the United States unintentionally brought the western corn rootworm, actually a beetle larva, to war-torn Yugoslavia. By 1995, the rootworms had spread to Croatia and Hungary, and today they are in Bulgaria and Italy, too. By the time the people noticed that their corn crops were disappearing, it was too late to eradicate the pest.

Plants may also be invasive. At least 5,000 nonindigenous plant species live in U.S. ecosystems. One example is hydrilla (*Hydrilla verticillata*), which chokes waterways in Florida, alters nutrient cycles, affects aquatic animals, and reduces recreational use of lakes and rivers (**figure 46.15**).

In Hawaii, the state government is addressing species invasions by creating refuges for native species. At the Kulani correc-

**FIGURE 46.15    Biological Invaders.** *Hydrilla verticillata* clogs this waterway in central Florida. This species came to the United States from southeast Asia as an aquarium plant and quickly became a nuisance when residents put the imported plant into local waters.

tional facility, for example, conservationists have gathered the Mauna Loa silversword plant and tree ferns, protecting them from voracious leaf-eating nonnative pigs. The refuge also houses several species of songbirds, whose numbers are plummeting from a triple attack—diminishing habitat, avian malaria, and invading species that eat their eggs.

The dangers of introduced species are twofold. On an ecosystem-by-ecosystem basis, natural selection will favor those species that grow and reproduce the fastest and are generalists—whatever their origin. This is the case for the red maple that is taking over the temperate deciduous forest in the eastern United States because it can tolerate a wide range of conditions. More philosophically, introducing species is homogenizing the biosphere, gradually transforming it from a collection of distinctive interacting areas whose species arrived by their own means—such as birds flying to islands—to a human-directed sameness.

● **temperate forests, p. 898**

## Reviewing Concepts

- Human-directed habitat destruction and invasion of nonnative species are hastening the shrinking of the biosphere; a once-diverse living landscape is becoming a disjointed collection of species-poor remnants.
- Diverse communities are more stable than those without diversity.
- We must preserve biodiversity.

# 46.9   The Resiliency of Life

Life on Earth has had many millions of years to adapt, diversify, and occupy nearly every part of the planet's surface. It would be very difficult to halt life on Earth completely, short of a global catastrophe such as a meteor collision or a nuclear holocaust. The biosphere has survived mass extinctions in the past, and it probably will in the future.

Life has prevailed through many localized challenges—from natural events such as hurricanes, volcanic eruptions, and forest fires to human-caused garbage heaps, oil spills, and pollution. Organisms may perish—many of them—but the plasticity of genetic material ensures that, in most cases, some individuals will inherit what in one environment is a quirk but in another is the key to survival. Even when biodiversity is severely challenged, some species usually persist. Life does not completely end; instead, it changes.

Can we continue to rely on adaptation and diversity to maintain life on Earth? Life may be resilient in an overall sense, yet it is fragile in its interrelationships. Intricate food webs collapse if a new species invades or if existing species decline or vanish, such as the chain of events that disrupted the Aleutian Island ecosystem when whales began to kill otters.

Maybe we have just been lucky, so far, that no single event has decimated so many key players in the game of life that other species could not replace them. But could multiple stresses combine to make Earth too inhospitable for life as we know it to continue? Will lifeless patches of oceans grow and coalesce, as ozone-depleted areas over the poles and stretches of dead desert

do? A single eutrophied or dried-up lake, an acid deposition-ravaged forest, a grassland turned cornfield and shopping mall, an oil-soaked shore—alone they are terrible; together they could begin to unravel the tangled threads that tie all life together.

On the other hand, humans may have the power to undo some of our past mistakes. For example, because of the Endangered Species Act of 1973, some species that faced extinction, such as the bald eagle, are slowly recovering. The last several decades have seen other major pieces of legislation (such as the Clean Air and Clean Water Acts) dramatically improve environmental quality in the United States. Worldwide, governments and private organizations have set aside natural areas to protect them from agriculture and urbanization. Major restoration projects, such as one planned for the Everglades, show that recovery, while costly and difficult, may yet reverse some of the sobering trends described in this chapter.

As the human population continues to grow, pressure on natural resources will increase. Ironically, in developing countries, families may respond by having more children to help obtain increasingly scarce basic necessities such as fuel, food, and clean water. In all parts of the world, local depletion of natural resources, pollution, and occasional natural disasters will continue to damage the environment. For these reasons, one key to reversing environmental decline will be to slow the growth of the human population.

Scientists and politicians, as well as ordinary citizens, carry the heavy burden of protecting the remaining resources for the future, while maintaining a reasonable standard of living for all people. Part of the solution lies within you and how you choose to live. This book has shown you the wonder of life, from its constituent chemicals to its cells, tissues, and organs, all the way up to the biosphere. Do whatever you can to preserve the diversity of life, for in diversity lies resiliency and the future of life on Earth.

## Reviewing Concepts

- Life has adapted to many changes over the millennia.
- Humans have the potential to profoundly affect life on Earth. If we are not careful, our actions may inadvertently cause the loss of much of life's diversity and, ultimately, affect human existence.

# Connections

As we humans have increased our abilities to manipulate our environment and manufacture goods on a mass level, we have had an increasing impact on that environment. If we are not careful, we will be affected as well. One of the lessons learned from studying nature is that we do rely more on other species than we had realized. Recent data has shown that the temperatures in treeless cities are significantly higher than in the surrounding areas. We use energy to lower the temperatures in buildings, adding to pollution and heating other areas of the region. Studies show that planting trees can have a positive effect and provide other benefits as well. The diversity-stability hypothesis needs to become more and more a motivation for our actions. The loss of even one species may be destructive to the entire planet. As we have the ability to alter the planet in ways no species has ever done, we have the responsibility to apply what we have learned to improve life of all kinds.

# Student Study Guide

## Key Terms

acid deposition  *914*
deforestation  *919*
desertification  *920*
diversity-stability
  hypothesis  *925*

global warming  *916*
greenhouse effect  *917*
industrial smog  *914*
ozone layer  *916*

persistent organic
  pollutants  *916*
photochemical smog  *914*
pollution  *913*

sustainability  *912*
ultraviolet radiation
  (UV)  *916*

## Chapter Summary

### 46.1 Sustainability

1. As we alter habitats, we may irreversibly alter ecosystems, with unpredictable consequences.

2. The concept of sustainability holds that we should use natural resources in a way that does not deplete them for future generations.

### 46.2 Air Pollution

3. Pollution is the presence of any substance that exceeds the capacity of an ecosystem or that is unknown to the system.

4. Atmospheric pollution affects breathing and vision.

### 46.3 Acid Deposition

5. Certain atmospheric pollutants combine with water, forming acids, which then accumulate in lakes and water tables.

6. Acid deposition can alter ecosystems and damage populations.

7. Laws are addressing the release of the pollutants.

### 46.4 Ozone Thinning and Global Warming

8. Chlorine-containing compounds can destroy ozone, removing the Earth's protection from radiation.

9. Carbon compounds in the atmosphere trap solar energy, warming the planet.

### 46.5 Deforestation and the Loss of Food Webs

10. Lumber harvests and practices have destroyed forests, eliminating habitats and leading to soil erosion and water pollution.

### 46.6 Desertification and the Expansion of Lifelessness

11. Short-sighted agricultural practices combined with natural weather events and climatic conditions have led to the conversion of fertile cropland to desertlike lands that are difficult to reclaim.

### 46.7 The Global Water Crisis

12. Human activities are fouling the waters and destroying and displacing populations before we have a chance to fully understand the scope of biodiversity in aquatic ecosystems.

13. Pollution is leading to the loss of usable water resources and affecting the world's oceans.

## 46.8 Loss of Biodiversity

14. Human-directed habitat destruction and invasion of non-native species are hastening the shrinking of the biosphere; a once-diverse living landscape is becoming a disjointed collection of species-poor remnants.

15. Diverse communities are more stable than those without diversity and must be preserved.

## 46.9 The Resiliency of Life

16. Life has adapted to many changes in the Earth over the millennia.

17. If we are not careful, human impacts may inadvertently cause the loss of much of life's diversity and, ultimately, affect human existence.

# What Do I Remember?

1. What are major sources of air pollution? Water pollution?

2. How can too much of a nutrient alter an ecosystem?

3. Give an example of an environmental problem that had an immediate effect and one that had a delayed effect.

4. What is sustainability, and how has human activity affected it?

5. Compare and contrast global warming, ozone depletion, and acid deposition and their effects on life.

6. Why is ozone considered a pollutant if its loss in the upper atmosphere is considered to be a great danger to life?

7. Describe the diversity-stability hypothesis.

### Fill-in-the-Blank

1. _____ is the conversion of fertile land to wasteland due to overgrazing.

2. The key component of photochemical smog is _____.

3. _____ is the chemical that causes most of the damage to the ozone layer.

4. The _____ is a nonnative invader of the Great Lakes that is causing great harm.

### Multiple Choice

1. How do introduced species disrupt ecosystems?
   a. They add more nutrients than were originally present.
   b. They displace other species in the community.
   c. They add more individuals than the ecosystem can support.
   d. They disrupt the balance between producers and consumers.

2. Burning coal releases _____ that, when combined with rain, become nitric and sulfuric acids.
   a. nitrogen dioxide and sulfur dioxide
   b. carbon monoxide and ozone
   c. ammonia and sulfur oxide
   d. radioactive wastes and ozone
   e. carbon dioxide and carbon monoxide

3. The loss of the ozone layer would result in
   a. more damage to the DNA of living organisms.
   b. more acid deposition in lakes.
   c. increased temperatures worldwide.
   d. accumulation of more carbon monoxide in the atmosphere.
   e. all of these.

4. Which of the following is considered to be the major greenhouse gas?
   a. ozone
   b. carbon monoxide
   c. chlorofluorocarbon
   d. carbon dioxide
   e. sulfur dioxide

5. What is the major impact of deforestation?
   a. the collapse of food webs
   b. an increase in carbon dioxide worldwide
   c. global warming increases
   d. an increase in photochemical smog
   e. a decrease in overall rainfall

# Appendix A

# Answers to Student Study Guide Problems

## Chapter 1

### What Do I Remember?

1. organization; energy use and metabolism; homeostasis; reproduction, growth and development; irritability (response to stimuli); and adaptation

2. Sexual reproduction produces nonidentical offspring.

   Asexual reproduction never uses more than one parent.

   Asexually reproduced offspring are genetically identical to the parents.

   Sexual reproduction adapts species to new environments.

   Asexual reproduction can never result in adaptation.

3. Natural selection gives an advantage to those with certain heritable differences in reproducing, obtaining food, or in overall survival. Over time, these differences accumulate to the point that the offspring are unable to mate with members of the original population, and a new species is born.

4. Taxonomic names for species reflect the evolutionary relationships between and among organisms. The names make it easier to distinguish and identify species that look very much alike, while also providing information regarding their relationship to one another. These names are universal and are not subject to the confusion seen with nonscientific naming schemes.

5. Domains describe the fundamental features of the cell type used by organisms. Kingdoms are subcategories of domains.

6. Scientific theories are explanations based upon years of amassed data. Science is based upon drawing the best conclusions possible from repeatable observations. Since it is never easy to determine when absolute truth has been reached, science uses theories to summarize the best understanding available. Likewise, a hypothesis is a prediction of the outcome of an experiment based upon the best observations and conclusions available at the moment. Hypotheses suggest the next experiment to try to gather additional data on a particular subject of inquiry.

7. A scientific experiment must have the necessary controls to eliminate alternative explanations. They must also have a measurable outcome that is unambiguous and be based on current understanding of the principles being explored. Some studies use "blind" controls to eliminate human influences on the outcomes.

8. Fire can reproduce and grow, it converts energy from one form to another in maintaining itself and a particular "internal" environment, it responds to stimuli such as wind or water, it has a rudimentary organization, but lacks cells or the ability to evolve.

9. As with the difference between a theory and a guess, science is based on centuries of discovery, with each generation building on the work of those before it. Sometimes information once thought to be fact is discarded in favor of better-founded ideas and new discoveries. There is no declared "end point" where science suddenly knows everything.

10. Science can apply only to things that can be tested and produce observable, repeatable results. Many of the realms of human endeavor do not lend themselves to scientific exploration because they do not yield to rigid, logical exploration with results that can be repeated by anyone wishing to know. Anything that is prone to subjective conclusions is likely to be outside the scope of science.

### Fill-in-the-Blank

1. DNA, or mutations or heritable differences
2. fungi
3. hypothesis
4. diversity or adaptation
5. metabolism

### Multiple Choice

1. a. the ability to convert sunlight energy into chemical energy
2. d. one species giving rise to a different species
3. b. summarizes many years of accumulated experimental results
4. b. a group of organisms that can reproduce
5. b. cell, tissue, organ, organism, population

## Chapter 2

### What Do I Remember?

1. a. atom—the smallest unit of matter; these combine to form molecules, and each has a unique character based on the element it represents

   b. element—different forms of atoms based on the number of protons they contain

   c. molecule—a complex of atoms bonded together

   d. compound—any substance that contains two or more elements

   e. isotope—different versions of a particular element having slightly different numbers of neutrons; these have the same chemical characteristics, but may break down to other elements and emit energy in the form of radioactivity

   f. ion—an atom with a mismatched set of protons and electrons, giving it an overall positive or negative charge and a very different chemical property than the element from which it originates.

2. $C_{10}H_{16}O_3N_2S$

3. Covalent bonds are the strongest type and involve the sharing of pairs of electrons. Of medium strength, ionic bonds are attractions between charged atoms, whether they are alone or part of a molecule. Weakest of the bonds, hydrogen bonds are charge-to-charge attractions between slightly charged portions of molecules. Van der Waals interactions involve attractions between transient oppositely charged regions of molecules. Water interactions occur when hydrophobic molecules combine to minimize direct interaction with water.

4. Any dissolved substance is a solute, and a solvent is the medium it is dissolved into, forming a solution.

5. Water can be thought of as equal parts of hydrogen ions and hydroxyl ($OH^-$) ions. If there is an unequal amount of hydrogen ions, this constitutes an acid or a base. The relative proportion of hydrogen ions to OH ions is indicated by pH. Each unit of pH represents a 10-fold difference in hydrogen ion concentration from that found in water.

6. All organisms generate acids as a by-product of metabolism. Since enzymes can be destroyed at low pH, organisms must have a method for reducing the amount of free hydrogen ions in a cell. Buffer systems absorb hydrogen ions.

### Fill-in-the-Blank

1. 5
2. $(12 \times 6) + (12 \times 1) + (16 \times 6) = 180$
3. isotope
4. ion
5. covalent

### Multiple Choice

1. b. protons
2. a. the octet rule
3. b. 100 $H^+$ for every $OH^-$
4. c. 7
5. a. high heat capacity

## Chapter 3

### What Do I Remember?

1. carbohydrates, lipids, proteins, and nucleic acids

2. Carbohydrates have the overall formula $(CH_2O)n$ and occur as monosaccharides, disaccharides, oligosaccharides, and polysaccharides. Starch and cellulose have the same formula but different chemical properties.

   Lipids are mostly carbon and hydrogen, with few charged groups. As a result they are mostly hydrophobic. Most are found as fatty acids or complexes of two or three fatty acids—phospholipids and triglycerides. Some are multiring complexes such as cholesterol.

   Amino acids are mostly carbon, hydrogen and oxygen plus nitrogen and sulfur. Proteins are polymers of amino acids and can be anywhere from a few to thousands of peptides in length. They can be hydrophobic or hydrophilic.

   Nucleic acids are polymers of nucleotides, which are composed of carbon, hydrogen, oxygen, and nitrogen. They contain phosphate groups that serve as energy storage for the cell or as an integral part of the backbone of long polymers such as DNA or RNA. They differ in the sugar—ribose or deoxyribose—and the nitrogenous base—A, C, T, G, or U.

3. Carbohydrates are structural elements and also serve to identify cells and store energy. Lipids form barriers to create compartments having unique environments, such as organelles. They also store energy.

   Proteins can be structural elements, enzymes, membrane channels, defensive molecules or organization elements.

   Nucleic acids are information storage, and nucleotides function as energy carriers.

4. Proteins are very diverse because they consist of variable combinations of 20 building blocks—amino acids. Carbohydrates and lipids are each based on essentially one type of molecule: a sugar or a fatty acid, respectively.

5. The shape of a protein determines its function.

6. DNA is information storage; RNA serves as a temporary copy and can function as an enzyme. DNA contains deoxyribose and the base thymine, whereas RNA contains ribose and the base uracil.

7. See table 3.1.

### Fill-in-the-Blank

1. monomer
2. $(CH_2O)n$
3. tertiary or 3°
4. ATP
5. a carboxylic acid group

### Multiple Choice

1. b. a triple covalent bond
2. a. storing information
3. c. ribose
4. d. all of these
5. a. glycerol and three fatty acids

## Chapter 4

### What Do I Remember?

1. A cell is a highly organized collection of molecules with all of the characteristics of life. A cell is the smallest unit of life, and cells come from existing cells.

2. A protein is a molecule, a polymer of amino acids. A molecule is composed of bonded atoms. Specific molecules in a specific organization comprise an organelle, which is a component of a cell. All organisms contain cells, which are composed of molecules.

3. See figure 4.4.

   Archaea—single-celled, found in hostile environments, membranes of isoprenes, no organelles

   Bacteria—single-celled, found everywhere, fatty acid membranes, cell wall of peptidoglycan, no organelles

   Eukaryotes—single-celled and multi-celled organisms, fatty acid membranes, some lack cell walls, others composed of chitin or cellulose, multiple organelles.

4. See figure 4.11.

5. Peroxisomes contain enzymes made in the cytoplasm, then transported to the organelles.

Lysosomal enzymes are made on the rough ER. Peroxisomal enzymes help the cell use oxygen; lysosomal enzymes are useful only for degrading biomolecules at low pH.

6. Organelles are formed by membranes enclosing a specific set of proteins, some embedded within the membrane itself. They create and maintain a unique chemical environment for chemical reactions to be separate from the rest of the cytoplasm.

### Fill-in-the-Blank

1. Archaeans
2. scanning probe microscope
3. nucleoid
4. lysosomes
5. oxygen free-radicals
6. cytoplasm

### Multiple Choice

1. a. ribosome
2. b. nucleolus
3. c. endosymbiont
4. a. vacuole
5. a. Archaea

## Chapter 5

### What Do I Remember?

1. Phospholipids, comprised of fatty acids joined by a phosphate plus some additional hydrophilic molecule form a barrier for the cell. Proteins, some that span the cell membrane, form channels and connections with the outside world for the cell, in addition to transporting molecules in and out of the cell. Both are also often modified by the addition of carbohydrates to form glycolipids and glycoproteins. Cholesterol is imbedded in the lipid bilayer and adds fluidity.

2. Through osmosis, water travels from regions of low salt concentrations toward those with higher concentrations. Since salt cannot pass through membranes, water moves in and out of cells and compartments, usually by way of protein channels.

3. Simple diffusion requires no energy input or protein channel assistance. Facilitated transport uses the gradient of a substance to provide movement, but the protein must help the substance across the membrane. Active transport moves substances against their concentration gradient, which requires energy and protein transporters.

4. An antiporter would use the energy of potassium moving out of the cell to drive glucose into the cell. An ATP pump would replenish the potassium gradient.

**anabolism** *an-AB-o-liz-um* Metabolic reactions that use energy to synthesize compounds. 93

**anaerobe** *AN-air-robe* Organism that can live in an environment lacking oxygen. 127

**anaerobic respiration** *an-air-RO-bic res-per-A-shun* Cellular respiration in the absence of oxygen. 127

**analogous structure** *ah-NAL-eh-ges STRUC-cher* Body part in different species that is similar in function but not in structure; evolved in response to a similar environmental challenge. 333

**anaphase** *AN-ah-faze* The stage of mitosis when centromeres split and two sets of chromosomes part. 151

**anapsid** *an-AP-sid* Reptile lacking holes on the side of its skull. 503

**aneuploid** *AN-you-ploid* A cell with one or more extra or missing chromosomes. 218

**angiosperm** *AN-gee-o-sperm* A group of plants that produce flowers and whose seeds are borne within a fruit. 423

**Animalia** The kingdom that includes the animals. 8

**antagonistic muscles** *an-tag-o-NIS-tik MUS-uls* Two muscles or muscle groups that flank a bone and move it in opposite directions. 695

**antenna complex** An interconnected series of membrane proteins in the thylakoid membranes that harvests light energy. 112

**anther** *AN-ther* Pollen-producing body at the tip of a stamen. 554

**antheridium** *an-ther-ID-ee-um* A sperm-producing structure in algae, bryophytes, and seedless vascular plants. 425

**antibody** Protein that B cells secrete that recognizes and binds to foreign antigens, disabling them or signaling other cells to do so. 783

**anticodon** *AN-ti-ko-don* A three-base sequence on one loop of a transfer RNA molecule; it is complementary to an mRNA codon and joins an amino acid and its mRNA instructions. 243

**antidiuretic hormone (ADH)** A hypothalamic hormone released from the posterior pituitary; acts on kidneys and smooth muscle cells of blood vessels to maintain the composition of body fluids. 667, 773

**antigen** The specific part of a molecule that elicits an immune response. 778

**antigen binding sites** Specialized ends of antibodies that bind specific antigens. 786

**antiparallelism** *AN-ti-PAR-a-lel-izm* The head-to-tail relationship of the two rails of the DNA double helix. 232

**antisense technology** Blocking gene expression by binding a complementary RNA sequence to a messenger RNA so that translation cannot take place. 266

**anus** Exit of digestive tract, where feces are released. 743

**aorta** The largest artery; it leaves the heart. 707

**apical complex** Microtubular structure that helps apicomplexa attach to or penetrate host cells. 411

**apical dominance** In plants, the suppression of growth of lateral buds by an intact terminal bud. 573

**apical meristem** The meristem (dividing cells) at the tip of a root or shoot. 516

**apicomplexa** A group of protista whose cells have apical complexes of specialized organelles. 411

**apoplastic pathway** In plants, the path that water takes along the cell walls and extracellular spaces. 545

**apoptosis** *ape-o-TOE-sis* Programmed cell death. 146

**appendicular skeleton** *AP-en-DEK-u-lar SKEL-eh-ten* In vertebrates, the limb bones and the bones that support them. 681

**appendix** A thin sac extending from the cecum in the human digestive system. 753

**aqueous humor** A fluid between the cornea and the lens that helps focus incoming light rays and maintains the shape of the eyeball. 647

**aqueous solution** A solution in which water is the solvent. 24, 68

**arachnid** A terrestrial, eight-legged, chelicerate arthropod, such as a spider. 480

**Archaea** *ar-KEE-a* One of the three domains of life. 386

**archegonium** *arch-eh-GO-nee-um* In bryophytes and some vascular plants, a multicellular egg-producing organ. 425

**arteriole** Small artery in which the smooth muscle layer predominates; important in regulation of blood pressure. 702

**artery** Vessel that carries blood away from the heart. 702

**artificial selection** Selective breeding. 277

**ascomycete** *ass-ko-MI-seet* Fungus that produces spores in sacs called asci. 444

**ascospore** A spore produced after meiosis in an ascus. 451

**ascus** (pl. asci) Saclike structure in which ascospores are produced. 451

**asexual reproduction** Any form of reproduction which does not require the fusion of gametes. 164, 552

**aster** Collection of microtubules that anchors the centriole and spindle apparatus to the cell membrane. 150

**atom** *AT-um* A chemical unit, composed of protons, neutrons, and electrons, that cannot be further broken down by chemical means. 3, 18

**atomic mass** Also called atomic weight; the mass of an atom. 18

**ATP** *See* adenosine triphosphate.

**ATP synthase** *ATP SIN-thaze* An enzyme complex that admits protons through a membrane, where they trigger phosphorylation of ADP to ATP. 114

**atrioventricular (AV) node** Specialized autorhythmic cells that control heartbeat. 707

**atrioventricular valve** Flap of heart tissue that prevents the backflow of blood from a ventricle to an atrium. 706

**atrium** An upper heart chamber that receives blood from veins. 703

**atrophy** *AT-tre-fee* Muscle degeneration resulting from lack of use or immobilization. 697

**auditory canal** The ear canal; funnels sounds from the pinna to the tympanic membrane. 650

**auditory nerve** Nerve fibers from the cochlea in the inner ear to the cerebral cortex. 651

**autoantibody** Antibody that attacks the body's tissues, causing autoimmune disease. 792

**autoimmunity** An organism's immune system attacking its own body. 792

**autonomic nervous system** Motor pathways that lead to smooth muscle, cardiac muscle, and glands. 623

**autopolyploid** *aw-toe-POL-ee-ploid* An organism with multiple identical chromosome sets. 307

**autosomal dominant** The inheritance pattern of a dominant allele on an autosome. The phenotype can affect males and females and does not skip generations. 190

**autosomal recessive** The inheritance pattern of a recessive allele on an autosome. The phenotype can affect males and females and skip generations. 190

**autosome** *AW-toe-soam* A nonsex chromosome. 168

**autotroph** Also known as a producer, an organism that produces its own nutrients by acquiring carbon from inorganic sources. 90, 107, 391, 875

**auxin** *AWK-zin* One of five classes of plant hormone. 570

**axial skeleton** In a vertebrate skeleton, the skull, vertebral column, ribs, and sternum. 681

**axon** An extension of a neuron that transmits messages away from the cell body and toward another cell. 592, 611

# B

**background extinction rate** The steady, gradual loss of a small percentage of existing species through natural competition or loss of diversity. 315

**Bacteria** *bac-TEAR-e-a* One of the three domains of life. 386

**bacteriophage** A virus that infects bacteria. 370

**balanced polymorphism** *BAL-anced POL-ee-MORF-iz-um* Stabilizing selection that maintains a genetic disease in a population because heterozygotes resist an infectious disease. 294

**bark** In woody plants, all tissues outside the vascular cambium. 532

**basal eukaryotes** Diverse protista that evolved earlier than crown eukaryotes. 422

**base** A molecule that releases hydroxide ions into a solution. 25

**basement membrane** A thin, noncellular layer that anchors epithelial tissues to other tissues. 591

**basidiomycete** *bass-ID-eo-MI-seet* Fungus that produces spores on basidia. 444

**basidiospore** A spore produced, after meiosis, on a basidium. 452

**basidium** *bass-ID-ee-um* Club-shaped structure on which basidiospores are produced externally. 451

**basilar membrane** *BA-sill-ar MEM-brane* The membrane beneath hair cells in the cochlea of the inner ear; vibrates in response to sound. 651

**B cell** Lymphocyte that produces antibodies. 779

**benthic zone** *BEN-thick ZONE* The bottom of an ocean or lake. 903, 906

**bilateral symmetry** A body form in which only one plane divides the animal into mirror image halves. 463

**bile** A digestive biochemical that emulsifies fats. 750

**binary fission** *BI-nair-ee FISH-en* A type of asexual reproduction in which a prokaryotic cell divides into two identical cells. 164, 394

**bioaccumulation** Higher concentration of a substance in cells compared to the surrounding environment. 882

**biochemical** Molecule that is important in biological systems. 4

**biodiversity** The spectrum of species or alleles in an area. 7

**bioenergetics** The study of energy in life. 88

**biogeochemical cycle** *bi-o-gee-o-KEM-i-kal SI-kull* Geological and biological processes that recycle chemicals vital to life. 878

**biogeography** The physical distribution of organisms. 275

**biological species** A type of organism that can sexually reproduce only with others of the same type, or a type of organism whose set of characteristics distinguishes it from all others. 302

**biomagnification** *bi-o-mag-nif-i-KAY-shun* Increasing concentrations of a chemical in higher trophic levels. 882

**biomass** Total dry weight of organisms in a given area at a given time. 878

**biome** *BI-ohm* One of several major types of terrestrial ecosystems. 875, 890

**bioremediation** *bi-o-ree-meed-e-AY-shun* Use of organisms that metabolize toxins to clean the environment. 399

**biosphere** The ecosystem of the entire planet. 4, 875

**biotic community** The interacting populations in a given area. 866

**biotic potential** The potential growth of a population under ideal conditions and with unlimited resources. 855

**bipedalism** *by-PEED-a-liz-m* The ability to move on two limbs. 361

**biramous** *bi-RAY-mus* Having two branches. 480

**bird** A tetrapod vertebrate that has feathers, wings, lungs, a four-chambered heart, and endothermy. 505

**bivalve** A mollusk that has a two-part shell. 473

**blade** The flattened region of a leaf. 524

**blastocyst** Stage of human prenatal development in which the morula hollows out and fills with fluid. 810

**blastomere** A cell in a preembryonic animal resulting from cleavage divisions. 810

**blastopore** An indentation in an animal embryo that develops into the mouth in protostomes and the anus in deuterostomes. 465

**blastula** The stage of early animal embryonic development that consists of a hollow ball of cells. 464

**blind spot** Point where the optic nerve exits the retina; it is devoid of photoreceptors. 644

**blood** A complex mixture of cells suspended in a liquid matrix that delivers nutrients to cells and removes wastes. 592, 711

**blood-brain barrier** Close-knit cells that form capillaries in the brain, limiting the substances that can enter. 633

**blood pressure** The force that blood exerts against blood vessel walls. 710

**bolus** *BO-lus* Food rolled into a lump by the tongue. 747

**bond energy** The energy required to form a particular chemical bond. 93

**bone** A connective tissue consisting of osteoblasts, osteocytes, and osteoclasts, embedded in a mineralized matrix. 592, 683

**bony fish** A fish with a skeleton reinforced with mineral deposits to form bone. 498

**boom-and-bust cycle** The repeated and regular increases and decreases in size of some populations. 858

**braced framework** A skeleton built of solid structural components strong enough to resist collapsing. 679

**bract** Floral leaf that protects a developing flower. 527

**brain** A distinct concentration of nervous tissue at the anterior end of an animal. 623

**brainstem** Part of the vertebrate brain closest to the spinal cord; controls vital functions. 626

**bronchiole** Microscopic branch of the bronchi within the lungs. 729

**bronchus** (pl. bronchi) *BRON-kus* One of a pair of tubules that branch from the trachea as it reaches the lungs. 729

**brown alga** Multicellular photosynthetic protist (stramenopile) with unique pigments. 415

**bryophyte** *BRI-o-fite* Collective term for plants that lack vascular tissue, including liverworts, hornworts, and mosses. 423

**buffer system** Pairs of weak acids and bases that maintain body fluid pH. 25

**bulbourethral gland** *BUL-bo-u-REE-thral GLAND* Small gland near the male urethra that secretes mucus. 804

**bulk element** An element that an organism requires in large amounts. 18

**bundle-sheath cell** Thick-walled plant cell surrounding veins; functions in $C_4$ photosynthesis. 118

**bursa** (pl. bursae) *BURR-sa* Small packet in joint that secretes lubricating fluid. 695

## C

**$C_3$ plant** Plant that uses only the Calvin cycle to fix carbon dioxide. 116

**$C_4$ photosynthesis** In plants, a biochemical pathway that helps prevent photorespiration. 118

**caecilian** A type of limbless amphibian. 501

**calcitonin** A thyroid hormone that decreases blood calcium levels. 669

**calorie** The energy required to raise the temperature of 1 gram of water by 1°C under standard conditions. 88

**calyx** The outermost whorl of a flower that contains the sepals. 554

**camouflage** Adaptation, usually coloration, that helps an organism blend in with its surroundings. 868

**CAM photosynthesis** Reactions that reduce the effect of photorespiration by storing carbon during nighttime for use during the day in hot, arid climates. 118

**canaliculus** (pl. canaliculi) *can-al-IK-u-lus* Passageway in bone that connects lacunae. 592

**capacitation** Activation of sperm cells in the female reproductive tract. 806

**capillary** Tiny vessel that connects an arteriole with a venule. 703

**capillary bed** Network of capillaries. 709

**capsid** The protective protein container of the genetic material of a virus. 370

**capsomer** The individual proteins that comprise a viral capsid. 370

**capsule** Firm, sticky layer surrounding some prokaryotic cells. 390

**carbohydrate** *KAR-bo-HI-drate* Compound containing carbon, hydrogen, and oxygen, with twice as many hydrogens as oxygens; sugar or starch. 32

**carbon** An element that is a main component of life forms on Earth and is capable of forming four bonds with other elements including other carbons. 30

**carbon fixation** The process of converting carbon from $CO_2$ into organic compounds. 116

**carbon reactions** Also known as the Calvin-Benson cycle, the reactions of photosynthesis that synthesize glucose from carbon dioxide. 106

**carboxyl group** *kar-BOX-ill GROOP* A carbon atom double-bonded to an oxygen and single-bonded to a hydroxyl group (OH). 36

**cardiac cycle** The sequence of contraction and relaxation that makes up the heartbeat. 707

**cardiac muscle** Type of muscle composed of branched, striated, involuntary, single-nucleated contractile cells. 594

**cardiac muscle cell** Branched, striated, involuntary, single-nucleated contractile cell in the mammalian heart. 687

**cardiac output** The volume of blood the heart ejects in 1 minute. 716

**cardiovascular system** The system of vessels and muscular pump that transports blood throughout the body. 597, 705

**carotid rete** *care-OT-id REE-tee* A configuration of blood vessels that cools the brain. 764

**carpel** *KAR-pel* Structure in a flower that encloses one or more ovules. 554

**carrying capacity** The theoretical maximum number of individuals an environment can support indefinitely. 856

**cartilage** A supportive connective tissue consisting of chondrocytes embedded in collagen and proteoglycans. 592, 683

**cartilaginous fishes** Group of jawed fishes that have a skeleton made of cartilage rather than bone; includes sharks, skates, and rays. 497

**Casparian strip** *kas-PAHR-ee-an STRIP* A waxy region of suberized cell walls in the endodermis of roots. 530

**caspase** Enzyme that triggers apoptosis. 158

**catabolism** *cah-TAB-o-liz-um* Metabolic degradation reactions, which release energy. 93

**catalysis** *kat-AL-i-sis* Speeding a chemical reaction. 38

**catastrophism** A theory of geological change championed by Cuvier that stated that new life comes to an area damaged or destroyed by a catastrophe. 274

**catecholamine** *kat-ah-KOLE-ah-meen* Hormone of the adrenal medulla. 670

**cavitation** A break in the water column in a xylem tube. 545

**cecum** Compartment in digestive tract where plant matter is absorbed. 744

**cell** The structural and functional unit of life. 2, 44

**cell body** The enlarged portion of a neuron that contains most of the organelles. 592, 611

**cell cycle** The sequence of events that occur in an actively dividing cell. 147

**cell membrane** Proteins embedded in a lipid bilayer, which forms the boundary of cells. 44

**cell theory** The ideas that all living matter consists of cells, cells are the structural and functional units of life, and all cells come from preexisting cells. 46

**cellular adhesion molecule (CAM)** A protein that enables cells to interact with each other. 81

**cellular immune response** The actions of T and B cells in the immune system. 780

**cellular respiration** Biochemical pathways that extract energy from nutrient molecules. 89, 722

**cellular slime mold** A protist that is unicellular when food is available; cells come together to form a mobile, multinucleated mass when food is scarce. 407

**cellulose** An insoluble polysaccharide that is a component of plant cell walls; indigestible by humans. 32

**cell wall** A rigid boundary surrounding cells of many prokaryotes, plants, and fungi. 50, 389

**central nervous system (CNS)** The brain and the spinal cord. 611

**centromere** *SEN-tro-mere* A characteristically located constriction in a chromosome. 148, 168

**centrosome** *SEN-tro-soam* A region near the cell nucleus that contains the centrioles. 149

**cephalization** Development of an animal body with a head end. 463

**cephalopod** A type of marine mollusk with a reduced or absent shell and well-developed brain and eyes; includes octopuses and squids. 474

**cerebellum** *ser-a-BELL-um* An area of the brain that coordinates muscular responses. 626

**cerebral cortex** *ser-EE-bral KOR-tex* The outer layer of the cerebrum. 628

**cerebral hemisphere** One of the two halves of the cerebrum. 628

**cerebrospinal fluid** Fluid similar to blood plasma that bathes and cushions the CNS. 633

**cerebrum** The region of the brain that controls intelligence, learning, perception, and emotion. 626

**cervical vertebra** Vertebra of the neck. 683

**cervix** The opening of the uterus into the vagina. 805

**charophyte** Group of green algae thought to be most closely related to terrestrial plants. 423

**checkpoints** Points in the cell cycle when signals halt cell division. 147

**chemical equilibrium** When a chemical reaction proceeds in both directions at the same rate. 94

**chemical reaction** Interactions in which atoms transfer or share electrons, forming new chemicals. 19

**chemiosmotic phosphorylation** Also known as oxidative phosphorylation, the reactions that produce ATP using the energy of a proton gradient. 114, 128

**chemoautotrophs** Organisms that obtain carbon from inorganic sources and energy from chemical bonds. 90

**chemoreception** *KEEM-o-ree-SEP-shun* Smell and taste. 640

**chemoreceptor** Receptor cell responsive to chemicals. 640

**chemotaxis** Directed movement toward or away from a chemical. 390

**chemotroph** *KEEM-o-trofe* An organism that obtains energy by oxidizing chemicals. 391

**chitin** An insoluable polysaccharide found in the exoskeleton of certain arthopods, including insects, and in the cell walls of fungi. 33

**chlorophyll a** *KLOR-eh-fill A* A green pigment plants use to harness the energy in sunlight. 109

**chloroplast** *KLOR-o-plast* An organelle housing the reactions of photosynthesis in eukaryotes. 59, 106

**cholesterol** A sterol that is vital to cell membranes and can be modified to form sex hormones and other lipids. 35

**chondrocyte** *KON-dro-site* A cartilage cell. 592, 683

**chordate** *KOR-date* Animal that at some time during its development has a notochord, hollow nerve cord, gill slits, and postanal tail. 490

**chorion** *KOR-ee-on* A membrane that develops into the placenta. 810

**chorionic villus** Fingerlike projection extending from the chorion to the uterine lining. 813

**choroid** *KOR-oid* The middle layer of the eyeball that is rich in blood vessels. 645

**chromatid** *KRO-mah-tid* A continuous strand of DNA comprising one-half of a replicated chromosome. 148, 168

**chromatin** *KRO-ma-tin* DNA and its associated proteins, which form a dark-staining material in the nuclei of eukaryotic cells. 149

**chromatophore** *kro-MAT-o-for* Structure in an amphibian's skin that provides pigmentation. 501

**chromosome** *KRO-mo-soam* A dark-staining, rod-shaped structure in the nucleus of a eukaryotic cell consisting of a continuous molecule of DNA wrapped in protein. Also, the genetic material of a prokaryotic cell. 208

**chyme** Semisolid food in the stomach. 748

**chytridiomycete** *KI-trid-eo-MI-seet* A microscopic fungus that produces motile zoospores. 444

**ciliate** *SIL-e-ate* Group of protista whose cells are covered with cilia. 413

**cilium (pl. cilia)** *SIL-ee-um* Protein projection from cell that beats, moving cells and substances. 75

**circadian rhythm** *sir-KA-dee-en RITH-um* Regular, daily rhythm of a biological function. 583

**clade** A group of similar organisms sharing a set of traits derived from a common ancestor. 313, 323

**cladogram** A diagram representing the evolutionary relationships of a group of organisms. 323

**class** A taxonomic category below phylum but above order. 7

**classical conditioning** A form of learning in which an animal responds to a formerly irrelevant stimulus. 830

**cleavage** A period of rapid cell division following fertilization but before embryogenesis. 810

**cleavage furrow** The initial indentation between two daughter cells in mitosis. 151

**climax community** A community that persists unless it is disturbed by environmental change. 870

**cline** A change in allele frequencies across a region, sometimes due to geographic barriers. 291

**clitoris** A small, highly innervated bit of tissue that is the female anatomical equivalent of the penis. 805

**cloaca** *klo-AY-ka* An opening common to the digestive, reproductive, and excretory systems in some animals. 744

**clone** A duplicate copy of DNA. Also an individual that is a genetic duplicate of another. 257

**cloning vector** A vector that is used to produce a huge number of DNA clones. 257

**closed circulatory system** A circulatory system that confines blood to vessels. 476, 702

**club moss** A type of seedless vascular plant. 430

**cnidocyte** *NID-o-site* Cell in cnidaria that contains nematocysts. 468

**coccyx** The final four fused vertebrae at the base of the human backbone. 683

**cochlea** The spiral-shaped part of the inner ear, where vibrations are translated into nerve impulses. 651

**coda** Patterns of sounds that are repeated as signals among groups of relatives, such as those of sperm whales. 841

**codominant** *ko-DOM-eh-nent* Alleles that are both expressed in the heterozygote. 197

**codon** *KO-don* A triplet of mRNA bases that specifies a particular amino acid. 243

**coelom** *SEE-loam* A fluid-filled animal body cavity that is completely lined with tissue derived from mesoderm. 465

**coelomate** *SEE-loam-ate* An animal with a coelom. 465

**coenzyme** Organic molecule required as part of a functional enzyme complex. 98

**coenzyme A (CoA)** A carrier molecule that combines with an acetyl group to form acetyl CoA in aerobic respiration. 131

**coevolution** The mutual, direct influence of two species on each others' evolution. 869

**cofactor** Inorganic molecule required for activity of an enzyme. 98

**cohesion** The attraction of water molecules to each other. 24

**cohesion-tension theory** Theory that explains the movement of water in xylem from roots through stems to leaves. 543

**cohort** A group of individuals that begin life at the same time. 853

**collagen** *COLL-a-jen* A connective tissue protein. 592, 683

**collecting duct** A structure in the kidney into which nephrons drain urine. 772

**collenchyma** *kol-LEN-kah-mah* Plant tissue composed of elongated, living cells with thickened walls that support growing regions of leaves and shoots. 518

**colon** The large intestine. 753

**columnar** Tall cells, as in epithelium. 591

**commensalism** *co-MEN-sal-izm* A symbiotic relationship in which one member benefits without affecting the other member. 868

**community** All the interacting organisms in a given area. 2, 850

**compact bone** A layer of solid, hard bone that covers spongy bone. 686

**complement** A group of proteins that assist other immune defenses. 780

**complementary base pairs** Bonding of adenine to thymine and guanine to cytosine in the DNA double helix. 39, 232

**complementary DNA (cDNA)** A DNA fragment produced by reverse transcriptase from an mRNA strand. 257

**compound** A molecule including different elements. 19

**compound eye** An eye formed from several ommatidia. 479, 644

**compression fossil** A fossil made by pressing an organism in layers of sediment. 327

**concentration gradient** Difference in solute concentrations between two adjacent compartments. 69

**concordance** *kon-KOR-dance* A measure of the inherited component of a trait. The number of pairs of either monozygotic or dizygotic twins in which both members express a trait, divided by the number of pairs in which at least one twin expresses the trait. 202

**conditioned stimulus** A new stimulus coupled to a familiar or unconditioned stimulus so that an animal associates the two. 830

**cone cell** Specialized receptor cell in the center of the retina that detects colors. 646

**conformation** *KON-for-MAY-shun* The three-dimensional shape of a protein. 37

**confusion effect** The indecision caused in a predator when viewing a tight grouping of prey. 839

**conidium** (pl. conidia) *kon-ID-ee-um* Asexually produced fungal spore. 448

**conifer** One of four divisions of gymnosperms. 432

**conjugation** *con-ju-GAY-shun* A form of gene transfer in prokaryotes. 166, 395

**connective tissue** Tissue type consisting of widely spaced cells in a matrix; includes loose and fibrous connective tissues, cartilage, bone, fat, and blood. 590

**conservative** DNA replication that retains the complete, original parental DNA while making a new copy. 235

**constant regions** Sequences of amino acids in the heavy and light chains that are the same for all antibodies. 784

**consumer** An organism that must obtain nutrients by consuming other organisms. 4, 875

**contact inhibition** A property of most non-cancerous eukaryotic cells that inhibits cell division when they contact one another. 155

**contraception** The use of practices, devices, or drugs to prevent conception; birth control. 809

**control** An untreated group used as a basis for comparison with a treated group in an experiment. 13

**convergent evolution** Organisms that have similar adaptations to a similar environmental challenge but that are not related by descent. 275

**coral reef** Underwater deposits of calcium carbonate formed by colonies of animals. 906

**cork cambium** A type of lateral meristem that produces the periderm of woody plants. 534

**cornea** A modified portion of the human eye's sclera that forms a transparent curved window that admits light. 645

**corolla** The next-to-outermost whorl of a flower that contains the petals. 554

**corona radiata** *ko-RONE-a raid-e-AH-ta* Layer of cells around the oocyte. 806

**corpus callosum** *KOR-pus kal-O-sum* Thick band of nerve fibers that interconnect the cerebral hemispheres. 629

**cortex** General anatomical term for the outer portion of an organ. Also, the ground tissue between the epidermis and vascular tissue in stems and roots. 523

**cotransport** Two substances crossing the cell membrane together through a single channel complex. 72

**cotyledon** *KOT-ah-LEE-don* Embryonic seed leaf in flowering plants. 561

**countercurrent exchange system** A system of parallel vessels in which fluid flows in opposite directions and maximizes the exchange of substances or heat. 763

**countercurrent flow** Fluid flow in different parts of a continuous tubule in opposite directions, which maximizes the amount of a particular substance that diffuses out of the tubule. 725

**coupled reactions** Two chemical reactions that occur simultaneously and have a common intermediate. 92

**covalent bond** Atoms sharing electrons. 21

**cranial nerve** Peripheral nerve that exits the vertebrate CNS from the brain. 623

**crista** (pl. cristae) *KRIS-ta* Fold of the inner mitochondrial membrane along which many of the reactions of cellular respiration occur. 59, 128

**critical period** The time during prenatal development when a structure is vulnerable to damage. 818

**crop** A part of the digestive tract in birds and some invertebrates that stores or digests food. 744

**crossing over** Exchange of genetic material between homologous chromosomes during prophase I of meiosis. 168

**crown eukaryotes** Eukaryotic organisms, including plants, animals, fungi, and many protista, thought to have diverged after basal eukaryotes. 405

**crustacean** *krus-TAY-shun* A segmented, aquatic arthropod with gills, two-part appendages, mandibles, and antennae. 481

**cuboidal** Cube shaped, as in some epithelial cells. 591

**cuticle** A waxy layer covering the aerial parts of a plant. 518

**cycad** *SI-kad* One of four divisions of gymnosperms. 432

**cyclic adenosine monophosphate (cAMP)** A second messenger formed from ATP by the activation of the enzyme adenyl cyclase. 658

**cyclin** *SI-klin* A type of protein that controls the cell cycle. 155

**cyst** A dormant form of a protist that has a cell wall and lowered metabolism. 406

**cytogenetics** *si-to-jen-ET-ix* Correlation of an inherited trait to a chromosomal anomaly. 209

**cytokine** Protein synthesized in certain immune cells that influences the activity of other immune cells. 779

**cytokinesis** *SI-toe-kin-E-sis* Distribution of cytoplasm, organelles, and macromolecules into two daughter cells in cell division. 147

**cytokinin** *SI-toe-KI-nin* One of five classes of plant hormone. 572

**cytoplasm** *SI-toe-PLAZ-um* The jellylike fluid in which cell structures are suspended. 44

**cytoskeleton** *SI-toe-SKEL-eh-ten* A framework of protein rods and tubules in cells. 52, 66

**cytotoxic T cell** Immune system cell that kills nonself cells by binding them and releasing chemicals. 787

## D

**day-neutral plant** Plant that does not rely on photoperiod to flower. 579

**decomposer** An organism that consumes feces and dead organisms, returning inorganic molecules to ecological cycles. 4, 875

**defensin** Antimicrobial peptide in arthropods. 779

**deforestation** Removal of tree cover from a previously forested area. 919

**dehydration synthesis** Formation of a covalent bond between two molecules by loss of water. 30

**demography** The statistical study of populations 850

**denaturation** A dramatic change in structure that destroys a protein's function. 38, 100

**dendrite** Thin neuron branch that receives neural messages and transmits information to the cell body. 592, 611

**denitrifying bacterium** *de-NI-tri-fy-ing bak-TEAR-e-um* A bacterium that reacts with nitrites and nitrates and releases nitrogen gas. 880

**dense connective tissue** Connective tissue with dense collagen tracts. 592

**density dependent factor** Condition that limits population growth when populations are large. 857

**density independent factor** Population-limiting condition that acts irrespective of population size. 858

**dentin** *DEN-tin* Bonelike substance beneath a tooth's enamel. 747

**deoxyribonucleic acid** *de-OX-ee-RI-bo-nu-KLAY-ic AS-id* (**DNA**) A double-stranded nucleic acid composed of nucleotides containing a phosphate group, a nitrogenous base (A, T, G, or C), and deoxyribose. 39

**deoxyribose** *de-OX-ee-RI-bose* A five-carbon sugar that is part of DNA. 39, 231

**deposit feeder** Animal that eats soil and strains out nutrients. 741

**derived character** The common distinguishing feature of a clade. 323

**dermal tissue** *DER-mal TISH-ew* Tissue that covers a plant. 518

**dermis** The layer of skin, derived from mesoderm, that lies beneath the epidermis in vertebrates. 601

**desert** One of several types of terrestrial biomes; very low precipitation. 896

**desertification** The encroachment of desert into surrounding areas. 920

**desmosome** A junction that anchors intermediate filaments of two adjoining cells in a single spot on the cell membrane. 80

**determinate cleavage** Cell division in early animal embryo in which cells commit to a particular developmental pathway. 464

**detritus** Collective term for feces and dead organic matter. 875

**deuterostome** An animal lineage with radial, indeterminate cleavage and an anus that forms from the blastopore. 464

**diabetes mellitus** Disease resulting from an inability of the body to produce or use insulin. 671

**diaphragm** A sheet of muscle separating the thoracic and abdominal cavities. 732

**diapsid** *di-AP-sid* Animal with two openings behind each eye orbit in its skull. 503

**diastole** *di-A-sto-lee* Relaxation of the heart muscle. 707

**diatom** *DI-a-tom* A photosynthetic aquatic protist with distinctive silica walls. 415

**dicotyledon (dicot)** *di-kot-ah-LEE-don* An angiosperm that has two cotyledons. 435, 523

**digestion** Chemical breakdown of food. 740

**digestive system** System of tubes where food is broken down into nutrient molecules that are absorbed into capillaries. 597, 742

**dikaryon** *di-KAR-e-on* A cell having two genetically distinct haploid nuclei. 445

**dilution effect** Formation of a group to reduce each individual's chance of being eaten. 839

**dinoflagellate** *di-no-FLADJ-el-et* Photosynthetic aquatic protist with flagellated cells. 410

**dioecious** *di-EE-shus* In plants, having separate male and female individuals of the same species. 555

**diploblastic** An animal whose tissues arise from two germ layers in the embryo. 464

**diploid cell** *DIP-loid SEL* A cell with two different copies of each chromosome. Also known as 2*n*. 164

**Diplomonadida** Group of basal eukaryotes whose cells lack mitochondria. 405

**direct development** The gradual development of a juvenile animal into an adult, without an intervening larval stage. 464

**directional selection** Changes in the prevalence of a characteristic that reflects differential survival of individuals better adapted to a particular environment. 294

**disaccharide** *di-SAK-eh-ride* A sugar that consists of two bonded monosaccharides. 32

**dispersive** Replication of DNA producing daughter molecules that contain parts of the original parent molecule and newly synthesized DNA. 235

**disruptive selection** A population in which two extreme expressions of a trait are equally adaptive. 294

**distal convoluted tubule** The region of the kidney distal to the nephron loop and proximal to a collecting duct. 772

**disulfide bond** Attraction between two sulfur atoms within a protein molecule. 37

**diversity-stability hypothesis** The more species in a community, the better the community can resist environmental change. 925

**division** A taxonomic category identical to phylum. 7

**dizygotic twins** *di-zi-GOT-ik TWINZ* Fraternal (non-identical) twins. 202

**DNA fingerprinting or profiling** The use of DNA patterns to identify individuals based on the vast variations in DNA sequence on the nucleotide level. 262

**DNA microarray** A collection of genes or gene pieces on a tiny grid allowing the monitoring of gene expression under different conditions. 266

**DNA polymerase** *DNA po-LIM-er-ase* An enzyme that inserts new bases and corrects mismatched base pairs in DNA replication. 237

**DNA sequencing** The technique used to read the sequence of nucleotides in a DNA molecule. 262

**DNA synthesizer** A laboratory machine that makes short sequences of DNA such as primers. 259

**domain** A taxonomic designation that supercedes kingdom. 7

**dominance hierarchy** A social ranking of members of a group of the same sex, which distributes resources with minimal aggression. 834

**dominant** An allele that masks the expression of another allele. 184

**dormancy** *DOR-man-see* A temporary state of lowered metabolism and arrested growth. 583

**dorsal hollow nerve cord** One of the four characteristics of chordates; derived from a plate of embryonic ectoderm. 490

**double blind** An experimental protocol where neither participants nor researchers know which subjects received a placebo and which received the treatment being evaluated. 13

**double fertilization** In angiosperms and gnetophytes, the fusion of one sperm cell with an egg, and another sperm cell with the polar nuclei to yield a triploid endosperm. 559

**duodenum** *doo-AH-de-num* The first section of the small intestine. 750

**dynamic equilibrium** A condition where equal concentrations of a substance exist across a semipermeable membrane. 69

**dynein** A type of motor protein found in cytoskeleton elements that uses ATP to produce movement of cilia or flagella. 75

**E**

**ecdysozoan** One of two protostome lineages, characterized by periodic molting. 465

**echolocation** Ability to locate objects by bouncing sound waves off them. 649

**ecological succession** Change in the species composition of a community over time. 870

**ecology** The study of relationships among organisms and their environments. 850

**ecosystem** All organisms and their nonliving environment in a defined area. 4, 866

**ecotone** A transition area between two adjacent ecological communities. 309

**ectoderm** *EK-toe-derm* In an animal embryo, the outermost germ layer, whose cells become part of the nervous system, sense organs, and the outer skin layer. 464

**ectotherm** *EK-toe-therm* An animal that obtains heat primarily from its external surroundings. 494, 762

**efferent arteriole** Arteriole that receives blood from the glomerular capillaries of a nephron. 770

**ejaculation** Discharge of sperm through the penis. 804

**elastin** *e-LAS-tin* A type of connective tissue protein. 592

**electrocardiogram (ECG)** A chart showing the electrical changes that accompany the contraction of the heart. 707

**electrolyte** *e-LEK-tro-lite* Solution containing ions. 25

**electromagnetic spectrum** A spectrum of naturally occurring radiation. 109

**electron** *e-LEK-tron* A negatively charged subatomic particle with negligible mass that orbits the atomic nucleus. 18

**electronegativity** The tendency of an atom to attract electrons. 22

**electron transport chain** Membrane-bound molecular complex that transfers energy from electrons to form proton gradients and energy-carrying molecules. 95, 113, 128, 133

**Electrophoresis** A technique that separates molecules in a mixture on the basis of electrical charge. 260-261

**element** A pure substance consisting of atoms containing a characteristic number of protons. 18

**embryonic disc** Flattened inner cell mass of the embryo. 810

**embryonic stage** In humans, the stage of prenatal development from the second through eighth weeks, when tissues and organs begin to differentiate. 804

**embryonic stem (ES) cell** Cells from an early stage of embryo development that are capable of specializing to become any cell type. 810

**embryo sac** The mature female gametophyte in angiosperms. 558

**emergent property** A quality that appears as complexity increases. 4

**empiric risk** Risk calculation based on prevalence. 201

**enamel** Hard substance covering a tooth. 747

**endergonic reaction** *en-der-GONE-ik re-AK-shun* An energy-requiring chemical reaction. 93

**endocrine gland** A concentration of hormone-producing cells in animals. 658

**endocrine system** Glands and cells that secrete hormones in animals. 597, 658

**endocytosis** *EN-doe-si-TOE-sis* The cell membrane's engulfing extracellular material. 73

**endoderm** In an animal embryo, the germ layer, whose cells become the organs and linings of the digestive and respiratory systems. 464

**endodermis** *en-do-DER-mis* The innermost cell layer in a root's cortex. 530

**endomembrane system** In eukaryotic cells, a series of highly folded membranes connected to the nucleus. 52

**endometrium** The inner uterine lining. 805

**endorphin** *en-DOOR-fin* Pain-killing protein produced in the nervous system. 667

**endoskeleton** An internal scaffolding type of skeleton in vertebrates and some invertebrates. 482, 680

**endosome** A membrane-bounded compartment containing the products of endocytosis. 73

**endosperm** *EN-do-sperm* In angiosperms, a triploid tissue that nourishes the embryo in a seed. 559

**endospore** *EN-doe-spor* A walled structure that enables some prokaryotic cells to survive harsh environmental conditions. 389

**endosymbiont theory** *EN-doe-SYM-bee-ont THER-ee* The idea that eukaryotic cells evolved from large prokaryotic cells that engulfed once free-living bacteria. 60

**endothelium** *en-doe-THEEL-e-um* Layer of single cells that lines blood vessels. 707

**endotherm** *EN-doe-therm* An animal that uses metabolic heat to regulate its temperature. 505, 762

**energy** The ability to do work. 18, 88

**energy shell** Levels of energy in an atom formed by electron orbitals. 20

**entropy** *EN-tro-pee* Randomness or disorder. 91

**envelope** The lipid layer around some viruses. 370

**environmental resistance** Factors that limit population growth. 856

**enzyme** *EN-zime* A protein that catalyzes a specific type of chemical reaction. 35, 38, 100

**epidemiology** *EP-eh-dee-mee-OL-o-gee* The analysis of data derived from real-life, nonexperimental situations. 10, 279

**epidermis** *ep-eh-DERM-is* The outer integumentary layer in several types of animals; also the outermost cell layer of young roots, stems, and leaves. 518, 600

**epididymis** *ep-eh-DID-eh-mis* In the human male, a tightly coiled tube leading from each testis, where sperm mature and are stored. 175, 804

**epigenesis** *ep-eh-JEN-eh-sis* The idea that specialized tissue arises from unspecialized tissue in a fertilized ovum. 799

**epiglottis** *ep-eh-GLOT-is* Cartilage that covers the glottis, routing food to the digestive tract and air to the respiratory tract. 729, 748

**epinephrine** *ep-eh-NEF-rin* **(adrenaline)** A hormone produced in the adrenal medulla that raises blood pressure and slows digestion. 670

**epistasis** *eh-PIS-tah-sis* A gene masking another gene's expression. 196

**epithelial tissue (epithelium)** Tightly packed cells that form linings and coverings. 590

**equational division** The second meiotic division, when four haploid cells form from two haploid cells that are the products of meiosis I. 168

**erythrocyte** Red blood cell. 592

**erythropoietin** A hormone produced in the kidneys that stimulates red blood cell production when oxygen is lacking. 713

**esophagus** A muscular tube that leads from the pharynx to the stomach. 748

**essential nutrient** Nutrient that must come from food because an organism cannot synthesize it. 755

**estrogen** A steroid hormone produced in ovaries of female vertebrates that helps regulate reproductive cycles. 35, 673

**estuary** *ES-tu-air-ee* An area where fresh water in a river meets salty water of an ocean. 904

**ethology** *eth-OL-o-gee* Study of how natural selection shapes adaptive behavior. 826

**ethylene** *ETH-eh-leen* One of five classes of plant hormone. 573

**euchromatin** *u-KROME-a-tin* Light-staining genetic material. 208

**Euglenida** Group of motile basal eukaryotes (protista), some of which are photosynthetic. 407

**Eukarya** *yoo-CARE-ee-a* One of the three domains of life, including organisms that have eukaryotic cells. 8

**eukaryotic cell** *yoo-CARE-ee-OT-ik SEL* A complex cell containing membrane-bounded organelles. 43

**euploid** *YOO-ployd* A normal chromosome number. 218

**eusocial** A population of animals that communicate, cooperate in caring for young, have overlapping generations, and divide labor. 838

**eustachian tube (auditory tube)** Tube that connects the middle ear with the air passageways; allows for the adjustment of pressure on the inside of the tympanic membrane. 651

**eutrophic** *yoo-TRO-fik* A lake containing many nutrients and decaying organisms, often tinted green with algae. 903

**evaporative cooling** Loss of body heat by evaporation of fluid from the body's surface. 764

**evolution** Changing allele frequencies in a population over time. Affected traits are said to **evolve.** 7, 273

**excision repair** *ex-SIZ-jhun ree-PARE* Cutting pyrimidine dimers out of DNA. 239

**exergonic reaction** *ex-er-GONE-ik re-AK-shun* An energy-releasing chemical reaction. 94

**exocrine gland** Structure that secretes substances through ducts. 659

**exocytosis** *EX-o-si-TOE-sis* Fusing of secretion-containing organelles with the cell membrane. 73

**exon** *EX-on* The bases of a gene that code for amino acids. 243

**exoskeleton** A braced framework skeleton on the outside of an organism. 478, 679

**experiment** A test to disprove a hypothesis. 8

**exponential growth** Population growth in which numbers double at regular intervals. 853

**expression vector** A molecule carrying a gene of interest along with a promoter that casues expression of the gene. 257

**expressivity** The variation of a trait's expression in different individuals. 195

**external respiration** Exchange of gases between respiratory surfaces and the blood. 722

**extracellular digestion** Dismantling of food by hydrolytic enzymes in a cavity within an organism's body. 742

**extracellular matrix** A nonliving complex of substances that surrounds cells of connective tissue. 591

**F**

**facilitated diffusion** Movement of a substance down its concentration gradient with the aid of a carrier protein. 71

**facultative anaerobe** Organism that can live in the presence or absence of oxygen. 391

**fall turnover** The seasonal mixing of the upper and lower layers of a lake. 903

fast-twitch fiber A msucle fiver that splits ATP quickly but is not well supplied with oxygen; it fatigues quickly. 693

**family** A taxonomic category below order and above genus. 7

**fast-twitch fiber** A muscle fiber that splits ATP quickly but is not well supplied with oxygen; it fatigues quickly. 693

**fatty acid** A hydrocarbon chain that is a part of a triglyceride. 34

**feather follicle** An extension of the epidermis from which a feather extends. 602

**fecundity** *fee-KUN-dit-ee* The number of offspring an individual produces in its lifetime. 853

**fern** *See* seedless vascular plant, true fern, whisk fern.

**fertilization** The union of two gametes. 165, 809

**fetal period** The final stage of prenatal development, when structures grow and elaborate. 800

**fiber** An elongated, plant cell that occurs in strands. 518

**fibrin** A threadlike protein that forms blood clots. 716

**fibroblast** A connective tissue cell that secretes collagen and elastin. 592

**fibrous root system** A root system composed of many similar-sized roots, as in grasses and many other monocots. 528

**filter feeder** An aquatic animal that uses ciliated tissue surfaces to feed on small food particles. 741

**filtration** Filtering of substances across a filtration membrane, as occurs in the glomerulus. 770

**fixed action pattern (FAP)** An innate, stereotyped behavior. 827

**flagellum** (pl. flagella) A long whiplike appendage a cell uses for motility; composed of microtubules. 75, 390

**flame cell system** A simple excretory system in flatworms. 767

**flower** The reproductive structure in angiosperms. 553

**follicle stimulating hormone (FSH)** A pituitary hormone that controls oocyte maturation, development of ovarian follicles, and their release of estrogen. 667

**food chain** The linear series of organisms that successively eat each other. 875

**food web** A network of interconnecting food chains. 875

**forebrain** The front part of the vertebrate brain. 626

**founder effect** Genetic drift that occurs after a small group founds a new settlement, partitioning a subset of the original population's genes. 291

**fountain effect** Splitting and regrouping of a school of fish, which confuses a predator. 839

**fovea centralis** An indentation in the retina opposite the lens; it has only cones and provides visual acuity. 646

**frameshift mutation** A mutation that adds or deletes one or two DNA bases, altering the reading frame. 251

**free energy** The usable energy in the bonds of a molecule. 93

**frond** Leaf of a true fern; also, a part of a lichen or alga that resembles a leaf. 431

**fruit** Seed-containing structure in flowering plants. 554

**fruiting body** Multicellular spore-bearing organ of a fungus. 446

**functional group** An atom or molecule added to a carbon skeleton that confers distinct chemical properties. 30

**fundamental niche** *fun-da-MEN-tal NEESH* All the resources that a species could possibly use in its environment. 867

**G**

**$G_0$ phase** Resting phase of the cell cycle where cells continue to function, but do not divide. 147

**$G_1$ phase** The gap stage of interphase when proteins, lipids, and carbohydrates are synthesized. 147

**$G_2$ phase** The gap stage of interphase when membrane components are synthesized and stored. 148

**gallbladder** An organ beneath the liver; stores bile. 754

**gamete** *GAM-eet* A sex cell. The sperm or ovum. 164

**gametogenesis** *ga-meet-o-JEN-eh-sis* Meiosis and maturation; making gametes. 166

**gametophyte** *gam-EET-o-fite* The haploid, gamete-producing stage of the plant life cycle. 165, 422, 553

**ganglion** Cluster of neuron cell bodies. 622

**ganglion cell** The first cell type in the visual pathway to generate action potentials. 646

**gap junction** A connection between two cells that allows cytoplasm to flow between both cells. 80

**gastrin** A hormone that stomach cells secrete; stimulates secretion of more gastric juice. 748

**gastrointestinal tract** A continuous tube along which food is physically and chemically digested. 744

**gastropod** A mollusk with a broad flat foot for crawling (snails and slugs). 473

**gastrovascular cavity** A digestive chamber with a single opening in cnidarians and flatworms. 470, 742

**gastrula** *GAS-troo-la* A three-layered embryo. 464, 810

**gemma** (pl. gemmae) Asexual reproductive structure in liverworts. 425

**gene** A sequence of DNA that specifies the sequence of amino acids of a particular polypeptide. 39

**gene flow** A change in a gene pool due to the random loss or addition of alleles through migration. 288, 291

**gene pool** All the genes and alleles in a population. 288

**generative cell** A haploid cell in a pollen grain that divides to form two sperm cells. 558

**gene therapy** Techniques that replace a nonfunctioning gene in somatic cells with a functioning one. 265

**genetic code** Correspondence between specific DNA base triplets and amino acids. 39, 244

**genetic drift** Changes in gene frequencies caused by separation of a small group from a larger population. 291

**genetic heterogeneity** Containing nonidentical components of genetic information. 199

**genetic load** Collection of deleterious alleles in a population. 294

**genotype** *JEAN-o-type* Genetic constitution of an individual. 187

**genus** A taxonomic category below family; many species may be grouped into a genus. 7

**geological timescale** A division of time into major eras of biological and geological activity, then periods within eras, and epochs within some periods. 327

**germ cell** Gamete or sex cell. 168

**germinal mutation** *JER-min-al mew-TAY-shun* A mutation in a sperm or oocyte. 252

**germination** The beginning of growth in a seed, spore, or other structure. 564

**gibberellin** *JIB-ah-REL-in* One of five classes of plant hormone. 571

**gill** A highly folded respiratory surface for gas exchange in aquatic animals. 490, 724

**ginkgo** *GEENG-ko* One of four divisions of gymnosperms. 432

**gizzard** A muscular part of the digestive tract that grinds food in some animals. 744

**global warming** An increase in average global temperature that appears to be a result of human activities. 916

**glomerular capsule** *glo-MARE-u-lar CAP-sool* The cup-shaped proximal end of the renal tubule that surrounds the glomerulus. 770

**glomerulus** *glo-MARE-u-lus* A ball of capillaries between the afferent arterioles and efferent arterioles in the proximal part of a nephron. 769

**glottis** Opening from the pharynx to the larynx. 729

**glucagon** A pancreatic hormone that breaks down glycogen into glucose, raising blood sugar levels. 671

**glucocorticoid** *glu-ko-KORT-eh-koyd* Hormone that the adrenal cortex secretes; enables the body to survive prolonged stress. 667, 670

**glycocalyx** *gli-ko-KAY-lix* A sticky layer outside a prokaryotic cell wall; consists of proteins and/or polysaccharides. 390

**glycolipid** A molecule made up of a oligosaccharide combined with a lipid. 32

**glycolysis** *gli-KOL-eh-sis* A catabolic pathway occurring in the cytoplasm of all cells. One molecule of glucose splits and rearranges into two molecules of pyruvic acid. 127

**glycoprotein** A molecule made up of an oligosaccharide combined with a protein. 32

**gnetophyte** *NEE-toe-fite* One of four divisions of gymnosperms. 433

**goiter** Swelling of the thyroid gland caused by lack of iodine in the diet. 669

**golden alga** Photosynthetic protist with unique pigments. 416

**Golgi apparatus** A system of flat, stacked, membrane-bounded sacs where cell products are packaged for export. 54

**gonad** Organ that manufactures gametes in animals. 673, 804

**gonadotropin-releasing hormone** A hormone produced by the hypothalamus that causes release of LH and FSH from the anterior pituitary. 673

**gradualism** Slow evolutionary change. 311

**Gram stain** Technique for classifying major groups of bacteria, based on cell wall structure. 389

**granum** *GRAN-um* A stack of flattened thylakoid discs that forms the inner membrane of a chloroplast. 59, 111

**gravitropism** *grav-eh-TROP-izm* A plant's growth response toward or away from the pull of gravity. 575

**gray matter** Nervous tissue in the CNS consisting of neuron cell bodies, unmyelinated fibers, interneurons, and neuroglial cells. 624

**green alga** Photosynthetic protist that has pigments, starch, and cell walls similar to those of land plants. 417

**greenhouse effect** Elevation in surface temperature caused by carbon dioxide and other atmospheric gases. 917

**ground tissue** The tissue that makes up most of the primary body of a plant; consists of parenchyma, collenchyma, and sclerenchyma cells. 517

**growth factor** A protein that binds a specific receptor type on certain cells, starting a cascade of molecular messages that signals the cell to divide. 155

**growth hormone (GH)** A pituitary hormone that promotes growth and development of tissues by increasing protein synthesis and cell division rates. 667

**guard cell** One of a pair of epidermal cells that open and close stomata in plants by gaining and losing turgor pressure. 519

**gustation** The sense of taste. 642

**gymnosperm** *JIM-no-sperm* A seed-producing plant whose seeds are not enclosed in an ovary. 423

## H

**habitat** The physical place where an organism lives. 866

**habituation** The simplest form of learning, in which an animal learns not to respond to irrelevant stimuli. 830

**hagfish** A jawless fish with a cranium and lacking supportive cartilage around the nerve cord. 496

**hair cell** Mechanoreceptor that initiates sound transduction in the cochlea. 651

**hair follicle** An epidermal structure anchored in the dermis, from which a hair grows. 602

**half-life** The time it takes for half the isotopes in a sample of an element to decay into a second isotope. 331

**haploid cell** A cell with one copy of each chromosome. Also called 1*n*. 164

**Hardy-Weinberg equilibrium** Maintenance of the proportion of genotypes in a population from one generation to the next. 288

**heart** Muscular organ that pumps blood or hemolymph. 705

**heartwood** Nonfunctioning wood in the center of a tree. 532

**heat capacity** The amount of heat necessary to raise the temperature of a substance. 26

**heavy chain** Large polypeptide of an antibody subunit. 784

**heliotrophism** The response in plants that turns flowers to face the sun; also called solar tracking. 583

**helper T cell** Lymphocyte that produces cytokines and stimulates activities of other immune system cells. 783

**heme group** *HEEM GROOP* An iron-containing complex that forms the oxygen-binding part of hemoglobin. 713

**hemizygous** *HEM-ee-ZY-gus* A gene on the Y chromosome in humans. 214

**hemocyanin** *HEEM-o-SI-a-nin* A respiratory pigment in mollusks that carries oxygen. 703

**hemoglobin** *HEEM-o-glo-bin* The iron-binding protein that carries oxygen in mammals. 703

**hemolymph** *HEEM-o-limf* The "blood" in animals with open circulatory systems. 478, 702

**hepatic portal system** A division of the circulatory system that enables the liver to rapidly harness chemical energy in digested food. 709

**heritability** The proportion of a trait attributable to heredity. 201

**hermaphroditic** Refers to an individual that produces both sperm and eggs. 463

**heterochromatin** *het-er-o-KROME-a-tin* Dark-staining genetic material. 208

**heterogametic sex** *HET-er-o-gah-MEE-tik SEX* The sex with two different sex chromosomes. 212

**heterotroph** *HET-er-o-TROFE* An organism that obtains carbon by eating another organism. 107, 391, 875

**heterozygous** *HET-er-o-ZI-gus* Possessing two different alleles for a particular gene. 187

**highly conserved** A protein or nucleic acid sequence that is very similar in different species. 336

**hindbrain** The lower portion of the vertebrate brain, which includes the brainstem and controls vital functions. 626

**hippocampus** Area of the cerebral cortex important in long-term memory. 631

**histamine** An allergy mediator that dilates blood vessels and causes allergy symptoms. 782

**histone** A cluster of proteins around which a length of DNA wraps to form a nucleosome. 234

**homeostasis** The ability of an organism to maintain constant body temperature, fluid balance, and chemistry. 5, 597

**homing** Returning to a given spot. 833

**hominid** *HOM-eh-nid* Animal ancestral to humans only. 361

**hominoid** *HOM-eh-noid* Animal ancestral to apes and humans. 361

**homogametic sex** *HO-mo-gah-MEE-tik SEX* The sex with two identical sex chromosomes. 212

**homologous pairs** *ho-MOL-eh-gus PAIRZ* Chromosome pairs that have the same sequence of genes. 168

**homologous structures** *ho-MOL-eh-gus STRUK-churs* Similar structures in different species; they have the same general components, indicating descent from a common ancestor. 333

**homozygous** *HO-mo-ZI-gus* Possessing two identical alleles for a particular gene. 187

**horizontal gene transfer** Transfer of genetic information between individuals, compared to vertical gene transfer between generations. 394

**hormone** A chemical synthesized in small quantities in one part of an organism and transported to another, where it affects target cells. 155, 570, 658

**hornwort** *HORN-wart* A type of bryophyte. 427

**horsetail** A type of seedless vascular plant. 430

**host range** The organisms a virus can infect. 376

**humoral immune response** Secretion of antibodies by B cells in response to a foreign antigen. 780

**humus** Partially decomposed organic matter in soil. 541

**hybridization** In DNA technology, matching a known DNA probe with its complementary sequence in genomic DNA to identify the location of the sequence. 262

**hydrocarbon** *HI-dro-kar-bon* A molecule containing mostly carbon and hydrogen. 30

**hydrogen bond** A weak chemical bond between oppositely charged portions of molecules. 23

**hydrolysis** *hi-DROL-eh-sis* Splitting a molecule by adding water. 30

**hydrophilic** *HI-dro-FILL-ik* Attracted to water. 23

**hydrophobic** *HI-dro-FOBE-ik* Repelled by water. 23

**hydrostatic skeleton** The simplest type of skeleton, consisting of flexible tissue surrounding a constrained liquid. 679

**hypertonic** *hi-per-TON-ik* The solution on one side of a membrane where the solute concentration is greater than on the other side. 69

**hypertrophy** *hi-PER-tro-fee* Increase in muscle mass, usually due to exercise. 695

**hypha** *HI-fa* A fungal thread, the basic structural unit of a multicellular fungus. 446

**hypoglycemia** A deficient level of glucose in the blood because of insufficient carbohydrate intake or excess insulin. 672

**hypothalamus** *hi-po-THAL-a-mus* A small structure beneath the thalamus that controls homeostasis and links the nervous and endocrine systems. 627, 664

**hypothesis** *hi-POTH-eh-sis* An educated guess based on prior knowledge. 8

**hypotonic** *hi-po-TON-ic* The solution on one side of a membrane where the solute concentration is less than on the other side. 69

## I

**idiotype** The part of an antibody molecule that binds to a specific antigen. 786

**ileum** The last section of the small intestine. 750

**imbibition** *IM-bih-BISH-un* The absorption of water by a seed. 24, 564

**immune system** System of specialized cells that defends the body against infections, cancer, and foreign cells. 597

**impact theory** Idea that mass extinctions were caused by impacts of extraterrestrial origin. 317

**implantation** Nestling of the blastocyst into the uterine lining. 810

**impression fossil** The preserved evidence of an organism, such as a footprint. 327

**imprinting** A type of learning that usually occurs early in life and is performed without obvious reinforcement. 830

**inclusive fitness** Fitness defined by combined reproductive success of an individual and its relatives. 838

**incomplete dominance** A heterozygote whose phenotype is intermediate between the phenotypes of the two homozygotes. 197

**incomplete penetrance** A genotype that does not always produce a phenotype. 194

**incus** *INK-us* A small bone in the middle ear. 650

**independent assortment** The random alignment of homologs during metaphase of meiosis I. 169

**indeterminate cleavage** Cell division in early animal embryo in which cells are not committed to a particular developmental pathway. 465

**industrial melanism** *in-DUS-tree-al MEL-an-iz-um* Coloration of an organism that is adaptive in a polluted area. 294

**industrial smog** Air pollution resulting directly from industrial and urban emissions. 914

**inferior vena cava** The lower branch of the largest vein that leads to the heart. 707

**inflammation** Increased blood flow and accumulation of fluid and phagocytes at the site of an injury, rendering it inhospitable to bacteria. 780

**inhibiting hormone** A hormone produced by one gland that inhibits another gland. 665

**innate** Instinctive; developing independently of experience. 827

**innate immunity** Cells and substances that provide preexisting defenses against infection without prior exposure to an antigen. 778

**inner cell mass** The cells in the blastocyst that develop into the embryo. 810

**insertion** The end of a muscle on the bone that moves. 695

**insight learning** Ability to apply prior learning to a new situation without trial-and-error activity. 831

**insulin** A pancreatic hormone that lowers blood sugar level by stimulating body cells to take up glucose from the blood. 671

**integrin** A type of protein that anchors cells to connective tissue. 81

**integument (integumentary system)** Outer covering of an animal's body. 597

**intercalated disks** *in-TER-kah-LAY-tid DISKS* Tight foldings in cardiac muscle cell membranes that join adjacent cells. 687

**interferon** A polypeptide produced by a T cell infected with a virus; stimulates surrounding cells to manufacture biochemicals that halt viral replication. 779

**interleukin** *in-ter-LOO-kin* A class of immune system biochemicals. 779

**intermediate filament** Cytoskeletal element intermediate in size between a microtubule and a microfilament. 75

**intermembrane compartment** The space between a mitochondrion's two membranes. 128

**internal respiration** Exchange of gases between the blood and the body cells. 722

**interneuron** A neuron that connects one neuron to another to integrate information from many sources and to coordinate responses. 611

**internode** Part of stem between nodes. 522

**interphase** *IN-ter-faze* The period when the cell synthesizes proteins, lipids, carbohydrates, and nucleic acids. 147

**interspecific competition** The struggle between members of a different species for vital resources. 867

**interstitial fluid** Liquid that bathes cells in a vertebrate's body. 717

**intertidal zone** The region along a coastline where the tide recedes and returns. 904

**intracellular digestion** Breakdown of molecules in food vacuoles in cells. 742

**intron** *IN-tron* Bases of a gene that are transcribed but are excised from the mRNA before translation into protein. 243

**invertebrate** Animal without a vertebral column. 462

**ion** *I-on* An atom that has lost or gained electrons, giving it an electrical charge. 22

**ionic bond** *i-ON-ik bond* Attraction between oppositely charged ions. 22

**iris** Colored part of the eye; regulates the size of the pupil. 645

**irritability** An immediate response to a stimulus. 6

**isotonic** *ice-o-TON-ik* When solute concentration is the same on both sides of a membrane. 69

**isotope** *I-so-tope* A differently weighted form of an element. 19

## J

**jejunum** *je-JOO-num* The middle section of the small intestine. 750

**joint** Where a bone contacts another bone. 681

**juvenile hormone** An insect hormone produced in larvae; controls metamorphosis. 662

## K

**karyokinesis** *KAR-ee-o-kah-NEE-sus* Division of the genetic material. 147

**karyotype** *KAR-ee-o-type* A size-order chart of chromosomes. 208

**keratin** *KER-a-tin* A hard protein that accumulates in the integument of many animals and forms specialized structures. 600

**keratinocytes** Epidermal cells that synthesize keratin. 600

**keystone species** Species that exert an effect on community structure that is disproportionate to their biomass. 869

**kidney** Organ consisting of millions of tubules that excrete nitrogenous waste and regulate ion and water levels. 768

**kilocalorie (kcal)** The energy required to raise 1 kilogram of water 1°C. One food calorie. 88, 755

**kinase** *KI-nase* A type of enzyme that activates other proteins by adding a phosphate. 155

**kinetic energy** *kin-ET-ik EN-er-gee* The energy of motion. 90

**kinetochore** *kin-ET-o-chore* Microtubule fibers anchored in the centromere that connect chromosomes to the spindle apparatus. 150

**kinetoplast** *kin-ET-o-plast* Unique, DNA-containing structure that forms part of the mitochondrion in Kinetoplastida. 408

**Kinetoplastida** Group of protista whose cells contain kinetoplasts. 408

**kingdom** A taxonomic category below the level of domain but above phylum. 7

**kin selection** Process by which an individual helps a relative survive or reproduce. 838

**knockout technology** Swapping or "knocking out" a normally functioning gene with a disabled form to study its function. 267

**Koch's postulates** Rules used to verify that an organism causes a particular disease. 397

**Krebs cycle** The stage in cellular respiration that completely oxidizes the products of glycolysis. 128

**K-selection** Selection for individuals that are long-lived, late-maturing, and have few offspring that each receive heavy parental investment. 859

## L

**labor** In mammals, the process involved in giving birth that includes strong contractions of the uterus to expel the offspring. 815

**lacteal** *LAK-tee-ul* A lymph vessel that absorbs fat in the small intestine. 753

**lactic acid fermentation** Reaction of pyruvic acid to produce lactic acid, occurring in some anaerobic bacteria and fatigued mammalian muscle cells. 139

**lacuna** (pl. lacunae) *la-KEW-na* Space in cartilage and bone tissue. 592

**lamprey** The first jawless fish with cartilage around the nerve cord; the simplest true vertebrate. 496

**lancelet** One of three types of invertebrate chordate. 493

**large intestine (colon)** Part of the digestive tract that extends from the small intestine to the rectum. 753

**larva** An immature stage of an animal; usually does not resemble the adult of the species. 464

**larynx** The "voice box" and a conduit for air. 729

**latent learning** Learning without reward or punishment; not apparent until after the learning experience. 831

**lateral line system** A network of canals that extends along the sides of fishes and houses receptor organs that detect vibrations. 497

**latent virus** A viral DNA that has become integrated with the host DNA and can remain dormant indefinitely. 375

**lateral meristem** *LAT-er-al MER-ih-stem* A meristem that gives rise to secondary plant tissue. 516

**law of independent assortment** Genes on different chromosomes are distributed independently of each other into gametes. 191

**law of segregation** Allele separation during meiosis. 186

**laws of thermodynamics** Physical laws that describe energy and energy use. 91

**leaf** The primary photosynthetic organ of most plants. 524

**learning** A persistent change in behavior as a result of experience. 829

**lens** The structure in the eye through which light passes and is focused. 645

**leukocyte** White blood cell. 592

**lichen** *LI-ken* An association of a fungus and an alga or cyanobacterium. 452

**life expectancy** Prediction of how long an individual will live, based on current age and epidemiology. 820

**life span** The longest a member of a species can live. 820

**life table** Data summarizing the probability of reproducing and dying at a given age. 853

**ligament** Tough band of fibrous connective tissue that connects bone to bone across a joint. 695

**ligand** *LI-gand* A messenger molecule that binds to a cell surface protein. 73

**ligase** *LI-gase* An enzyme that catalyzes formation of covalent bonds in the DNA sugar-phosphate backbone. 237, 257

**light chain** Small polypeptide chain in an antibody subunit. 784

**light reactions** Photosynthetic reactions that harvest light energy and store it in molecules of ATP or NADPH. 106

**lignin** *LIG-nin* A tough, complex molecule that strengthens the walls of some plant cells. 423

**limnetic zone** *lim-NET-ik ZONE* The layer of open water in a lake or pond where light penetrates. 902

**linkage map** Diagram of gene order on a chromosome based on crossover frequencies. 211

**linked genes** Genes on the same chromosome. 209

**lipid** *LIP-id* Hydrophobic compound that contains carbon, hydrogen, and oxygen, but with much less oxygen than carbohydrates have. 33

**littoral zone** *LIT-or-al ZONE* The shallow region along the shore of a lake or pond where sufficient light reaches to the bottom for photosynthesis. 902

**liver** The organ that detoxifies blood, stores glycogen and fat-soluble vitamins, synthesizes blood proteins, and monitors blood glucose level. 754

**liverwort** *LIV-er-wart* A type of bryophyte. 425

**lobe-finned fish** A type of fish with limblike fins. 499

**logistic growth** Population growth that levels off as the carrying capacity is approached. 856

**long-day plant** Plant that requires dark periods shorter than some critical length to flower. 579

**loose connective tissue** Connective tissue with widely spaced fibroblasts and a few fat cells. 592

**lophotrochozoan** One of two protostome lineages, characterized by distinct developmental patterns and/or specialized feeding structures. 465

**lumbar vertebra** Vertebra of the small of the back. 683

**lungfish** A type of fish with air bladders adapted as lungs. 499

**lungs** Paired structures that house the bronchial tree and the alveoli; the sites of gas exchange in some vertebrates. 725

**luteinizing hormone (LH)** A hormone made in the anterior pituitary; promotes ovulation. 667

**lymph** Blood plasma minus some large proteins, which flows through lymph capillaries and lymph vessels. 717

**lymphatic system** A circulatory system that consists of lymph capillaries and lymph vessels that transport lymph. 717

**lymph node** Structure in the lymphatic system that contains white blood cells and fights infection. 717

**lymphocyte** A type of white blood cell. T and B cells. 718

**lysogenic infection** *li-so-JEN-ik in-FEK-shun* A bacteriophage infection in which the phage DNA is incorporated in the host chromosome. 372

**lysogeny** *li SAW jen ee* The process by which a virus incorporates its DNA into a bacterial genome. 375

**lysosome** *LI-so-soam* A sac in a eukaryotic cell in which molecules and worn-out organelles are enzymatically dismantled. 54, 57

**lytic infection** *LIT-ik in-FEK-shun* A viral infection that bursts the host cell. 372

# M

**macroevolution** Large-scale evolutionary changes, such as speciation and extinction. 272, 302

**macromolecule** A very large molecule. 20

**macronutrients** In plants, nine elements required in large amounts. 538 In human nutrition, carbohydrates, fats, and proteins obtained from food. 755

**macrophage** A phagocyte that destroys bacteria and cell debris and presents antigens to T cells. 780

**major histocompatibility complex (MHC)** A cluster of genes that code for cell surface proteins. 783

**malleus** *MAL-e-us* A small bone in the middle ear. 650

**Malpighian tubule** An insect excretory structure. 767

**mammal** A tetrapod vertebrate with hair, mother's milk, endothermy, a four-chambered heart, and a muscular diaphragm. 508

**mammary gland** Milk-producing sweat gland derivative in mammals. 509

**mangrove swamp** Tropical wetland dominated by salt-tolerant trees. 904

**mantle** A dorsal fold of tissue that secretes a shell in mollusks. 473

**marrow cavity** Space in a bone shaft that contains yellow marrow. 686

**marsupial** A pouched mammal. 508

**marsupium** A pouch in which the immature young of marsupial mammals nurse and develop. 509

**mass extinction** The abrupt loss of many species over a wide area. 315

**mast cell** Immune system cell that releases allergy mediators when stimulated. 782

**matrix** The inner compartment of a mitochondrion. Also, the nonliving part of connective tissue. 128

**matter** Any material that takes up space and has mass. 18

**mechanoreceptor** Receptor cell sensitive to mechanical energy. 640

**medulla oblongata** *med-OOL-a ob-long-AT-a* The part of the brainstem nearest the spinal cord; regulates breathing, heartbeat, blood pressure, and reflexes. 626

**medusa** The free-swimming form of a cnidarian. 470

**megagametophyte** *MEG-ah-gah-MEE-toe-fight* The female gametophyte in a plant. 555

**megaspore** *MEG-ah-spor* Structure in plants that develops into the female gametophyte. 555

**meiosis** *mi-O-sis* Cell division that halves the genetic material. 165

**Meissner's corpuscles** A type of nerve receptor in the skin that detects light touch. 652

**melanocyte** Cell that produces melanin pigment. 600

**melanocyte-stimulating hormone (MSH)** A pituitary hormone that controls skin pigmentation in some vertebrates. 668

**melatonin** A hormone produced in the pineal gland that may control other hormones by sensing light and dark cycles. 672

**memory B cell** Mature B cell, specific to an antigen already met, that responds quickly by secreting antibodies when that antigen is encountered again. 782

**meninges** *meh-IN-geez* The three membranes that cover and protect the CNS. 633

**meristem** *MER-ih-stem* Undifferentiated plant tissue that gives rise to new cells. 516

**mesoderm** *MEZ-o-derm* The middle embryonic germ layer, whose cells become bone, muscle, blood, dermis, and reproductive organs. 464

**mesophyll cell** *MEZ-o-fill SEL* Thin-walled plant cell that takes part in photosynthesis. 118, 524

**messenger RNA (mRNA)** A molecule of ribonucleic acid that is complementary in sequence to the sense strand of a gene. 239

**metabolic pathway** A series of connected, enzymatically catalyzed reactions in a cell that build up or break down substances. 92

**metabolism** *meh-TAB-o-liz-um* The biochemical reactions of a cell. 4, 92

**metacentric** *met-a-SEN-trik* A chromosome whose centromere divides it into two similarly sized arms. 209

**metamorphosis** A developmental process in which an animal changes drastically in body form between the juvenile and the adult. 464

**metaphase** *MET-ah-faze* The stage of mitosis when chromosomes align down the center of a cell. 150

**metastasis** *meh-TAH-stah-sis* Spreading of cancer. 159

**microevolution** Subtle, incremental single-trait changes that underlie speciation. 272

**microfilament** Actin rods abundant in contractile cells. 75

**microgametophyte** *MIK-ro-gah-MEE-toe-fight* The male gametophyte in a plant. 555

**micronutrients** In plants, seven elements required in very small amounts. 528 In human nutrition, vitamins and minerals required in small amounts. 755

**microspore** *MIKE-ro-spor* Structure in plants that develops into the male gametophyte. 555

**microtubule** Long tubule of tubulin protein that moves cells. 75

**microvillus (pl. microvilli)** Tiny projection on the surface of an epithelial cell; part of an intestinal villus. 752

**midbrain** Part of the brain between the forebrain and hindbrain; important in vision and hearing. 626

**middle lamella** The region where cell walls of adjacent plant cells meet. 80

**migration** A regularly repeated journey from one specific geographic region to another. 831

**mimicry** Similar appearances of different species. 868

**mineralocorticoid** Adrenal hormone that helps maintain blood volume and electrolyte balance. 670

**mismatch repair** A DNA repair system that recognizes and corrects a noncomplementary base. 239

**missense mutation** A mutation that changes a codon specifying a certain amino acid into a codon specifying a different amino acid. 251

**mitochondrion** *MI-toe-KON-dree-on* Organelle that houses the reactions of cellular respiration. 59

**mitosis** *mi-TOE-sis* A form of eukaryotic cell division in which two genetically identical cells form from one. 146

**mitotic spindle** *mi-TOT-ik SPIN-del* A structure of microtubules that aligns and separates chromosomes in mitosis. 148

**modal action pattern** Same as *fixed action pattern,* but recognizes that individual behavior patterns vary. 828

**molecular clock** Application of the rate at which DNA mutates to estimate when two types of organisms diverged from a shared ancestor. 339

**molecular tree diagram** A depiction of hypothesized relationships among species based on molecular sequence data. 302

**molecular mass** The sum of the masses of the atoms that make up a molecule. 20

**molecule** *MOL-eh-kuel* Atoms joined by chemical bonds. 4, 19

**molting hormone** An insect hormone produced in the larva; triggers molting. 662

**monoclonal antibodies (MAB)** *MON-o-KLON-al AN-tee-bod-eez* Identical antibodies provided by a single B cell or hybridoma. 787

**monocotyledon (monocot)** *MON-o-kot-ah-LEE-don* An angiosperm that has one cotyledon, or seed leaf. 435, 523

**monoecious** *mon-EE-shus* In plants, having male and female reproductive parts on the same plant. 555

**monogamy** Formation of a permanent male-female pair. 835

**monomer** *MON-o-mer* A single link in a polymeric molecule. 30

**monosaccharide** *MON-o-SAK-eh-ride* A sugar that is one five- or six-carbon unit. 32

**monosomy** *MON-o-SOAM-ee* Absence of one chromosome. 218

**monotreme** An egg-laying mammal. 508

**monozygotic twins** *mon-o-zi-GOT-ik TWINZ* Identical twins resulting from the splitting of a fertilized ovum. 202

**morphogenesis** *morf-o-GEN-eh-sis* The series of events during embryonic development that leads to formation of distinct structures. 810

**morula** *MOR-yoo-la* The preembryonic stage consisting of a solid ball of cells. 810

**moss** A type of bryophyte. 432

**motif** A common part of a transcription factor. 37

**motor neuron** A neuron that transmits a message from the CNS toward a muscle or gland. 611

**motor unit** A neuron and all the muscle fibers it contacts. 694

**M phase** The part of the cell cycle when genetic material divides; mitosis. 149

**multicellular** *mull-tee-SEL-u-lar* An organism that consists of many cells. 6

**multifactorial** *mull-tee-fac-TORE-e-al* Traits molded by one or more genes and the environment. 200

**muscle fiber** Skeletal muscle cell. 688

**muscle spindle** Receptor in skeletal muscle fiber that monitors tension. 694

**muscle tissue** Tissue consisting of contractile cells that provide motion. 590

**muscular foot** A ventral organ in mollusks that is used in locomotion. 473

**muscular system** Organ system of muscles whose contractions form the basis of movement. 597

**musculoskeletal system** A combined system of muscle attached to a skeleton, either of bone in the case of vertebrates or other types of skeletons in invertebrates. 678

**mutagen** *MUTE-a-jen* An agent that causes a mutation. 250

**mutant** A phenotype or allele that is not the most common for a certain gene in a population or that has been altered from the "normal" condition. 187

**mutation** A change in a gene or chromosome. To mutate is to undergo such a change. 7, 187, 249

**mutualism** *MU-chu-a-lism* Symbiosis that benefits both partners. 868

**mycelium** *my-SEAL-ee-um* An assemblage of hyphae that forms an individual fungus. 446

**mycorrhiza** *mi-cor-IZ-a* An association of a fungus and the roots of a plant. 452, 542

**myelin sheath** A fatty material that insulates some nerve fibers in vertebrates, allowing rapid nerve impulse transmission. 616

**myofibril** A cylindrical subunit of a muscle fiber. 689

**myofilament** Actin or myosin "string" that is part of a myofibril. 689

**myosin** The protein that forms thick filaments in muscle tissue. 594, 691

## N

**nastic movement** Plant growth or movement that is not oriented toward the provoking stimulus. 576

**natural selection** The differential survival and reproduction of organisms whose genetic traits better adapt them to a particular environment. 7, 272

**navigation** The following of a specific course. 831

**negative feedback** An action that counters an existing condition; important in homeostatic responses. 100, 598, 661

**nematocyst** A stinging structure contained within cnidocytes of cnidarians. 468

**nephridium** *nef-RID-e-um* Network of tubules in some invertebrates that has an excretory function. 767

**nephron** A microscopic tubular subunit of a kidney, consisting of a renal tubule and peritubular capillaries. 768

**nephron loop** Also called loop of Henle; part of a nephron that lies between the proximal and distal convoluted tubules, where water is conserved and urine is concentrated by a countercurrent exchange system. 772

**neritic zone** *ner-IT-ik ZONE* The region of an ocean from the coast to the edge of the continental shelf. 906

**nerve** Bundle of axons. 611

**nerve net** Diffuse network of neurons, as in cnidarians. 622

**nervous system** Interconnected network of neurons and supportive neuroglia that transmits information rapidly throughout the body. 597

**nervous tissue** A tissue whose cells (neurons and neuroglia) form a communication network. 590

**net primary production** Energy available to consumers in a food chain, after cellular respiration by producers. 875

**neural impulse** The electrical message that a neuron conducts. 613

**neural tube** Embryonic precursor of the CNS. 811

**neuroglia** Cells that support, nourish, or assist neurons. 592, 610

**neuromodulator** Peptide that alters a neuron's response to a neurotransmitter or blocks the release of a neurotransmitter. 617

**neuromuscular junction** A chemical synapse of a neuron onto a muscle cell. 692

**neuron** *NUHR-on* A nerve cell, consisting of a cell body, a long "sending" projection (axon), and numerous "receiving" projections (dendrites). 592, 610

**neurosecretory cell** Cell that functions as a neuron at one end but as an endocrine cell at the other by receiving neural messages and secreting hormones. 659

**neurotransmitter** A chemical passed from a neuron to receptors on another neuron or on a muscle or gland cell. 617

**neurulation** Interaction between the notochord and nearby ectoderm that triggers formation of the nervous system in an early animal embryo. 811

**neutron** *NEW-tron* A particle in an atom's nucleus that is electrically neutral and has one mass unit. 18

**neutrophil** A type of white blood cell that, like a macrophage, can engulf bacteria. 780

**niche** All resources a species uses for survival, growth, and reproduction. 867

**nitrifying bacterium** *NI-tri-fy-ing bac-TEAR-e-um* Bacterium that converts ammonia into nitrites and nitrates. 880

**nitrogen fixation** A microbial process that reduces atmospheric nitrogen gas to ammonia. 386

**nitrogen fixing bacterium** Bacterium that reduces atmospheric nitrogen to nitrogen-containing compounds that plants can use. 542, 880

**nitrogenous base** A nitrogen-containing compound that forms part of a nucleotide in a nucleic acid. 39

**node** Area of leaf attachment on a stem. 522

**node of Ranvier** A short region of exposed axon between Schwann cells on neurons of the vertebrate peripheral nervous system. 616

**nodule** Swelling, inhabited by symbiotic nitrogen-fixing bacteria, on the roots of certain types of plants. 542

**nondisjunction** *NON-dis-JUNK-shun* Unequal partition of chromosomes into gametes during meiosis. 218

**nonpolar covalent bond** *non-POE-lar co-VAY-lent BOND* A covalent bond in which atoms share electrons equally. 21

**nonsense mutation** A point mutation that alters a codon encoding an amino acid to one encoding a stop codon. 251

**nonseptate hypha** *non-sep-tate HI-fa* A multicellular hypha lacking crosswalls, or septa. 446

**norepinephrine** An adrenal hormone that raises blood pressure, constricts blood vessels, and slows digestion. 670

**notochord** A semirigid rod running down the length of a chordate's body. 490, 811

**nuclear envelope** A two-layered structure bounding a cell's nucleus. 55

**nuclear pore** A hole in the nuclear envelope. 55

**nucleic acid** *new-CLAY-ic AS-id* DNA or RNA. 39

**nucleoid** *NEW-klee-oid* The part of a prokaryotic cell where the DNA is located. 51, 388

**nucleolus** *new-KLEE-o-lis* A structure within the nucleus where RNA nucleotides are stored. 55

**nucleosome** *NEW-klee-o-some* DNA wrapped around eight histone proteins as part of chromosome structure. 234

**nucleotide** *NEW-klee-o-tide* The building block of a nucleic acid, consisting of a phosphate group, a nitrogenous base, and a five-carbon sugar. 39

**nucleus** *NEW-klee-is* The central region of an atom, consisting of protons and neutrons. Also, a membrane-bounded sac in a eukaryotic cell that contains the genetic material. 18, 46

**nyctinasty** *NIK-tin-asty* A nastic response in plants to light and dark. 578

## O

**obligate aerobe** An organism that requires oxygen to live. 391

**obligate anaerobe** An organism that must live in the absence of oxygen. 391

**oceanic zone** The open sea beyond the continental shelf. 906

**ocellus** *o-SELL-us* Eyespot; visual organ in flatworms. 644

**octet rule** The tendency of an atom to fill its outermost shell. 20

**olfaction** The sense of smell. 642

**oligodendrocyte** Cell that produces myelinated neurons in the brain. 616

**oligonucleotide** A segment of nucleic acid, usually synthetic, about 20 bases in length. 266

**oligosaccharide** A medium-length complex carbohydrate formed by linking together several monosaccharides. 32

**oligotrophic** *OL-eh-go-TRO-fik* A lake with few nutrients; usually very blue. 903

**ommatidium** *oh-ma-TID-ee-um* The visual unit of a compound eye. 644

**oncogene** *ON-ko-jean* A gene that normally controls cell division but when overexpressed leads to cancer. 160

**oogenesis** *oh-oh-GEN-eh-sis* The differentiation of an egg cell from a diploid oogonium, to a primary oocyte, to two haploid secondary oocytes, and finally, after fertilization, to a mature ovum. 176

**oogonium** *oh-oh-GO-nee-um* The diploid cell in which egg formation begins. 176

**open circulatory system** A circulatory system in which hemolymph circulates freely through the body cavity. 478, 702

**operant conditioning** Trial-and-error learning, in which an animal voluntarily repeats any behavior that brings success. 830

**operon** *OP-er-on* A series of genes with related functions and their controls. 241

**opsin** A component of the photopigment rhodopsin. 647

**optic disc** The point on the retina of the eye where the optic nerve exists. 646

**optic nerve** Nerve fibers that connect the retina to the brain; formed of ganglion cell axons. 646

**orbital** The most likely location of an electron relative to an atom's nucleus. 19

**order** A taxonomic category below the level of class but above family. 7

**organ** A structure of two or more tissues that functions as an integrated unit. 2, 590

**organelle** *or-gan-NELL* Specialized structure in eukaryotic cells that carries out specific functions. 4, 44

**organic molecule** Any carbon-based molecule found in living cells. 20

**organogenesis** *or-GAN-o-GEN-eh-sis* Development of organs in an embryo. 811

**organ system** System of physically or functionally linked organs. 2, 590

**orgasm** A pleasurable sensation associated with sexual activity. 804

**orientation** Movement in a specific direction. 831

**origin** The end of a muscle on an immobile bone. 695

**origin of replication** A site on a chromosome where DNA replication begins with a helicase breaking hydrogen bonds that link a particular base pair. 237, 257

**osmoconformer** *oz-mo-con-FORM-er* An organism whose ion concentrations match those of its surroundings. 765

**osmoregulation** *OZ-mo-REG-u-LAY-shun* The control of water and ion balance in an organism. 765

**osmoregulator** *oz-mo-REG-u-LAY-tor* An organism that actively controls its ion concentrations in a changing environment. 765

**osmosis** *oz-MO-sis* Passive diffusion of water through a semipermeable membrane. 69

**osteoblast, osteoclast** Bone cells that respectively secrete and break down bone matrix. 683

**osteocyte** A mature bone cell in a lacuna. 592, 683

**osteon** Concentric circles of osteocytes in bone. 685

**osteonic canal** *oss-tee-ON-ik ka-NAL* Portal that houses blood vessels in bone. 685

**ostracoderm** *oss-TRAK-o-derm* An extinct, jawless, bottom-dwelling, filter-feeding fish. 497

**otolith** *OH-toe-lith* Calcium carbonate granule in the vestibule of the inner ear whose movements provide information on changes in velocity. 652

**outcrossing** The fertilization of the egg cell of one plant by pollen from a different plant. 558

**oval window** A membrane between the middle ear and the inner ear. 651

**ovary** One of the paired female gonads that house developing oocytes. Also, in a flowering plant, the enlarged basal portion of a carpel. 176, 554, 673, 804

**oviparous** Egg-laying (in animals). 494

**ovoviviparous** Retaining eggs inside the birth canal and giving birth to live offspring (in animals). 494

**ovulation** The release of an oocyte from an ovarian follicle. 805

**ovule** *OV-yul* In seed plants, a structure that contains a megagametophyte and egg cell. 554

**oxidation** The loss of one or more electrons by a participant in a chemical reaction. 95

**oxytocin** *ox-ee-TOE-sin* A hypothalamic hormone released from the posterior pituitary; stimulates muscle contraction in the mammary glands and the uterus. 668

**ozone layer** Atmospheric zone rich in ozone gas, which absorbs the sun's ultraviolet radiation. 916

## P

**pacemaker** Specialized cells in the wall of the right atrium that set the pace of the heartbeat. 707

**Pacinian corpuscle** *pah-SIN-ee-en KOR-pus-el* A receptor in the skin; senses vibration. 652

**paedomorphosis** *pay-doe-MORF-o-sis* When adults of a species have features of the larval stages of their ancestors. 501

**pain receptor** Specialized receptor cell that serves a protective function in detecting mechanical damage and temperature extremes. 652

**paleontology** *PAY-lee-on-TOL-ah-gee* Study of evidence of past life. 322

**palindrome** A sequence that reads the same forward and backward, like the word "radar." Restriction endonucleases often recognize palindromes in DNA. 256

**palisade mesophyll cell** A columnar mesophyll cell in leaves that is specialized for light absorption. 524

**pancreas** An organ with an endocrine part that produces somatostatin, insulin, and glucagon and a digestive part that produces pancreatic juice. 671, 754

**pancreatic islet** Also, islet of Langerhans; cluster of cells in the pancreas; secretes hormones that control nutrient use. 671

**Parabasalia** Group of protista whose cells lack mitochondria. 405

**parapatric speciation** The formation of a new species at the boundary zone between two species. 309

**parasitism** A symbiotic relationship in which one member derives nutrients or resources at the expense of the other. 868

**parasitoid** A wasp whose larvae live within a host and kill it, but whose adults are free-living. 868

**parasympathetic nervous system** Part of the autonomic nervous system; controls vital functions such as respiration and heart rate. 623

**parathyroid gland** One of four small groups of cells behind the thyroid gland; secretes parathyroid hormone. 669

**parathyroid hormone (PTH)** A hormone that maintains calcium levels in the blood. 669

**parenchyma** *pah-REN-kah-mah* Plant tissue composed of living, thin-walled cells of variable function. 518

**parental generation (P₁)** The first generation in a genetic cross. 210

**parental investment** The time and resources a parent spends on producing and raising its offspring. 836

**parthenogenesis** *par-then-o-GEN-eh-sis* Female reproduction without fertilization. 164

**passive immunity** Immunity generated when an organism receives antibodies from another organism. 791

**passive transport** Movement of substances across a membrane down a concentration gradient. 71

**pectoral girdle** Collarbones and shoulder blades in the vertebrate skeleton. 682

**pedigree** A chart showing relationships of relatives and which ones have a particular trait. 190

**pelagic zone** *pah-LA-gik ZONE* Water above the ocean floor. 906

**pelvic girdle** Bones that support a vertebrate's hind limbs. 682

**penetrance** The percentage of individuals with a particular genotype who express the associated phenotype. 194

**penis** Male organ of copulation and, in mammals, urinary excretion. 804

**pepsin** A stomach enzyme that chemically digests protein. 748

**peptide bond** *PEP-tide BOND* A chemical bond between amino acids; results from dehydration synthesis. 37

**peptide hormone** A water-soluble, amino acid–based hormone that cannot freely diffuse through a cell membrane. 659

**peptidoglycan** A complex, cross-linked polysaccharide in a bacterial cell wall. 389

**per capita rate of increase** The difference between the birthrate and death rate in a population. 853

**perception** An animal's interpretation of a sensation. 638

**pericardium** *pear-ih-KAR-dee-um* Connective tissue sac that houses the heart. 706

**pericycle** *PEAR-ee-si-kel* A ring of cells in a root's cortex that produces branch roots. 530

**periderm** *PEAR-ih-derm* Outer covering of woody stems and roots. Includes cork, cork cambium, and phelloderm. 532

**periodic table** Chart that lists naturally occurring elements according to their properties. 18

**peripheral nervous system** Neurons that transmit information to and from the CNS. 617

**peristalsis** *pear-ih-STAL-sis* Waves of muscle contraction that propel food along the digestive tract. 747

**peritubular capillaries** Capillaries that surround renal tubules in kidney nephrons. 769

**permafrost** Permanently frozen part of the ground in the tundra. 901

**peroxisome** *per-OX-eh-soam* A membrane-bounded sac that buds from the smooth ER and that houses enzymes important in oxygen use. 57

**persistent organic pollutants** Thousands of carbon-based chemicals that pollute ecosystems, including chlorofluorocarbons and certain pesticides. 916

**petal** A flower part whose color often lures pollinators. 554

**petiole** *PET-ee-ol* The stalklike part of a leaf. 524

**petrifaction** The preservation of tissue structure by replacing biomolecules with minerals. 328

**P_fr, P_r** The far-red and red forms, respectively, of the plant pigment phytochrome. $P_r$ absorbs red light and converts to $P_{fr}$, and vice versa. 580

**phagocyte** White blood cell that engulfs and digests foreign material and cell debris. 778

**pharyngeal pouch** Where endoderm and ectoderm grow toward each other in the throat region of a chordate embryo. 490

**pharynx** A muscular tube that connects the mouth and esophagus. 490, 729, 748

**phenocopy** *FEEN-o-kop-ee* An environmentally caused trait that resembles an inherited trait. 198

**phenotype** *FEEN-o-type* Observable expression of a genotype. 187

**pheromone** *FER-o-moan* Biochemical an organism secretes that elicits a response in another member of the species. 638, 658

**phloem** *FLOW-m* Plant tissue that transports photosynthetic products and other dissolved chemicals. 423, 519

**phloem sap** Solution of water, simple sugars, and other dissolved substances in phloem. 546

**phosphate** A functional group composed of phosphorous and oxygen, having the formula $PO_4$. 34

**phospholipid** *FOS-fo-LIP-id* A molecule consisting of two fatty acids and a phosphate; hydrophobic at one end and hydrophilic at the other end. 34, 67

**phosphorylation** The addition of a phosphate to a molecule. 98

**photochemical smog** *fo-to-KEM-i-kal SMOG* Air pollution resulting from chemical reactions among pollutants in sunlight. 914

**photon** *FOE-ton* A packet of light energy. 89, 109

**photoperiodism** *fo-toe-PER-ee-o-diz-um* A plant's ability to measure seasonal changes by the length of day and night. 579

**photoreactivation** *fo-toe-re-ak-ti-VAY-shun* A type of DNA repair in which an enzyme uses light energy to break pyrimidine dimers. 238

**photoreceptor** Receptor cell sensitive to light energy. 644

**photorespiration** *fo-toe-res-per-A-shun* A process that counteracts photosynthesis. 118

**photosynthate** *fo-toe-SIN-thate* The product of photosynthesis. 106

**photosynthesis** *fo-toe-SIN-the-sis* The series of biochemical reactions that enable plants to harness sunlight energy to manufacture organic molecules. 106

**photosystem** *FO-toe-sis-tum* A cluster of pigment molecules that enables photosynthetic organisms to harness solar energy. 112

**phototroph** *FO-toe-trofe* An organism that derives energy from the sun. 391

**phototropism** *fo-toe-TROP-iz-um* A plant's growth towards unidirectional light. 574

**pH scale** A measurement of how acidic or basic a solution is. 25

**phylogeny** *fi-LODJ-ah-nee* Depiction of evolutionary relationships among species. 304

**phylum** *FI-lum* A major taxonomic group, just beneath kingdoms. 7

**phytochrome** *FI-toe-krome* A pale blue plant pigment involved in timing flowering, seed germination, and other processes. 580

**pilus** (pl. pili) Short projection on bacterial cells; attaches to objects or other cells. 390

**pineal gland** A small structure in the brain that produces melatonin in response to light and dark periods. 672

**pioneer species** The first species to colonize an area devoid of life. 871

**pistil** *PIS-til* The female reproductive structure in a flower. 555

**pith** Ground tissue in the center of a stem or root, within the vascular cylinder. 524

**pituitary gland** A pea-sized gland attached to the vertebrate brain; releases several types of hormones. 664

**placebo** *pla-SEE-bo* An inert substance used as an experimental control. 13

**placenta** A structure that connects the developing fetus to the maternal circulation in many mammals. 509, 813

**placental mammal** A mammal in which the developing fetus is nourished by a placenta. 508

**placoderm** An extinct line of giant fishes with jaws, paired fins, and notochord with some bone. 497

**Plantae** *PLAN-tay* The plant kingdom. 8

**plaque** With respect to viral infections, a zone of killed cells in a layer of host cells. 368

**plasma** A watery, protein-rich fluid that forms the matrix of blood. 592, 711

**plasma cells** Mature B cells that secrete large quantities of a single antibody type. 782

**plasmid** Small, circular DNA apart from an organism's chromosome. 257, 388

**plasmodesma** *plaz-mo-DEZ-ma* Connection between plant cells that allows cytoplasm to flow between them. 80

**plasmodium** *plaz-MO-dee-um* A multinucleated mass of an acellular slime mold. 407

**plastid** *PLAS-tid* A double-membraned plant organelle. 111

**platelet** A cell fragment that is part of the blood and orchestrates clotting. 592, 713

**plate tectonics** The movement of landmasses resting on plates that float on molten rock. 308

**pleiotropic** *PLY-o-TRO-pik* A genotype with multiple expressions. 195

**point mutation** A change in a single DNA base. 251

**polar** Used to describe a difference that exists between two poles, such as a difference in charge between two ends of a single molecule. 21

**polar body** A small cell generated during female meiosis, enabling cytoplasm to be partitioned into just one of the four meiotic products, the ovum. 176

**polar covalent bond** *PO-lar co-VAY-lent BOND* A covalent bond in which electrons are attracted more toward one atom's nucleus than to the other. 21

**polar nuclei** *PO-lar NU-klee-i* The two nuclei in a cell of a plant's megagametophyte that fuse with sperm nucleus to form the triploid endosperm nucleus. 558

**pollen grain** Immature male gametophyte in seed plants. 431, 558

**pollen sac** One of four cavities in an anther, in which pollen grains are produced. 555

**pollen tube** A tube, formed upon germination of a pollen grain, that carries sperm to the ovule. 435

**pollination** Transfer of pollen from an anther to a receptive stigma. 558

**pollution** Any change in air, land, or water that adversely affects organisms. 913

**polyacrylamide** A compound used in electrophoresis to separate DNA fragments with tiny differences such as a single base. 261

**polygamy** A mating system in which a member of one sex associates with several members of the opposite sex. 836

**polygenic** *pol-ee-JEAN-ik* A trait caused by more than one gene. 200

**polymer** *POL-eh-mer* A long molecule composed of similar subunits. 30

**polymerase chain reaction (PCR)** A series of steps by which a DNA sequence can be amplified to produce millions of copies of the sequence. 259

**polyp** The sessile form of a cnidarian. 470

**polyploidy** *POL-ee-PLOID-ee* A cell with extra chromosome sets. 220, 307

**polysaccharide** A complex carbohydrate consisting of hundreds of linked monosaccharides. 32

**pons** An oval mass in the brainstem where white matter connects the medulla to higher brain structures and gray matter helps control respiration. 626

**population** A group of interbreeding organisms living in the same area. 4, 288, 850

**population bottleneck** A type of genetic drift. An event kills many members of a population, and a small number of individuals restore its numbers, restricting the gene pool. 292

**population density** The number of individuals of a species per unit area or volume of habitat. 850

**population dispersion** The pattern in which individuals are scattered throughout a habitat. 851

**positive feedback** Mechanism by which the products of a process stimulate that process. 101, 598, 661

**postanal tail** One of the four characteristics of chordates in which the notochord and associated muscles extend posteriorly beyond the anus. 490

**postsynaptic neuron** *post-sin-AP-tik NUHR-on* One of two adjacent neurons that receives a message. 618

**postzygotic reproductive isolation** The separation of species due to nonviability of a hybrid embryo or offspring. 305

**potential energy** The energy stored in the position of matter. 90

**prebiotic simulation** Experiment that attempts to recreate the conditions on early Earth that gave rise to the first cell. 346

**precocial** Babies that are born capable of independent behavior. 506

**predator** An organism that kills another for food. 868

**preembryonic stage** Prenatal development before the organism folds into layers. 804

**preformation** The idea that a gamete or fertilized ovum contains an entire preformed organism. 798

**pressure flow theory** Theory that explains how phloem sap moves from photosynthetic tissues to nonphotosynthetic plant parts. 547

**presynaptic neuron** *pre-sin-AP-tic NUHR-on* Neuron that releases neurotransmitters into a synaptic cleft. 618

**prey** An organism killed by another for food. 868

**prezygotic reproductive isolation** The separation of species due to factors that prevent the formation of a zygote. 305

**primary growth** Lengthening of a plant due to cell division in the apical meristems. 516

**primary immune response** The immune system's response to its first encounter with a foreign antigen. 784

**primary motor cortex** A band of cerebral cortex extending from ear to ear across the top of the head that controls voluntary muscles. 628

**primary oocyte** *PRI-mare-ee oh-oh-site* An intermediate in ovum formation. 176

**primary producer** Species forming the base of a food web or the first link in a food chain. 875

**primary spermatocyte** *PRI-mare-ee spur-MAT-o-site* An intermediate in sperm formation. 174

**primary somatosensory cortex** A part of the cerebral cortex that receives sensory input from the skin, muscles, joints and bones. 628

**primary (1°) structure** The amino acid sequence of a protein. 37

**primary succession** Appearance of life in an area previously devoid of life. 870

**primer** A short DNA segment at which replication begins, especially in the polymerase chain reaction (PCR). 259

**primitive streak** Pigmented band along the back of an embryo; develops into the notochord. 811

**principle of competitive exclusion** The idea that two or more species cannot indefinitely occupy the same niche. 867

**principle of superposition** The idea that lower rock layers are older than those above them. 274

**prion** *PREE-on* Infectious protein particle. 380

**producer** Also called an autotroph; organism that produces its own nutrients from inorganic carbon sources. 4

**product** The result of a chemical reaction. 20

**product rule** The chance of two events occurring equals the product of the chances of either event occurring. 192

**profundal zone** *pro-FUN-dal ZONE* The deep region of a lake or pond where light does not penetrate. 902

**progenote** *pro-JEAN-note* Collection of nucleic acid and protein; forerunner to cells. 349

**progeny virions** Newly synthesized viral particles assembled inside a host cell. 375

**progesterone** A steroid hormone produced by the ovaries that controls secretion patterns of other reproductive hormones. 673

**prolactin** A pituitary hormone that stimulates milk production. 667

**prometaphase** *pro-MET-a-faze* The stage of mitosis just before metaphase, when condensed and paired chromosomes approach the center of the cell. 150

**promoter** *pro-MOW-ter* A control sequence near the start of a gene; attracts RNA polymerase and transcription factors. 240

**prophase** *PRO-faze* The first stage of mitosis, when chromosomes condense and become visible. 149

**prostaglandin** Lipid released locally and transiently at the site of a cellular disturbance. 658

**prostate gland** A small gland that produces a milky, alkaline fluid that activates sperm. 804

**protein** *PRO-teen* A polymer of amino acids. 35

**Protista** *pro-TEES-ta* The kingdom that includes mostly unicellular, eukaryotic organisms. 404

**proton** *PRO-ton* A particle in an atom's nucleus carrying a positive charge and having one mass unit. 18

**protonephridium** (pl. protonephridia) *pro-toe-nef-RID-ee-um* Structure in flatworms that helps maintain internal water balance. 472

**protoplasm** *PRO-tow-plaz-m* Living matter. 31

**protostome** An animal lineage with spiral, determinate cleavage and a blastopore that develops into a mouth. 464

**proximal convoluted tubule** Region of the nephron, adjacent to the glomerular capsule, where selective reabsorption of useful components of the glomerular filtrate occurs. 771

**pseudocoelom** *soo-doe-SEAL-ohm* A fluid-filled animal body cavity lined by endoderm and mesoderm. 465

**pseudocoelomate** *soo-doe-SEAL-oh-mate* An animal with a pseudocoelom. 465

**pulmonary artery** The artery that leads from the right ventricle to the lungs. 707

**pulmonary circulation** Blood circulation through the heart and lungs. 704

**pulmonary vein** Vein that leads from the lungs to the left atrium. 707

**pulp** The soft inner part of a tooth, consisting of connective tissue, blood vessels, and nerves. 747

**punctuated equilibrium** The view that life's history has had periods of little change interrupted by bursts of rapid change. 311

**Punnett square** A device used to diagram the various possible genetic results of combining gametes. 188

**pupil** The opening in the iris that admits light into the eye. 645

**purine** *PURE-een* A type of organic molecule with a double ring structure, including the nitrogenous bases adenine and guanine. 232

**pyloric sphincter** Circular muscle at the stomach's exit. 749

**pyramid of biomass** A diagram depicting dry weight of organisms at each trophic level. 878

**pyramid of energy** A diagram depicting energy stored at each trophic level at a given time. 875

**pyramid of numbers** A diagram depicting number of organisms at each trophic level. 875

**pyrimidine** *pie-RIM-eh-deen* A type of organic molecule with a single ring structure, including the nitrogenous bases cytosine, thymine, and uracil. 232

**pyruvic acid** *pie-ROO-vik AS-id* The product of glycolysis. 128

## Q

**quaternary (4°) structure** *QUAT-eh-nair-ee STRUK-sure* The organization of polypeptide chains of a protein. 37

## R

**radial cleavage** The pattern of directly aligned blastomeres in the early deuterostome embryo. 465

**radial symmetry** An animal body form in which any plane passing from one end to the other divides the body into mirror images. 463

**radiometric dating** Using measurements of natural radioactive decay as a clock to date fossils. 331

**radula** *RAD-yew-la* A chitinous, tonguelike structure that mollusks use to eat. 473

**ray-finned fish** Group of bony fishes with fins supported by parallel bony rays connected by webs of thin tissue. 499

**reabsorption** *re-ab-SORP-shun* The kidney's return of useful substances to the blood. 770

**reactant** *re-AK-tant* A starting material in a chemical reaction. 19, 38

**reaction chain** A sequence of releasers that joins several behaviors. 829

**realized niche** The resources in a species' environment that it can actually use, considering competition and other limitations. 867

**receptacle** The area where a flower attaches to a floral stalk. 554

**receptor-mediated endocytosis** *re-CEP-ter ME-dee-a-ted en-do-ci-TOE-sis* Binding of a ligand by a cell surface protein stimulates the cell to draw in the ligand in a vesicle. 73

**receptor potential** A change in membrane potential in a neuron specialized as a sensory receptor. 640

**recessive** *re-SESS-ive* An allele whose expression is masked by the activity of another allele. 184

**reciprocal translocation** *re-SIP-ro-kal tranz-lo-CAY-shun* Two nonhomologous chromosomes exchanging parts. 223

**recombinant** Containing genetic material from two or more sources. Recombinant chromosomes include genes from both parents. 210

**recombinant DNA** DNA assembled by putting together a DNA fragment cut by restriction enzymes with plasmid DNA and ligase to create a new plasmid carrying a gene of interest. 257

**rectum** A storage region leading from the large intestine to the anus. 753

**red alga** A type of multicellular photosynthetic protist. 416

**red blood cell** A disc-shaped cell, lacking a nucleus, that contains hemoglobin. 592, 712

**reduction** The gain of one or more electrons by a reactant. 95

**reduction division** Meiosis I, when the diploid chromosome number is halved. 168

**reflex arc** A neural pathway that links a sensory receptor and an effector. 625

**regulatory protein** A protein that controls the activity of a specific enzyme by binding to it. 100

**relative dating** Determining the age of a fossil by comparisons to known ages of adjacent fossils in rock strata. 331

**releaser** A sign stimulus that carries information between members of the same species. 828

**releasing hormone** A hormone produced by the hypothalamus that stimulates another gland. 665

**renal cortex** The outer portion of a kidney. 769

**renal medulla** The middle part of a kidney. 769

**renal pelvis** The inner part of a kidney, where urine collects. 770

**renal tubule** The tubule part of a nephron that contains filtrate. 769

**renin** A hormone that stimulates release of aldosterone. 773

**replication fork** A locally unwound portion of DNA where replication occurs. 237

**repressor** A regulatory protein that inhibits transcription. 240

**reproductive system** A system of organs that produces and transports gametes and may nurture developing offspring. 597

**reptile** A tetrapod vertebrate with an amniote egg, ectothermy, and a tough scaly body covering; the first vertebrate to reproduce independent of water. 502

**reservoir** An organism that can harmlessly harbor a virus that infects a different species. 378

**resource partitioning** Specialization of resource use; allows species with similar niches to coexist. 867

**respiratory cycle** The cycle of one inhalation and one exhalation in vertebrates. 733

**respiratory surface** Site of external respiration. 722

**respiratory system** Organ system that acquires oxygen gas and releases carbon dioxide. 597

**resting potential** The electrical potential inside of a neuron not conducting a nerve impulse. 614

**restriction endonuclease** One of a group of enzymes that cut DNA in highly specific locations. 256

**reticular formation** A diffuse network of nerve tracts that extends through the brainstem and into the thalamus; screens sensory input to the cerebrum. 627

**retina** *RET-eh-na* A sheet of photoreceptors at the back of the human eye. 645

**retinal** Photosensitive portion of rhodopsin molecule. 647

**retrovirus** Viruses with a genome consisting of single-stranded RNA. 257

**reuptake** The reabsorption of a neurotransmitter by the presynaptic axon after release. 619

**reverse transcriptase** *re-VERS tran-SCRIPT-aze* An enzyme that uses RNA as a template to construct a DNA molecule. 257, 348

**R group** An amino acid side chain. 36

**rhizoid** *RI-zoyd* Rootlike extension on gametophytes of some nonvascular plants; anchors the plant and absorbs water and minerals. 425

**rhizome** *RI-zome* Fleshy, horizontal underground stem. 430

**rhizosecretion** The ability of roots to secret proteins; may be exploited for manufacture of commerical protein products. 542

**rhodopsin** *roe-DOP-sin* A pigment that transduces light into an electrochemical signal in photoreceptor cells. 644

**ribonucleic acid** *RI-bo-nu-KLAY-ik AS-id* (**RNA**) A single-stranded nucleic acid consisting of nucleotides containing a phosphate, ribose, and nitrogenous bases adenine, guanine, cytosine, and uracil. 39

**ribose** *RI-bose* The five-carbon sugar that is a structural component of RNA. 39, 231

**ribosomal RNA** *RI-bo-SOAM-el RNA* (**rRNA**) RNA that, along with proteins, forms a ribosome. 239

**ribosome** *RI-bo-soam* A structure built of RNA and protein where mRNA anchors during protein synthesis. 51, 55, 389

**ribulose bisphosphate** *RIH-byu-los bis-FOS-fate* The five-carbon intermediate of the carbon reactions of photosynthesis. 116

**RNA polymerase** *RNA poe-LIM-er-ase* An enzyme that takes part in DNA replication and RNA transcription. 239

**RNA primer** A small piece of RNA, inserted at the start of a piece of DNA to be replicated, which attracts DNA polymerase and is later removed. 237

**RNA world** The theory that the first genetic material was RNA. 348

**Robertson translocation** A chromosome aberration in which the short arms of two nonhomologs break, leaving sticky ends that then join the two long arms into a new, large chromosome. 222

**rod cell** Specialized receptor cell in the retina that provides black-and-white vision. 646

**root** The underground part of a plant. 516

**root cap** A thimble-shaped mass of cells that protects a growing root tip. 528

**root hair** Outgrowth of root epidermal cell; increases the surface area for absorbing water and minerals. 519

**rough endoplasmic reticulum** The ribosome-studded organelle where secreted proteins are synthesized. 52

**r-selection** Selection for individuals that are short-lived, reproduce early, and have many offspring that each get little parental investment. 859

**rubisco** The enzyme that adds carbon to ribulose bisphosphate. 116

**ruga** (pl. rugae) Fold in the mucosa of the stomach. 748

**rumen** A bacteria-rich part of the stomach of large grazing animals. 744

**ruminant** An animal that has a rumen. 744

## S

**saccule** *SAK-yul* A pouch in the vestibule of the inner ear, containing a jellylike fluid and calcium carbonate granules that move in response to changes in velocity. 652

**sacrum** Fused pelvic vertebrae. 683

**salivary amylase** An enzyme produced in the mouth; begins chemical digestion of starch. 747

**salt** A molecule composed of cations and anions. 22

**saltatory conduction** Jumping of an action potential between nodes of Ranvier in myelinated axons. 616

**saprotroph** An organism that absorbs nutrients from dead organic matter. 444

**sapwood** Outermost wood that actively transports water and dissolved nutrients within a plant. 532

**sarcolemma** *sar-ko-LEM-a* Cell membrane of a skeletal muscle cell. 689

**sarcomere** *SAR-ko-meer* A pattern of repeated bands in skeletal muscle. 692

**sarcoplasmic reticulum** *sar-ko-plaz-mic ret-IK-u-lum* The endoplasmic reticulum of a skeletal muscle cell. 689

**saturated** (fat) A triglyceride with single bonds between the carbons of its fatty acid tails. 34

**savanna** One of several terrestrial biomes; a grassland with scattered trees. 895

**Schwann cell** A type of neuroglia that forms a sheath around certain neurons. 594, 616

**scientific method** A systematic approach to interpreting observations; involves reasoning, predicting, testing, and drawing conclusions and then putting them into perspective with existing knowledge. 10

**sclera** The outermost layer of the eye; the white of the eye. 645

**sclereid** A type of plant cell that may contain hard inclusions. 518

**sclerenchyma** *sklah-REN-kah-mah* Supportive plant tissue composed of cells with thick, nonstretchable secondary cell walls. 518

**scrotum** *SKRO-tum* The sac of skin containing the testes. 804

**sebaceous gland** *se-BAY-shis GLAND* Gland in human skin that secretes a mixture of oils that soften hair and skin. 605

**secondary growth** Thickening of a plant due to cell division in lateral meristems. 516

**secondary immune response** The immune system's response to subsequent encounters with a foreign antigen. 784

**secondary oocyte** *SEC-un-derry OH-OH-site* A haploid cell that is an intermediate in ovum formation. 176

**secondary spermatocyte** *SEC-un-derry sper-MAT-o-site* A haploid cell that is an intermediate in sperm formation. 174

**secondary (2°) structure** The shape a protein assumes when amino acids close together in the primary structure chemically attract. 37

**secondary succession** Change in a community's species composition following a disturbance. 871

**second messenger** A biochemical activated by an extracellular signal; transmits a message inside a cell. 660

**secretion** A cell's release of a biochemical. Also, the addition of substances to the material in a kidney tubule. 770

**seed** In seed plants, a dormant sporophyte within a protective coat. 561

**seed coat** A tough outer layer of a seed; protects a dormant plant embryo and its food supply. 561

**seedless vascular plant** A plant with specialized vascular tissues but that does not produce seeds. 423

**segmentation** In digestion, localized muscle contractions in the small intestine that provide mechanical digestion. Also, division of an animal body into repeated subunits. 750

**selectin** A type of cellular adhesion protein that slows white blood cells down in the vicinity of an injury. 81

**semicircular canal** Fluid-filled structure in the inner ear; provides information on the position of the head. 652

**semiconservative replication** Mode of DNA replication in which each new double helix has one parental and one new strand. 234

**semilunar valve** Tissue flaps in the artery just outside each ventricle; maintain one-way blood flow. 706

**seminal vesicle** *SEM-in-el VES-eh-kel* In the human male, one of a pair of structures that adds fructose and prostaglandins to sperm. 804

**seminiferous tubule** Tubule within the testis where sperm form and mature. 175, 804

**senescence** The cessation of growth and subsequent aging of tissues. 582

**sensory adaptation** Lessening of sensation with prolonged exposure to the stimulus. 640

**sensory neuron** A neuron that transmits information from a stimulated body part to the CNS. 611

**sensory receptor** Sensory neuron or specialized epithelial cell that detects and passes stimulus information to a sensory neuron. 638

**sepal** *SEE-pel* A leaflike floral structure that encloses and protects inner floral parts. 554

**septate hypha** A hypha with crosswalls that partition off individual cells. 446

**seta** (pl. setae) *SEE-ta* Bristle on the side of an earthworm; also, hairlike, vibration-sensitive structure at base of insect antenna. 649

**sex chromosome** A chromosome that carries genes that determine sex. 168

**sex pilus** *SEX PILL-us* A prokaryotic cell outgrowth that transfers DNA from one cell to another. 395

**sexual dimorphism** The difference in appearance between males and females of the same species. 836

**sexual reproduction** The combination of genetic material from two individuals to create a third individual. 6, 164, 553

**sexual selection** Natural selection of traits that increase an individual's reproductive success. 279

**shoot** The aboveground part of a plant. 516

**short-day plant** Plant that requires dark periods longer than some critical length to flower. 579

**sieve cell** Relatively unspecialized phloem cell, found in gymnosperms and seedless vascular plants. 521

**sieve tube member** Specialized conducting cell in phloem of flowering plants. 521

**sign stimulus** A feature of an object that triggers a behavior. 828

**signal transduction** The biochemical transmission of a message from outside the cell to inside. 82

**signature sequence** DNA sequence unique to the members of specific taxonomic groups. 392

**sign stimulus** A feature of an object that triggers a behavior. 828

**single nucleotide polymorphism (SNP)** A single-base site in DNA that tends to vary in a population. SNPs play a role in DNA fingerprinting. 262

**sink** In plants, areas where products of photosynthesis are unloaded from phloem. 547

**sinoatrial (SA) node** Specialized cells in the wall of the right atrium that set the pace of the heartbeat; the pacemaker. 707

**skeletal muscle** Voluntary muscle type composed of unbranched, multinucleated cells that connect to skeletal elements. 594

**skeletal muscle cell** Single, multinucleated cell that contracts when actin and myosin filaments slide. Makes up voluntary, striated muscle. 687

**skeletal system** System of bones and ligaments that support body structures and attach to muscles. 597

**sliding filament model** Sliding of protein myofilaments past each other to shorten muscle cells, leading to contraction. 692

**slime layer** A loose, sticky layer surrounding some prokaryotic cells. 390

**slow-twitch fiber** A muscle fiber that splits ATP slowly and resists fatigue because it is well supplied with oxygen. 693

**small intestine** Connects the stomach with the colon; important site of chemical digestion and absorption. 750

**smooth endoplasmic reticulum** The organelle where lipids are synthesized. 54

**smooth muscle** Type of muscle consisting of involuntary, nonstriated, spindle-shaped cells. 594

**smooth muscle cell** Cell that makes up involuntary, nonstriated contractile tissue that lines the digestive tract and other organs. 687

**sodium-potassium pump** A mechanism that uses energy released from splitting ATP to transport $Na^+$ out of cells and $K^+$ into cells. 72, 613

**solute** SOL-yoot A chemical that dissolves in another, forming a solution. 24, 68

**solution** A homogenous mixture of a substance (the solute) dissolved in a solvent. 24, 63

**solvent** SOL-vent A chemical in which others dissolve, forming a solution. 24, 68

**somatic cell nuclear transfer** The transfer of the nucleus of a differentiated cell into an oocyte whose nucleus has been destroyed or removed. The new organism that develops is a clone of the somatic cell nucleus donor. 801

**somatic mutation** so-MAT-ik mew-TAY-shun A mutation in a body (nonsex) cell. 252

**somatic nervous system** Motor pathways that lead to skeletal muscles. 623

**somatostatin** so-MAT-owe-STAT-in A pancreatic hormone that controls the rate of nutrient absorption into the bloodstream. 671

**source** In plants, areas where products of photosynthesis are loaded into phloem. 547

**Southern blotting** A technique for identifying the location of DNA sequences in a genome by using a known DNA probe to base-pair with fragments blotted on a membrane. 262

**speciation** SPEE-she-AY-shun Appearance of a new type of organism. 303

**species** A group of similar individuals that breed sexually in nature only among themselves, or a microorganism with a distinctive set of features. Also the lowest level in the taxonomic hierarchy 6

**spermatid** SPER-ma-tid An intermediate stage in sperm development. 174

**spermatogenesis** sper-MAT-o-JEN-eh-sis The differentiation of a sperm cell from a diploid spermatogonium, to primary spermatocyte, to two haploid secondary spermatocytes, to spermatids, and finally to mature spermatozoa. 174

**spermatogonium** sper-mat-o-GO-nee-um A diploid cell that divides, yielding cells that become sperm cells. 174

**spermatozoon (pl. spermatozoa)** sper-mat-o-ZO-en Mature sperm cell. 175

**S phase** The synthesis phase of interphase, when DNA replicates and microtubules assemble from tubulin. 148

**sphincter** Muscular ring that controls passage of a substance from one part of the body to another. 748

**spicule** Glassy or limy material that makes up sponge skeletons. 467

**spinal cord** Tube of nervous tissue that extends through the vertebral column. 623

**spinal nerve** Peripheral nerve that exits the vertebrate CNS from the spinal cord. 623

**spiracle** Opening in the body wall of arthropods; used for breathing. 724

**spiral cleavage** Pattern of early cleavage cells in protostomes, resembling a spiral. 465

**spleen** An abdominal organ that produces and stores lymphocytes and red blood cells. 718

**spongy bone** Flat bones and tips of long bones with large spaces between a web of bony struts. 685

**sporangium** A structure in which spores are produced. 425

**spontaneous process** A process that occurs without the input of energy. 92

**spore** Reproductive structure in prokaryotes and fungi that may be specialized for survival or dissemination. In plants, the spore develops into a gametophyte. 442, 446

**sporophyte** SPOR-o-fite Diploid, spore-producing stage of the plant life cycle. 165, 422, 553

**spring turnover** The seasonal mixing of upper and lower layers of a lake. 903

**squamous** SKWAY-mus Flat, as in epithelium. 591

**stabilizing selection** When extreme phenotypes are less adaptive than an intermediate phenotype. 294

**stamen** STAY-men Male reproductive structure in flowers; consists of a stalklike filament with a pollen-producing anther at its tip. 554

**stapes** STAY-peez A small bone in the middle ear. 650

**starch** A polysaccharide used as a storage molecule by plants; a major food source for humans and other animals. 32

**statocyst** STAT-o-sist A fluid-filled cavity that contains minerals and sensory hairs that control balance and equilibrium. 649

**statolith** STAT-o-lith Structure in plant or animal cells that helps the organism detect the direction of gravity. 575, 649

**stem** Central axis of a plant's shoot system. 522

**stem cell** An undifferentiated cell that divides to give rise to additional stem cells and cells that specialize. 45, 155, 590

**steroid hormone** A lipid hormone that can cross a target cell's membrane and bind to intracellular receptors. 659

**sterol** Lipid molecule based on a complex molecule of four interconnected carbon rings. 35

**sticky end** A single-stranded fragment at the end of a DNA double strand that has been cut by a restriction enzyme. Sticky ends can bind together through base pairing. 256

**stigma** STIG-mah The tip of a flower's style; receives pollen grains. 554

**stoma** (pl. stomata) STO-mah Pore in a plant's epidermis through which gases are exchanged between the plant and the atmosphere. 116, 519

**stomach** J-shaped compartment that receives food from the esophagus. 748

**stramenopile** stra-MEEN-o-pile Diverse group of protista including brown algae, diatoms, and water molds. 414

**stroma** STRO-ma The nonmembranous inner region of the chloroplast. 59, 111

**style** A stalk that arises from the top of an ovary in a flower. 545

**submetacentric** A chromosome in which the centromere sets off a short and a long arm. 208

**substrate** A reactant an enzyme acts upon. 38

**substrate-level phosphorylation** ATP formation from transferring a phosphate group to ADP. 128

**superior vena cava** The upper branch of the largest vein that leads to the heart. 707

**superposition** The principle that older fossils lay in lower layers of rock strata. 273

**survivorship** The fraction of a group of individuals that survive to a particular age. 842

**suspension feeder** Animal that captures food particles suspended in the water. 475, 493

**sustainability** Use of resources while not depleting them for the future. 912

**sweat gland** Epidermal invagination that produces sweat. 603

**swim bladder** An organ that allows bony fishes to adjust their buoyancy. 498

**symbiosis** *sim-bi-O-sis* One type of organism living in or on another. 867

**sympathetic nervous system** Part of the autonomic nervous system; mobilizes the body to respond to environmental stimuli. 623

**sympatric speciation** The formation of a new species within the boundaries of a parent species. 309

**symplastic pathway** In plants, the path that water takes moving from cell to cell through the cytoplasm via plasmodesmata. 545

**synapse** Area where two neurons meet and transmission of electrochemical information occurs. 617

**synapsid** *sin-AP-sid* Vertebrate with a single opening behind each eye orbit; mammals and their immediate ancestors. 503

**synaptic cleft** The space between two neurons at a chemical synapse. 618

**synaptic integration** A cell's overall response to many incoming neural messages. 621

**synaptic knob** Enlarged tip of an axon; contains synaptic vesicles. 618

**synonymous codons** Different codons that encode the same amino acid. 251

**synovial capsule** A lining of fibrous connective tissue between bones in a moveable joint. 695

**synovial joint** A freely moveable joint between bones; these joints have synovial capsules filled with fluid that make movement low in friction. 695

**synteny** *SIN-ten-ee* Comparison of gene order on chromosomes between species. 337

**systematics** *sis-te-MAT-ix* Study of the evolutionary relationships among species. 322

**systemic circulation** Blood circulation through the heart and the body. 704

**systole** *SIS-toll-ee* The contraction of the ventricles of the heart. 707

# T

**taiga** *TIE-gah* One of several terrestrial biomes; the northern coniferous forest. 900

**taproot system** A root system in which lateral root branches arise from a tapering main root. 528

**target cell** A cell that a hormone binds and directly affects. 659

**taxis** *TAX-iss* An orientation movement made toward or away from a stimulus. 833

**taxonomy** *tax-ON-o-mee* Classification of organisms on the basis of evolutionary relationships. 7

**T cell** T lymphocyte; a component of the immune system. 780

**tectorial membrane** *tek-TOR-ee-al MEM-brane* The membrane above hair cells in the cochlea of the inner ear; pressed by hair cells responding to the basilar membrane's vibration in the presence of sound waves. 651

**telocentric** *tell-o-SEN-trik* A chromosome with the centromere at the tip. 208

**telomerase** *tell-OM-er-ase* An enzyme that extends chromosome tips using RNA as a template. 154

**telomere** *TELL-o-meer* A chromosome tip. 154

**telophase** *TELL-o-faze* The final stage of mitosis, when two nuclei form from one and the spindle is disassembled. 151

**temperate coniferous forest** One of several terrestrial biomes; coniferous trees dominate. 899

**temperate deciduous forest** *TEMP-er-et de-SID-yoo-us FOR-est* One of several terrestrial biomes; dominated by deciduous trees. 899

**temperate grassland** One of several terrestrial biomes; grazing, fire, and drought restrict tree growth. 898

**temporal caste** A group in a society with a role that changes with time. 839

**tendon** A band of fibrous connective tissue attaching a muscle to a bone. 688, 695

**teratogen** *teh-RAT-eh-jen* Something that causes a birth defect. 818

**territoriality** Behavior that defends one's area. 834

**tertiary (3°) structure** *TER-she-air-ee STRUK-sure* The shape a protein assumes when amino acids far apart in the primary structure chemically attract one another. 37

**test cross** Breeding an individual of unknown genotype to a homozygous recessive individual to reveal the unknown genotype. 188

**testis** Male gonad containing seminiferous tubules, where sperm are manufactured. 175, 673, 804

**testosterone** A male steroid hormone that regulates sperm production and development of male characteristics. 35, 673

**tetanus** Maximal muscle contraction caused by continual stimulation. 694

**tetrapod** A four-limbed vertebrate. 490

**thalamus** A tight grouping of nerve cell bodies beneath the cerebrum; relays sensory input to the cerebrum. 627

**thallus** *THALL-us* Relatively undifferentiated body form lacking roots, stems, and leaves. 446

**theory** An explanation for observations and experimental evidence of natural phenomena. 10

**thermodynamics** *THERM-o-die-NAM-ix* Study of energy transformations in nature. 91

**thermogenesis** *therm-o-JEN-eh-sis* Generating heat metabolically. 763

**thermoreceptor** A receptor cell that responds to temperature changes. 652

**thermoregulation** Ability of an animal to balance heat loss and gain with the environment. 762

**thigmonasty** *THIG-mo-NAS-tee* A nastic response to touch. 576

**thigmotropism** *thig-mo-TRO-piz-um* A plant's growth response toward touch. 576

**thoracic vertebra** Vertebra of the upper back. 683

**threshold potential** The point at which the membrane of a neuron is depolarized sufficiently to trigger an action potential. 614

**thylakoid** *THI-lah-koyd* Disclike structure that makes up the inner membrane of a chloroplast. 59

**thylakoid membrane** The membrane of the thylakoid, the location of the light reactions of photosynthesis. 111

**thylakoid space** The inner compartment of the thylakoid. 111

**thymus** A lymphatic organ in the upper chest where T cells learn to distinguish foreign from self antigens. 718

**thyroid gland** In humans, a gland in the neck that manufactures thyroxine, a hormone that increases energy expenditure. 668

**thyroid-stimulating hormone (TSH)** A pituitary hormone that stimulates the thyroid gland to release two types of hormones. 667

**thyrotropin-releasing hormone** A hormone secreted by the hypothalamus that stimulates release of TSH from the pituitary. 669

**thyroxine** A thyroid hormone that increases the rate of cellular metabolism. 668

**tidal volume** Volume of air inhaled or exhaled during a normal breath. 733

**tight junction** A connection between two cells that prevents fluid from flowing past the cells. 79

**tissue** Group of cells with related functions. 2, 590

**tonsil** Collection of lymphatic tissue in the throat. 718

**topsoil** The uppermost layer of soil, often rich in organic matter. 541

**totipotent** Capable of giving rise to a complete organism or to cells that can specialize in any way. 800

**trace element** An element an organism requires in small amounts. 18

**trachea (pl. tracheae)** In vertebrates, the respiratory tube just beneath the larynx; the "windpipe." In invertebrates, a branched tubule that brings the external environment in close contact with cells, facilitating gas exchange. 724, 729

**tracheid** *TRAY-kee-id* Relatively unspecialized conducting cell in xylem. 521

**tracheole** Tiny fluid-filled branch of arthropod trachea, contacts individual cells. 724

**transcription** *tranz-SKRIP-shun* Using DNA as a template to manufacture RNA. 239

**transcription factor** *tranz-SCRIPT-shun fac-TOR* A protein that turns on and off different genes in a particular cell. 241

**transcytosis** Movement of a substance into and across a cell for release at the opposite side. 73

**transduction** Conversion of energy from one form to another. Also, a method of horizontal gene transfer in which a virus transfers DNA from one host cell to another. 106, 397

**transfer RNA (tRNA)** A small RNA molecule that binds an amino acid at one site and an mRNA codon at another site. 239

**transformation** Method of horizontal gene transfer in which an organism takes up naked DNA from the environment. 395

**transgenic technology** The use of recombinant DNA technology in multicellular organisms. 265

**translation** Assembly of an amino acid chain according to the sequence of base triplets in a molecule of mRNA. 239

**transmissible spongiform encephalopathy** A disease, caused by the ingestion of prions, that riddles the brain with characteristic holes and lesions. 380

**transpiration** The movement of water vapor from plant parts to the atmosphere through open stomata. 519, 543

**transplantation** Replacing a diseased or damaged organ with one from a donor. 60, 600

**transposable element** *tranz-POSE-a-bull EL-e-ment* Jumping gene. 261

**transverse (T) tubule** Part of the sarcolemma that juts from the sarcoplasmic reticulum of a skeletal muscle cell. 689

**trichome** A hairlike outgrowth of plant epidermis. 519

**triglyceride** *tri-GLI-sir-ide* A type of fat that consists of one glycerol bonded to three fatty acids. 34

**triiodothyronine** *tri-i-o-do-THI-ro-neen* A thyroid hormone that increases the rate of cellular metabolism. 668

**triploblastic** An animal whose adult tissues arise from three germ layers in the embryo. 464

**trisomy** *TRI-som-mee* A cell with one extra chromosome. 218

**trophic level** *TRO-fik LEV-l* An organism's position along a food chain. 875

**trophoblast** *TRO-fo-blast* A layer of cells in the preembryo; develops into the chorion and then the placenta. 810

**tropical dry forest** One of several terrestrial biomes; yearly dry and rainy seasons in tropics. 894

**tropical rain forest** One of several terrestrial biomes; year-round high temperatures and precipitation. 891

**tropic hormone** A hormone that affects another hormone's secretion. 665

**tropism** *TRO-piz-um* Plant growth toward or away from an environmental stimulus. 574

**tropomyosin** *TRO-po-MI-o-sin* A type of protein in thin myofilaments of skeletal muscle cells. 689

**troponin** *tro-PO-nin* A type of protein in thin myofilaments of skeletal muscle cells. 689

**true fern** A type of seedless vascular plant. 431

**trypanosome** Protist (Kinetoplastida) that causes human diseases transmitted by biting flies. 408

**tube (vegetative) cell** A haploid cell in a pollen grain that gives rise to the pollen tube. 558

**tube foot** Cuplike structure connected to the water vascular system in echinoderms; provides locomotion. 483

**tumor necrosis factor** An immune system biochemical with varied functions in cancer, infection, and inflammation. 779

**tumor suppressor gene** A gene which, when inactivated or suppressed, causes cancer. 160

**tundra** *TUN-drah* One of several terrestrial biomes; low temperature and short growing season. 901

**tunicate** *TOON-i-cat* A type of invertebrate chordate. 492

**turgor pressure** *TER-ger PRESH-er* Rigidity of a cell caused by water pressing against the cell wall. 70

**twitch** Rapid contraction and relaxation of a muscle cell following a single stimulation. 693

**tympanal organ** *tim-PAN-al OR-gan* A thin part of an insect's cuticle; detects vibrations and therefore sound. 649

**tympanic membrane** The eardrum, a structure upon which sound waves impinge. 650

## U

**ultraviolet (UV) radiation** Portion of the electromagnetic spectrum with wavelengths shorter than 400 nm. 916

**umbilical cord** *um-BIL-ik-kel KORD* A ropelike structure that contains one vein and two arteries; connects a pregnant female placental mammal to unborn offspring. 805

**unconditioned stimulus** A stimulus that normally triggers a particular response. 830

**unicellular** Consisting of a single cell. 6

**uniformitarianism** The idea that geological changes are the result of steady, gradual processes whose overall rates remain constant. 273

**uniramous** *yoon-i-RAY-muss* Unbranched. 479

**unsaturated** *un-SAT-yur-RAY-tid* (fat) A triglyceride with at least one double bond between the carbons in its fatty acid tails. 34

**upwelling** Upward movement of cold, nutrient-rich lower layers of a body of water. 906

**urea** A nitrogenous waste derived from ammonia. 766

**ureter** A muscular tube that transports urine from the kidney to the bladder. 768

**urethra** Tube that transports urine from the bladder out of the body. 768, 804

**uric acid** A nitrogenous waste derived from ammonia. 766

**urinary bladder** A muscular sac where urine collects. 768

**urinary system** Organ system that filters blood and helps maintain concentrations of body fluids. 768

**uterine tube** Also, falopian tube. In the human female, one of a pair of tubes leading from near the ovaries to the uterus. 804

**uterus** The muscular, saclike organ in a female placental mammal where the embryo and fetus develop. 804

**utricle** *YOO-trah-kel* A pouch in the vestibule of the inner ear filled with a jellylike fluid that contains calcium carbonate granules that move in response to changes in velocity. 652

## V

**vacuole** *VAK-yoo-ole* A storage sac in a cell. 52

**vagina** Conduit from the uterus to the outside of the body in female mammals. 805

**valence shell** The outermost shell of an atom. 20

**van der Waals attractions** Dynamic attractions within or between oppositely charged molecules. 23

**variable** A portion of an experiment that is changed to test a hypothesis. 13

**variable region** Amino acid sequence that forms the upper part of heavy and light chains; varies for different antibodies. 784

**vasa recta** Portion of peritubular capillary in which blood flows in the opposite direction of the filtrate in the nephron loop. 772

**vascular bundle** *VAS-ku-ler BUN-del* A strand of vascular tissue containing primary xylem, primary phloem, and associated cells. 523

**vascular cambium** *VAS-ku-ler KAM-bee-um* A type of lateral meristem in plants; produces secondary xylem and phloem. 531

**vascular tissue** *VAS-ku-ler TISH-ew* Specialized conducting tissue in plants; xylem and phloem. 519

**vas deferens** In the human male, a tube from the epididymis that joins the urethra in the penis. 175, 804

**vasoconstriction** Decrease in the diameter of a vessel. 707

**vasodilation** Increase in the diameter of a vessel. 707

**vector** A molecule, such as a plasmid, that can carry a DNA fragment into a cell. 257

**vein** In plants, a strand of vascular tissue in leaves. In animals, a vessel that returns blood to the heart. 703

**ventilation** Any mechanism that increases the flow of air or water across a respiratory surface. 725

**ventricle** Space in the brain into which cerebrospinal fluid is secreted. Also, a muscular heart chamber beneath the atria. 703

**venule** Vessel that drains blood from capillaries. 703

**vernal pool** A small body of water formed from snowmelt in the spring that may persist until summer; home to a variety of organisms. 884

**vertebra** One unit of the vertebral column. 494

**vertebral column** *ver-TEE-bral COL-um* Bone or cartilage that supports and protects the spinal cord. 682

**vertebrate** An animal that has a vertebral column. 462, 494

**vertical gene transfer** Inheritance of genetic information from one generation to the next. 394

**vertical stratification** Layering of different plant species in the forest canopy. 893

**vesicle** *VES-i-cal* A membrane-bounded sac in a cell. 52

**vessel element** Specialized conducting cell in xylem. 521

**vestibule** *VES-teh-bule* A structure in the inner ear; provides information on the position of the head with respect to gravity and changes in velocity. 652

**vestigial organ** *ves-TIJ-e-el OR-gan* A structure that seems not to have a function in an organism but that resembles a functional organ in another species. 334

**villus** (pl. villi) Tiny projection on the inner lining of the small intestine; greatly increases surface area for nutrient absorption. 752

**virion** A complete, infectious virus particle. 370

**viroid** *VIE-royd* Infectious RNA molecule. 379

**virus** An infectious particle consisting of a nucleic acid (DNA or RNA) wrapped in protein. 368

**visceral mass** Part of the molluscan body; contains the digestive and reproductive systems. 473

**vital capacity** The maximal volume of air that can be forced in and out of the lungs during one breath. 733

**vitreous humor** *VIT-ree-us HU-mer* A jellylike substance behind the lens that makes up most of the volume of the eye. 647

**viviparous** Bearing live young without retaining eggs in the birth canal (in animals). 494

**vocal cord** Elastic tissue band that covers the glottis and vibrates as air passes, producing sound. 729

## W

**warning coloration** Bright or distinctive display that advertises an organism's defenses. 868

**water mold** Filamentous, heterotrophic protist, also called an oomycete. 415

**water vascular system** System of canals in echinoderms; provides locomotion and osmotic balance. 483

**wavelength** The distance a photon moves during a complete vibration. 109

**whisk fern** A type of seedless vascular plant. 430

**white blood cell** A cell that helps fight infection. 592, 713

**white matter** Nervous tissue in the CNS consisting of myelinated fibers. 624

**wild type** The most common phenotype or allele for a certain gene in a population. 187

## X

**X inactivation** Turning off one X chromosome in each cell of a female mammal at a certain point in prenatal development. 216

**X-linked** Traits, usually recessive, whose alleles reside on the X chromosome. 214

**xylem** *ZI-lem* Plant tissue that transports dissolved ions and water. 423, 519

**xylem sap** Dilute solution of water and dissolved minerals in xylem. 543

## Y

**Y-linked** A gene carried on the Y chromosome. 214

**yeast** A unicellular fungus. 446

**yolk sac** An extraembryonic membrane that forms beneath the embryonic disc and manufactures blood cells. 814

## Z

**zona pellucida** A thin, clear layer of proteins and sugars that surrounds a secondary oocyte. 806

**zoonosis** An infection in humans that originates in another animal. 378

**zoospore** A flagellated spore produced by chytrids and water molds. 448

**zygomycete** *zi-go-MI-seet* A fungus that produces zygospores. 444

**zygospore** *ZI-go-spoor* Diploid resting spore produced by fusion of haploid gametangia in zygomycetes. 450

**zygote** *ZI-goat* The fused egg and sperm that develops into a diploid individual. 166, 798

# *Credits*

## Photographs

### Chapter 1

Opener: Tennessee Aquarium; 1.1/top left: Corbis Eye on Earth CD; 1.1/middle left: © Carl R. Sams II/Peter Arnold, Inc.; 1.1/top right: © Lynn Stone/Animals Animals/Earth Scenes; 1.1/bottom right: © Rod Planck/Photo Researchers, Inc.; 1.1/bottom left: © Rod Planck/Photo Researchers, Inc.; 1.4a: © R. Kessel-G. Shih/Visuals Unlimited; 1.4b: ©Runk/Schoenberger/Grant Heilman; 1.4c: Corbis Animals in Action CD; 1.5a: © Michael and Patricia Fogden/Animals Animals/Earth Scenes; 1.5b: © Michael and Patricia Fogden/Animals Animals/Earth Scenes; 1.6a-c: © Steven Vogel; 1.7/left: © Kwangshin Kim/Photo Researchers, Inc.; 1.7/middle left: © James King-Holmes/Science photo Library/Photo Researchers; 1.7/center left: © Ken W. David/Tom Stack & Associates, Inc.; 1.7/center right: © Dwight Kuhn; 1.7/middle right: © David Dennis/Animals Animals/Earth Scenes; 1.7/right: © Michael Fogden/Animals Animals/Earth Scenes; 1.A: © Katy Payne; 1.9a: © Inga Spence/Tom Stack & Associates; 1.9b: Dr. James G. Sikarskie; 1.9c: Planet Earth Pictures The Innovation Centre; 1.9d: © Scimat/Photo Researchers, Inc.; 1.10a: © Philip Gould/Corbis; 1.10b: © Galen Rowell/Corbis; 1.11: Courtesy National Library of Medicine

### Chapter 2

Opener: © Kevin Krajick; 2.4b: © Dan McCoy/Rainbow; 2.6a-c: 2.11: Courtesy of Diane R. Nelson

### Chapter 3

Opener: © Kenneth Eward/BioGrafix/Photo Researchers, Inc.; 3.2b: © Darylyne A. Murawski/Peter Arnold, Inc; 3.2/inset: © BioPhoto Associates/Photo Researchers, Inc.; 3.5: © Stephen J. Krasemann/DRK Photo; 3.7a: © Richard R. Hansen/Photo Researchers, Inc.; 3.7b: © Gerald Lacz/Peter Arnold, Inc.; 3.7c: © John Cancelosi/Peter Arnold, Inc; 3.10b: Div. of Computer Research & Technology/Photo Researchers, Inc.

### Chapter 4

Opener: © Nancy Kedersha/Immunogen/SPL/Photo Researchers, Inc.; 4.A: Schematic Courtesy Michael Schmid, Institut F. Allgemeine Physik, TU Wien; 4.1a: © Manfred Kage/Peter Arnold, Inc.; 4.1b: © Edwin A. Reschke/Peter Arnold, Inc; 4.2a: © Kathy Talaro/Visuals Unlimited; 4.2b: Science VU/Visuals Unlimited; 4.A/top left: © BioPhoto Assoc./Photo Researchers, Inc.; 4.A/top right: © Doug Martin/Photo Researchers, Inc.; 4.A/middle left: © Don Fawcett/Visuals Unlimited; 4.A/middle right : © SIU/Visuals Unlimited; 4.A/bottom: © Science VU-IBMRL/Visuals Unlimited; 4.C(a): © Fred Hossler/Visuals Unlimited; 4.C(b): Boehringer Ingelheim International Gmbtt, photo by Lennart Nilsson, The Body; 4.C(d): © Aaron H. Swihart and Robert C. Mac Donald; 4.6b: CNRI/Science Photo Library/Photo Researchers, Inc.; 4.7a: © David M. Phillips/Visuals Unlimited; 4.7b: © Thomas Tottleben/Tottleben Scientific Company; 4.7c: © T. E. Adams/Visuals Unlimited; 4.8: From J.T. Staley, M.P. Bryant, N. Pfenning, and J.G. Holt, (Eds.); 4.9/left: © K.R. Porter / Photo Researchers, Inc.; 4.9/right: © Biophoto Assoc./Science Source/Photo Researchers, Inc.; 4.9/top: © David M. Phillips/The Population Council/Science Source / Photo Researchers, Inc.; 4.10: © Biophoto Assoc./Science Source/Photo Researchers, Inc.; 4.12c: © David M. Phillips/The Population Council/Science Source/Photo Researchers, Inc.; 4.13: © R. Bolender - D. Fawcett/Visuals Unlimited; 4 14, 4.15: Prof. P Motta & T. Naguro/Science Photo Library/Photo Researchers, Inc.; 4.16a: © D. Friend - D. Fawcett / Visuals Unlimited; 4.16b: From S.E. Frederick and E.H. Newcomb, Journal of Cell Biology, 4:343—53, 1969; 4.17: © Bill Longcore/Photo Researchers, Inc.; 4.18: © Biophoto Assoc./Photo Researchers, Inc.

### Chapter 5

Opener:Christine Ortlepp; 5.1: © NIBSC/SPL/Photo Researchers, Inc.; 5.2b: Gordon Leedale/BioPhoto Associates; 5.6a: © Stanley Flegler/Visuals Unlimited; 5.6b: © Veronika Burmeister/Visuals Unlimited; 5.6c: © Science VU/Visuals Unlimited; 5.7b-c: © Cabisco/Visuals Unlimited; 5.13a: © M. Schiwa/Visuals Unlimited; 5.13b : Biology Media/Photo Researchers, Inc.; 5.13/top, middle top, middle bottom, bottom: The Company of Biologist, Ltd.; 5.14: © K.G. Murti/Visuals Unlimited; 5.15a: © D.W. Fawcett/Photo Researchers, Inc.; 5.15b: © CNRI/SPL/Photo Researchers, Inc.; 5.16b: © I. Gibbons, Don Fawcett/Visuals Unlimited; 5.16c: © David M. Phillips/Visuals Unlimited; 5.17: Mary Reedy; 5.18b: AP/Wide World Photos; 5.19/top: © David M. Phillips/Visuals Unlimited; 5.19/middle: © D.W. Fawcett/Photo Researchers, Inc.; 5.19/bottom: © D. Albertini - D. Fawcett/Visuals Unlimited; 5.20b: © C. Gerald Van Dyke/Visuals Unlimited

### Chapter 6

Opener: © Gunter Ziesler/Bruce Coleman Collection; 6.1: © Stephen Dalton/Photo Researchers, Inc.; 6.2/top: © Jeremy Walker/Tony Stone Images/Getty Images; 6.2/bottom: Des & Jen Bartlett/NGS Image Collection; 6.3: © D. Foster/Visuals Unlimited; 6.4a: © Stephen Dalton/NHPA; 6.4b: © Bryan Yablosky/Duomo; 6.6: © Manoj Shah/Tony Stone Images/Getty Images; 6.7: Dr. E.R. Degginger; 6.9a: © Blair Seitz/Photo Researchers, Inc.; 6.9b: © Gamma Liason/Getty Source; 6.A: © Ivan Polunin/Bruce Coleman Inc.; 6.A/inset: © E.R. Degginger/Photo Researchers, Inc.

### Chapter 7

Opener: NOAA; 7.1: Steve Raymer/NGS Image Collection; 7.7: Courtesy of Dr. Eldon Newcomb, University of Wisconsin, Madison; 7.13: © BioPhot; 7.14a: © Geoff Bryan 1996/Photo Researchers, Inc.; 7.14b-c: © Colin Weston/Planet Earth Pictures

### Chapter 8

Opener: Willard R. Chappell; 8.1a: © Lori Adamski Peek/Tony Stone Images/Getty Images; 8.1b: © Ed Reschke; 8.2: Courtesy of Bruce Parker; 8.4: © SPL/Custom Medical Stock Photos; 8.9: PhotoDisc Vol. Series 06 #6066 NA004439; 8.11a: Courtesy of Peter Hinkle; 8.13: PhotoDisc Food & Dining Vol. 12#12282; 8.14a: © Vance Henry/Nelson Henry

### Chapter 9

Opener: © Jennifer W. Shuler/Photo Researchers; 9.1: © Lennart Nisson; 9.3: AP/Wide World Photos; 9.4b: Dr. A. T. Sumner, "Mammalian chromosomes/© Springer Verlag; 9.5: © Conley Rieder/Biolgical Photo Service; 9.6: © Ed Reschke; 9.7/interphase, prophase, prometaphase, metaphase, anaphase: Ed Reschke; 9.7/telophase: © Peter Arnold Inc./Ed Reschke; 9.7/cytokinesis: © Ed Reschke; 9.8b: Dr. Mark W. Kirschner Dept. of Cell Biology Harvard Med. School; 9.9: © R. Calentine/Visuals Unlimited; 9.15/top: © Lennart Nisson; 9.15a: Robert Maier/Animals Animals Earth Scenes; 9.15b: © Gordon & Cathy Illg/Animals Animals/Earth Scenes; 9.16: Dr. John Heuser of Washington University School of Medicine, St Louis

### Chapter 10

Opener: © J. Luke/PhotoLink/Getty Images; 10.1a/top left, middle left, bottom left, top right, middle right, bottom right: © Carolina Biological Supply Company/Photo Take; 10.1b: © Francis Gohier/Photo Researchers, Inc.: 10.3: Dr. L. Caro/Science Photo Library/Photo Researchers, Inc.; 10.4b: © R. Kessel - G. Shih/Visuals Unlimited; 10.5: CNRI/Science Photo Library/Photo Researchers, Inc.; 10.8: © John/Cabisco/Visuals Unlimited; 10.11: © Francis LeRoy/BioCosmos/SPL/Photo Researchers, Inc.; 10.12b: © David M. Phillips/Visuals Unlimited; 10.14: Larry Johnson, Dept. of Veterinary Anatomy and Public Health; 10.16: © David M. Phillips/Visuals Unlimited; 10.A © CNRI/Phototake

### Chapter 11

Opener: © MedioImages; 11.7a: Dr. M. G. Neuffer University of Missouri-Columbia; 11.7b: © Joe McDonald/Visuals Unlimited; 11.14: © Porterfield - Chickering/Photo Researchers; 11.15: © Lester Bergman; p. 195/TA11.3/gray: © Norvia Behling/Animal Animals Earth Scenes; TA11.3/chinchilla: © Grant Heilman/Grant Heilman Photography; TA11.3/light gray: © Robert Maier/Animals Animals Earth Scenes; TA11.3/himalayan: © Hans Burton/Bruce Coleman, Inc.; TA11.3/albino: © Jane Burton/Bruce Coleman, Inc.; 11.16a: North Wind Picture Archives; 11.20/left, middle, right: Courtesy Thomas Kaufman. Photo by Phil Randazzo and Rudi Turner; 11.22: Library of Congress

## Chapter 12

Opener: © Brand X Pictures/Punchstock; 12.1: © Science Vu/Visuals Unlimited; 12.8: © BioPhoto Associates/Photo Reasearchers; 12.11b: © Cabisco/Visuals Unlimited; 12.12a: © Horst Schaefer/Peter Arnold, Inc; 12.12b: © William E. Ferguson; 12.14a: Courtesy Colleen Weisz; 12.14b: Courtesy of Integrated Genetics; 12.16b: Courtesy of Lawrence Livemore National Laboratory; 12.17a: © David M. Phillips/Visuals Unlimited; 12.17b/top left, top right, bottom left, bottom right: From R. Simensen, R. Curtis Rogers, Courtesy of Integrated Genetics

## Chapter 13

Opener left: © A.M. Siegelman/Visuals Unlimited; Opener right: © Harvey Lloyd/Peter Arnold; 13.1: © Dr. Gopal Murti/SPL/Photo Researchers, Inc.; 13.4: © Oliver Meckes/MPI - Tubingen/Photo Researchers; 13.6: © A.C. Barrington Brown / Photo Researchers; 13.7c: © Div. Of Cumputer Research & Technology, National Institute of Health/SPL/Photo Researchers; 13.10c: © 1948 M-C Escher Foundation-Baarn-Holland, All Rights Reserved; 13.11/top: © Ada Olins/Biological Photo Service; 13.11/bottom: Visuals Unlimited; 13.24c: © Tripos Associates/Peter Arnold, Inc.; 13.30b: © Kiseleva-Fawcett/Visuals Unlimited; 13.31/left, right: © Bill Longcore/Photo Researchers, Inc.

## Chapter 14

Opener: Dr. Rainer H. Koehler; 14.2: © K.G. Murti/Visuals Unlimited; 14.5b: Coutesy of Cellmark Diagnostics, Germantown, Maryland; 14.9/left: © Corbis #CB011809; 14.9/middle: © Norbert Wu/Peter Arnold, Inc.; 14.9/right: © Corbis #CB038761; 14.9/bottom: PhotoDisc Vol. Series 44

## Chapter 15

Opener: © C&M Denis Huot/Peter Arnold, Inc.; 15.2b: © Jeff Greenberg/Peter Arnold, Inc.; 15.3: © Galen Rowell/Peter Arnold, Inc.; 15.4a/left,right: © Tom McHugh/Photo Researchers, Inc.; 15.4b/top: © A.J. Copley/Visuals Unlimited; 15.4b/middle: Peter Scoones/Planet Earth Pictures; 15.4b/bottom: © D. Holden Bailey/Tom Stack & Associates; 15.5/top, bottom: © Peter Grant/Princeton University; 15.6a: © Tom McHugh/Photo Researchers, Inc.; 15.6b: © Robert and Linda Mitchell; 15.6c: © Tom McHugh/Photo Researchers, Inc.; 15.6d: Courtesy of John Alcock; 15.7a: © David M. Phillips/Visuals Unlimited; 15.7b: © A.B. Dowsett/SPL/Photo Researchers, Inc.; 15.7c: © Stanley F. Hayes, National Institutes of Health/SPL/Photo Researchers, Inc.; 15.7d: © Ken Greer/Visuals Unlimited; 15.7/inset: © Cabisco/Phototake; 15.8: Corbis/The Bettmenn Archives

## Chapter 16

Opener: © Digital Vision/Getty Images; 16.2b: Courtesy of Dr. Victor A. McKusick/Johns Hopkins Hospital; 16 A/topleft: © Gerard Lacz/Peter Arnold, Inc.; 16 A/top right: © D. Cavagnaro/Visuals Unlimited; 16.A/bottom left: © Scott Camazine/Photo Researchers, Inc.; 16 A/bottom right: © Barb Zurawski

## Chapter 17

Opener: © AP/Wide World; 17.1a/left: S. Lowry/Univ. Ulster/© Tony Stone Images; 17.1a/middle: Corbis id #CB030324; 17.1a/right: © Robert Fried/Stock Boston; 17.1b/left: Corbis AWI029 Animals & Wildlife; 17.1b/middle: Corbis #CB009677; 17.1b/right: PhotoDisc id#WL003983; 17.1/bottom: David Liittschwager & Susan Middeton/© 1973 Reprinted with permission of Discover Magazine; 17.3a: © Tom McHugh/Photo Researchers, Inc.; 17.4/toad: © John R. MacGregor/Peter Arnold, Inc.; 17 4/toad : © Allan Morgan/Peter Arnold, Inc.; 17.4/abalone: © Tom McHugh/Photo Researchers, Inc.; 17.4/dog-cat: © Renee Lynn/Photo Researchers, Inc.; 17.4/cow-moose: © Sandy Macys/Gamma Liaison; 17.4/tiger: © Portefield/Chickering/Photo Researchers, Inc.; 17.6a-b: Courtesy of Dr. Tom Smith, San Francisco State University; 17.7: WL Perry, DM Lodge, and JL Feder, Dept. of Biology Univeristy of Notre Dame; 17.8/top: © Melvin Grey; 17.8/bottom: © Melvin Grey; 17.10a/top: © Kjell B. Sandved/Visuals Unlimited; 17.10a/bottom: © E. R. Degginger/Photo Researchers, Inc.; 17.10b/top, bottom : © J. Wm. Schopf/UCLA; 17.11: © Discover Magazine; 17.12a-d: © Kevin de Queiroz; 17.12e: Jonathan B. Losos; 17.14a: Photo by Mario Maier, Eflangen/Germany

## Chapter 18

Opener: Natural History Museum/London; 18.1a: Reprinted with permission from SCIENCE elect. Vol. 282 No. 5394 pages 1601-1772 27 November 27 © 1998 / David Dilcher and Ge Sun; 18.1b: Reprinted with permission from SCIENCE elect. Vol. 282 No. 5399 pages 133-288 January 8 © 199 / Gifford H. Miller; 18.1c: Photo by Dr. Hendrik Poinar/Max-Planck Institute; 18.3/left: Photo courtesy of Terry D. Jones and John A. Ruben; 18.3/middle: O. Louis Mazzatenta/NGS Image Collection; 18.3/right: Corbis / Animals in Action #AIA0079; 18.4a-c: Dr. Luis M. Chiappe; 18.7a-b: © William E. Ferguson; 18.7c: © Robert Gossington/Bruce Coleman; 18.7d: © John Reader/SPL/Photo Researchers, Inc.; 18.7e: Dr. Shuhai Xiao/Botanical Museum Harvard University; 18.12a: © E.R. Degginger/Animals Animals; 18.12b: © Visuals Unlimited/Science VU; 18.13/top-all: Courtesy Michael Richardson; 18.13/bottom-all: © Bodleian Library; 18.14a-b: Courtesy Dr. H. Hameister, Universitatulm Ulm, Abteilung Medizinische Genetik; 18.14c: Courtesy Ricki Lewis; 18.17/left: From James H. Asher and Thomas B. Friedman, Journal of Medical Genetis 27:618-626, 1990; 18.17/middle: © Vickie Jackson; 18.17/right: From James H. Asher and Thomas B. Friedman, Journal of Medical Genetis; 18.10: From Kevin Padian and Luis M. Chiappe, "The Origin of Birds and Their Flight" in Scientific American, vol. 278, No. 2, February 1998. Reprinted by permission of Ed Heck, Illustrator; 18.11: From Patterns in Evolution by Roger Lewin, © 1997 by Scientific American Library

## Chapter 19

Opener: © 2001 David L. Brill/Brill Atlanta; 19.A: Corbis/Bettmann; 19.10b: © M. Abbey/Visuals Unlimited; 19.10c: © W. Burgdorfer/Visuals Unlimited; 19.10d: Charles O'Kelly, Bigelow Laboratory for Ocean Sciences; 19.11: © Dr. Malcolm Walter; 19.12, 19.13: J.G. Gehling, all rights reserved; 19.13: From R. Raff, The Shape of Life, c The University of Chicago Press. Reprinted by permission; 19.14: Milwaukee Public Museum Silurian Reef Diorama; 19.15: Photo by W.A. Taylor/permission NATURE 9/4/97 vol 389 p. 35 fig 2d; 19.17A: Field Museum of Natural History Chicago (Geo 75400C); 19.17b (1-3): Bruce Parker; 19.23a: © G. Hinter/Gamma Liaison; 19.23b: © Burt Silverman/Silverman Studios, Inc.

## Chapter 20

Opener: © Biophoto Associates/Photo Researchers, Inc.; 20.1a: © Runk/Schoenberger/Grant Heilman Photography, Inc; 20.1b: © Dennis Kunkel/Phototake; 20.2: © E.C.S Chan/Visuals Unlimited; 20.3/top left: © Science VU/Visuals Unlimited; 20.3/middle left : © Dr. O. Bradfute/Peter Arnold, Inc.; 20. 3/bottom left: © Peter Arnold, Inc.; 20.3/top right: © Oliver Mexkes/MPI - Tubingen/Photo Researchers, Inc.; 20.3/middle right: © E.O.S./Gelderblom/Photo Researchers, Inc.; 20.3/bottom right: © Hans Gelderblom/Visuals Unlimited; 20.5a: © Oliver Meckes/MPI - Tubingen/Photo Researchers, Inc.; 20.9: © Science Source/Photo Researchers, Inc.; 20.A: © Culver Pictures, Inc.; 20.10: © R.F. Ashley/Visuals Unlimited; 20.11: Theodore Diener/USDA/Plant Virology Laboratory; 20.13: © Ralph Eagle Jr./Photo Researchers, Inc.

## Chapter 21

Opener: From L. Tijuis, W.A.J. van Benthum, M.C.M. van Looschrecht and J.J. Heinen, "Solids Retention Time in Spherical Biofilms in a Biofilm Airlift Suspension Reactor," Biotechnology and Bioengineering, 44: 867-879, 1994 Reprinted by permission of John Wiley & Sons; 21.1a: © Sylvan Wittwer/Visuals Unlimited; 21.1b: © C.P. Vance/Visuals Unlimited; 21.3a : © G. Murti/SPL/Photo Researchers, Inc.; 21.3b: American Society for Microbiology; 21.4: © Raymond B. Otero/Visuals Unlimited; 21.5a: Courtesy of Charles C. Brinton, Jr. and Judith Camaham; 21.5b: © Fred Hossler/Visuals Unlimited; 215c: © George Musil/Visuals Unlimited; 21.6a: © David M. Phillips/Visuals Unlimited; 21.6b: © George J. Wilder/Visuals Unlimited; 21.6c: © Thomas Tottleben/Tottleben Scientific Company; 21.7: Esther R. Angert & Norman R. Pace; 21.8: Dr. Walther Stoeckenius; 21.9b: © Dr. Kari Lounatmaa/Science Photo Library/Photo Researchers, Inc.; 21.12: Courtesy of E.L. Wollman; 21.13a: © Arthur M. Siegelman/Visuals Unlimited; 21.13b: © Dr. Dennis Kunkel/Phototake; 21.14: © Joe Munroe/Photo Researchers, Inc.

## Chapter 22

Opener: Whirling Disease Foundation; 22.1: Serguei Karpov; 22.2: © A.M. Siegelman/Visuals Unlimited; 22.3: © M. Abbey/Visuals Unlimited; 22.4a-b: From M. Schaechter, G. Medoff, and D. Schlessinger (Eds.) Mechanism of Microbial Disease, 1989 Williams and Wilkins; 22.5: © Tony Stone Images/Robert Brons/BPS; 22.6: © Beatty/Visuals Unlimited; 22.7/left, right : © Carolina Biological Supply/Phototake; 22.9b:

© Authur M. Siegelman/Visuals Unlimited; 22.9c: Armed Forces Institute of Pathology; 22.10a: © David M. Phillips/Visuals Unlimited; 22.10c: © Bill Bachman/Photo Researchers, Inc.; 22.11a: Courtesy of Dr. JoAnn Burkholder; 22.11b: NCSU Aquatic Botany Lab; 22.11c: Courtesy of Dr. JoAnn Burkholder; 22.12: DJP Ferguson, Oxford University with permission from The Biologist 47: 234238, 2000.; 22.A: © Lennart Nilsson, The Body Victorious, Dell Publishing Company; 22.14: © M. Abbey/Visuals Unlimited; 22.15: © Biophoto Associates/Science Source/Photo Researchers, Inc.; 22.16a: M.S. Fuller; 22.16b: W.E. Fry, Plant Pathology, Cornell University; 22.17: © John D. Cunnigham/Visuals Unlimited; 22.18: © Gregory Ochocki/Photo Researchers, Inc.; 22.19: © Daniel W. Gatshall/Visuals Unlimited; 22.20: © Andrew J. Martinez/Photo Researchers, Inc.

## Chapter 23

Opener: © Mitch Hrdlicka/Getty Images; 23.3/left : Brigitte Meyer-Berthaud permissions NATURE 398, p 700 4/22/99; 23.3/middle : Reprinted with permission from SCIENCE vol. 282 No. 5394 pages 1601-1772 27 November 27 © 1998 / David Dilcher and GE Sun; 23.3/right: Maria A. Gandolfo/Nature; 23.5: © Jane Burton/Bruce Coleman, Inc.; 23.6a-b: Courtesy of Howard Crum, University of Michigan Herbarium; 23.6c: © John D. Cunningham/Visuals Unlimited; 23.7, 23.8a: © William E. Ferguson; 23.8b: © Dave Schiefelbein/Tony Stone Images; 23.8c: © Kingsley Stern; 23.10/left: © Kjell B. Sandved/Bufferfly Alphabet; 23.10/right: © David M. Dennis/Tom Stack & Associates; 23.10/bottom: © Stan Elems/Visuals Unlimited; 23.11a: © Bud Lehnhausen/Photo Researchers, Inc.; 23.11b: © Joe McDonald/Visuals Unlimited; 23.11c: © W. Ormerod/Visuals Unlimited, Inc.; 23.11d: © Rod Planck/Tom Stack & Associates; 23.12a-b: © Robert & Linda Mitchell; 23.14a: Dr. Dennis Stevenson; 23.14b: © Biophoto Associates/Photo Researchers, Inc.; 23.14c: © Jeff Gnass; 23.14d: © Doug Sokell/Visuals Unlimited; 23.A(A) : © Fred Bavendam; 23.A(B): © Michael Ederegger/Peter Arnold, Inc.; 23.A(C ): © Kingsley Stern; 23.B(A-D): © W. Barthlott; 23.17a: © Richard Weiss/Peter Arnold, Inc.; 23.17b: © Dwight Kuhn; 23.17c: © James L. Castner; 23.17d: © Hans Reinhard/OKAPIA/Photo Researchers, Inc.

## Chapter 24

Opener/left: Coutesy Dr. Mats Wedin, photo by Heidi Doring; Opener/right: Courtesy Dr. Mats Wedin, photo by Mats Wedin; 24.1a: © R.S.Husse/Visuals Unlimited; 24.1b: Thomas J. Volk University of Wisconsin-La Crosse (www.wisc.edu/botany/fungi/volkmyco.html); 24.3: Courtesy of N. Allin and G.L. Barron, University of Guelph; 24.4: © Dwight Kuhn; 24.5a: Runk/Schoenberger From Grant Heilman; 24.5b: Courtesy of Dr. Garry T. Cole/The Univ. of Texas at Austin; 24.6a: © Carolina Biological/Phototake; 24.7a: © J. Robert Waaland/Biological Photo Service; 24.7b: M.S. Fuller; 24.8/left: Runk/Schoenberger From Grant Heilman Photography, Inc.; 24.8/right: © Ed Reschke; 24.9: © Dwight Kuhn; 24.10a: © John D. Cunningham/Visuals Unlimited; 24.10b: © Doug Sherman/Geofile; 24.11b: © Bruce

Iverson; 24.12a: © Bill Keogh/Visuals Unlimited; 24.12b: © Hans Reinhard/Bruce Coleman, Inc.; 24.12c-d: Courtesy of G.L. Barron, The University of Guelph; 24.13: © Stanley Flegler/Visuals Unlimited; 24.A: © E. Chan/Visuals Unlimited; 24.B: © Holt Studios, Ltd/Animals Animals/Earth Scenes; 24.14a: John Dennis/Canadian Forest Service; 24. 14b: © R.L. Peterson/Biological Photo Service; 24.15b: © V. Ahmadijian/Visuals Unlimited; 24.16: © Steve & Sylvia Sharnoff/Visuals Unlimited, Inc.

## Chapter 25

Opener: © Dennis Kunkel Microscopy, Inc.; 25.7: © James C. Amos/Photo Researchers, Inc.; 25.9b: H. Armstrong Roberts; 25.9c: © Science VU/Visuals Unlimited; 25.12a: © L. Newman & A. Flowers/Photo Researchers, Inc.; 25.12b: © David Mechlin/Phototake NYC; 25.13: © CNRI/SPL/Photo Researchers, Inc.; 25.14: © Carolina Biological/Phototake; 25.15b: © Franklin J. Viola; 25.15c: © Chesher/Photo Researchers, Inc.; 25.15d: © Fred Bavendam/Peter Arnold, Inc.; 25.17b: © E. R. Degginger/Bruce Coleman, Inc.; 25.17c: © Kjell B. Sandved/Visuals Unlimited; 25.17d: © Bill Beatty/Visuals Unlimited; 25.19: Craig Cary/Univ. of Delaware; 25.20b: © A.M. Siegelman/Visuals Unlimited; 25.21b: © William E. Ferguson; 25.22a: © Stephen Krasemann/Photo Researchers, Inc.; 25.22b: © Luis C. Marigo/Peter Arnold, Inc.; 25. 22c: Philip Brownell/Oregon State University; 25.23a: © (Zigmond) Carmella Leszczynski/Animals Animals/Earth Scenes; 25.23b: © G/C Merker/Visuals Unlimited; 25.23c: Dr. James L. Castner; 25.23e: © Susan Beatty/Animals Animals; 25.23d: Davies & Starr/© Tony Stone Images; 25.23e: George K. Bryce/Animals Animals; 25.24b: © Andrew Martinez/Photo Researchers, Inc.; 25.24c: © Norbert Wu/Peter Arnold, Inc.; 25.24d: © Rick M. Harbo; 25.25a: © David B. Fleetham/Visuals Unlimited; 25.25b: © Nancy Sefton/Photo Researchers, Inc.; 25.26a: © Peter Parks/Animals Animals/Earth Scenes; 25.26b: © E.R. Degginger/Animals Animals/Earth Scenes

## Chapter 26

Opener: MAR-ECO; 26.3b: © Nancy Sefton/Photo Researchers, Inc.; 26.4b: © Runk/Schoenberger/Grant Heilman Photography; 26.6b: Illustration by: Rob Wood - Wood Ronsaville Harlin, Inc; 26.5: Reprinted by permission from Nature from S. Kumar and S.B. Hedge, "A Molecular Time Scale for Vertebrate Evolution" in Nature, vol. 392, p. 919, 4/30/98, fig. 3, c 1998 Macmillan Magazines Limited; 26.6c: © Russ Kinne/Photo Researchers, Inc.; 26.8b: © Hal Beral/Visuals Unlimited; 26.8c: © W. Gregory Brown/Animals Animals/Earth Scenes; 26.10b: © Tom McHugh/Photo Researchers, Inc.; 26.10c: © Peter Scoones/Plant Earth Pictures; 26.10d: © Dr. E.R. Degginger; 26.12b: © Mark Moffet/Minden Pictures; 26.12c: © Suzanne L. Collins & Joseph T. Collins/Photo Researchers, Inc.; 26.12d: E.D. Brodie, Jr.; 26.14b: © Ed Reschke/Peter Arnold, Inc.; 26.14c: © Mark Moffett/Minden Pictures; 26.14d-e: © Joe McDonald/Animals Animals/Earth Scenes;

26.15: © Joe McDonald/Bruce Coleman, Inc.; 26.18b: © Meckes/Ottawa/Photo Researchers, Inc; 26.20b: © Fritz Prenzel/Animals Animals/Earth Scenes; 26.20c: © Dalton, S./Animals Animals/Earth Scenes; 26.20d: © Tim Davis/Photo Researchers, Inc.; 26.20e: © Ansel Horn/Phototake; 26.22b: © Tom Ulrich/Visuals Unlimited; 26.22c: © Roland Seitre/Peter Arnold, Inc.; 26.22d: © John Giustina/Bruce Coleman, Inc.

## Chapter 27

Opener: © PhotoLink/Getty Images; 27.2: © Jack M. Bostrack/Visuals Unlimited; 27.3a: © Randy Moore/BioPhot; 27.3b: © George Wilder/Visuals Unlimited; 27.3c: © BioPhot; 27.3d: © Bruno P. Zehnder/Peter Arnold, Inc.; 27.4a: © BioPhot; 27.4b: © Dwight Kuhn; 27.4c: From Yolanda Heslop-Harrison, "SEM of Fresh Leaves of Pinguicula" Science 667:173, © 1970 by the AAAS; 27.4d: © Olver Meckes/Eye of Science/Photo Researchers, Inc.; 27.4e: © Patti Murray/Animals Animals/Earth Scenes; 27.5/left, right: © Ray Simon/Photo Researchers, Inc.; 27.6a: © John D. Cunningham/Visuals Unlimited; 27.6b, 27.7: © George Wilder/Visuals Unlimited; 27.8a/left: © Cabisco/Visuals Unlimited; 27.8a/right: © Dwight Kuhn; 27.8b/right: © Science VU/Visuals Unlimited; 27.9a: © Charles Gurche; 27.9b: © Kenneth W. Fink/Photo Researchers, Inc.; 27.9c: © G.C. Kelley/Photo Researchers, Inc.; 27.9d: © Franz Krenn/Photo Researchers, Inc.; 27.9e-f: © Dwight R. Kuhn; 27.10a-b: © Kjell Sandved/Butterfly Alphabet; 27.10c: © Robert Maier/Animals Animals/Earth Scenes; 27.11a-b: © Dwight R. Kuhn; 27.12: © M.I. Walker/Photo Researchers, Inc.; 27.13: © Ed Reschke; 27.15a: © John D. Cunningham/Visuals Unlimited; 27.15b-c: © Stan Elems/Visuals Unlimited; 27.15e: © John D. Cunningham/Visuals Unlimited; 27.16: © Runk/Schoenberger/Grant Heilman Photography; 27.18a-b: © W.H. Hodge/Peter Arnold, Inc.; 27.18c: © William E. Ferguson; 27.20b: © Manfred Kage/Peter Arnold, Inc.; 27.20c: © A.J. Copley/Visuals Unlimited, Inc.

## Chapter 28

Opener: © Dennis Flaherty/Photo Researchers, Inc.; 28.3a: © 1991 Regents University of California Statewide IPM Project; 28.3b: © 1991 Regents University of California Statewide IPM Project; 28.4: © Deborah A. Kopp/Visuals Unlimited, Inc.; 28.9a: Dr. E.R. Degginger

## Chapter 29

Opener: © Getty Images; 29.1a: © Nardin/Jacana/Photo Researchers, Inc.; 29.1b: © David Sieren/Visuals Unlimited, Inc.; 29.1c: Dr. Jeffery B. Milton/University of Colorado; 29.A(1-2): © Davis Littschwager, Susan Middleton/Discover Magazine; 29.B(1-3): © Davis Littschwager, Susan Middleton/Discover Magazine; 29.6a: © William E. Ferguson; 29.6b(1-2): © Robert A. Tyrrell; 29.6c: © Joe McDonald/McDonald Wildlife Photography; 29.6d: © Merlin D. Tuttle/Bat Conservation International/Photo Researchers, Inc.; 29.7a-b: © Leonard Lessin/Photo Researchers, Inc.; 29.8: © Angelina

Lax/Photo Researchers, Inc.; 29.10b: © John D. Cunningham/Visuals Unlimited; 29.10c: Biodisc.com; p. 563/Table 29.1(1 top): © Inga Spence/Visuals Unlimited; Table 29.1(2): © Alan & Linda Detrick/Photo Researchers, Inc.; Table 29.1(3): © Zig Leszczynski/Earth Scenes/Animals Animals/Earth Scenes; Table 29.1(4): © Dwight Kuhn; Table 29.1(5): © Phil Degginger; Table 29.1(6): © Dwight Kuhn; Table 29.1(7): Dr. E.R. Degginger; Table 29 1(8 bottom): © R.J. Erwin/Photo Researchers, Inc.; 29.12a: © Adam Hart-Davis/SPL/Photo Researchers, Inc.; 29.12b: © O.S.F./Earth Scenes/Animals Animals/Earth Scenes; 29.12c: © Rod Planck/Photo Researchers, Inc.; 29.12d: © W.H. Hodge/Peter Arnold, Inc.; 29.12e: © Richard Shiell/Earth Scenes/Animals Animals/Earth Scenes; 29.13a: © Dwight Kuhn; 29.13b: © Ed Reschke

## Chapter 30

Opener: Courtesy of University of California, Statewide Integrated; 30.3a: © Robert E. Lyons/Visuals Unlimited; 30.3b: © Sylvan H. Wittwer/Visuals Unlimited; 30.5: David G. Clark; 30.6: © Leonard Lessin/Peter Arnold, Inc.; 30.7a: © C. Calentine/Visuals Unlimited; 30.7b: BioPhot; 30.9: © William E. Ferguson; p.577/30.A(1-2): © Dr. Randy Moore/BioPhot; 30.10a: © John Kaprielian/Photo Researchers, Inc.; 30.10b: © Richard H. Gross; 30.10c: © John Kaprielian/Photo Researchers, Inc.; 30.11/both: © Tom McHugh/Photo Researchers, Inc.; 30.11: © Tom McHugh/Photo Researchers, Inc.; 30.16: © John D. Cunningham/Visuals Unlimited, Inc.; 30.17: © L. West/Photo Researchers, Inc; 30.18: © Tim Thompson/Tony Stone Images

## Chapter 31

Opener: © Lennart Nilsson, Behold Man, Little Brown & Co./Bonnierforlagen; 31.7/zebra: © Roland Seitre/Peter Arnold, Inc.; 31.7/toucan: © Michael Sewell/Peter Arnold, Inc.; 31.7/iguana: © Werner H. Muller/Peter Arnold, Inc.; 31.7/frog: © Rob & Ann Simpson; 31.B/left: © Vaughn Youtz/Sipa Press; 31.B/right: Jewish Hospital, Kleinert, Kutz and Associates Hand Core Center, and University of Louisville; 31.9b: © CNRI/SPL/Photo Researchers, Inc.

## Chapter 32

Opener: The New England Journal of Medicine vol. 338 p. 1672-1676, 06/04/1998 © Massachusetts Medical Society All Rights reserved. Courtesy Dr. David W. Nierenberg; 32.1: © Leonard Lee Rue III; 32.2b: © Secchi-League/Roussel - UCLAF/CNRI/SPL/Photo Researchers, Inc.; 32.6b: © Fawcett/Coggeshall/Photo Researchers, Inc.; 32.8b: © Don Fawcett/Photo Researchers, Inc.; 32.10: © E.R. Lewis, Y.Y. Zaevi, T.E.Evenhart/University of California, Berkeley; 32.14d: © Manfred Kage/Peter Arnold, Inc.

## Chapter 33

Opener: Jerry Chunn Photography; 33.1a-b: © Thomas Eisner; 33.4a: © Hans Pfletschinger/Peter Arnold, Inc.; 33.4b: © R.A. Steinbrecht; 33.7b: © Thomas Eisner; 33.7c: © Fred Bavendam/Peter Arnold, Inc.; 33.8c: © Frank S. Werblin

## Chapter 34

Opener/both: Courtesy of Eli Lilly & Company; 34.6a: © Dan Kline/Visuals Unlimited; 34.9a-c: Clinical Pathological Conference on Acromegaly, Diabetes, Hypemetabolism, Protein Use & Heart Failure, American Journal of Medicine 20: 133 (1956)

## Chapter 35

Opener: Copyright Sea Studios, Inc., Cameraman Bob Cranston; 35.1a: Photo © Lois Greenfield, 1994; 35.1b: © Dwight Kuhn; 35.3/top: © David Scharf/Peter Arnold, Inc.; 35.3/bottom: © Zigmund Leszezynski/Animals Animals/Earth Scenes; 35.4a: © Nancy Stefton/Photo Researchers, Inc.; 35.4b: © Gregory Ochocki/Photo Researchers, Inc.; 35.4c: © J&B Photo/Earth Scenes/Animals Animals/Earth Scenes; 35.5: From Kenneth R. Gordon, PH.D., "Adaptive Nature of Skeletal Design," BioScience 39(11): 784-790, Dec.1989; 35.7a: © Chuck Brown/Photo Researchers, Inc.; 35.7b: © Ed Reschke; 35.B/p.685: © Michael Klein/Peter Arnold, Inc.; 35.10a: © Ed Reschke; 35.10b: © Manfred Kage/Peter Arnold, Inc.; 35.10c: © Ed Reschke; 35.19a: © Custom Medical Stock Photos; 35.19b: © B.S.I.P./Custom Medical Stock Photos

## Chapter 36

Opener: © Brand X Pictures/Punchstock; 36.8b: © Lennart Nilsson, Behold Man, Little Brown & Co.; 36.12: © Boehringer Ingelheim International GmbH, Photo: Lennart Nilsson, *The Body Victorious,* Dell Publishing; 36.A/p.714: © J & L Weber/Peter Arnold, Inc.; 36.14b: SPL/Photo Researchers, Inc.

## Chapter 37

Opener: © National Geographic/Getty; 37.1: © Douglas M. Munnecke/Biological Photo Service; 37.9b: © D.W. Fawcett/Photo Researchers, Inc.; 37.10: © John Watney Photo Library; 37.A/p.730: © USCS; 37.B/p.731 both: © Martin Rotker/Martin Rotker Photography

## Chapter 38

Opener: © Getty Images; 38.1a: © Dwight Kuhn; 38.1b: © James D. Watt/Planet Earth Pictures; 38.1c: © Tom McHugh/Photo Researchers, Inc.; 38.12a: © David Scharf/Peter Arnold, Inc.; 38.12c: Courtesy of David H. Alpers, M.D.

## Chapter 39

Opener: © Neil Beer/Getty Images; 39.3 Courtesy of Eric Toolson; 39.6: © Roger Eriksson; 39.8: © Fred Bavendam/Peter Arnold

## Chapter 40

Opener: North Wind Picture Archives; 40.5: © Biology Media/Photo Researchers, Inc.; 40.8: © Manfred Kage/Peter Arnold, Inc.; 40.11b: © Len Lessin/Peter Arnold, Inc.; 40.15b: © Dr. A. Liepins/SPL/Photo Researchers, Inc.; 40.18/top: © David Scharf/Peter Arnold, Inc.; 40.18/bottom: © Phil Harrington/Peter Arnold, Inc.

## Chapter 41

Opener A-D: Jill A. Helms; 41.1a: © Professor P. Motta/Dept. of Anatomy/University of La Sapienza, Rome/SPL/Photo Researchers, Inc.; 41.3: Furnished through the Courtesy of C.L. Markert; p. 803/41.B/top left: Courtesy of Christine Nusslein-Volhard; 41.B/top right, bottom left, bottom right: Courtesy of Jim Langeland, Steve Paddock, Sean Carroll/Howard Hughes Medical Institute, University of Wisconsin; p. 803/41.C/left, right: Courtesy of F.R. Turner, Indiana University, Bloomington, IN; 41.12: Courtesy Kathryn Tosney; 41.13: © Petit Format/Nestle/Science Source/Photo Researchers, Inc.; 41.17: © Erika Stone/Peter Arnold, Inc.; 41 E/p. 818 : © Topham/The Image Works; 41.18: © J.L. Bulcao/Gamma Liaison

## Chapter 42

Opener: Bryan D. Neff; 42.1: © J.A.L. Cooke; 42.2: © K&K Ammann/Bruce Coleman; 42.3a: © Maslowski Wildlife Productions; 42.4 : © Frans Lanting/Minden Pictures; 42.5: © Willard Luce/Animals Animals/Earth Scenes; 42.6: Bonnie Bird; 42.7: Nina Leen/Life Magazine; 42.A/p. 832: © Marty Snyderman; 42.9: © Lynn Rogers; 42.10: © Rick Sullivan/Bruce Coleman, Inc.; 42.11: © Tom & Pat Leeson/Photo Researchers, Inc.; 42.12a: © Lady Phillppa/Scott/NHPA; 42.12b: © Tom Brakefield/Bruce Coleman, Inc.; 42.13: © Franklin J. Viola; 42.15: David W. Pfenning; 42.16: © E.R. Degginger/Color-Pic. Inc.; 42.17: From Mace A. Mack and Daniel Rubenstein, "Zebra Zones" in Natural History, March 1998, p. 79, American Museum of Natural History. Reprinted by permission of Joe LeMonnier, Illustrator; 42.17b: © Gregory G. Dimijian/Photo Researchers, Inc.; 42.19a: © Emmett Duffy, College of William & Mary, Virginia Institute of Marine Science; 42.19b: © Raymond A. Mendez/Animals Animals/Earth Scenes; 42.19c: © Donald Specker/Animals Animals/Earth Scenes; 42.20: From Animal Communication by Judith Goodenough from Carolina Biological Reader Series No. 143. Copyright c 1984 Carolina Biological Supply Company. Used by permission; 42.20c: © Scott Camazine/Photo Researchers, Inc.

## Chapter 43

Opener: Gordon H. Rodda/USGS; 43.1a: Wayne P. Armstrong; 43.1b: © John A. Novak/Animals Animals/Earth Scenes; 43.1c: © NOAA/NGDC DMSP Digital Archive/Science Source/Photo Researchers, Inc.; p. 852/43.A: © C. Gable Ray; p. 852/43.B: Courtesy of Montana Dept. of Fish, Wildlife, and Parks; 43.9: Based on a study by Joel Berger, University of Nevada-Reno/NYT Pictures

## Chapter 44

Opener: OAR/National Undersea Research Program (NURP); College of William & Mary; 44.1: © Janis Burger/Bruce Coleman, Inc.; 44.3: © J. Sneesby/B. Wilkins/Tony Stone Images; 44.5a: © Edward S. Ross; 44.5b: © James L. Castner; 44.5c: © Simon D. Pollard/Photo Researchers, Inc.; 44.6: © Norm Smith and Jim Dawson; 44.8a-b: USDA Forest Service - IITF Library; 44.9a-d: Courtesy, Dr. Stuart Fisher, Department of Zoology, Arizona State University; 44.17: